SUPERNOVA ENVIRONMENTAL IMPACTS

IAU SYMPOSIUM No. 296

COVER ILLUSTRATION:

The cover illustration is a montage of the Supernova Remnant Simeis 147 over a view of River Ganga near Kolkata.

IAU SYMPOSIUM PROCEEDINGS SERIES

INTERNATIONAL ASTRONOMICAL UNION

UNION ASTRONOMIQUE INTERNATIONALE

International Astronomical Union

IAU

SUPERNOVA ENVIRONMENTAL IMPACTS

PROCEEDINGS OF THE 296th SYMPOSIUM OF THE INTERNATIONAL ASTRONOMICAL UNION HELD AT RAICHAK ON GANGES NEAR CALCUTTA, INDIA JANUARY 7 – 11, 2013

Edited by

ALAK K. RAY
Tata Institute of Fundamental Research, Mumbai

and

RICHARD McCRAY
University of Colorado, Boulder

Shaftesbury Road, Cambridge CB2 8EA, United Kingdom

One Liberty Plaza, 20th Floor, New York, NY 10006, USA

477 Williamstown Road, Port Melbourne, VIC 3207, Australia

314–321, 3rd Floor, Plot 3, Splendor Forum, Jasola District Centre, New Delhi – 110025, India

103 Penang Road, #05–06/07, Visioncrest Commercial, Singapore 238467

Cambridge University Press is part of Cambridge University Press & Assessment, a department of the University of Cambridge.

We share the University's mission to contribute to society through the pursuit of education, learning and research at the highest international levels of excellence.

www.cambridge.org
Information on this title: www.cambridge.org/9781107044777

First published 2014

A catalogue record for this publication is available from the British Library

ISBN 978-1-107-04477-7 Hardback

Table of Contents

Preface

As the title of this volume indicates, the unifying theme of this Symposium was the interaction of supernovae with their environments, both interstellar and circumstellar. New telescopes spanning the entire electromagnetic spectrum have caused the study of supernovae (SNe) and supernova remnants (SNRs) to advance at a breathtaking pace. In only a decade, automated synoptic surveys have increased the detection rate of supernovae by more than an order of magnitude and have enabled discovery of highly unusual supernovae. Observations of gamma rays SNRs with ground-based Cherenkov telescopes and the Fermi telescope have given us new insights into particle acceleration in supernova shocks. Far-infrared observations from the Spitzer and Herschel observatories are telling us much about the properties and fate of dust grains in SNe and SNRs. Multi-wavelength surveys have yielded new insights into the influence of SNe on the ecosystems of galaxies.

Core-collapse SNe have a great diversity of light curves and spectral evolution. We now understand that their diversity is determined largely by the history of mass loss from the progenitor star, which determines the distribution of circumstellar matter and the structure of the star immediately before it explodes. The remarkable SN2009ip shows that the some supernova progenitors have major episodes of mass loss only a few years before they finally explode. The impact of the explosion debris with the extended stellar envelope may result in a supernova of exceptional luminosity, such as SN2006gy. Explosions of stars with extended circumstellar envelopes give rise to X-ray and radio supernovae. Such systems blur the distinction between SNe and SNRs. Alternatively, if the supernova progenitor has lost most of its envelope and has become a relatively compact star, the explosion may be sub-luminous, as is the case of SN1987A. A strong stellar wind from a blue giant progenitor to a supernova may create a cavity in the circumstellar environment, displacing the interstellar matter to great distance from the supernova. Thus, the evolution of the supernova progenitor may have a great influence on the subsequent evolution of X-ray and radio emission from the SNR. Moreover, supernova explosions can stimulate star formation by compressing interstellar gas and can also help to terminate star formation by dispersing gas in star-forming molecular clouds. Supernovae in low-density regions, such as superbubbles or galactic bulges, cannot effectively radiate energy therefore may drive global outflows, affecting the galactic ecosystem.

In 2011, the Scientific Organizing Committee proposed that an IAU Symposium be held in India during January 7 - 11, 2013. After some initial search for sites within India, the organizers selected the venue of the Symposium at The Fort Raichak on the banks of river Ganga about 65 km south of the city of Kolkata (formerly Calcutta). In hindsight, the IAU's decision was a good fortune for the supernova community. Most speakers that the organizers proposed as Invited Speakers were able to attend, and many young researchers presented exciting new results in oral presentations and poster sessions. Apart from professional astronomers (including PhD students), graduate students and advanced undergraduates with strong interest in astronomy took part in the meeting. Students and professional astronomers participated from many countries of north and south as well as from all regions of India.

The scientific program started with a session on SN 1987A, the nearest SN in recent times in the Large Magellanic Cloud and continued with recent developments of supernova models and surveys to find new supernovae. Almost all aspects of research on different types of core-collapse supernovae and their interactions with the surrounding medium, including observational studies of light curves and spectra in the radio, mm, optical, UV, X-ray and gamma-ray bands using many ground and space based telescopes and facilities were presented at the Symposium. There was also an impromptu session on SN 2009ip where research in progress was discussed on the first evening. Research on various aspects of Supernova Remnants (as a young supernova ages and develops full-scale interaction with the interstellar medium), how they affect galactic outflow, star formation and superbubbles and large scale structure of the host galaxies, acceleration of high energy cosmic rays were presented in the five days of the meeting. The unifying theme of the Symposium was the physics of shocks and the observations of their radiation in both SNe and SNRs. The website of the conference at: http://www.tifr.res.in/ iau296/ has many details of the scientific programme and rationale, members of the Local and Scientific Organizing Committees, talks and posters presented at the Symposium etc. It also has a link to the videographic recordings of the talks presented at the meeting arranged according to the scientific sessions. These recordings will remain deployed at this site for a period up to the end of 2015.

On the second evening of the Symposium a classical Indian Dance programme of the Odissi genre was presented by Sharmila Biswas and her disciples from the Odissi Vision and Movement Centre, Kolkata. The Symposium ended with a cruise up the River Ganga to take the delegates from Raichak to Calcutta.

All delegates to the Symposium stayed in the same Hotel Complex throughout the meeting, which significantly enhanced the interactions both formal and informal among the participants. Many individuals and organizations helped in multiple ways to make the IAU Symposium 296 a successful scientific meeting in a rural setting in the state of West Bengal. Generous funding was received from the International Astronomical Union, Tata Institute of Fundamental Research, Mumbai, Indian Institute of Astrophysics, Bangalore and the Indian National Science Academy, New Delhi. The organizers thank the Director, the Dean, Natural Sciences Faculty and the Registrar, Tata Institute of Fundamental Research, Mumbai for administrative support and help. The organizers appreciate the conference work carried out by Messers Surendra Kulkarni, Nassim Khan and Ms. Magnes Johny and the help rendered by the members of the Local Organizing Committee and the student volunteers. The organizers thank Prof. Malabika Sarkar, Vice Chancellor, Presidency University, Calcutta and Prof Somak Raychaudhury for critical help in liaising with local agencies and representatives of the Governments of West Bengal and India.

Richard McCray and Alak Ray,
On behalf of the Scientific Organizing Committee:

Roger Chevalier (USA)
Evgeny Berezhko (Russia)
Catherine Cesarsky (France)
Yang Chen (China)
Claes Fransson (Sweden)
Marianne Lemoine-Goumard (France)
Fangjun Lu (China)
Virginia Trimble (USA)
Massimo Turatto (Italy)
Jacco Vink (The Netherlands)
Daniel Wang (USA)

CONFERENCE PHOTOGRAPH

Photo Credits: Tata Institute of Fundamental Research.

Photo Credits: Tata Institute of Fundamental Research.

Photo Credits: Tata Institute of Fundamental Research.

Photo Credits: Tata Institute of Fundamental Research.

Photo Credits: Tata Institute of Fundamental Research.

Photo Credits: Tata Institute of Fundamental Research.

Photo Credits: Tata Institute of Fundamental Research.

Photo Credits: Tata Institute of Fundamental Research.

Photo Credits: R. Sankrit.

Photo Credits: R. Sankrit.

Photo Credits: R. Sankrit.

Photo Credits: R. Sankrit.

Photo Credits: R. Sankrit.

Photo Credits: R. Sankrit.

Supernova Environmental Impacts
Proceedings IAU Symposium No. 296, 2013
A. Ray & R. A. McCray, eds.

doi:10.1017/S1743921313009150

Recent *Hubble Space Telescope* Observations of SN 1987A: Broad Emission Lines

Kevin France

Center for Astrophysics and Space Astronomy, University of Colorado, 389 UCB, Boulder, CO 80309; kevin.france@colorado.edu

Abstract. Observations with the Hubble Space Telescope (HST), conducted since 1990, have allowed us to create a "movie" of the evolution of the core-collapse supernova SN 1987A from 3–25 years after the explosion. Critical to understanding the late time evolution of SN 1987A was the successful HST Servicing Mission 4 in May 2009. The repair of the STIS instrument and the installation of the WFC3 imager and COS spectrograph have provided crucial data points for understanding the temporal variability in the physical structure and energy sources for SN 1987A, as well as measurements of the chemical abundances of the ejecta. In this proceeding, I will focus on two topics that have made use of the expanded capability of HST and highlight the importance of access to a UV/optical space observatory for the studies of local supernovae: 1) 2) The decreasing maximum velocity of neutral hydrogen crossing the reverse shock front and the role of soft X-ray/EUV heating in the outer supernova debris and 2) The detection of metals (N^{4+} and C^{3+} ions) crossing the reverse shock front and CNO processing in the progenitor star.

Keywords. circumstellar matter, shock waves, supernovae: individual: SN 1987A

1. Introduction

Borkowski, Blondin, & McCray (1997) predicted that the spectrum of SN 1987A should display very broad ($\Delta v \sim \pm$ 12,000 km s^{-1}) emission lines of Lyα, Hα, NV λ1240, and HeII λ1640, produced where the freely expanding supernova debris crosses a reverse shock located inside the equatorial circumstellar ring. In September 1997, using the *Hubble Space Telescope*-Space Telescope Imaging Spectrograph (*HST*-STIS), Sonneborn *et al.* (1998) detected broad Lyα emission, and Michael *et al.* (1998) showed how observations of this emission can be used to map the shape of the reverse shock and the flux of HI atoms crossing it. Michael *et al.* (2003) and Heng *et al.* (2006) analyzed subsequent (February 1999 – October 2002) STIS observations of both Lyα and Hα to map the increasing flux of HI atoms across the reverse shock.

In this proceeding, I will summarize recent analyses of the broad emission features coming from the outer debris and the reverse shock region of SN 1987A. We refer readers to France *et al.* (2010) and France *et al.* (2011) for additional details related to the observations, spectral reduction, and data analysis of the *HST* observations of SN 1987A.

2. Observations

SN 1987A was observed with the medium resolution far-UV modes of *HST*-COS (G130M and G160M) on 2011 February 11 and March 14 for a total of 7 spacecraft orbits (18555 s; Table 1) as part of the Supernova 1987A INTensive Study (SAINTS - GO12241; PI - R. Kirshner). A description of the COS instrument and on-orbit performance characteristics can be found in Osterman *et al.* (2011). Star 3 contributes negligible flux to the one-dimensional spectrum. The custom data extractions were then reprocessed with the COS calibration pipeline, CALCOS v2.13.6, and combined with the custom IDL

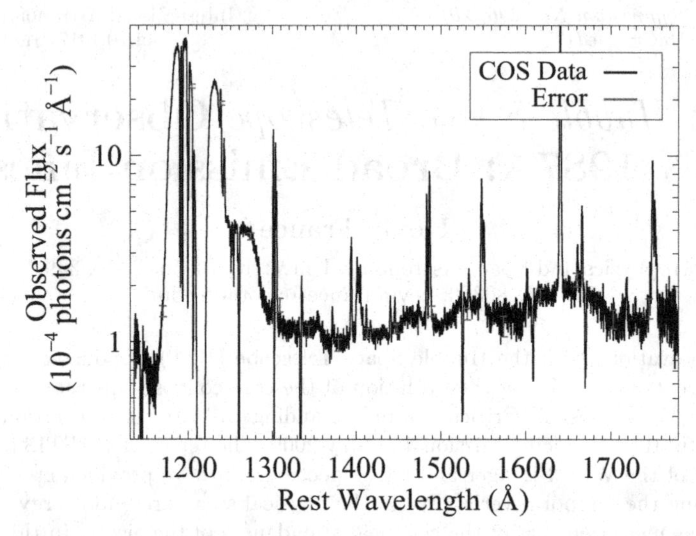

Figure 1. Full far-UV spectrum of SN 1987A, obtained with the Cosmic Origins Spectrograph, aboard *Hubble*. Broad emission from the SN 1987A reverse shock, narrow emission lines attributable to circumstellar hotspots, as well as underlying continuum emission are observed. Representative 1-σ error bars (a combination of photon statistics and flux calibration uncertainties) are shown in red.

coaddition procedure described by Danforth *et al.* (2010). Data were obtained in four central wavelength settings in each far-UV grating mode (λ1291, 1300, 1309, and 1318 with G130M and λ1577, 1589, 1600, and 1611 with G160M) at the default focal-plane split position. Observations at multiple wavelength settings provide continuous spectral coverage over the 1136 – 1782 Å bandpass and minimize the residual fixed pattern noise from the detector grid wires and the MCP pores. The point source resolving power of the medium resolution COS far-UV modes is $R \equiv \Delta\lambda/\lambda \approx 18{,}000$ (Δv = 17 km s^{-1}); however, the filled-aperture resolving power is $R \sim 1500$ ($\Delta v \sim 200$ km s^{-1}; France *et al.* 2009).

The STIS observations were obtained in the G140L mode, through the 52$''$ × 0.2$''$ slit, on 31 January 2010. These data were the first ultraviolet observations of SN 1987A following the *Hubble* Servicing Mission 4. Approximately three *HST* orbits worth of data were acquired at a resolving power $R \sim 1500$, at $\sim$0.1$''$ angular resolution across the 1150 - 1710 Å bandpass. STIS G750L observations in the red were also obtained with somewhat lower spectral resolution, although with a factor of $\sim$2 higher angular resolution.

3. Lyα, 2 – photon Continuum, and Heating of the Outer Ejecta

Figures 1, 2, and 4 show the broad emission from H I; Lyα and Hα line emission. The velocity distribution of the neutral hydrogen emission in 2011 (Figure 1) extends from −13000 – +8000 km s^{-1}. The velocity maxima are much smaller than the initial observations by Sonneborn *et al.* (1998; ± 20000 km s^{-1}) and are consistent with the decrease in maximum projected velocity observed from 2004 to 2010 (France *et al.* 2010). Previous work noted that the Lyα/Hα ratios from the reverse shock exceed the 5:1 photon production ratio expected for a Balmer-dominated shock (Heng *et al.* 2006; Heng & McCray 2007). France *et al.* (2010) found Lyα/Hα ratios $\geqslant$ 30 from −8000 – −2500 km s^{-1} and $\geqslant$ 20 from +3000 – +7000 km s^{-1} for isolated cuts across the northern

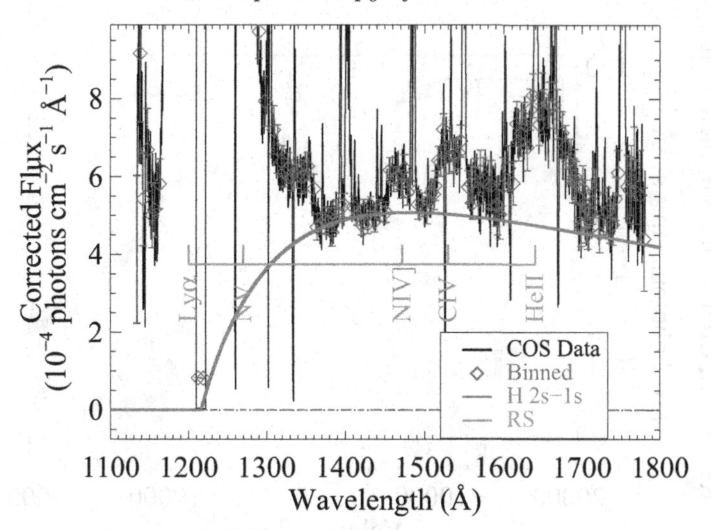

Figure 2. The extracted, one-dimensional spectrum of SN 1987A (*black*) and the spectrum averaged over 0.8 Å bins chosen to avoid narrow features from hotspot emission and interstellar absorption. These data have been corrected for interstellar neutral hydrogen absorption ($\log_{10}(\mathrm{N(H)}) = 21.43$) and interstellar dust extinction, assuming $E(B-V) = 0.19$ and $R_V = 3.1$. The broad spectral features are produced by a combination of hydrogen 2-photon emission ($2s \rightarrow 1s$) and ionic emission from the reverse shock. A theoretical hydrogen 2-photon spectrum is shown overplotted in orange. Reverse shock emission features are identified in green.

and southern sides of the reverse shock, respectively. They attributed the enhanced Lyα emission to a second source of Lyα photons. They argued that Lyα photons from the hotspots are resonantly scattered by onrushing hydrogen with a distribution of velocities spanning a width $\Delta v_H \sim 3000 - 9000$ km s^{-1} (unlike Lyα, Hα is not a resonance line and therefore hotspot Hα photons pass through the debris freely).

While the Lyα enhancement at the largest negative velocities can likely be explained by this mechanism, a more likely physical scenario is that the lower velocity ($\pm$7000 km s^{-1}) Lyα excess is attributable to the X-ray heating of the outer supernova debris. The low-density gas in the outer debris ($n_H \sim 100$ cm^3; Smith *et al.* 2005; Heng *et al.* 2006) is heated by X-rays emitted from the shocked gas near the hotspots. In particular, the energy deposition of the soft X-ray/EUV photons from the shocked hotspots will be concentrated in a layer near the reverse shock (Fransson *et al.* 2011). When the ionization fraction is $\gtrsim 3 \times 10^{-2}$, the majority of the X-ray energy heats the gas through Coulomb stopping of fast photoelectrons (Xu *et al.* 1992; Kozma & Fransson 1992). The primary coolants for this gas are Lyα and 2-photon emission.

In this scenario, we expect to detect H I 2-photon emission from the same regions of enhanced Lyα production. The high sensitivity of the COS observations reveal this emission for the first time (Figure 2). The total integrated (1216 Å – 6 μm) H I 2-photon flux is $9.1 \pm 0.6 \times 10^{-1}$ photons cm^{-2} s^{-1}, and the observed ratio of Lyα line emission to 2-photon is $F(\mathrm{Ly}\alpha)/F(2s) = 1.96 \pm 0.23$. Since this ratio is consistent with the $2p/2s$ ratio of 2.1 (observed as $F(\mathrm{Ly}\alpha)/F(2s)$) expected for recombination at $T \approx 10^4$ K (Spitzer 1978) and the $(1s \rightarrow 2p)/(1s \rightarrow 2s)$ ratio of 2.05 – 2.10 expected for excitation by thermal electrons (Callaway 1988), we infer that it is likely that the Lyα and the 2-photon emission come from the same source. At the densities of the recombination regions in the hotspots ($n_H > 4 \times 10^6$ cm^{-3}; Groningsson *et al.* 2008), the 2-photon emission

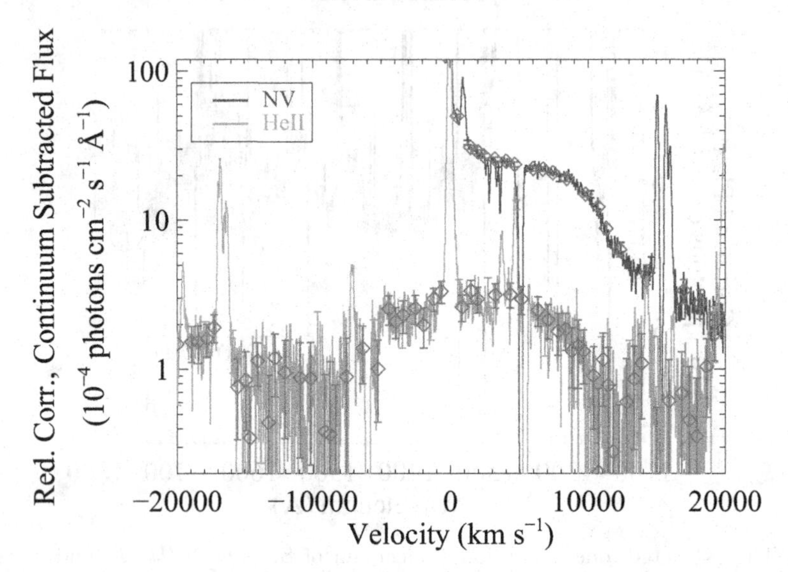

Figure 3. 2-photon continuum-subtracted NV and HeII in the extracted, one-dimensional spectra. The blue side of the NV distribution is lost under the RS Lyα emission. The HeII velocity profile appears qualitatively similar to that of NV.

should be dramatically suppressed (France *et al.* 2011) and therefore we conclude that the 2-photon emission continuum is dominated by the lower-density outer debris.

Is it reasonable to expect this level of X-ray heating of gas near the reverse shock? We calculate the total X-ray flux from the circumstellar ring using the two-component model spectrum of Zhekov *et al.* (2006), scaled to the total 0.5 – 2.0 keV luminosity observed by *Chandra* near day 8000 ($L(0.5 - 2.0 \text{ keV}) \approx 1.5 \times 10^{36}$ erg s^{-1}; Racusin *et al.* 2009). The total 2-photon luminosity is $L(2s) \approx 2.2 \times 10^{36}$ erg s^{-1}, therefore the 0.5 – 2.0 keV X-ray flux is insufficient to power the far-UV continuum. However, the total shock luminosity is most likely dominated by emission in the soft X-ray/EUV band (0.01 – 0.5 keV) that is attenuated by the neutral hydrogen in the interstellar media of the Milky Way and LMC (Fransson *et al.* 2011). The luminosity in this band inferred from the model by Zhekov *et al.* is $L(0.01 - 0.5 \text{ keV}) \approx 3 \times 10^{38}$ erg s^{-1}. Assuming that roughly half of this emission intersects the outer ejecta, we find that approximately 1.4% of the soft X-ray/EUV luminosity from the shocked ring must be reprocessed into H I 2-photon emission.

4. High-velocity Metal Emission Lines and CNO Processing in the Progenitor

Figure 2 shows the broad emission lines of N V, N IV], C IV, and He II imposed on the 2-photon continuum. Prior to the first STIS observations of SN 1987A, Borkowski *et al.* (1997) predicted that strong reverse shock emission from Li-like N V λ1240 would be detectable. This emission was not apparent, however, in the first deep far-UV STIS spectra presented by Sonneborn *et al.* (1998) and Michael *et al.* (1998). In recent (January 2010) STIS observations, we tentatively detected broad, redshifted N V emission (Figure 4), but low S/N precluded a detailed analysis. Now, in our COS observations, we unambiguously detect this emission. While the blue wing of the line is lost under the bright Lyα emission, we observe the complete red wing of the N V λ1240 velocity profile. In Figure 3, we compare the line profile of N V with that of He II, observing that the red

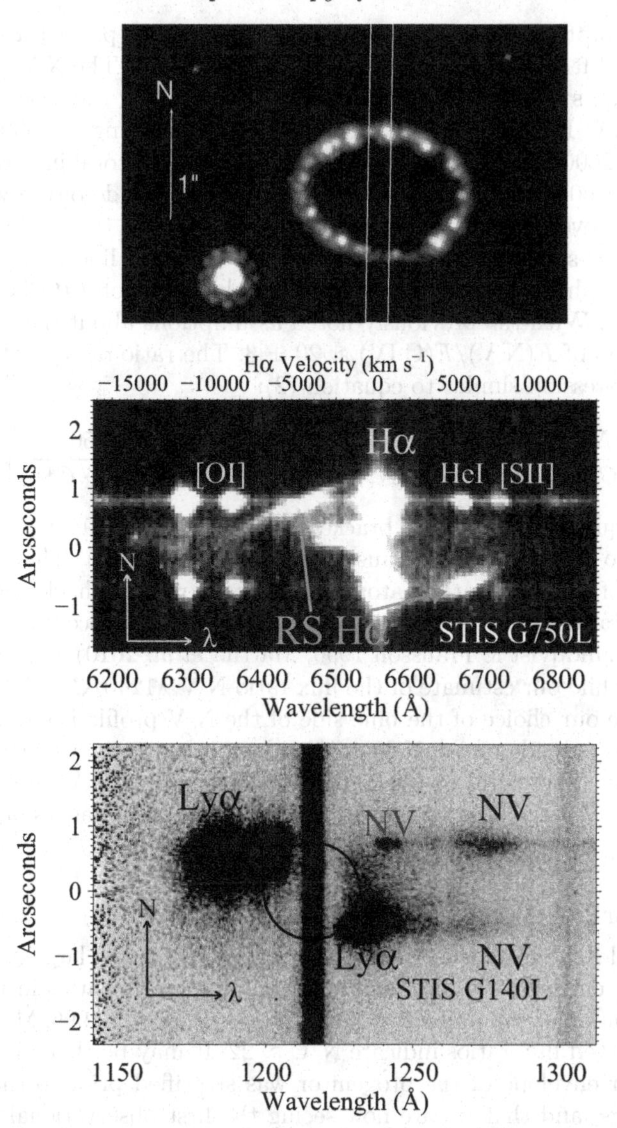

Figure 4. Long-slit imaging spectroscopy from the January 2010 *HST* observing campaign. *top*) *HST*-Advanced Camera for Surveys image obtained on 28 November 2003 in the F625W filter with an exposure time of 800s, illustrating the slit orientation used. (*middle*) *STIS* STIS G750L spectrum of SN 1987A, centered about the Hα emission line. The vertical bar at the center of the image is stationary Hα emission from interstellar or circumstellar gas. The bright spots at the north and south of this bar are the emissions from Hα + [N II] $\lambda\lambda$6548,6583 Å from hotspots on the equatorial ring. [O I] $\lambda\lambda$6300,6364, He I λ6678, and [S II] $\lambda\lambda$6716,6731 Å emission from the hotspots is also observed. The blueshifted streaks near the center are Hα emission excited by radioactivity in the interior of the supernova debris. The curved, blueshifted streak extending from the north side of the vertical bar and the redshifted streak on the south side (noted with orange arrows) are Hα emission from the brightest parts of the reverse shock. *bottom* STIS G140L observations of SN 1987A. The bright vertical stripe is the slit image in geocoronal Lyα. The blue ellipse approximates the location of the circumstellar emission ring. Broad, faint features seen on the north and south sides between ∼1260–1290 Å (labeled N V in blue) are produced by nitrogen ions in the reverse shock, and the shock-excited N V hotspot emission is labeled in magenta.

wings of the two profiles are qualitatively similar. Both ions present a boxy line profile, with a flat top and fall off between +9000 – +10000 km s^{-1}. The N V profile extends to at least +14000 km s^{-1}, where the data are contaminated by the geocoronal O I triplet.

We also detect C IV λ1550 from the reverse shock covering $\sim$ 1500 – 1580 Å, with emission from −12000 – −1500 km s^{-1} having 2.2 times the total integrated photon flux as the +1500 – +8000 km s^{-1} interval (Table 2). The red side of the velocity profile is only marginally above the noise level at $v_{CIV} > +3000$ km s^{-1}.

Interpolating the spectrum through the narrow emission lines, we measure an integrated broad line photon flux (−12000 – +8000 km s^{-1}) of F(C IV) = 7.7 × 10^{-3} photons cm^{-2} s^{-1}. With the previously-noted assumptions about the N V flux in mind, we find a flux ratio of F(N V)/F(C IV) ≈ 20 ± 3. The ratio of N V λ1240/C IV λ1550 is given by an expression similar to equation (2):

$$\frac{F(NV)}{F(CIV)} = \frac{x_N}{x_C}\left[\frac{R(1240)}{R(N^{4+} \rightarrow N^{5+})}\right]\left[\frac{R(1550)}{R(C^{3+} \rightarrow C^{4+})}\right]^{-1} \tag{4.1}$$

The ratio of the quantities in square brackets is very close to 0.9 for v_s = (5 – 12) × 10^3 km s^{-1} (the ratio is near unity because both are Li-like ions). That implies that the abundance ratio of nitrogen/carbon atoms crossing the reverse shock is $x_N/x_C \approx 22 \pm 3$, greater than the ratio $x_N/x_C \approx 8.5 \pm 3.5$ inferred from spectroscopic observations of the equatorial ring (Lundqvist & Fransson 1996; Mattila *et al.* 2010).

We note that while our estimate of the flux ratio N V λ1240/C IV λ1550 is somewhat uncertain because our choice of the blue side of the N V profile is speculative, the integrated flux in the red side of the N V profile alone is more than 10 times greater than the *total* C IV flux. Therefore, we are confident that the N V/CIV ratio is enhanced well beyond what can be attributed to the circumstellar ring abundances of the two species.

5. Future Work

One unresolved issue from the studies described above is the large N V/C IV ratio in the reverse shock emission. Carbon and nitrogen abundance ratios in the circumstellar ring suggest this number should be ≈ 8 (Lundqvist & Fransson 1996; Mattila *et al.* 2010), however the observed flux ratios indicate N/C ≈ 22. It may be that the N/C abundance ratio in the outer envelope of the progenitor was stratified prior to the ejection of the circumstellar rings, and that we are now seeing the first observational evidence of that stratification. A second possibility to account for this discrepancy is that ongoing thermonuclear processing continued to convert C to N in the supernova progenitor following the circumstellar ring ejection. The CNO bi-cycle will enrich the ^{14}N abundance at the expense of the abundances of ^{12}C and ^{16}O and, in equilibrium, will convert almost all of the primordial C and O into N (Caughlan & Fowler 1962). CNO processing has been invoked to explain the fact that the observed nitrogen abundance in the circumstellar ring is elevated by a factor ~10 over its value in the LMC (Fransson *et al.* 1989). Our observation that the He abundance does not change between the ring and reverse shock is qualitatively consistent with the 14E1 model presented by Shigeyama & Nomoto (1990). They show that the high-velocity material we observe crossing the reverse shock front is only a small fraction of the total ejected mass, and does not probe deep enough into the interior of the ejecta to observe significantly elevated He abundances.

If CNO processing continued near the stellar surface following the ejection of the circumstellar rings, it could have in principle converted most of the remaining C and O abundances seen in the equatorial ring into N. Heng *et al.* (2008) note reduced oxygen abundances, possibly related to the high N V/C IV ratio observed in our observations. For this explanation to be viable, the timescale to reach equilibrium in the CNO cycle must be $\lesssim$ 20,000 years, the interval since the ejection of the equatorial ring. Additional modeling of the progenitor, ejecta, and reverse shock would be useful for better understanding the large N V/C IV ratio that has been observed in our deep ultraviolet spectra of SN 1987A.

Acknowledgements

I would like to thank Dick McCray for the opportunity to study SN 1987A; Dick's enthusiasm and vast knowledge of supernova physics have made this collaboration a highlight of my research program. I also acknowledge enjoyable collaboration with the SAINTS team, in particular Bob Kirshner, Pete Challis, and Claes Fransson, during the course of this work. I appreciate the hard work invested by Steve Penton to produce the custom extractions from these challenging COS observations. The SAINTS team thanks Svetozar Zhekov for making his X-ray shock model available, and I also thank James Green for enjoyable discussions about the spectroscopic imaging capabilities of COS.

References

Callaway, J. 1988, PRA, 37, 3692

Caughlan, G. R. & Fowler, W. A. 1962, ApJ, 136, 453

Danforth, C. W., Keeney, B. A., Stocke, J. T., Shull, J. M., & Yao, Y. 2010, ApJ, 720, 976

France, K., Beasley, M., Keeney, B. A., Danforth, C. W., Froning, C. S., Green, J. C., & Shull, J. M. 2009, ApJl, 707, L27

France, K., McCray, R., Heng, K., Kirshner, R. P., Challis, P., Bouchet, P., Crotts, A., Dwek, E., Fransson, C., Garnavich, P. M., Larsson, J., Lawrence, S. S., Lundqvist, P., Panagia, N., Pun, C. S. J., Smith, N., Sollerman, J., Sonneborn, G., Stocke, J. T., Wang, L., & Wheeler, J. C. 2010, Science, 329, 1624

France, K., McCray, R., Penton, S. V., Kirshner, R. P., Challis, P., Laming, J. M., Bouchet, P., Chevalier, R., Garnavich, P. M., Fransson, C., Heng, K., Larsson, J., Lawrence, S., Lundqvist, P., Panagia, N., Pun, C. S. J., Smith, N., Sollerman, J., Sonneborn, G., Sugerman, B., & Wheeler, J. C. 2011, ApJ, 743, 186

Fransson, C., Cassatella, A., Gilmozzi, R., Kirshner, R. P., Panagia, N., Sonneborn, G., & Wamsteker, W. 1989, ApJ, 336, 429

Fransson, C., Larsson, J., Spyromilio, J., Chevalier, R., Groningsson, P., Jerkstrand, A., Leibundgut, B., McCray, R., Challis, P., Kirshner, R. P., Kjaer, K., Lundqvist, P., & Sollerman, J. 2012, ArXiv e-prints

Heng, K., Haberl, F., Aschenbach, B., & Hasinger, G. 2008, ApJ, 676, 361

Heng, K. & McCray, R. 2007, ApJ, 654, 923

Heng, K., McCray, R., Zhekov, S. A., Challis, P. M., Chevalier, R. A., Crotts, A. P. S., Fransson, C., Garnavich, P., Kirshner, R. P., Lawrence, S. S., Lundqvist, P., Panagia, N., Pun, C. S. J., Smith, N., Sollerman, J., & Wang, L. 2006, ApJ, 644, 959

Kozma, C. & Fransson, C. 1992, ApJ, 390, 602

Lundqvist, P. & Fransson, C. 1996, ApJ, 464, 924

Mattila, S., Lundqvist, P., Gröningsson, P., Meikle, P., Stathakis, R., Fransson, C., & Cannon, R. 2010, ApJ, 717, 1140

Michael, E., McCray, R., Borkowski, K. J., Pun, C. S. J., & Sonneborn, G. 1998, ApJ, 492, L143

Michael, E., McCray, R., Chevalier, R., Filippenko, A. V., Lundqvist, P., Challis, P., Sugerman, B., Lawrence, S., Pun, C. S. J., Garnavich, P., Kirshner, R., Crotts, A., Fransson, C., Li, W., Panagia, N., Phillips, M., Schmidt, B., Sonneborn, G., Suntzeff, N., Wang, L., & Wheeler, J. C. 2003, ApJ, 593, 809

Osterman, S., Green, J., Froning, C., Béland, S., Burgh, E., France, K., Penton, S., Delker, T., Ebbets, D., Sahnow, D., Bacinski, J., Kimble, R., Andrews, J., Wilkinson, E., McPhate, J., Siegmund, O., Ake, T., Aloisi, A., Biagetti, C., Diaz, R., Dixon, W., Friedman, S.,

Ghavamian, P., Goudfrooij, P., Hartig, G., Keyes, C., Lennon, D., Massa, D., Niemi, S., Oliveira, C., Osten, R., Proffitt, C., Smith, T., & Soderblom, D. 2011, ApS & S, 157
Racusin, J. L., Park, S., Zhekov, S., Burrows, D. N., Garmire, G. P., & McCray, R. 2009, ApJ, 703, 1752
Shigeyama, T. & Nomoto, K. 1990, ApJ, 360, 242
Sonneborn, G., Pun, C. S. J., Kimble, R. A., Gull, T. R., Lundqvist, P., McCray, R., Plait, P., Boggess, A., Bowers, C. W., Danks, A. C., Grady, J., Heap, S. R., Kraemer, S., Lindler, D., Loiacono, J., Maran, S. P., Moos, H. W., & Woodgate, B. E. 1998, ApJl, 492, L139+
Spitzer, L. 1978, Physical processes in the interstellar medium
Xu, Y., McCray, R., Oliva, E., & Randich, S. 1992, ApJ, 386, 181
Zhekov, S. A., McCray, R., Borkowski, K. J., Burrows, D. N., & Park, S. 2006, ApJ, 645, 293

Discussion

PODSIADLOWSKI: You argued that the high N/C ratio in the reverse shock indicates that there must have been significant CNO processing after the inner ring ejection. This of course strongly constrains evolutionary models. I have not looked at this specifically in my own models, but I know that the temperature in the H-burning shell in a post-merger model is high. So the conditions will be right somewhere, but to make this quantitative one needs the total amount that has been processed and also how this is mixed into the region where it is observed (as the star will have a radiative envelope in that phase). What are you going to do about this?

FRANCE: The authors have discussed this problem with Alex Heger, and have discussed the possibility of making a follow-up study comparing stellar models and HST observations. Admittedly, this has not gone anywhere in over a year. I would be happy to work with stellar modelers who have definitive predictions for N and C abundances so that we could compare them to observations.

SANKRIT: Can you detect any spatial variation of NV/CIV, or YN/XC, (or XN/YN) around the Ring/Ejecta?

FRANCE: With the limited angular resolution of HST-COS ($\sim 0.8'' - 1.0''$), we can separate the blue vs. red sides of the reverse shock emission; albeit with some overlap. We do see quantitatively and qualitatively different profiles on the red and blue sides in the different lines. However, the broad Reverse Shock Lines (except for Ly α and the red NII) are quite faint. Attempts to look at variations when breaking up the data into subregions results in plots too noisy to be useful.

CHEVALIER: Where do we stand with regard to determining the geometry of the RS, particularly out of the ring plane.

FRANCE: At present, we have slices of the RS. To understand the true, 3D geometry of the RS, one would want to step the $52 \times 0.2''$ STIS slit (6140L) across the circumstellar ring. The problem is that the ring is $\sim 2''$ in diameter. So that is 10 STIS pointings X 3-4 orbits per pointing, and you have 30-40 orbits. A political challenge for an HST Panel. However, we do have good evidence that the RS extends beyond / out of the ring plane. At the time of the HST-COS observations, the expected RS velocity was $\sim$ 7000 km/s, however we observed Ly α (blue) and NII (red) in excess of 12,000 km/s. Since $V \propto r/t$, this suggests that part of the emission comes from $r_{RS} > r_{CS}$, or out of the ring plane.

Supernova Environmental Impacts
Proceedings IAU Symposium No. 296, 2013
A. Ray & R. A. McCray, eds.

doi:10.1017/S1743921313009162

9500 Nights of Mid-Infrared Observations of SN 1987A: the birth of the remnant

Patrice Bouchet[1] and John Danziger[2]

[1]DSM/IRFU/Service d'Astrophysique, CEA-Saclay
L'Orme des Merisiers, F91191 Gif-sur-Yvette Cedex, France
email: Patrice.Bouchet@cea.fr

[2]INAF, Osservatorio Astronomico di Trieste, I-34143 Trieste, Italy
email: danziger@oats.inaf.it

Abstract. The one-in-a-life-time event Supernova SN 1987A, the brightest supernova seen since Kepler's in 1604, has given us a unique opportunity to study the mechanics of a supernova explosion and now to witness the birth of a supernova remnant. A violent encounter is underway between the fastest-moving debris and the circumstellar ring: shocks excite "hotspots". ATCA/ANTF, Gemini, VLT, HST, Spitzer, Chandra, and recently ALMA observations have been so far organized to help understanding the several emission mechanisms at work. In the mid-infrared SN 1987A has transformed from a SN with the bulk of its radiation from the ejecta to a SNR whose emission is dominated by the interaction of the blast wave with the surrounding interstellar medium, a process in which kinetic energy is converted into radiative energy. Currently this remnant emission is dominated by material in or near the inner equatorial ring (ER). We give here a brief history of our mid-infrared observations, and present our last data obtained with the *SPITZER* infrared satellite and the ESO VLT and Gemini telescopes: we show how together with Chandra observations, they contribute to the understanding of this fascinating object. We argue also that our imaging observations suggest that warm dust is still present in the ejecta, and we dispute the presence of huge amount of very cold dust in it, as it has been claimed on the basis of data obtained with the *HERSCHEL1* satellite.

Keywords. supernovae: SN 1987A, ISM: supernova remnants, infrared: ISM

1. Introduction

There are several ways in which the presence of dust may be shown to be associated with supernovae. Each individually may not be unequivocal concerning its precise location and time of origin. This is mainly due to the fact that dust can form in the ejecta, but it may also form or be already present in the CSM. SN 1987A provided the means of examining these methods. Possible indicators for dust in the ejecta are: (i) the early presence of molecules, (ii) an apparent blueward line shifts, and (iii) light curves showing a decrease in the visual together with an increase in the infrared emission, while no effect is seen on the bolometric light curve. Some of these methods have been used to infer the presence of dust in or near other supernovae but never all of them simultaneously, but for SN 1987A.

2. Dust in the ejecta of SN 1987A

Our ground based observations showed an infrared emission excess starting at day ~ 76 after outburst (Suntzeff & Bouchet 1990, and Bouchet, Danziger, & Lucy 1991a), as well as an early (<150 days after outburst) SiO and CO molecules formation (Danziger & Bouchet 1989). However, these data were not judged conclusive at the time as far as the dust origin was concerned, and it was not until March 1989 (day 736) that Danziger *et al.* 1989 reported from observations carried out in August - September 1988, that the emission line profiles of OI (630.0 nm and 636.3 nm) and CI (982.4 nm and 985.0 nm) had become asymmetric with their peak emission blueshifted by 500-600 km/s, and that similar behavior had been seen later on in the Na I and H-alpha profiles. These authors attributed this effect to extinction by dust, and were then the first ones to report on the discovery of dust condensation within the metal-rich ejecta. It is worth noting that they even gave a simple modeling of this newly discovered dust in their IAU circular, which was further on elaborated in subsequent papers (Lucy *et al.* 1989, Lucy *et al.* 1991). The spectral energy distribution longward of 5 μm has evolved continuously in time. After day $\sim$ 300 a clear dustlike component appeared. SN 1987A was then regularly observed in the mid-infrared (where >90% of the total energy was radiated) until day $\sim$ 2000 (Danziger *et al.* 1991, and Bouchet & Danziger 1993), when it had faded so much that it was not observable any longer with the 4m-class ground based telescopes at ESO/La Silla and CTIO/NOAO, and their associated infrared instruments. SN 1987A was observed at day 4100 with ISOCAM, on board the ESA ISO satellite (Fischera & Tuffs 2000), and Bouchet 2004 reported a possible weak detection of the supernova at day 4300 with OSCIR at the CTIO 4-m Blanco telescope. Except for these two observations, there have been no other detections of the mid-IR emission from the ejecta/ring region of SN 1987A between 1998 and 2004.

We had to wait for the advent of T-ReCS and VISIR, the mid-IR instruments at the 8-m Gemini South telescope, and at the ESO/VLT respectively, to resume our observations. They have not been interrupted since. Bouchet *et al.* 2004 together with the first ever resolved image of a supernova environment in the mid-IR reported a detection of the central ejecta at 10 μm at day 6067, obtained during the commissioning time of T-ReCS at Gemini/South; subsequent observations showed that this faint emission was still present at days 7241, 7565, 8720 and most probably at day 9375, as it can be seen in Figure 1 (Bouchet, Danziger, & de Buizer 2013). Our measurements together with the current theoretical models set a temperature of $80K < T_{Dust,Ejecta} < 100K$ and a mass range $M_{Dust,Ejecta}$= 3 x 10^{-4} - 2 x 10^{-3} $M_{\odot}$ (Bouchet *et al.* 2006). The warm dust discovered at day 430 had not been destroyed and that it is what we are still observing now.

3. Dust in the inner equatorial ring

3.1. *ESO/VLT and Gemini-South ground based observations*

The observed mid-IR flux in the region of SN 1987A is dominated by emission from dust in the inner equatorial ring (ER), which we have detected and resolved with T-ReCS and VISIR (Bouchet *et al.* 2004; Bouchet *et al.* 2006). So-called Hot spots similar to those found in the optical and near-IR are clearly present, and the morphology of the 10 μm

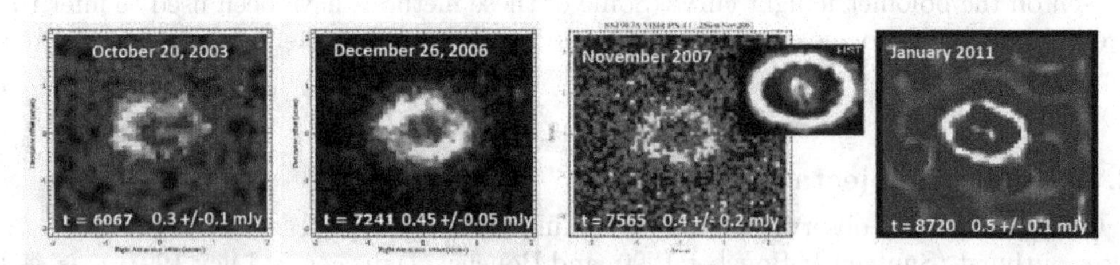

Figure 1. 10 μm imaging of SN 1987A at different epochs. Note the faint emission near but not exactly at the center of the ER.

emission is globally similar to the morphology at other wavelengths from X-rays to radio (Figure 2). We do not have space here to discuss the changes in morphology, but see Bouchet, Danziger, & de Buyzer 2013. The mid-IR emission in the ER originates from 180 K dust collisionally-heated by X-ray emitting gas of T $\sim 5 \times 10^6$ K, with density $(2-4) \times 10^4$ cm^{-3}. The mass of this radiating dust is $\sim 1.2 \times 10^{-6}$ M$_\odot$ (at day 7554) and it scales linearly with the IR flux. Note that there is no significant increase of the mid-IR flux between days 8708 and 9375.

3.2. *SPITZER Observations*

We have used the *SPITZER* satellite to monitor the mid-IR evolution of SN 1987A over a 5 year period spanning the epochs between days $\sim$ 6000 and 8000 since the explosion. Its radiative output during this epoch is dominated by the interaction of the SN blast wave with the pre-existing ER. The main results of our study can be briefly summarized as follows (Dwek *et al.* 2010):

(*a*) The dust grains are mainly silicates and their emission increased as $t^{0.87}$ up to day 8708, consistent with X-rays observations suggesting that the blast wave has transitioned from a free expansion to the Sedov phase (now expanding into the main body of the ER)(Figure 3a).

(*b*) A secondary emission component, the nature of which remains a mystery, dominates the spectrum in the 5 - 8 μm region: its intensity and spectral shape rule out any possible gas or synchrotron emission mechanism as the source of this emission. It must therefore be attributed to a secondary dust component radiating at temperatures above 350 K (Dwek *et al.* 2010) (Figure 3b). The question of whether this could be a clue to a binary coalescence origin of the progenitor should be addressed. However, it has to be stressed that the low angular resolution of the IRS (InfraRed Spectrograph) on board *SPITZER* (1.8 arcsec/pixel at most for the shortest wavelengths) is too low to argue that this second component lies inside the ER.

(*c*) The grain radii or IR emissivities of this secondary component must be significantly smaller than those of the silicates. Their sputtering lifetime could therefore be significantly shorter than, and their evolution quite different from, that of the silicates.

(*d*) The overall shape of the $\sim 5-8\mu$m dust spectrum has not changed during the observations, suggesting that the density and temperature of the soft X-ray emitting gas have not significantly changed during more than 5 years of observations. The spectral shape of this IR emission is remarkably constant, suggesting also that the mass ratio of the silicate to the secondary dust component remained roughly constant during that period.

(*e*) The infrared-to-X-ray ratio (IRX) is constant at 2.5 throughout this epoch. This value shows that the cooling of the shocked gas is dominated by IR emission from the

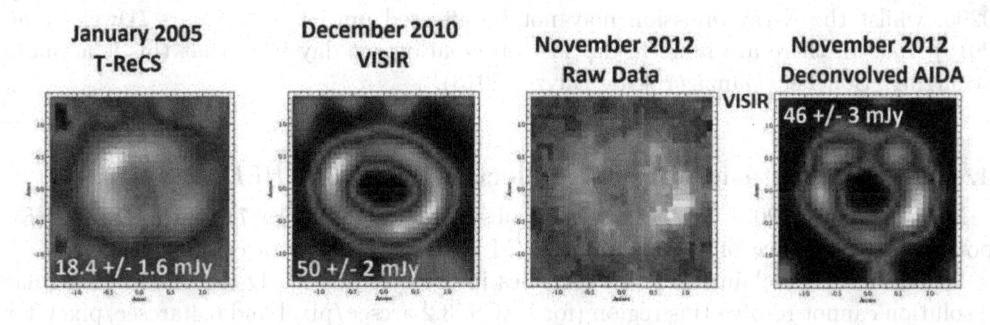

Figure 2. 10 μm imaging of the ER around SN 1987A at different epochs.

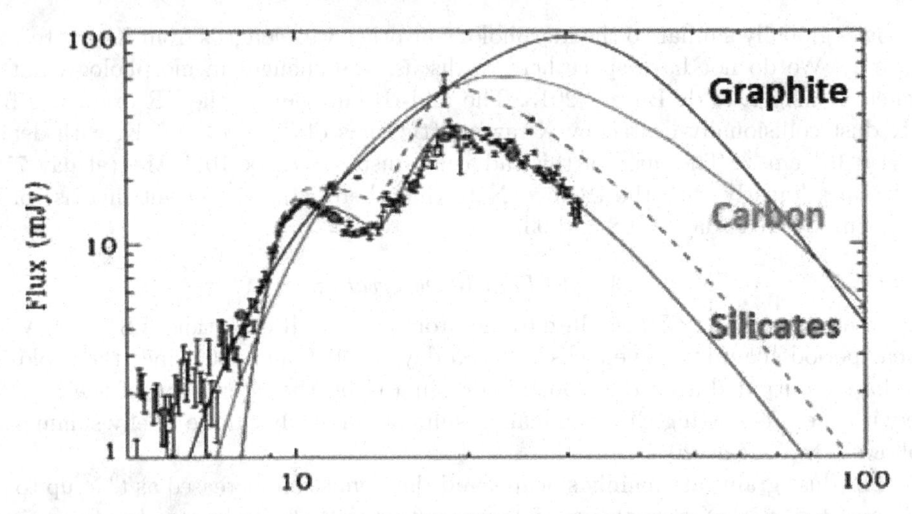

Figure 3. The low resolution spectrum of SN 1987A obtained with the InfraRed Spectrograph (IRS) on board the infrared satellite *SPITZER* with theoretical fitting for three grains species. It is clear that silicates are the main dust component.

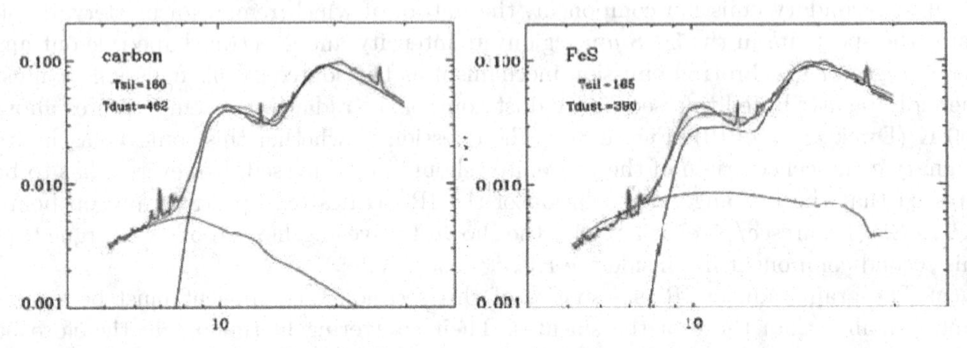

Figure 4. The low resolution IRS spectrum of the SN 1987A environment with two fits of the blue region excess emission corresponding to the second dust component. Carbon species at T = 462 K give the best fit for a corresponding temperature of the silicates T = 180 K. The temperature would be slightly lower (T $\sim$ 39 K) were FeS grains dominate.

collisionally-heated dust with radii >0.30 μm, and that a significant fraction of the refractory elements in the ER is depleted onto dust.

(*f*) The constancy of the IRX means that neither grain destruction by sputtering nor cooling of the shocked gas has played a significant role during the epoch of the observations. It was expected that grain destruction becomes important only at day 9200, whilst the X-ray emission may not be affected until t $\sim$ 30 years (Dwek *et al.* 2010). Indeed, there are hints in our last observations at day 9375 that this is actually occurring (Bouchet, Danziger & de Buyzer 2013).

4. The Interpretation of data collected by *HERSCHEL*

Matsuura *et al.* 2011 argue that their observations of SN 1987A with *HERSCHEL* point to the presence of a huge amount (M = 0.4 - 0.7 $M_\odot$) of cold dust (17 K < T < 23 K), and they claim that this dust lies in the ejecta. Clearly the Herschel angular resolution cannot resolve this region (for PACS, 3.2 arcsec/pixel and 6.4 arcsec/pixel, for the short and long wavelengths respectively, while - I would like to remind the audience

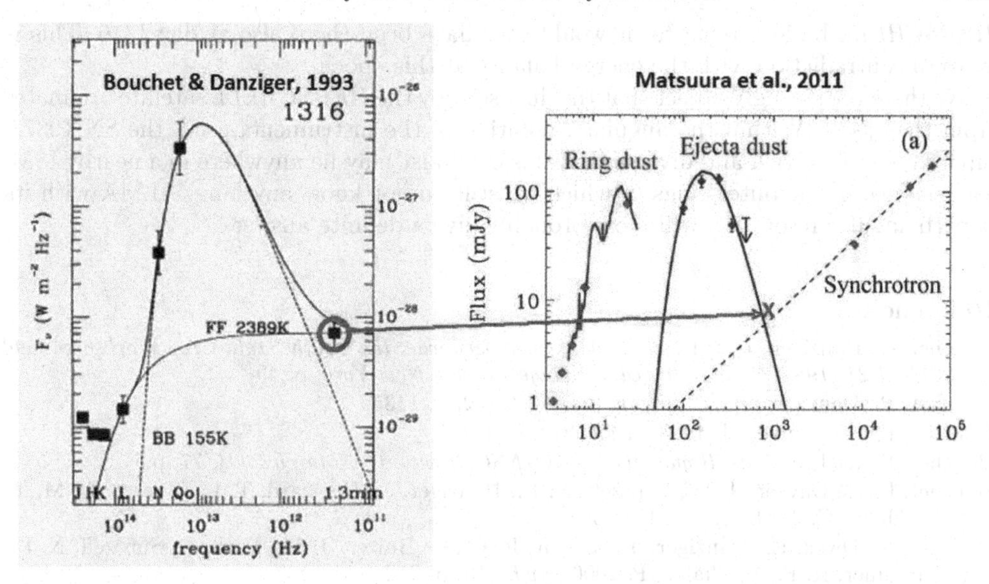

Figure 5. The SED of SN 1987A at day 1316, and the one derived from *Herschel* observations. Note the data point at 1.3 mm obtained with the SEST telescope at La Silla.

- the ER is about 1 arcsecond diameter wide), and there are strong arguments against this interpretation. Among them, let me quote the following ones:

- When would have this dust appeared in the ejecta? The observed energy budget up to day 2100, more than 90% of which is radiated in the mid-IR by the warm dust discovered at day 430, was perfectly consistent with a powering by the radioactive decays of ^{56}Co, ^{57}Co, ^{44}Ti and ^{22}Na, and some theoretical considerations as time-dependent effects due to long recombination and cooling time leading to a frozen-in structure of the ejecta (Fransson & Kozma 1993). The observed bolometric light curve agrees remarkably well with these theoretical models and there was then no room from the '*HERSCHEL*' dust at that time.
- We strongly believe that our mid-IR observations show that the warm dust is still present in the ejecta at the time of the *HERSCHEL* observations: although the issue of whether both warm and cold dust could cohabit should be addressed, it seems highly improbable.
- We could envisage that a fraction of the 'warm' dust has cooled down. However, how a few 'warm' 10^{-4} $M_\odot$ could lead to 0.4 - 0.7 $M_\odot$ of cold dust?
- According to Matsuura *et al.*'s (2011) model, all refractive elements must go into dust. Nonetheless, there is no evidence of the enormous absorption in the visible which would be expected; moreover, there is emission from the refractive elements in the debris so not all by far has gone into grains.
- The temperature of the '*HERSCHEL*' dust is the same as the general ISM 20 K and inside the debris heated by strong shocks, X-rays and radioactivity, which seems highly suspicious to us.
- We have been monitoring regularly the supernova at 1.3mm with the ESO/Swedish SEST radio telescope at La Silla between 1989 and 1997. Bouchet, Danziger & Lucy 1991) report a data point of $\sim$ 9 mJy at this wavelength for day 1316. Figure 4 shows that this value is surprisingly quite consistent with the spectral energy density shown in Matsuura *et al.*'s (2011) figure 2: this suggests that the flux density at this wavelength has remained constant for more than $\sim$ 8000 days, and should the dust detected by

HERSCHELL be in the ejecta, it would thus have been there also at day 1316. This is in total contradiction with the energy balance at this epoch.

We therefore strongly object that the dust seen by the *HERSCHELL* satellite originates from the ejecta. Within the angular resolution of the instruments used, the SN 1987A environment is so rich and diverse that this cold dust may lie anywhere in a nearby ISM, for instance in the outer rings of which we still do not know anything. ALMA with its superb angular resolution will most probably give a definite answer.

References

Bouchet, P., Danziger, J., & Lucy, L 1991, *in Supernovae: the Tenth Santa Cruz Workshop, held July 9 21, 1989; S. Woosley ed., Springer-Verlag New-York*, p. 49

Bouchet, P., Danziger, J., & Lucy, L 1991, *AJ*, 102, p. 1135

Bouchet, P. & Danziger, J. 1993, *A&A*, 273, p. 451

Bouchet, P. 2004, *Science Highlights NOAO/NSO Newsletter, March 2004*, 77, p.5

Bouchet, P., de Buyzer, J. M., Suntzeff, N. B., Danziger, J., Hayward, T. L., Telesco, C. M., & Packham, C. 2004, *ApJ*, 611, p. 394

Bouchet, P., Dwek, E., Danziger, J., Arendt, R. G., de Buizer, J. M., Park, S., Suntzeff, N. B., Kirsshner, R. P., & Challis, P. 2006, *ApJ*, 650, p. 212

Bouchet, P., Danziger, J., & de Buyzer, J. 2013, *in preparation*

Danziger, J., Gouiffes, C., Bouchet, P., & Lucy, L. 1989, *IAU Circ. 4746; Green D.W.E. ed., March 1989*

Danziger, J. & Bouchet, P. 1989, *in Evolutionary Phenomena in Galaxies, CUP*, p.283

Dwek, E., Arendt, R. G., Bouchet, P., Burrows, D. N., Challis, P., Danziger, J., de Buizer, J. M., Gehrz, R. D., Kirshner, R. P., McCray, R., Park, S., Polomski, E. F., & Woodward, C. E. 2008, *ApJ*, 676, p. 1029

Dwek, E., Arendt, R. G., Bouchet, P., Burrows, D. N., Challis, P., Danziger, J., de Buizer, J. M., Gehrz, R. D.,Park, S., Polomski, E., Slavin, J., & Woodward, C. E. 2010, *ApJ*, 722, p. 425

Fransson, C. & Kozma, C. 1993, *ApJ*, 408, L25

Lucy, L., Danziger, J., Gouiffes, C., & Bouchet, P. 1989, *in Structure and Dynamics of the Interstellar Medium, IAU Colloq. 120, G. Tenorio-Tagle, M. Moles & J. Melnick eds., Lect. Notes in Phys.*, 350, p. 164

Lucy, L., Danziger, J., Gouiffes, C., & Bouchet, P. 1991, *in Supernovae: the Tenth Santa Cruz Workshop; S. Woosley ed., Springer-Verlag New-York*, p. 82

Matsuura, M., *et al.* 2011, *Science, Volume 333*, Issue 6047, pp. 1258

Suntzeff, N. & Bouchet, P. 1990, *AJ*, 99(2), p. 650

Discussion

CESARSKY: So, you do not believe that *Herschel* detected that amount of cold dust in the ejecta?

BOUCHET: Absolutely and definitely not! These data are indeed very valuable, and I agree that most probably this far-infrared emission originates from dust, but I object that this dust be in the ejecta. As I said, there are many places where it could lie, and my guess goes toward the outer rings. But this is just a guess

Supernova Environmental Impacts
Proceedings IAU Symposium No. 296, 2013
A. Ray & R. A. McCray, eds.

doi:10.1017/S1743921313009174

Radio Observations of Supernova 1987A

L. Staveley-Smith[1,2], T. M. Potter[1], G. Zanardo[1], B. M. Gaensler[2,3] and C.-Y. Ng[4]

[1]International Centre for Radio Astronomy Research, University of Western Australia, Crawley, WA 6009, Australia
email: Lister.Staveley-Smith@icrar.org

[2]ARC Centre of Excellence for All-sky Astrophysics (CAASTRO)

[3]Sydney Institute of Astronomy, School of Physics, The University of Sydney, NSW 2006, Australia

[4]Department of Physics, The University of Hong Kong, Pokfulam Road, Hong Kong

Abstract. Supernovae and their remnants are believed to be prodigious sources of Galactic cosmic rays and interstellar dust. Understanding the mechanisms behind their surprisingly high production rate is helped by the study of nearby young supernova remnants. There has been none better in modern times than SN1987A, for which radio observations have been made for over a quarter of a century. We review extensive observations made with the Australia Telescope Compact Array (ATCA) at centimetre wavelengths. Emission at frequencies from 1 to 100 GHz is dominated by synchrotron radiation from an outer shock front which has been growing exponentially in strength from day 3000, and is currently sweeping around the circumstellar ring at about 4000 km s^{-1}. Three dimensional models of the propagation of the shock into the circumstellar medium are able to reproduce the main observational features of the remnant, and their evolution. We find that up to 4% of the electrons encountered by the shock are accelerated to relativistic energies. High-frequency ALMA observations will break new ground in the understanding of dust and molecule production.

Keywords. Supernovae, SN1987A, radio astronomy

1. Overview

As the brightest supernova since Kepler's of 1604, SN1987A has been a Rosetta Stone in our understanding of the physics of Type II supernova explosions. It was the first supernova with a known progenitor, and the brightest since the invention of the telescope. It has therefore allowed a number of unprecedented studies including: detailed comparison with models of the very final stages of stellar evolution; the detection of the first neutrinos from an extrasolar source, providing evidence for the formation of a neutron star (Vissani *et al.* 2010); measurement of absolute distance using the light travel time method (Panagia *et al.* 1991); the probing and analysis of the circumstellar medium excited by the ultraviolet flash; measurement of the radioactive decay products of the explosion; study of the interaction of the expanding blast wave with the anisotropic circumstellar medium laid out by the stellar winds from various stages of the progenitor star; and formation of dust and molecules in the cooling ejecta. The fact that, to date, over 2000 refereed papers have been written about SN1987A, or have mentioned SN1987A in their abstracts, is testament to the significance of the event.

In this paper, we discuss radio observations in the two and a half decades following the explosion. These observations are particularly sensitive to detailed shock physics, magnetic field amplification and circumstellar structure, and have allowed us to clearly

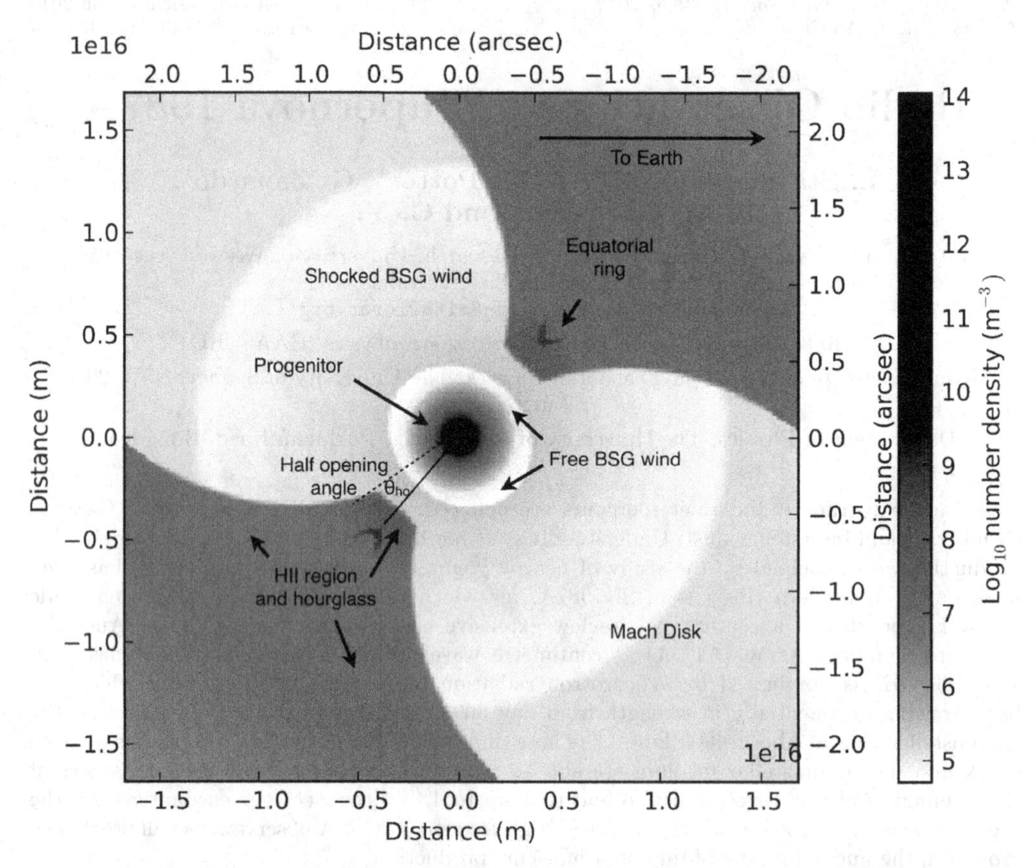

Figure 1. A model of the initial 'hourglass' environment prior to explosion. The environment was created by the progenitor star in its evolution from red supergiant to blue supergiant (BSG). The supernova shock is currently at the position of the equatorial ring the ring is perpendicular to the plane of the paper (Potter 2012). A Mach disk is formed at the point where the BSG wind again goes supersonic. The axes are labelled in arcsec and in units of 10^{16} m.

follow the evolution of a Type II supernova into the important supernova remnant (SNR) phase.

2. Environment

The initial interaction of the SN1987A blast wave with the circumstellar medium resulted in the generation of a burst of synchrotron radiation lasting a few days (Turtle *et al.* 1987; Storey & Manchester 1987). Radio (and X-ray) observations of other, more distant, type II supernovae have been used to compare progenitor mass-loss rates using the theoretical framework laid out by Chevalier (1982), and to probe the uniformity of the mass loss. Similar studies on SN1987A were consistent with the observation that the progenitor was a blue supergiant, with a fast but tenuous stellar wind at the time of explosion. However SN1987A was notable in that the luminosity of this initial radio burst, which peaked at about day 3, was around 10^4 times weaker than powerful type II supernovae such as SN1993J. Nevertheless, following the discovery by the ESO New Technology Telescope (Wampler *et al.* 1990) and the HST of a ring-like structure around SN1987A, there was some expectation of a re-brightening of the radio emission. This was seen at day 1200 by the MOST and ATCA telescopes (Staveley-Smith *et al.* 1992).

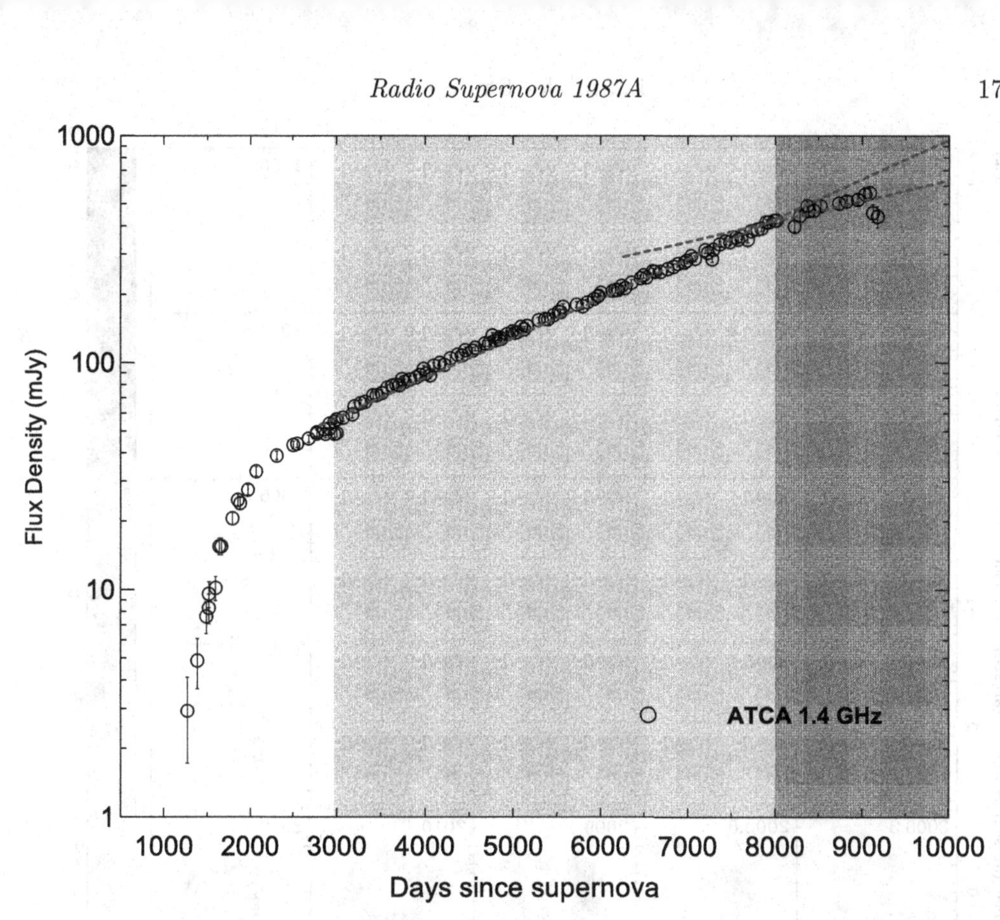

Figure 2. Radio observations at 1.4 GHz have been conducted with ATCA at intervals of 4-6 weeks since the remnant was re-detected around day 1200. Its evolution is divided into three phases: (a) a linear phase up to day 3000; (b) an exponential phase to day 8000 (light shading); and (c) a slower exponential phase from day 8000 (darker blue shading). The red dotted line is the fit of Zanardo *et al.* (2010).

As a result of early optical and radio observations, and studies of the light echoes of the explosion itself (Crotts, Kunkel & McCarthy 1989), a detailed picture of the environment around the supernova has been constructed. The main features of the circumstellar environment are illustrated in Fig. 1. The well-known equatorial ring, which defines the waist of an hourglass feature, is perpendicular to the plane of the paper. The northernmost edge of the ring is closest to Earth. Within the hourglass, there are (pre-supernova) regions which signify both freely expanding wind and shocked wind. It is at the interface of the two where it is believed that the first radio emission was generated (Chevalier & Dwarkadas 1995).

3. Evolution

After re-emergence of the radio remnant at day 1200, the flux density was seen to increase linearly with time until day 4500, when small departures were evident (Manchester *et al.* 2002). In datasets from day 3000 to 8000, a better description is an exponential increase (Zanardo *et al.* 2010), perhaps reflecting the exponential nature of the particle acceleration process and/or the increasing area of the expanding shock front. Observational data from the ATCA at 1.4 GHz is shown in Fig. 2. Similar exponential evolution was seen in the X-rays, though the detailed temporal behaviour has been quite different

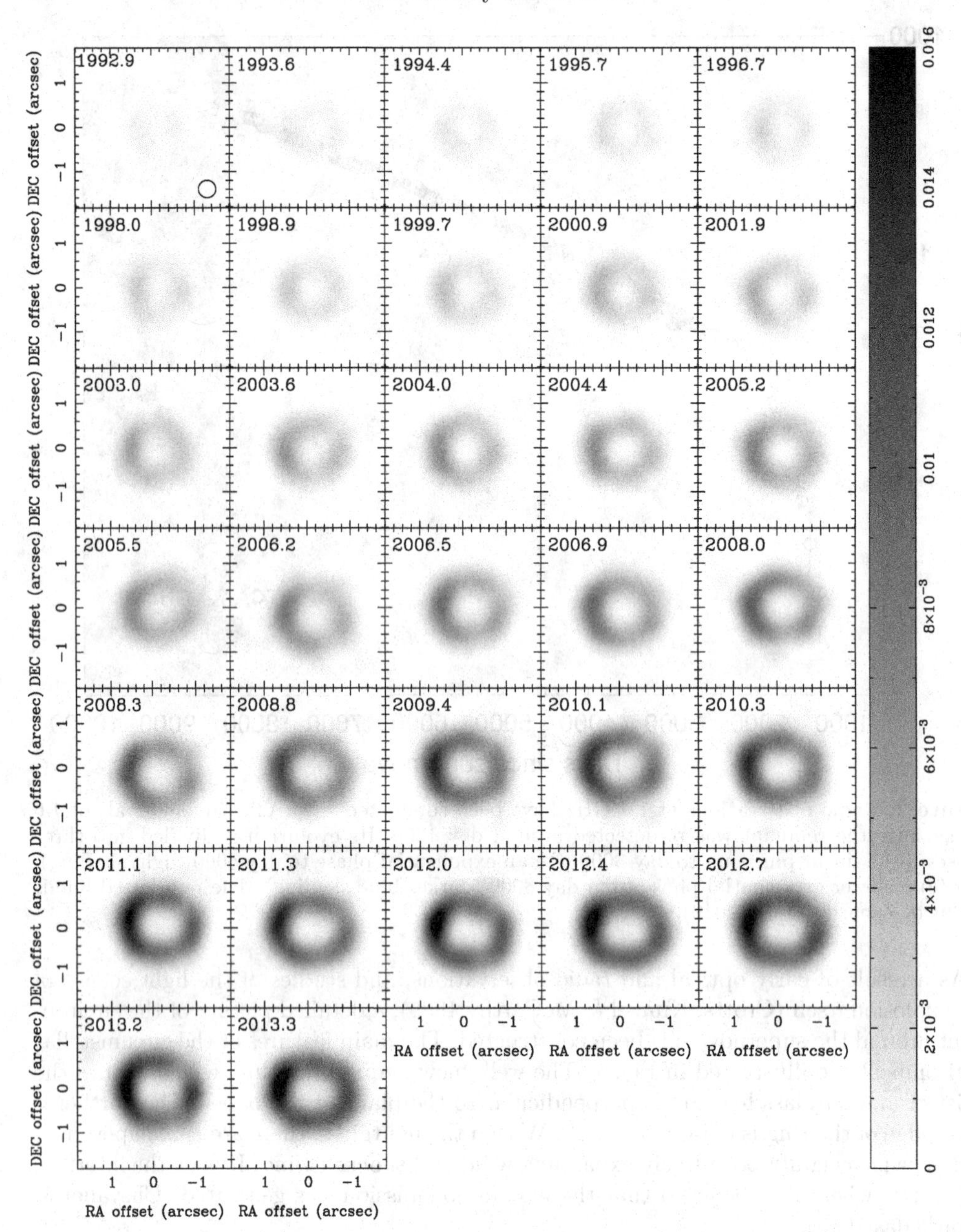

Figure 3. Radio observations at 9 GHz have been conducted with ATCA at half-yearly or yearly intervals since the SN1987A remnant was first resolved in 1992. The above panels illustrate: (a) the strengthening radio emission; (b) the change from a circular (torus) to an elliptical (equatorial ring) morphology; and (c) the increasing diameter of the remnant. The year (CE) of the observation is listed in the top-left corner of each panel. Note the asymmetric nature of the remnant. Data from Ng *et al.* (2013, in preparation).

(Helder *et al.* 2013). This is now understood to be due to the different emission regions and the dominance of X-ray thermal radiation over synchrotron radiation dominant at radio wavelengths.

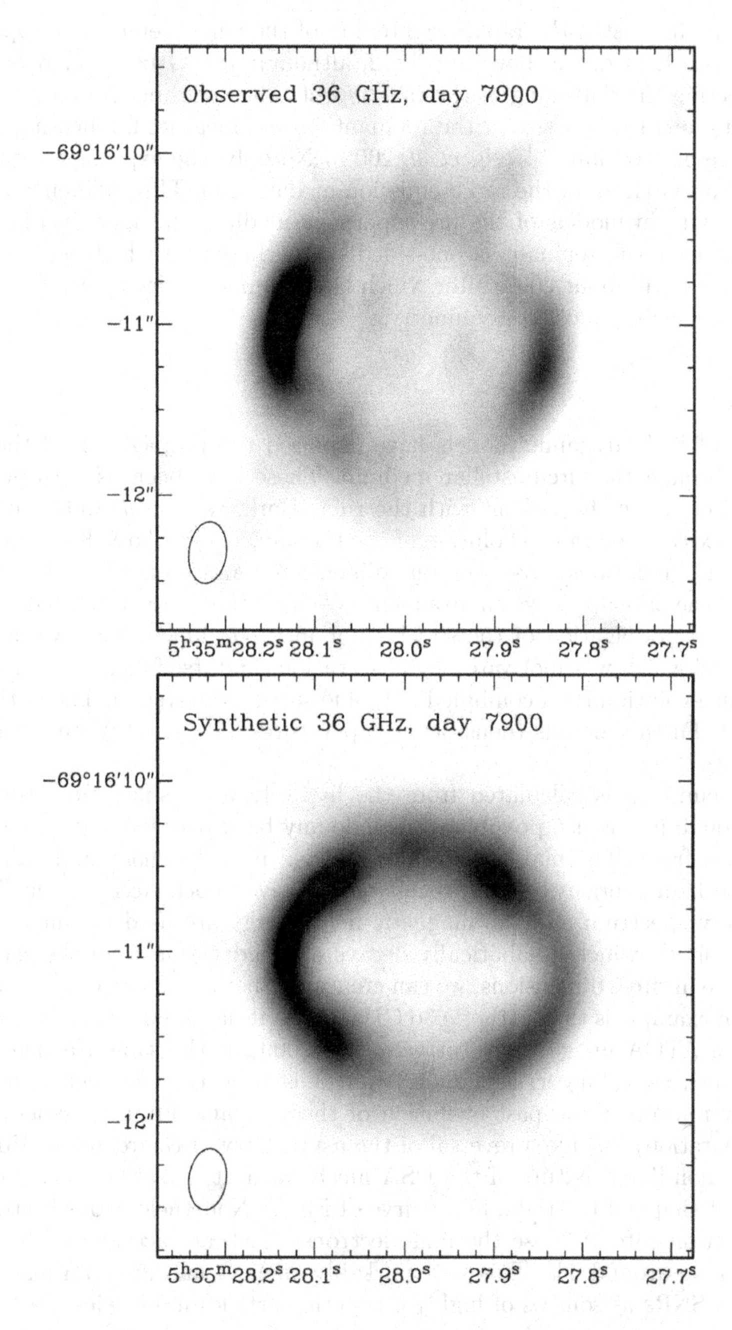

Figure 4. *Top:* A 36-GHz ATCA image of the radio remnant of SN1987A on day 7900 (Potter *et al.* 2009). *Bottom:* multi-dimensional hydrodynamic DSA model of the remnant at the same frequency and on the same day (Potter 2012).

Interpretation of the radio light curve has been assisted by high-resolution radio observations with the Australia Telescope Compact Array (ATCA), particularly when the remnant became strong enough to apply the technique of 'super-resolution'. This allows a resolution increase by a factor of 1.5 to 2 for a compact source like SN1987A with the available signal-to-noise ratio. A compilation of all 9 GHz full uv-coverage ATCA images is shown in Fig. 3. Firstly, the rapid brightening of the whole remnant is apparent. This reflects the radio light curve shown in Fig. 2, although at 9 GHz the increase is slightly steeper, reflecting the flattening of the radio spectrum over time. Secondly, on close inspection, the growth in the size of the remnant can be measured, reflecting the current expansion rate of 4000 km s^{-1} (Ng *et al.* 2008). Notably, the expansion rate was much higher prior to switch-on of the radio emission at day 1200. This difference is explained as an inner cavity in models of the pre-supernova medium as shown in Fig. 1. Finally, it is apparent that the remnant is one-sided, with the eastern half being consistently $\sim 30\%$ brighter until about 2007, after which the asymmetry decreases. The most likely explanation of this is explosion asymmetry.

4. Models

A number of hydrodynamic models have explored the propagation of the SN1987A shock front through the circumstellar medium. These have been used to predict dates for the collision of the shock front with the ring (Borkowski *et al.* 1997) and have also been used to explain the radio evolution of the remnant (Berezhko & Ksenofontov 2006). However, most simulations have been one-dimensional and do not capture the inherent asymmetry of the pre-supernova environment. We have therefore attempted to model the hydrodynamical propagation of the shock front in three dimensions. We do this using an initial model which was motivated by observations and itself created from a plausible hydrodynamic evolution for a combined red/blue supergiant system. The initial model is that of Fig. 1. Further details (densities, temperatures, dimensions) are given in Potter (2012).

The radio emission is calculated from the hydrodynamic simulation using sub-grid physics, although it was not possible to include any back-reaction from the cosmic rays onto the shock front. Thermal electron are injected into the shock and assumed to be accelerated to high energies by the process of Diffusive Shock Acceleration (DSA). The electron energy spectrum and the magnetic field energy are used to calculate the synchrotron emissivity, which adiabatically decays after advection from the shock. As the simulations are in three dimensions, we can create synthetic radio images at any time and frequency. An example is the synthetic 36 GHz image at day 7900, which is plotted alongside an actual ATCA image from Potter *et al.* (2009) at the same frequency in Fig.4. Parameters such as diameter, asymmetry and axis ratio can be calculated, and seem to accurately reproduce the past evolution of the remnant. Future predictions (Potter *et al.* in preparation) include a reversal of the asymmetry of the remnant. However, due to the highly non-linear nature of the DSA mechanism, it is more challenging to reproduce the exact shape of the radio light curve of Fig. 2. Nonetheless, our best fit suggests that the injection rate of those thermal electrons which are accelerated to cosmic ray energies must be around 4%. This is remarkably efficient and gives an insight into the importance of SNRs as sources of highly energetic particles in galaxies.

5. Future

SNRs are believed to be important sources of cosmic rays in galaxies. These particles are important in facilitating the conversion of atomic gas into molecular form by providing charge to dust grains deeply embedded in these clouds. They also provide interstellar pressure that slows the accretion of cool gas onto galaxies. Cosmic ray acceleration processes can be further studied by tracking the evolution of the radio emission at the forward shock during the ongoing transition of SN1987A into an SNR.

High energy particles are also created by magnetised spinning neutron stars. Although no pulsar has yet been detected, it is believed that a neutron star was created during the initial explosion. Evidence for this comes from the $\sim$ 21 non-background neutrinos seen by the Kamiokande II, IMB and Baksan detectors about 2 hrs before the visible flash (Vissani *et al.* 2010). Zanardo *et al.* (2013) suggest that the conditions may now be favourable for detection of a pulsar. Indeed, tentative signs of a flat spectrum central component have been found in previous multi-frequency radio studies (Potter *et al.* 2009; Zanardo *et al.* 2013).

Finally, sub-mm observations indicate the presence of large amounts of cold dust in the ejecta of SN1987A (Matsuura *et al.* 2010). This is exciting as it implies that dust formation in supernovae is probably more important than previously believed, and perhaps resolves the puzzle of the rapid appearance of large quantities of dust at high redshift. Although not the subject of this paper, multi-frequency and spectral-line observations with ALMA will be crucial in the measurement of dust and gas mass and temperature, gas kinematics and in-situ dust and molecule formation.

Acknowledgements

Parts of this research were conducted by the Australian Research Council Centre of Excellence for All-sky Astrophysics (CAASTRO), through project number CE110001020.

References

Berezhko, E. G. & Ksenofontov, L. T. 2006, *ApJ*, 650, 59

Borkowski, K. J., Blondin, J. M., & McCray, R. 1997, *ApJ*, 476, 31

Chevalier, R. A. 1982, *ApJ*, 259, 302

Chevalier, R. A. & Dwarkadas, V. V. 1995, *ApJ*, 452, L45

Crotts, A. P. S., Kunkel, W. E., & McCarthy, P. J. 1989, *ApJ*, 347, 61

Helder, E. A., Broos, P. S., Dewey, D., Dwek, E., McCray, R., Park, S., Racusin, J. L., Zhekov, S. A., & Burrows, D. N. 2013, *ApJ*, 764, 11

Manchester, R. N., Gaensler, B. M., Wheaton, V. C., Staveley-Smith, L., Tzioumis, A. K., Bizunok, N. S., Kesteven, M. J., & Reynolds, J. E. 2002, *PASA*, 19, 207

Matsuura, M. *et al.* 2010, *Science*, 333, 1258

Ng, C.-Y., Gaensler, B. M., Staveley-Smith, L., Manchester, R. N., Kesteven, M. J., Ball, L., & Tzioumis, A. K. 2008, *ApJ*, 684, 481

Panagia, N., Gilmozzi, R., Macchetto, F., Adorf, H.-M., & Kirshner, R. P. 1991, *ApJ*, 380, 23

Potter, T. M. 2012, *Radio Observations and Multi-dimensional Simulations of the Expanding Remnant of SN1987A*, Ph.D thesis, University of Western Australia

Potter, T. M., Staveley-Smith, L., Ng, C.-Y., Ball, L., Gaensler, B. M., Kesteven, M. J., Manchester, R. N., Tzioumis, A. K., & Zanardo, G. 2009, *ApJ*, 705, 261

Staveley-Smith, L., Manchester, R. N., Kesteven, M. J., Reynolds, J. E., Tzioumis, A. K., Killeen, N. E. B., Jauncey, D. L., Campbell-Wilson, D., Crawford, D. F., & Turtle, A. J. 1992, *Nature*, 355, 147

Storey, M. C. & Manchester, R. N. 1987, *Nature*, 329, 421

Turtle, A. J., Campbell-Wilson, D., Bunton, J. D., Jauncey, D. L., Kesteven, M. J., Manchester, R. N., Norris, R. P., Storey, M. C., & Reynolds, J. E. 1987, *Nature*, 327, 38

Vissani, F., Costantini, M. L., Fulgione, W., Ianni, A., & Pagliaroli, G. 2010, in *Frontier Objects in Astrophysics and Particle Physics*, 2010 Vulcano workshop, arXiv:1008.4726

Wampler, E. J., Wang, L., Baade, D., Banse, K., D'Odorico, S., Gouiffes, C., & Tarenghi, M. 1990, *ApJ*, 362, 13

Zanardo, G., Staveley-Smith, L., Ball, L., Gaensler, B. M., Kesteven, M. J., Manchester, R. N., Ng, C.-Y., Tzioumis, A. K., & Potter, T. M. 2010, *ApJ*, 710, 1515

Zanardo, G., Staveley-Smith, L., Ng, C.-Y., Gaensler, B. M., Potter, T. M., Manchester, R. N., & Tzioumis, A. K. 2013, *ApJ*, 767, 98

Discussion

CHAKRABORTI: Do you see the spectral index change with time?

STAVELEY-SMITH: The spectral index has flattened considerably from an initial value of -1 to the current value of -0.7 ($S \propto \nu^{\alpha}$) - see Zanardo *et al.* (2010).

BARTEL: What is your current limit on the spectral luminosity of any pulsar wind nebula that may exist in the center of SN1987A, e.g. in terms of the spectral luminosity of the Crab Nebula?

STAVELEY-SMITH: The limit at frequencies of 8 GHz and higher for a Central Nebula or pulsar is of the order of 1 mJy. This corresponds to a small fraction ($\sim$1%) of the spectral power of the Crab Nebula.

WANG: Do you have any handle on the total mass of the ring?

STAVELEY-SMITH: Not really. The synchrotron component arises in the lower density components. There are some constraints on the ionized gas mass.

Supernova Environmental Impacts
Proceedings IAU Symposium No. 296, 2013
A. Ray & R. A. McCray, eds.

doi:10.1017/S1743921313009186

The radio remnant of Supernova 1987A at high frequencies and high resolution

G. Zanardo[1], L. Staveley-Smith[1,4], C. -Y. Ng[2], B. M. Gaensler[3,4] T. M. Potter[1], R. N. Manchester[5] and A. K. Tzioumis[5]

[1]International Centre for Radio Astronomy Research (ICRAR), M468, The University of Western Australia, Crawley, WA 6009, Australia.
email: giovanna.zanardo@icrar.org

[2]Department of Physics, The University of Hong Kong, Pokfulam Road, Hong Kong

[3]Sydney Institute for Astronomy (SIfA), School of Physics, The University of Sydney, NSW 2006, Australia

[4]Australian Research Council Centre of Excellence for All-sky Astrophysics (CAASTRO)

[5]CSIRO Astronomy and Space Science, Australia Telescope National Facility, PO Box 76, Epping, NSW 1710, Australia

Abstract. As the remnant of Supernova (SN) 1987A has been getting brighter over time, new observations at high frequencies have allowed imaging of the radio emission at unprecedented detail. We present a new radio image at 44 GHz of the supernova remnant (SNR), derived from observations performed with the Australia Telescope Compact Array (ATCA) in 2011. The diffraction-limited image has a resolution of 349×225 mas, which is the highest achieved to date in high-dynamic range images of the SNR. We also present a new image at 18 GHz, also derived from ATCA observations performed in 2011, which is super-resolved to $0''.25$. The new 44 and 18 GHz images yield the first high-resolution spectral index map of the remnant. The comparison of the 44 GHz image with contemporaneous X-ray and Hα observations allows further investigations of the nature of the remnant asymmetry and sheds more light into the progenitor hypotheses and SN explosion. In light of simple free-free absorption models, we discuss the likelihood of detecting at 44 GHz the possible emission originating from a pulsar wind nebula (PWN) or a compact source in the centre of the remnant.

Keywords. circumstellar matter, ISM: supernova remnants, radio continuum: general, supernovae: individual (SN 1987A), acceleration of particles, radiation mechanisms: nonthermal

---

1. Introduction

Supernova 1987A in the Large Magellanic Cloud, as the only nearby core-collapse supernova observed to date, has provided a unique opportunity to study the evolution of the interaction between the propagating blast wave and the progenitor's circumstellar medium (CSM). The complex CSM distribution in the supernova remnant (SNR) is believed to have originated from a red supergiant (RSG), which has evolved into a blue supergiant (BSG) about 20,000 years before the explosion (Crotts & Heathcote 2000). Models of the progenitor evolution suggest that the equatorially denser, i.e. slower, RSG wind (Blondin & Lundqvist 1993, Martin & Arnett 1995), was swept up by the faster BSG wind (Morris & Podsiadlowski 2007), thus forming high density rings. In particular, beside the central circular ring in the equatorial plane (*equatorial ring*, ER), observations have also revealed two outer rings that formed from the mass loss of the progenitor star, located on either side of the equatorial plane (Jakobsen *et al.* 1991, Plait *et al.* 1995), which confer to SNR 1987A a peculiar triple-ring nebula structure.

Since the radio detection of the remnant in mid-1990 (Turtle *et al.* 1990), the synchrotron emission has been generated by the shock wave propagating into the ring-shaped distribution of the CSM in the equatorial plane. Monitoring of the flux density has been regularly undertaken with the Molonglo Observatory Synthesis Telescope (MOST) at 843 MHz (Ball *et al.* 2001) and at 1.4, 2.4, 4.8 and 8.6 GHz with the Australia Telescope Compact Array (ATCA) (Staveley-Smith *et al.* 1992). ATCA observations have been ongoing for ~25 years (Staveley-Smith *et al.* 1993, Gaensler *et al.* 1997, Manchester *et al.* 2002, Staveley-Smith *et al.* 2007, Zanardo *et al.* 2010). An exponential increase of the flux density has been measured at all frequencies since day ~5000 after the explosion, which is likely due to an increasing efficiency of the acceleration process of particles by the shock front (Zanardo *et al.* 2010). The morphology of the nonthermal radiation emitted by relativistic electrons accelerated in the remnant has been investigated using images at 9 GHz since 1992 (Staveley-Smith *et al.* 1992), with a spatial resolution of 0.″5 achieved via maximum entropy super-resolution (Gaensler *et al.* 1997, Ng *et al.* 2008). These images have provided the first insight into the marked east-west asymmetry of the radio emission. The first imaging observations at 18 GHz were undertaken in 2003 July, at an effective resolution of 0.″45 (Manchester *et al.* 2005). Very long baseline interferometry (VLBI) observations of the SNR were successful in 2007 October (Tingay *et al.* 2009) and 2008 November (Ng *et al.* 2011) at 1.4 and 1.7 GHz, respectively. These observations, while with low sensitivity and dynamic range, at a resolution of ~ 0.″1 captured the presence of structures smaller than 0.″2 in bright regions (Ng *et al.* 2011). At the same time, ATCA observations at 36 GHz in 2008 April and October resulted in high-dynamic range images with an angular resolution of 0.″3 (Potter *et al.* 2009). After the ATCA upgrade in mid-2009 with the Compact Array Broadband Backend (Wilson *et al.* 2011), the remnant was imaged at higher frequencies. The first resolved image at 94 GHz was produced from observations between 2011 June and August by Lakićević *et al.* (2012).

2. New observations

The first image of the SNR at 44 GHz was derived from ATCA observations in 2011 January and November, with diffraction-limited resolution of 349 × 225 mas, which is the highest achieved to date at high-dynamic range (Zanardo *et al.* 2013). This new image has been analysed in conjunction with that derived from contemporaneous ATCA observations at 18 GHz (Zanardo *et al.* 2013) to investigate the spectral index distribution across the remnant. In Fig. 1, the 44 GHz image, slightly super-resolved to 0.″25, is shown together with the super-resolved 18 GHz image.

The high resolution of the 44 GHz image has allowed a comparison between the morphology of the radio and Hα emission, as seen in contemporaneous observations with the Hubble Space Telescope (*HST*)†, in terms of both emission around the ER and emission from the centre of the SNR, where the ejecta can be located (Larsson *et al.* 2011). Comparison with the X-ray observations performed with the *Chandra* Observatory (Helder *et al.* 2013) has provided further insight into the structure of the nonthermal emission.

3. Findings

Emission morphology and asymmetry. Consistent with previous radio observations, the 44 GHz image shows a marked asymmetry in the emission distribution. More specifically, the east-west asymmetry ratio is ~ 1.5 from the ratio of the brightness peaks in the radial profiles at PA ~90°, is ~ 1.6 from the integrated flux densities over the eastern and

† STScI-2011-21, NASA, ESA, & Challis P. (Harvard–Smithsonian Center for Astrophysics) http://hubblesite.org/newscenter/archive/releases/2011/21/image/

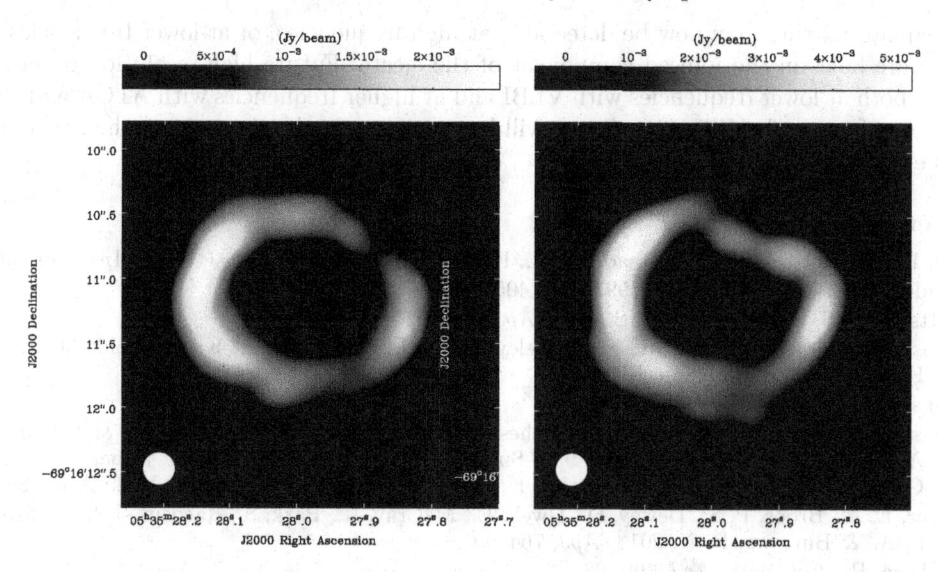

Figure 1. *Left:* Super-resolved Stokes-I continuum image of SNR 1987A at 44 GHz made by combining observations performed with the ATCA on 2011 January 24 and November 16. The diffraction-limited image is restored with a 0.″25 circular beam. *Right:* Super-resolved Stokes-I continuum 18 GHz image obtained from observations performed in 2011 January using a 0.″25 restoring circular beam. For details on both images see Zanardo *et al.* (2013).

western halves of the image. These values are higher than the ~1.4 ratio derived for the new 18 GHz image and the ratio previously measured with images at lower frequencies.

Remnant expansion and local velocities. The comparison between the new images at both 18 and 44 GHz with corresponding observations performed in earlier epochs, specifically the 2003 observations at 17 and 19 GHz and the 2008 observations at 36 GHz, highlights an asymmetric expansion of the remnant, with expansion velocities on the eastern lobe significantly higher than what measured on the western lobe.

Spectral index variations. The 18–44 GHz spectral index distribution is measured at an angular resolution of 0.″4. The spectral indices in SNR 1987A primarily range between −1.1 and −0.3, with a mean of −0.8. Spectral indices associated with the brightest sites over the eastern lobe are steeper than the mean value. The steeper spectrum on the eastern lobe implies compression ratios slightly lower than on the western bright sites, and could be correlated with the higher expansion rate measured on the eastern side of the remnant. Two regions of flatter spectral indices are identified, one approximately located in the centre of the SNR, which extends over the SN location (Reynolds *et al.* 1995), and the other located further north. These two features lie at PA ~ 30°.

Structure of the shock and progenitor explosion. There is a strong correspondence between major features of the emission at 44 GHz, and the arrangement of the hot spots shown in the Hα emission. The direction of the east-west asymmetry of the X-ray and radio emission, is opposite to that of the Hα emission. This fact supports the hypothesis that the remnant asymmetric morphology might be due to an asymmetric explosion, rather than to an asymmetric distribution of the CSM.

Likelihood of detecting the radiation emitted by a compact source. At 44 GHz, a central feature of fainter emission appears to extend over the SN site, and to overlap with the western side of the ejecta as seen by *HST*. This feature corresponds to a region of flatter spectral indices in the 18–44 GHz spectral map, which could indicate the presence of a compact source or a PWN. The origin of this emission is unclear. However, simple free-free absorption models suggest that the radiation emitted by a compact source inside the equatorial ring may now be detectable at high frequencies, or at lower frequencies if there are holes in the ionised component of the ejecta. Future high-resolution observations, both at lower frequencies with VLBI and at higher frequencies with ATCA and the Atacama Large sub-Millimeter Array, will be crucial to further investigate the nature of this emission.

References

Ball, L., Crawford, D. F., Hunstead, R. W., Klamer, I., & McIntyre, V. J. 2001, *ApJ*, 549, 599

Blondin, J. M. & Lundqvist, P. 1993, *ApJ*,405, 337

Crotts, A. P. S. & Heathcote, R. S. 2000, *ApJ*, 528, 426

Gaensler, B. M., Manchester, R. N., Staveley-Smith, L., Tzioumis, A. K., Reynolds, J. E., & Kesteven, M. J. 1997, *ApJ*, 479, 845

Gaensler, B. M. 1998, *ApJ*, 493, 781

Gaensler, B. M., Staveley-Smith, L., Manchester, R. N., Kesteven, M. J., Ball, L., & Tzioumis, A. K. 2007, in AIP Conf. Proc. 937, Supernova 1987A: 20 Years After: Supernovae and Gamma-Ray Bursters, ed. S. Immler, K. W. Weiler, & R. McCray (New York: AIP), 86

Helder, E. A., Broos, P. S., Dewey, D., Dwek, E., McCray, R., Park, S., Racusin, J. L., Zhekov, S. A., & Burrows, D. N. 2013, *ApJ*, 764, 11

Jakobsen, P. *et al.* 1991, *ApJ*, 369, 63

Lakićević, M., Zanardo, G., van Loon, J. Th., Staveley-Smith, L., Potter, T., Ng, C. -Y., & Gaensler, B. M. 2012, *A&A*, 541, L2

Larsson, J. *et al.* 2011, *Nature*, 474, L484

Manchester, R. N., Gaensler, B. M., Wheaton, V. C., Staveley-Smith, L., Tzioumis, A. K., Bizunok, N. S., Kesteven, M. J., & Reynolds, J. E. 2002, *PASA*, 19, 207

Manchester, R. N., Gaensler, B. M., Staveley-Smith, L., Kesteven, M. J., & Tzioumis, A. K. 2005, *ApJ*, 628, L131

Martin, C. L. & Arnett, D. 1995 *ApJ*, 447, 378

Morris, T. & Podsiadlowski, P. 2007, *Science*, 315, 1103

Ng, C.-Y., Gaensler, B. M., Staveley-Smith, L., Manchester, R. N., Kesteven, M. J., Ball, L., & Tzioumis, A. K. 2008, *ApJ*, 684, 481

Ng, C. -Y. *et al.* 2011, *ApJ* (Letters), 728, L15

Plait, P. C., Lundqvist, P., Chevalier, R. A., & Kirshner, R. P. 1995, *ApJ*, 439, 730

Podsiadlowski, Ph., Morris, T. S., & Ivanova, N. 2007, AIP Conf. Proc., 937, 125–133

Potter, T. M., Staveley-Smith, L., Ng, C.-Y., Ball, Lewis, Gaensler, B. M., Kesteven, M. J., Manchester, R. N., Tzioumis, A. K., & Zanardo, G. 2009, *ApJ*, 705, 261

Reynolds, J. E. *et al.* 1995, *A&A*, 304, 116

Staveley-Smith, L. *et al.* 1992, *Nature*, 355, 147

Staveley-Smith, L. *et al.* 1993, *Nature*, 366, 136

Staveley-Smith, L., Gaensler, B. M., Manchester, R. N., Ball, L., Kesteven, M. J., & Tzioumis, A. K. 2007, in AIP Conf. Proc. 937, Supernova 1987A: 20 Years After: Supernovae and Gamma-Ray Bursters, ed. S. Immler, K.W. Weiler, & R. McCray, (New York: AIP), 96

Tingay, S. *et al.* 2009, in 8th International e-VLBI Workshop, Proc. Sci., 100.

Turtle, A. J., Campbell-Wilson, D., Manchester, R. N., Staveley-Smith, L., & Kesteven, M. J. 1990, IAU Circ. 5086, 2

Wilson W. E. *et al.*, *MNRAS*, 416, 832

Zanardo, G., Staveley-Smith, L., Ball, Lewis, Gaensler, B. M., Kesteven, M. J., Manchester, R. N., Ng, C.-Y., Tzioumis, A. K., & Potter, T. M. 2010, *ApJ*, 710, 1515

Zanardo, G., Staveley-Smith, L., Ng, C.-Y., Gaensler, B. M., Potter, T. M. Manchester, R. N., & Tzioumis, A. K. 2013, *ApJ*, 767, 98

Supernovae Environmental Impacts
Proceedings IAU Symposium No. 296, 2013
A. Ray & R. A. McCray eds.

doi:10.1017/S1743921313009198

Core Collapse Supernova Models and Nucleosynthesis

Ken'ichi Nomoto

Kavli Institute for the Physics and Mathematics of the Universe (WPI)
The University of Tokyo, Kashiwa, Chiba 277-8583, Japan
email: `nomoto@astron.s.u-tokyo.ac.jp`

Abstract. After the Big Bang, production of heavy elements in the early Universe takes place in the first stars and their supernova explosions. The nature of the first supernovae, however, has not been well understood. The signature of nucleosynthesis yields of the first supernovae can be seen in the elemental abundance patterns observed in extremely metal-poor stars. Interestingly, those abundance patterns show some peculiarities relative to the solar abundance pattern, which should provide important clues to understanding the nature of early generations of supernovae. We review the recent results of the nucleosynthesis yields of massive stars. We examine how those yields are affected by some hydrodynamical effects during the supernova explosions, namely, explosion energies from those of hypernovae to faint supernovae, mixing and fallback of processed materials, asphericity, etc. Those parameters in the supernova nucleosynthesis models are constrained from observational data of supernovae and metal-poor stars.

Keywords. stellar evolution, supernova, nucleosynthesis

---

1. Introduction

In the Big Bang Universe, the first heavier elements, such as C, O, Ne, Mg, Si and Fe, must be synthesized in the evolution and explosion of the first stars (metal-free = Population III = Pop III stars) early in the history of the universe. The massive first stars evolve to explode as the first supernovae (SNe), which release large explosion energies and eject nucleosynthetically-enriched materials.

In the early Universe when the metal content was extremely low, the enrichment by a single supernova (SN) can dominate the pre-existing metal contents. Then the abundance pattern of the enriched gas may reflect nucleosynthesis in the individual SN. The next generation of stars formed from the enriched gas and the long-lived low mass stars may be observed as extremely metal-poor (EMP) stars (Beers & Christlieb 2005). Thus the abundance patterns of EMP stars can constrain the nucleosynthetic yields of the Pop III SN and thus the mass range of the first stars. Beers & Christlieb (2005) have defined the metal-poor stars with metallicity [Fe/H]† to coin as follows: Very metal-poor (VMP) stars for $-3 \leqslant$ [Fe/H] < -2, EMP stars for $-4 \leqslant$ [Fe/H] < -3, Ultra metal-poor (UMP) stars for $-5 \leqslant$ [Fe/H] < -4, Hyper metal-poor (HMP) stars for $-6 \leqslant$ [Fe/H] < -5, and Mega metal-poor (MMP) stars for [Fe/H] < -6.

Actually, recent observations discovered several EMP stars, whose abundance patterns are quite unusual, such as carbon enhanced metal-poor (CEMP) and hyper metal-poor (HMP) stars (e.g., Beers & Christlieb 2005), being significantly different from previously known nucleosynthesis yields of massive stars. These new observations have raised important challenges to the stellar evolution and nucleosynthesis theory.

† $[A/B] = \log_{10}(N_A/N_B) - \log_{10}(N_A/N_B)_\odot$, where the subscript $\odot$ refers to the solar value and N_A and N_B are the abundances of elements A and B, respectively.

Interestingly, there is another challenge to the conventional stellar evolution and supernova models. That is the establishment of the Gamma-Ray Burst (GRB)-Supernova Connection (e.g., Woosley & Bloom 2006). Four GRB-associated SNe have been confirmed spectroscopically so far. They are all very energetic supernovae, whose kinetic energy E exceeds 10^{52} erg, more than 10 times the kinetic energy of normal core-collapse SNe. (We use the explosion energy E for the final kinetic energy of the explosion.) In the present paper, we use the term 'Hypernova (HN)' to describe such a hyper-energetic supernova with $E_{51} = E/10^{51}$ erg $\gtrsim 10$.

Motivated by these challenges, we briefly review the recent results of core-collapse supernova models and their nucleosynthesis. The comparison between such stellar nucleosynthesis yields and the abundance patterns of EMP/UMP/HMP stars can provide a new approach to find out the individual supernova mechanism, especially for Pop III supernovae (Nomoto *et al.* 2013).

2. Progenitor's Mass and Explosion Energy

The fates of Pop III stars depend on the mass to which the initially low-mass stars grow through mass accretion. For various cases of feedback and mass accretion rates, Ohkubo *et al.* (2009) calculated the evolution of accreting Pop III stars to show that massive stars may form if the mass accretion is not much reduced during the main-sequence evolution (Fig. 1 (left), Ohkubo *et al.* 2009). It is possible that Pop III stars were even more massive than $\sim 300 M_\odot$, if rapid mass accretion continues during the whole main-sequence phase of Pop III stars (Ohkubo *et al.* 2006). Here the models of stellar evolution, supernova explosions, and nucleosynthesis are described as a function of the main-sequence mass M. These models are constrained from the comparison of theoretical supernova light curves and spectra with observations.

As mentioned in the Introduction, the explosion energies of core-collapse supernovae are fundamentally important quantities, and an estimate of $E \sim 1 \times 10^{51}$ ergs has often been used for nucleosynthesis calculations. A good example is SN1987A.

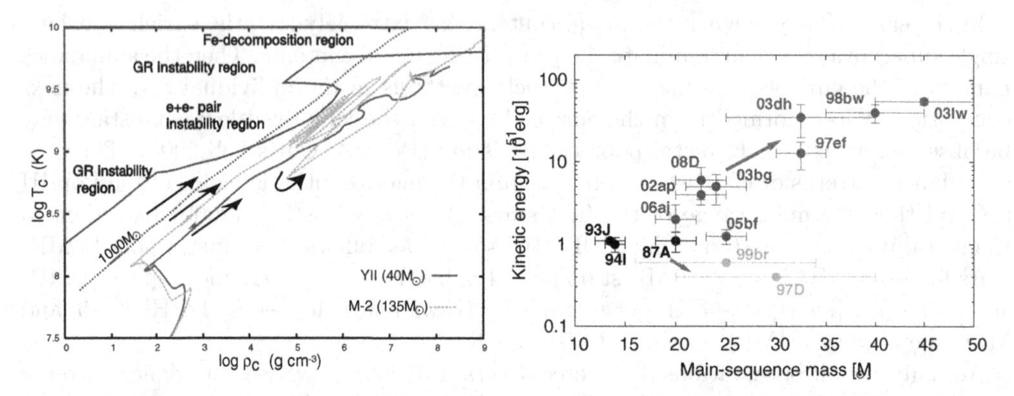

Figure 1. (left:) Evolutionary tracks of the central temperature and central density of Pop III massive stars with mass accretion (Ohkubo *et al.* 2009). The arrows indicate the direction of the evolution. The numbers in brackets are the final masses for various accretion rates and radiative feedback effects. The evolutionary track without accretion is shown for $M = 1000 M_\odot$. **(right:)** The explosion energy as a function of the main-sequence mass of the progenitors for several supernovae/hypernovae. Explosions of $13 - 25 M_\odot$ stars cluster at normal SNe, while explosions of $25 - 40 M_\odot$ stars have a large variety ranging from hypernovae to faint SNe.

Important change has come from the establishment of the connection between long GRBs and core-collapse SNe from GRB 980425/SN 1998bw, GRB 030329/SN 2003dh, GRB 031203/SN 2003lw, and GRB120422A/SN2012bz (Melandri *et al.* 2012 and references therein). These GRB-SNe are of Type Ic, showing the broad-line spectra (SNe BL-Ic). The properties of these GRB-SNe, such as the ejected mass (the main-sequence mass of the progenitor) and the kinetic energy of explosion, have been estimated from the comparison between the observed light curve and spectra and their theoretical models (Nomoto *et al.* 2006). As summarized in Figure 1 (right), these GRB-SNe have similar properties; they are all hypernovae with $E_{51} \sim 30$ - 50 and synthesize 0.3 - 0.5 $M_\odot$ of ^{56}Ni. The mass estimates, obtained from fitting the optical light curves and spectra, place hypernovae at the high-mass end of SN progenitors.

An X-Ray Flash-SN connection has also been found: GRB 060218/SN 2006aj and GRB 100316D/SN 2010bh. Compared with the above GRB-SNe, SN 2006aj is less energetic ($E_{51} \sim 2$) and its progenitor mass is smaller, $\sim 20 M_\odot$, while SN 2010bf may be as energetic as $E_{51} \sim 10$ (Bufano *et al.* 2012 and references therein).

In contrast, SNe II 1997D and 1999br were very faint SNe with very low E (e.g., Turatto *et al.* 1998). In the $E - M$ diagram (Fig. 1 (right)), therefore, we propose that SNe from stars with $M \gtrsim 20 - 25 M_\odot$ have different E, with a bright, energetic "hypernova branch" at one extreme and a faint, low-energy SN branch at the other (Nomoto *et al.* 2003). For the faint SNe, the explosion energy was so small that most ^{56}Ni fell back onto the compact remnant. Thus the faint SN branch may become even a "failed" SN branch at larger M. Between the two branches, there may be a variety of SNe.

This trend might be interpreted as follows. Stars more massive than ~ 25 $M_\odot$ form a black hole at the end of their evolution. Stars with non-rotating black holes are likely to collapse "quietly", ejecting a small amount of heavy elements (Faint supernovae). In contrast, stars with rotating black holes are likely to give rise to hypernovae. The hypernova progenitors might form the rapidly rotating cores by spiraling-in of a companion star in a binary system.

3. 8 - 10 $M_\odot$ Stars and Electron Capture Supernovae

Stars in the mass range of $\sim 8 M_\odot < M < 10 M_\odot$ can become "electron capture supernovae", if the electron-degenerate O-Ne-Mg core mass grows to 1.38 $M_\odot$ and the central density reaches 4×10^9 g cm^{-3} before the whole super-AGB envelope is lost by mass loss. (Thus the lower limit of this mass range depends on the mass loss rate.) At such a high central density, the electron Fermi energy exceeds the threshold for electron captures ^{24}Mg(e$^-$, ν)^{24}Na(e$^-$, ν)^{20}Ne and ^{20}Ne(e$^-$, ν)^{20}F(e$^-$, ν)^{20}O. The resultant decrease in Y_e triggers collapse (Nomoto 1987).

The resultant explosion is induced by neutrino heating, and is weak with the kinetic energy as low as $E \sim 10^{50}$ erg (Kitaura *et al.* 2006). These stars produce little α-elements and Fe-peak elements, but might be another sources of Zn and light p-nuclei.

Nucleosynthesis in the supernova explosion of a 9 $M_\odot$ star (Wanajo *et al.* 2009) shows that the largest overproduction is shared by ^{64}Zn, ^{70}Se, and ^{78}Kr. The ^{64}Zn production provides an upper limit to the occurrence of exploding O-Ne-Mg cores at about 20% of all core-collapse supernovae. This SN may produce a significant amount of weak r-process elements as well as ^{48}Ca from a neutron-rich blob (Wanajo *et al.* 2013). The ejecta mass of ^{56}Ni is $0.002 - 0.004 M_\odot$, much smaller than the $\sim 0.1 M_\odot$ in more massive progenitors. The light curve of electron capture supernova can be consistent with SN 1054 (the Crab Nebula's supernova) (Tominaga, Blinnikov, & Nomoto 2013).

4. 10 - 13 $M_\odot$ Stars and Faint Supernovae

The stars of 10 - 13 $M_\odot$ undergo off-center neon ignition due to the efficient neutrino cooling in the semi-degenerate O+Ne+Mg core (Nomoto & Hashimoto 1988); this is in contrast to the 13 $M_\odot$ star where neon is ignited at the center. The neon flame propagates inward due to core contraction. For 10 - 12 $M_\odot$, whether or not the neon flame reaches the center needs further study. If the neon flame is quenched, the central region could form an electron-degenerate O-Ne-Mg core surrounded by a layer with the neon-burning products, i.e., a Si-Fe-rich layer. It would become an electron capture supernova as in lower mass stars.

For stars with $M \gtrsim 12\ M_\odot$, the neon flame reaches the center (Nomoto & Hashimoto 1988). Then, the core would evolve into an Fe core smaller than 1.4 $M_\odot$. The explosion of a star with such a small mass Fe core could be powered by neutrino heating, becoming a weak SN as in the electron capture supernovae with O-Ne-Mg cores (Müller *et al.* 2012). They would eject only a small amount of heavy elements.

The Fe core collapse of these stars would certainly lead to the formation of a neutron star (NS) because there exists a steep density gradient around $\sim 1.4\ M_\odot$, and the outer envelope is too extended to accrete and further increase the mass of the collapsing core beyond 1.4 $M_\odot$. The resultant SNe tend to be faint because a negligibly small amount of ^{56}Ni is ejected (Müller *et al.* 2012). Such an SN may correspond to faint supernovae (Smartt 2009). A possible case of such a faint supernova from this mass range is the Type Ib SN 2005cz, which is unusually faint and rapidly fading (Kawabata *et al.* 2010).

5. 13 - 25 $M_\odot$ Stars and Normal Supernovae

These stars undergo Fe-core collapse to form a NS, and produce significant amount of heavy elements from α-elements to Fe-peak elements. The boundary mass between the NS and black hole (BH) formation, $M_{\rm NS/BH} \sim 20 - 25 M_\odot$, is only tentative.

For this mass range, SN 1987A in the LMC has provided the most detailed constraints on the explosion model (Arnett *et al.* 1989). The modeling of nearby supernovae suggest that the stars in the mass range of 13 $M_\odot$ $- M_{\rm NS/BH}$ undergo neutron star-forming Fe core-collapse and induce normal core collapse supernovae. These SNe produce significant amounts of heavy elements from α-elements to Fe-peak elements.

6. 25 - 140 $M_\odot$ Stars and Hypernovae & Faint Supernovae

These stars undergo Fe-core collapse to form a BH. As seen in Figure 1 (right), the resulting black hole-forming SNe seem to be bifurcate into two branches, Hypernovae and Faint SNe. If the BH has little angular momentum, little mass ejection would take place and it would be observed as a Faint SN. On the other hand, a rotating BH could eject matter in the form of jets to make a hypernova. Hypernovae produce a large amount of heavy elements from α-elements to Fe-peak elements.

6.1. Hypernovae

The right panel of Figure 2 shows the composition of the ejecta for a 25 $M_\odot$ hypernova model ($E_{51} = 10$). The nucleosynthesis in a normal 25 $M_\odot$ SN model ($E_{51} = 1$) is also shown for comparison in the left panel of Figure 2 (Umeda & Nomoto 2002).

We note the following characteristics of nucleosynthesis with very large explosion energies (Nakamura *et al.* 2001, Nomoto *et al.* 2001):

(1) Both the complete and incomplete Si-burning regions shift outward in mass compared with normal supernovae, so that the mass ratio between the complete and incomplete Si-burning regions becomes larger. As a result, higher energy explosions tend to produce larger [(Zn, Co, V)/Fe] and smaller [(Mn, Cr)/Fe], which can explain the trend observed in very metal-poor stars.

(2) In complete Si-burning of hypernovae, elements produced by α-rich freeze-out are enhanced. Hence, elements synthesized through capturing α-particles, such as ^{44}Ti, ^{48}Cr, and ^{64}Ge (decaying into ^{44}Ca, ^{48}Ti, and ^{64}Zn, respectively) are more abundant.

(3) Oxygen burning takes place in more extended regions for larger E. More O, C, and Al are burned to produce a larger amount of burning products such as Si, S, and Ar. Therefore, hypernova nucleosynthesis is characterized by large abundance ratios of [Si,S/O], which can explain the abundance feature of M82 (Umeda *et al.* 2002).

6.2. **Faint Supernovae**

In contrast to Hypernovae, faint supernovae undergo extensive fallback of processed materials. The ejecta of fallback supernovae have large [C/Fe] – [Al/Fe] at low metallicity. These patterns could explain the abundance patterns observed in CEMP stars.

Note that, in spherical explosions, substantial fallback occurs for relatively low E. However, in the jet-induced explosions, fallback occurs even for high E explosions. Note also the stellar mass dependence of the abundance pattern is quite weak. In comparing with the observed patterns of CEMP stars, it is difficult to identify the progenitor's mass.

6.3. **Pulsational Instability in Pre-Collapse Stars**

Stars more massive than $M \sim 90M_\odot$ undergo nuclear instabilities and associated pulsations (ϵ-mechanism) at various nuclear burning stages depending on the mass loss and thus metallicity (Heger & Woosley 2002).

(1) **Oxygen Burning**: For the $M = 137M_\odot$ Pop III star, the evolutionary track for the evolution of the central density and temperature is very close to (but outside of) the "e$^-$e$^+$ pair-instability region" of $\Gamma < 4/3$, where Γ denotes the adiabatic index (Fig. 1 (left)). During oxygen burning, the central temperature and density of such a massive

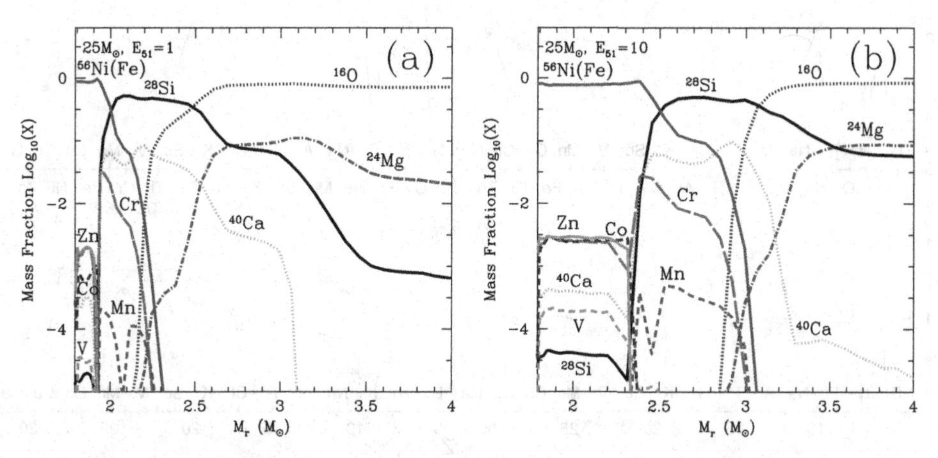

Figure 2. Abundance distribution against the enclosed mass M_r after the explosion of Pop III 25 $M_\odot$ stars with $E_{51} = 1$ (a) and $E_{51} = 10$ (b) (Umeda & Nomoto 2002, Tominaga, Umeda, & Nomoto 2007b). The high explosion energy in hypernovae leads to shift the complete and incomplete Si-burning regions outward in mass and to enhance α-rich freeze-out in the complete Si burning layer.

star oscillate several times (Woosley *et al.* 2007, Ohkubo *et al.* 2009). This is because in such massive stars radiation pressure is so dominant that Γ is close to 4/3, and thus the inner core of the stars easily expands with the nuclear energy released by O-burning. Once it expands, the temperature drops suddenly, the central O-burning is weakened, and the stellar core in turn shrinks. Since only a small amount of oxygen is burnt for each cycle, these pulsations occur many times. In extreme cases, the pulsation could induce dynamical mass ejection and optical brightening as might be observed in the brightest SN 2006gy (Woosley *et al.* 2007).

(2) **Silicon Burning**: $M \sim 90 M_\odot$ stars undergo nuclear instability due to silicon burning and pulsates several times (Umeda & Nomoto 2008). The amplitude of the pulsation due to Si-burning in the central density and temperature is smaller than O-burning (Ohkubo *et al.* 2009).

(3) **Core-Collapse and ^{56}Ni Production**: Eventually, these $\sim$ 90 - 140 $M_\odot$ stars undergo Fe-core collapse to form BHs. Then hypernova-like energetic SNe could occur to produce large amount ^{56}Ni. The synthesized ^{56}Ni mass increases with the increasing E and M. For $E = 3 \times 10^{52}$ ergs, ^{56}Ni masses of up to 2.2, 2.3, 5.0, and 6.6 $M_\odot$ can be produced for low metallicity ($Z = 0.0001$) progenitors with M = 30, 50, 80 and 100 $M_\odot$ (Umeda & Nomoto 2008). Thus the upper limit to the mass of ^{56}Ni produced by core-collapse SNe ($M \lesssim 140\ M_\odot$) would be $\sim$10 $M_\odot$. The abundance pattern of the ejecta does not depend much on the stellar masses.

Because of the large ejecta mass, the expansion velocities may not be high enough to form broad line features, as has been observed in SN Ic 1999as (Nomoto 2012). Thanks to the large E and ^{56}Ni mass, however, the SNe could be super-luminous supernovae (SLSNe) (Moriya *et al.* 2010).

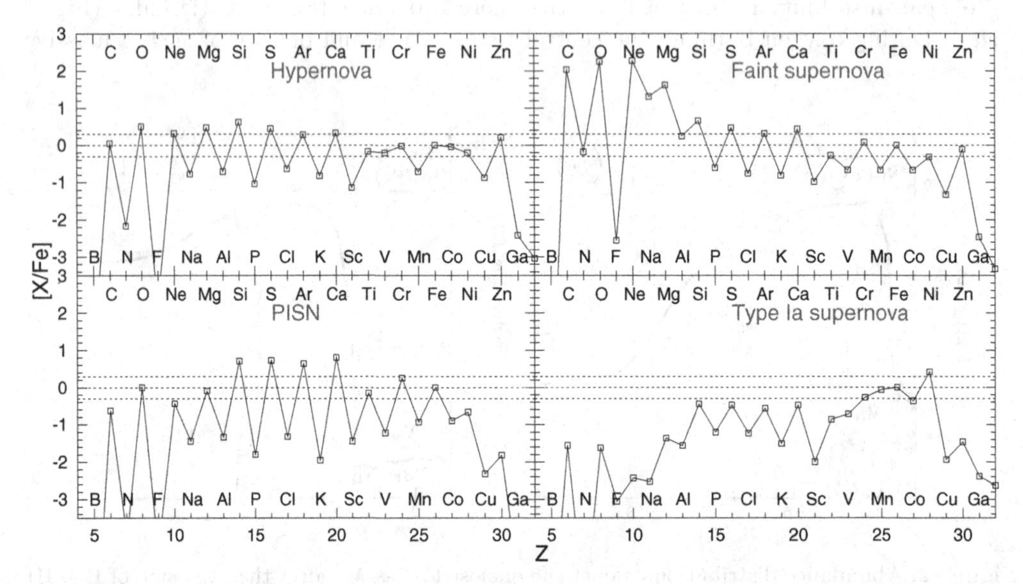

Figure 3. Yields of the core-collapse hypernova (HN), pair-instability supernova (PISN), faint SN, and Type Ia supernova are compared (Nomoto *et al.* 2013). The faint SN, PISN, and Type Ia supernova are characterized with high [C/Fe], low [Zn/Fe] and large odd-even effect, and low [α/Fe], respectively, compared to the hypernova.

7. Very Massive Stars

7.1. *Pair-Instability Supernovae of 140 - 300 $M_\odot$ Stars*

If very massive stars ($M > 140\ M_\odot$) do not lose much mass, they undergo thermonuclear explosions triggered by pair creation instability (pair-instability supernovae: PISN) (Barkat *et al.* 1967). The star is completely disrupted without forming a BH and thus ejects a large amount of heavy elements, especially ^{56}Ni (e.g., Umeda & Nomoto 2002, Heger & Woosley 2002). The largest mass of ^{56}Ni obtained in the PISN models amounts to $\sim 40\ M_\odot$ (Heger & Woosley 2002). The resultant radioactive decays of ^{56}Ni and ^{56}Co could produce SLSNe.

Figure 3 shows the abundance patters in the ejecta of typical models of the core-collapse hypernova, PISN, faint SN, and Type Ia supernova (SN Ia). The abundance features of PISNe to be compared with the observed abundances are as follows:

- Abundance ratios of iron-peak elements: [Zn/Fe] < -0.8 and [Co/Fe] < -0.2. Such small Zn and Co productions (relative to Fe) are due to the low central temperature at the bounce of the collapsing core. This abundance feature of yields is intrinsic to PISNe, because, if the central temperature would become higher, the core collapse would continue due to photodisintegration effects.
- Explosive O-burning leads to large [(Si, S, Ca)/O] (~ 0.8).
- The odd-even effect is significantly larger in PISNe than core-collapse SNe.

7.2. *Stars with $M \gtrsim 300 M_\odot$ and Intermediate Mass Black Holes*

Stars with $300 M_\odot < M < 3.5 \times 10^5 M_\odot$ enter the pair-instability region but are too massive to be disrupted by PISNe but undergo core collapse, forming intermediate-mass black holes (IMBHs). If such stars formed rapidly rotating black holes, jet-like mass ejection could possibly occur and produce processed material. In fact, for moderately aspherical explosions, the patterns of nucleosynthesis show small [O/Fe] and [Ne/Fe], and large [Mg/Fe], [Si/Fe] and [S/Fe]. Also [C/Si] is not so small compared with PISNe. These patterns match the observational data of both the intracluster medium and M82 better than PISNe (Ohkubo *et al.* 2006).

This result suggests that core-collapse explosions of very massive stars could contribute to chemical enrichment in clusters of galaxies. This might support the view that Pop III core-collapse very massive stars could be responsible for the origin of IMBHs.

Stars more massive than $\sim 3.5 \times 10^5 M_\odot$ (super-massive stars) collapse owing to general relativistic instability before reaching the main-sequence.

8. Extremely Metal Poor Stars and Abundance Profiling

8.1. *VMP Stars: Comparison of with Normal Core-Collapse Supernova Yields*

In Figure 4 (left), the averaged abundance pattern of VMP stars ($-2.7 <$ [Fe/H] < -2.0, Cayrel *et al.* 2004) are compared with the SN and HN yields integrated over the progenitors of $M = 10 - 50 M_\odot$ with the Salpeter's IMF (black solid, Tominaga, Umeda, & Nomoto 2007b). The observed and theoretical patterns are in reasonable agreement for many elements, although N, K, and Sc are largely underproduced in the model.

8.2. *Extremely Metal-Poor (EMP) Stars*

Figure 4 (right) shows the comparison between the averaged abundances of EMP stars ($-4.2 <$ [Fe/H] < -3.5) and normal [C/Fe] ~ 0 (Cayrel *et al.* 2004)) with the normal Pop III SN yield (*blue dashed*: 15 $M_\odot$, $E_{51} = 1$; Tominaga *et al.* 2007a). It is seen that the SN yield is in reasonable agreement with the observations for [(Mg, Al, Si)/Fe], but gives too small [(Mn, Co, Zn)/Fe].

If we compare the observations with the hypernova yield (black solid; $M = 20M_\odot$, $E_{51} = 10$) in Figure 4 (right), [(Ti, Co, Zn)/Fe] are larger and thus in much better agreement with observations than the SN yield. Another example of good agreement between the EMP star (BS 16467-062) and an HN model is seen in Figure 5a (Tominaga, Iwamoto, & Nomoto 2013).

The difference between the SN and HN models is seen in the abundance distribution of the SN and HN ejecta (Fig.2). Both Co and Zn are synthesized in complete Si burning in a high temperature region, which is more extended in the mass coordinate in the higher E model. Figure 2 also shows that the mass fractions of Zn, Co, and V in the complete Si-burning region are larger because of higher entropy in the HN model than the SN model. As a result, the integrated ratios of Co/Fe and Zn/Fe are larger in higher energy explosions.

8.3. *Carbon-Enhanced Metal-Poor (CEMP) Stars*

A significant fraction of the metal-poor stars show such a large carbon enhancement as [C/Fe] $\gtrsim +1$, thus being called CEMP stars (e.g., Beers & Christlieb 2005 and references therein).

The faint SNe occur as a result of fallback of a large amount of radioactive ^{56}Ni whose decay into Fe powers the SN light curve (Fig. 1 (right)). Since the fallback of carbon is less than Fe because of the outer location of C, large [C/Fe] results.

Figure 5b shows that the abundance pattern of a CEMP star (CS 29498-043: Aoki *et al.* 2004) is well reproduced by a *single* faint SN ($M = 25M_\odot$, Tominaga, Iwamoto, & Nomoto 2013).

The physical mechanism of the mixing-fallback that can yields [C/Fe] > 1 would be the jet-like explosion (Tominaga *et al.* 2007a) rather than the Rayleigh-Taylor instability resulting in [C/Fe] $\lesssim +0.5$.

Most CEMP stars show [O/Mg] > 1. Faint SNe enhance [O/Fe] more than [Mg/Fe]. This is because Mg is synthesized in the inner region, so that more Mg falls back onto the central remnant than O.

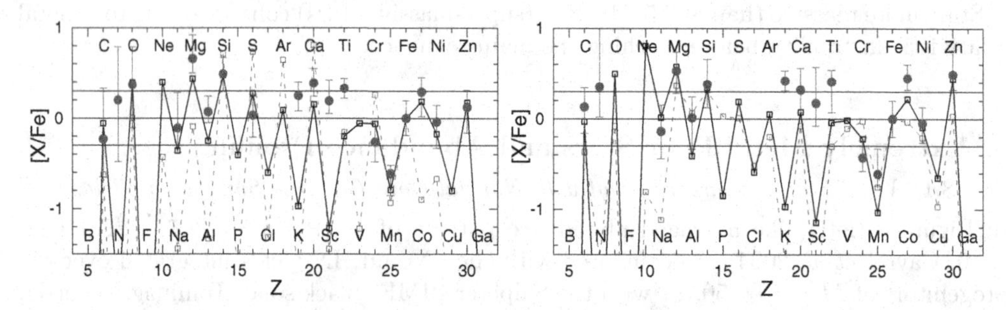

Figure 4. **(left:)** The abundance pattern of VMP stars with $-2.7 <$ [Fe/H] < -2.0 *filled circles with bars* is shown to be in good agreement with the IMF integrated yield of Pop III SNe and HNe from $10M_\odot$ to 50 $M_\odot$ (black solid, Tominaga, Umeda, & Nomoto 2007b), but not with the $200M_\odot$ PISN yield (blue dashed, Umeda & Nomoto 2002). **(right:)** Averaged elemental abundances of stars with $-4.2 <$ [Fe/H] < -3.5 compared with the normal SN yield (*blue dashed*: 15 $M_\odot$, $E_{51} = 1$) and the HN yield (20 $M_\odot$, $E_{51} = 10$) with the mixing and fallback (*black solid*). The HN yield have larger [(Ti, Co, Zn)/Fe] and thus is in better agreement with the EMP stars than the SN yield.

8.4. *Ultra Metal-Poor (UMP) and Hyper Metal-Poor (HMP) Stars*

The elemental abundance patterns of these HMP stars provide a key to the answer to the above questions. The abundance patterns of HE1327–2326 (Frebel *et al.* 2005) and HE0107–5240 (Collet *et al.* 2006) are quite unusual. The striking similarity of [Fe/H] (= −5.4 and −5.2 for HE1327–2326 and HE0107–5240, respectively) and [C/Fe] (∼ +4) suggests that similar chemical enrichment mechanisms operated in forming these HMP stars. However, the N/C and (Na, Mg, Al)/Fe ratios are more than a factor of 10 larger in HE1327–2326. In order for the theoretical models to be viable, these similarities and differences should be explained self-consistently.

9. Concluding Remarks

We report on the properties and nucleosynthesis of the two distinct classes of massive SNe: 1) very energetic Hypernovae, whose kinetic energy is more than 10 times that of normal core-collapse SNe, and 2) very faint and low energy SNe (Faint SNe). These two new classes of SNe are likely to be "black-hole-forming" SNe with rotating or non-rotating black holes. Nucleosynthesis in Hypernovae is characterized by larger abundance ratios (Zn,Co,V,Ti)/Fe than normal SNe, which can explain the observed ratios in EMP stars. Nucleosynthesis in Faint SNe is characterized by a large amount of fall-back, which explains the abundance pattern of the most Fe-poor stars.

These comparisons suggest that black-hole-forming SNe made important contributions to the early Galactic (and cosmic) chemical evolution. We discuss how nucleosynthetic properties resulted from such unusual supernovae are connected with the unusual

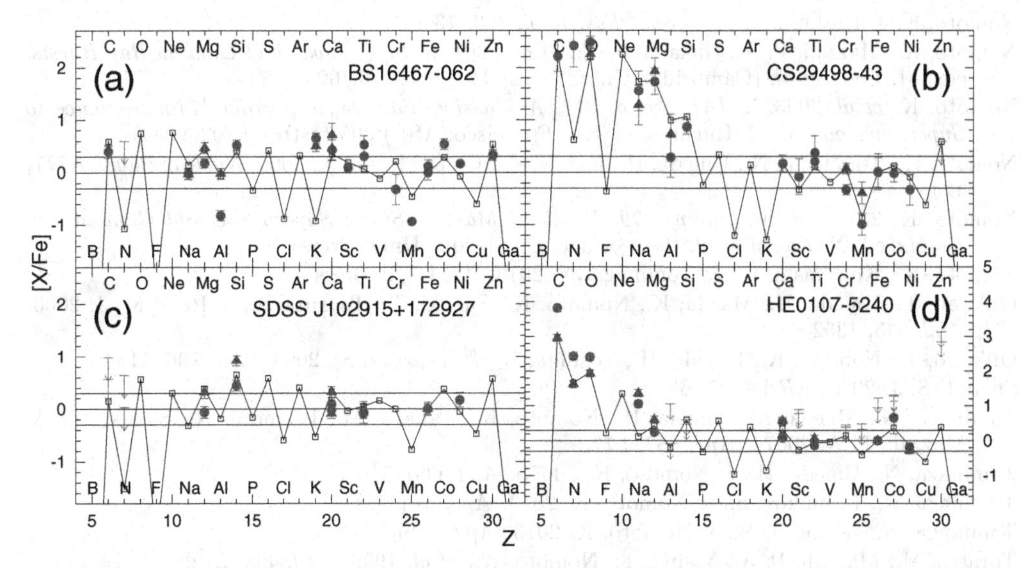

Figure 5. Comparison of the observed abundance patterns (1D-LTE: red filled circles with bars and 3D-(N)LTE: blue filled triangles with bars if available) of EMP (BS 16467-062, Cayrel *et al.* 2004), CEMP (CS 29498-43, Aoki *et al.* 2004), UMP (SDSS J102915+172927, Caffau *et al.* 2011), and HMP stars (HE 0107-5240, Collet *et al.* 2006) with relevant SN models (Tominaga, Iwamoto, & Nomoto 2013). The SN models are constructed for the $25M_\odot$ progenitor (Iwamoto *et al.* 2005). The explosion energies and ejected ^{56}Ni masses of SN models are $E_{51} = 20$ and $M(^{56}\mathrm{Ni}) = 0.044M_\odot$ (EMP star), $E_{51} = 20$ and $M(^{56}\mathrm{Ni}) = 9.1 \times 10^{-4} M_\odot$ (CEMP star), $E_{51} = 20$ and $M(^{56}\mathrm{Ni}) = 1.2 \times 10^{-1} M_\odot$ (UMP star), and $E_{51} = 5$ and $M(^{56}\mathrm{Ni}) = 8.0 \times 10^{-5} M_\odot$ (HMP star).

abundance patterns of extremely metal-poor stars. Such connections may provide important constraints on the properties of first stars.

This research has been supported by World Premier International Research Center Initiative, MEXT, Japan.

References

Aoki, W., Norris, J. E., & Ryan, S. G., *et al.* 2004, *ApJ*, 608, 971

Arnett, W. D., Bahcall, J. N., Kirshner, R. P., & Woosley, S. E. 1989, *ARAA*, 27, 629

Barkat, Z., Rakavy, G., & Sack, N. 1967, *Phys. Rev. Letters*, 18, 379

Beers, T. C. & Christlieb, N. 2005, *ARAA*, 43, 531

Bufano, F., *et al.* 2012, *ApJ*, 753, 67

Caffau, E., *et al.* 2011, *Nature*, 477, 67

Cayrel, R., *et al.* 2004, *A&A*, 416, 1117

Christlieb, N., *et al.* 2002, *Nature*, 419, 904

Collet, R., Asplund, M., & Trampedach, R. 2006, *ApJ*, 644, L121

Frebel, A. *et al.* 2005, *Nature*, 434, 871

Heger, A. & Woosley, S. E. 2002, *ApJ*, 567, 532

Iwamoto, N., Umeda, H., Tominaga, N., Nomoto, K., & Maeda, K., 2005, *Science*, 309, 451

Kawabata, K., Maeda, K., Nomoto, K., *et al.* 2010, *Nature*, 465, 326

Kitaura, F. S., Janka, H.-Th., & Hillebrandt, W. 2006, *A&A*, 450, 345

Melandri, A., *et al.* 2012, *A&A*, 547, 82

Moriya, T., Tominaga, N., Tanaka, M., Maeda, K., & Nomoto, K. 2010, *ApJ*, 717, 83

Müller, B., Janka, T., & Heger, A. 2012, *ApJ*, 761, 72

Nakamura, T., Umeda, H., Iwamoto, K., Nomoto, K., *et al.* 2001, *ApJ*, 555, 880

Nomoto, K. 1987, *ApJ*, 322, 206

Nomoto, K. & Hashimoto, M. 1988, *Phys. Rep.*, 163, 13

Nomoto, K., Mazzali, P. A., Nakamura, T., *et al.* 2001, in *Supernovae and Gamma Ray Bursts*, eds. M. Livio *et al.* (Cambridge Univ. Press) 144 (astro-ph/0003077)

Nomoto, K. *et al.* 2003, in *IAU Symp. 212, A Massive Star Odyssey, from Main Sequence to Supernova*, eds. V. D. Hucht *et al.*(San Francisco: ASP) 395 (astro-ph/0209064)

Nomoto, K., Tominaga, N., Umeda, H., Kobayashi, C., & Maeda, K. 2006, *Nuclear Phys.*, A777, 424

Nomoto, K. 2012, in *IAU Symp. 279, Death of Massive Stars: Supernovae and Gamma-Ray Bursts*, ed. N. Kawai *et al.* (Cambridge: Cambridge Univ. Press) 1

Nomoto, K., Kobayashi, C., & Tominaga, N. 2013, *ARAA*, in press

Ohkubo, T., Umeda, H., Maeda, K., Nomoto, K., Suzuki, T., Tsuruta, S., & Rees, M. J. 2006, *ApJ*, 645, 1352

Ohkubo, T., Nomoto, K., Umeda, H., Yoshida, N., & Tsuruta, S., 2009, *ApJ*, 706, 1184

Smartt, S. J. 2009, *ARAA*, 47, 63

Tominaga, N., Maeda, K., Umeda, H., Nomoto, K., Tanaka, M., Iwamoto, N., Suzuki, T., & Mazzali, P. A. 2007a, *ApJ*, 657, L77

Tominaga, N., Umeda, H., & Nomoto, K. 2007b, *ApJ*, 660, 516

Tominaga, N., Blinnikov, S., & Nomoto, K. 2013, *ApJ*, in press

Tominaga, N., Iwamoto, N., & Nomoto, K. 2013, *ApJ*, submitted

Turatto, M., Mazzali, P. A., Young, T., Nomoto, K., *et al.* 1998, *ApJ*, 498, L129

Umeda, H. & Nomoto, K. 2002, *ApJ*, 565, 385

Umeda, H., Nomoto, K., Tsuru, T., & Matsumoto, H. 2002, *ApJ*, 578, 855

Umeda, H. & Nomoto, K. 2003, *Nature*, 422, 871

Umeda, H. & Nomoto, K. 2008, *ApJ*, 673, 1014

Wanajo, S., Nomoto, K., Janka, H.-T., Kitaura, F. S., & Müller, B. 2009, *ApJ*, 695, 208

Wanajo, S., Janka, H.-T., & Müller, B. 2013, *ApJ* (Letters), 767, L26

Woosley, S. E. & Bloom, J. S. 2006, *ARAA*, 44, 507

Woosley, S. E., Blinnikov, S., & Heger, A. 2007, *Nature*, 450, 390

Supernova Environmental Impacts
Proceedings IAU Symposium No. 296, 2013
A. Ray & R. A. McCray, eds.

doi:10.1017/S1743921313009204

Supernova searches and rates

Enrico Cappellaro

INAF, Osservatorio Astronomico di Padova, vicolo dell'Osservatorio 5 I-35122 Padova, Italy
email:enrico.cappellaro@0apd.inaf.it

Abstract. Supernova statistics, establishing a direct link between stellar populations and explosion scenarios, is a crucial test of stellar evolution theory. Nowadays, a number of SN searches in the local Universe and at high redshifts are allowing observational probes of long standing theoretical scenarios. I will briefly review some of the most interesting results in particular for what concern the evolution with cosmic time of the SN rate, which is one of the topic that in the last few years had a most rapid development.

Keywords. supernovae: general, stars: evolution, galaxies: stellar content

1. Introduction

The use of supernova (SN) statistics to constrain the progenitor scenarios and explosion mechanisms began 50 yr ago with the basic observations that different SN types are linked to different stellar populations (Minkowki 1964). For a long time, the progresses were limited by the small number statistics until, with the progress of instrumentation and of the tools for data mining, searching for SNe has become a routine process that, nowadays, is producing several hundred events per year (Fig. 1). The best known outcome of these efforts is the discovery of the accelerated expansion of the Universe but, as a by product, we could obtain accurate measurements of the SN rates as a function of parent stellar population, at least in the local Universe, and cosmic age. In the following I will briefly review some of the most significant results and their implications.

2. Modern supernova searches

The rich SN harvest of modern SN searches has minimized the problem of the sample size, but we should be aware that using SN statistics is not only a question of sample numerosity, it requires also a careful consideration of the search and selection biases. Indeed, while each search has different characteristics, all have biases. One basic distinction is between targeted and un-targeted SN searches.

Typically, a targeted SN search monitors a well defined sample of bright, nearby galaxies, preferentially giant spirals which are known to have a higher SN rate. The most successful experiment of this sort is probably the Lick Observatory SN search (LOSS, Leaman *et al.* 2011) but in this category are also to be included the many SN searches conducted by amateur astronomers that today contribute with a significant number of SN discoveries in the local Universe.

Un-targeted SN searches make use of wide field imagers to monitor large sky areas, e.g. the Sloan Digital Sky Survey II (SDSSII) surveyed 300 sq. deg. in five filters (Frieman *et al.* 2008), but we should be aware that, contrary to a usual assumption, this does not guarantee that they are unbiased. In many wide field surveys the candidate selection algorithm is designed to strongly reduce the number of false detections, e.g. favoring the candidates associated with galaxies or neglecting those in the nucleus of bright galaxies.

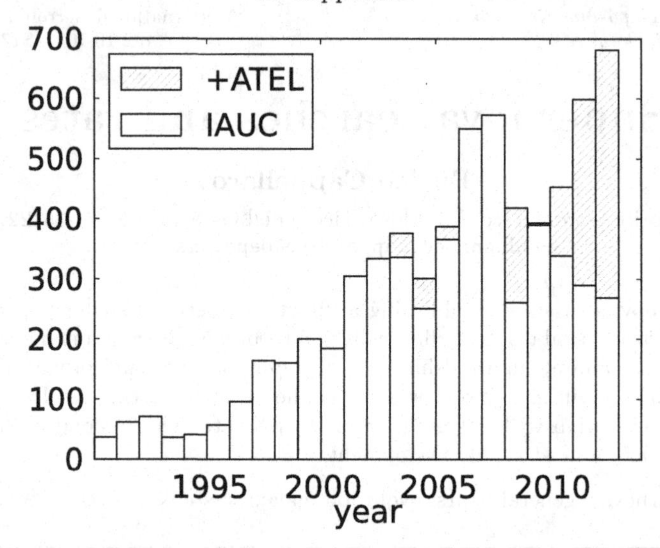

Figure 1. Number of SN discoveries per year (updated at 2012 Dec 31). The white histogram are the SNe reported through the *IAUC/CBET* (`www.cbat.eps.harvard.edu`) that, to date, are over 6100, whereas the hatched histogram shows those reported **only** via ATEL, The Astronomer's Telegram (`www.astronomerstelegram.org`), that are just over 1100. The ATELs are allowing the community for immediate dissemination of reports and comments upon new astronomical observations that is crucial to assure prompt follow-up. On the other hand because ATELs are not filtered or edited the reports are not standardized and there are more chances for duplications and typos. The fraction of SNe reported only in ATELs is growing, being over 50% in the last two years.

In addition, a specific observing strategy may be adopted to maximize the detection of selected transients: in most cosmological SN searches observations were spaced by 15 days to maximise the detection of type Ia SNe near maximum. Indeed many of the breakthroughs of the Palomar Transient Factory (PFT, Rau *et al.* 2009) are the result of the monitoring of un-explored time domain, with either very frequent cadence or otherwise long term survey duration.

The presence of specific biases affecting different searches was emphasised by Quimby *et al.* (2012) by comparing the SN discovery list of LOSS, SDSSII and their own SN search based on the ROTSE experiment (`www.rotse.net`). It turns out that the luminosity function of the SN Ia found by LOSS includes a higher fraction of faint events than either SDSS II and ROTSE, while the latter is peculiar with respect to the luminosity function of SN host galaxies, with an excess of faint galaxies compared with both SDSS II and LOSS. None of these two biases have an obvious explanation though likely they are the result of a combination of many different search features, including the instrument characteristics (field of view, pixel scale), search strategy (depth, frequency of monitoring) and detection/selection algorithms.

Despite the presence of biases, the plain statistics of SN discoveries can still lead to fundamental conclusions for what concern the origin of SNe. A good example in this respect is the statistics of the SN types as a function of the host galaxy properties, in particular the galaxy morphological type. Fig. 2 shows the number of supernovae of type Ia against the number of those of type II grouped according to the Hubble type of their host galaxies. It is well known that the Hubble sequence in the Local Universe is also a sequence of stellar population ages (eg. Buzzoni 2005): early type galaxies (*E, S0*) have completed their stellar assembly long time ago and therefore they host only low mass stars with an average age of several billion years. At the other extreme, in the late

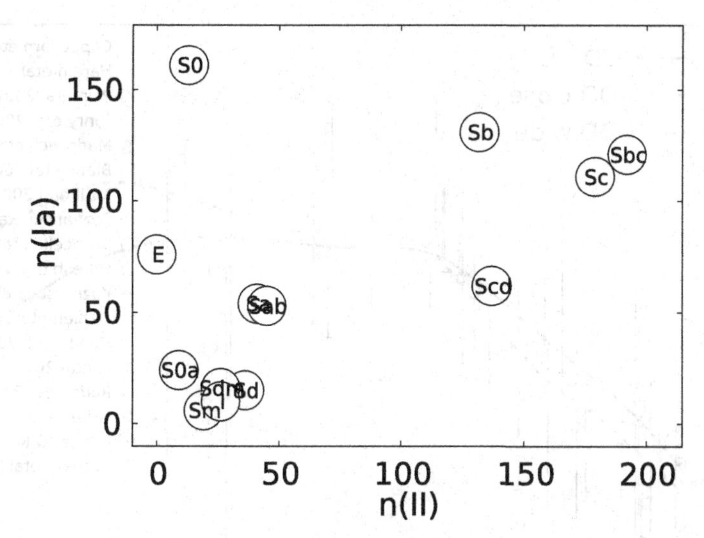

Figure 2. Number of SN Ia vs. number of SN II grouped according to the morphological type of the host galaxies. The sample is limited to host recession velocities $< 10000\,\mathrm{km}\,s^{-1}$ and the numbers where **not** normalized to the relative fraction of different type of galaxies in the Universe. Data were extracted from the Asiago Supernova Catalogue (`http://graspa.oapd.inaf.it`).

spirals (*Sc,Sd,Sm*) star formation is still ongoing and the luminosity is dominated by young massive stars. Intermediate Hubble types host a mixed stellar population with a comparable fraction of old, low mass stars and of young, massive stars.

Starting from this knowledge interpreting the distribution of points in Fig. 2 is straightforward. Type Ia SNe that strongly dominate in early type galaxies need to be associated to low mass progenitors and instead type II that are most frequent in late spirals are to be linked to young massive stars. We may notice that the number of SN Ia is high also in late spirals where old stars are rare. If we maintain that the type Ia progenitor mass range is the same in all galaxies, the similar rate of events in early and late spirals indicates a wide range of progenitor ages at explosions, from few 10^9 yr in ellipticals to few 10^5 yr in late spirals. The latter however are not the dominant fraction: for the same global on-going star formation rate, measured from the number of detected SN II, the number of type SN Ia in intermediate *Sb* spirals is much higher that in late *Scd* spirals arguing in favour of a significant fraction of SN Ia progenitors with intermediate ages. As shown in the next section, these qualitative conclusions based on the plain event statistics, are confirmed by the results of the detailed quantitative analysis.

3. SN rates and galaxy types

For a quantitative statistical analysis first of all we need to compute SN rates, that is not just count the discovered events but also account for the surveillance time and for the observational biases, at least the known ones, and normalise to the size of the stellar population under investigation. This approach requires a detailed knowledge of the SN search characteristics which usually limits the analysis to the SNe discovered by a specific search, then reducing the available SN statistics. The obvious solution, i.e. to combine the data of different searches, was attempted only in one case (Cappellaro *et al.* 1999). With this approach, by comparing the SN rate per unit blue luminosity (SNu) to the star formation rate (SFR) for galaxies of different colours (as derived by galaxy evolutionary model, Kennicutt 1998) it was possible to constrain the masses for core collapse SN

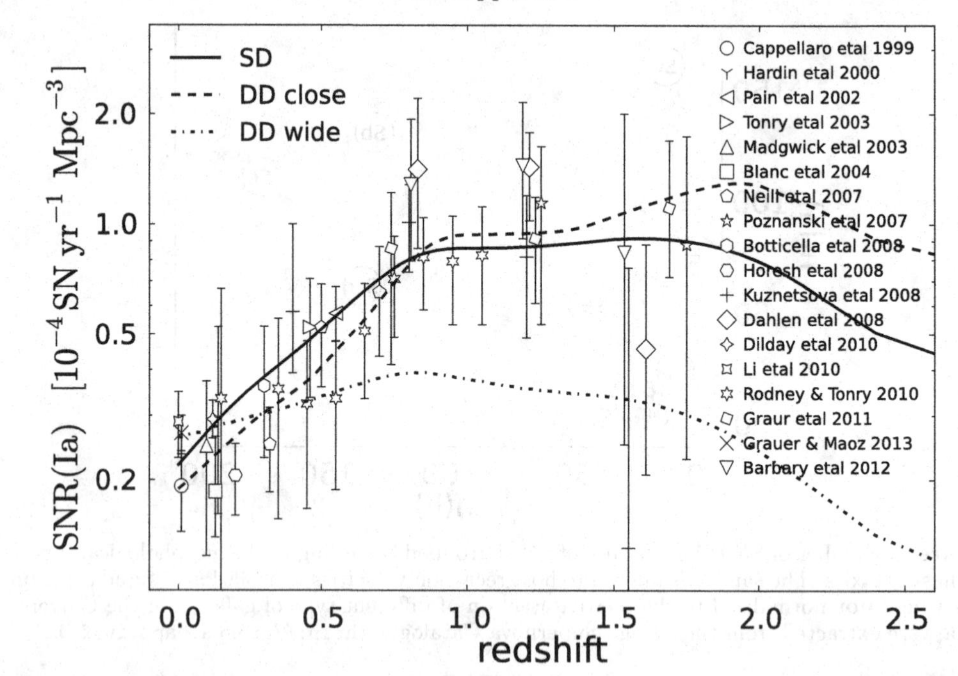

Figure 3. Evolution with redshift of the type Ia SN rate per unit volume. All measurements available in literature have been collected excluding early estimates that later have been revised and a few measurements at low redshift with very large error-bars. The predicted rate evolution assuming the current estimate of SFR (Cucciati *et al.* 2012) and three different SN Ia progenitor scenarios, namely single degenerate (SD), and two flavors of double degenerate (DD) with close and wide distribution of the binary separation (Greggio 2005) are also shown.

progenitors to the range $10 < M < 40\,\mathrm{M}_{\odot}$. Instead the rate of type Ia in SNu was found approximately constant in galaxies of different Hubble type. Because the M/L ratio is significantly lower in late spirals compared with early type galaxy, this implies that the SN Ia rate per unit mass is much higher in star forming galaxies, supporting the presence of a significant fraction of relatively young stars among SN Ia progenitors. This finding was confirmed by a direct calculation of the SN rate per unit mass, where the galaxy mass was derived from the K-band infrared luminosity and the B-K colours (Mannucci *et al.* 2005). More recently a detailed analysis of the data of LOSS (Li *et al.* 2011) lead to the identification of the dependence of the SN rate per unit mass on the galaxy mass: galaxies with higher mass have a lower SN rate per unit mass. In the context of galaxy evolution, this can be understood considering the downsizing effect for which, in the local Universe, the SFR per unit mass is higher in galaxies of lower mass. This effect is found also for SN Ia that again is an evidence for a significant fraction of SN Ia having young progenitors. Eventually, the direct evidence that the SN Ia rate increases with the specific SFR, that is the SFR per unit mass, has been derived by Smith *et al.* (2012) using the SDSS-II SN database.

The fact that SN Ia show a wide range of progenitor ages or, how it is usually refereed to, of delay time from star formation to explosion, is consistent with the consolidated stellar evolution paradigms. Indeed, while there are still fundamental uncertainties in the standard scenarios for SN Ia explosions, double degenerate or single degenerate or both, in all cases it is expected that the delay times distribution ranges from few 10^5 yr to several 10^9 yr, with more power at young ages (Greggio 2005, 2010).

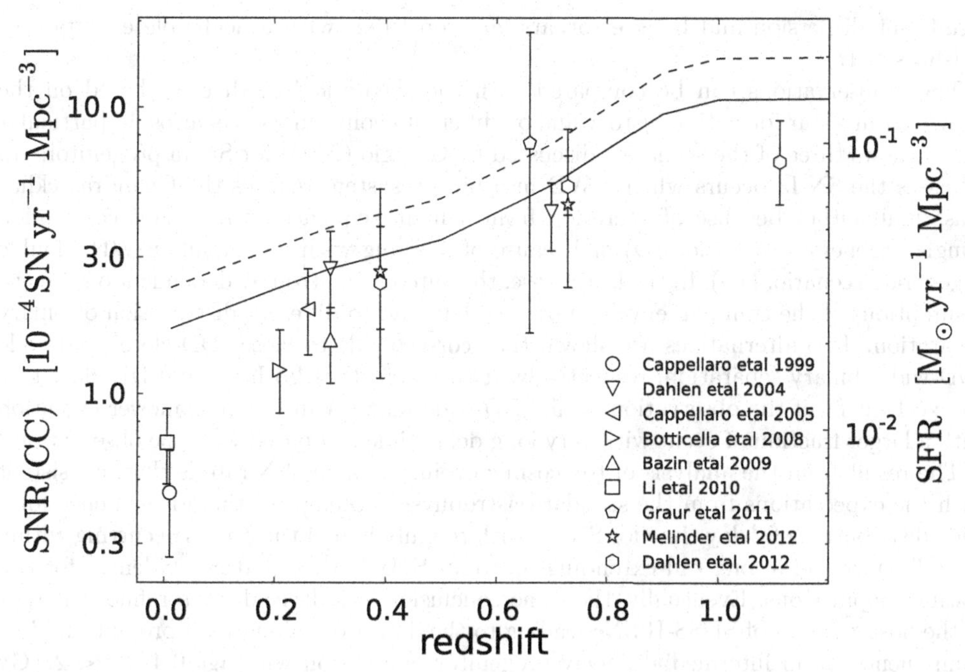

Figure 4. The evolution with redshift of the core collapse SN rate per unit volume is compared with the predicted evolution assuming the current estimate of the SFR (Cucciati *et al.* 2012) and assuming $10 < M < 40\,\mathrm{M}_\odot$ for the progenitor mass range (solid line) or adopting a lower limit of $8\,\mathrm{M}_\odot$ (dashed line). The measurements shown in the plot do not include the correction for the core collapse SNe that remain hidden in the nucleus of starburst galaxies (cf. Mattila *et al.* 2012).

4. Evolution of the SN Ia rate with redshift

Following the early attempts to measure the cosmic evolution of the SFR, Madau *et al.* (1998) were among the first to argue that measurements of the rate of SN Ia with redshift could be used to constrain their delay time distribution and in turn the progenitor scenario. Some time later, the first results based on the Great Observatories Origins Deep Survey (GOODS) lead to a surprising claim (Strolger *et al.* 2004): the SN Ia rate per unit volume shows a rapid increase up to redshift ~ 1 then start to decrease, and by redshift 1.7 the rate is a factor 3/4 lower than at peak. This seemed to imply that the SN Ia delay time distribution is limited to the range 2-4 billion yr.

The need to reconcile this finding with the evidences from the local Universe that a good fraction of SN Ia occurs after a short delay time gave raise to a number of speculations about the possible co-existence of two distinct populations of SN Ia (Scannapieco & Bildsten 2005, Mannucci *et al.* 2006) with very short and very long delay time, respectively. However, not even the two population model could match the GOODS observations: fitting simultaneously the high rate at redshift ~ 1 and the decline at higher redshift would need an ad hoc mechanism to suppress the prompt population at high redshift that otherwise would be dominating. Therefore the debate remained open while a number of groups spent their efforts to add new measurements of SN Ia rate at high redshift. The results of this effort are summarized in Fig. 3 showing that the picture changed significantly with respect to the early claims. First of all the new measurements along with a re-analysis of some of the early data (cf. Dahlen *et al.* 2008, Rodney & Tonry 2010) show that the raise of the SN Ia rate up to redshift ~ 1 is more gentle than initially believed. In addition, the high redshift measurements, while showing a significant dispersion and large errorbars, are consistent with a negligible evolution at redshifts > 1.

These observations can be compared with the theoretical predictions based on the known cosmic star formation history and on different evolutionary scenarios. In particular Fig. 3 shows three of the scenarios discussed by Greggio (2005) for SN Ia progenitors. In all cases the SN Ia occurs when a WD in a binary system reaches the Chandraseckhar mass limit either because of accretion from a main sequence or red giant companion (single degenerate scenario, SD) or because of merging with a companion WD (double degenerate scenario, DD). In the latter case, the outcome is strongly dependent on specific assumptions of the common envelope phases that lead to different distribution of binary separation. Two alternatives are shown that corresponds to close (DD-close) and wide (DD-wide) binary separation, respectively. It turns out that both SD and DD-close give an excellent fit of the observations while DD-wide fails because of a shallower evolution with a larger fraction of event with very long delay time compared with the observations.

The result from the analysis of the cosmic evolution of the SN rate is that, consistent with the expectations from the standard astrophysical scenarios, there is evidence for a wide distribution of delay time for SN Ia, with roughly half of the events occurring within 0.5 billion yr and a long tail extending up to an Hubble time and no evidences for two distinct populations. Eventually, the same conclusion was derived, by a refined analysis of the host galaxies of SDSS-II SNe leading to the detection, along with prompt and late components, of an intermediate-delay progenitor population with age $0.4 < t < 2.4\,\mathrm{Gy}$ (Maoz *et al.* 2012).

5. Core collapse SN rate and star formation history

The rate of core collapse SNe (CC SNe) in a given stellar stellar population is a direct measure of the on going SFR, and hence the CC SN rate evolution as a function of redshift can be used to track the cosmic star formation history. This is shown in Fig. 4 where the published estimates of the CC SN rates are compared with the SFR derived from measurements of the FUV galaxy luminosity (Cucciati *et al.* 2012). To match the two scales (SN rate vs. SFR) we need two basic assumptions:

(*a*) the initial mass function (IMF). While the fraction of massive stars for a given SFR depends from the IMF, the same quantity determines the luminosity/SFR conversion factor. Therefore, the actual choice of the IMF is insignificant as soon as it is consistent in all the comparison. Here we adopt a standard Salpeter IMF as in Cucciati *et al.* (2012).

(*b*) the mass range for core collapse SN progenitors. A good match between CC SN rate and SFR is obtained adopting $10 < M_{CC} < 40\,\mathrm{M}_\odot$ (solid line in Fig 4). While the uncertainty on the upper limit has a modest influence (for an upper limit of $100\,\mathrm{M}_\odot$ the expected SN rate increases by only $\sim$10%), the choice of the lower limit is more important.

For the latter, we remind that in recent years it has become possible to obtain direct measurements of the mass of CC SN progenitors by the analysis of deep, high resolution pre-discovery images. The method has still large uncertainties but the available data on a dozen events suggest that the lower limit for core collapse explosion is lower than that adopted above, i.e. $8 \pm 1\,\mathrm{M}_\odot$ (Smartt 2009). With this number the expected SN rate increases by $\sim 40\%$ (dotted line in the Fig. 4) causing a significant mismatch with respect to the observed rates.

As a possible explanation for the observed discrepancy it has been claimed that the observed rate are lower limit of the true value because a significant fraction of events would remain hidden in the nuclear regions of luminous infrared galaxies due to very high dust extinction (Mannucci *et al.* 2007, Mattila *et al.* 2012). Some support to this idea come from the results of infrared SN searches in starburst galaxies (Miluzio *et al.* 2013) although we would expect that the effect is negligible at low redshift where the fraction of LIRs is small. Instead, as we see in Fig. 4, the discrepancy is significant also at low redshift where SN searches are most complete. We should not forget that the issue of the uncertain extinction correction affects not just SN searches but also galaxy luminosity measurements, in particular in the (rest frame) UV band. Therefore the issue remains open and further more accurate estimates of the CC SN rates are definitely needed not least for a consistent check of more conventional SFR indicators.

6. Conclusion

SN rates as a function of redshift have been measured mainly as a by product of SN searches aimed to measure cosmological parameters. Despite some observational biases they have been able to set useful constraints on the progenitor scenarios and, when extinction biases are fully understood, also on the cosmic star formation history. We may notice however that for a further discrimination among competing scenarios, SN rate per unit volume appears insufficient and what we need now is a detailed investigation of the link between SNe and their parent stellar population as a function redshift. This requires a characterization of the monitored galaxies in term of luminosity, mass and star formation history.

With this in mind we started a new SN search at intermediate redshift using the VLT Survey Telescope (VST) equipped with the OMEGACAM wide field camera (Botticella *et al.* 2013). Differently from other projects, our emphasis is on measuring un-biased SN rates as a function of galaxy properties and, for this reason we focused on two well known sky fields, COSMOS and the Chandra Deep Field South, for which an outstanding amount of ancillary data allows the recovery of the galaxy physical properties. With our project, named SUDARE from Supernova Diversity and Rate Evolution, we aim to provide a first investigation of the statistics of SN diversity as a function of galaxy evolution. While statistics is not expected to be enormous (just few hundred SNe at the completion of the four years search) the project is also intended to prepare the ground for the analysis of future massive transient search, in particular LSST and EUCLID.

References

Barbary, K., Aldering, G., Amanullah, R., *et al.* 2012, *ApJ*, 745, 31
Bazin, G., Palanque-Delabrouille, N., Rich, J., *et al.* 2009, *A&A*, 499, 653
Blanc, G., Afonso, C., Alard, C., *et al.* 2004, *A&A*, 423, 881
Botticella, M. T., Riello, M., Cappellaro, E., *et al.* 2008, *A&A*, 479, 49
Botticella *et al.* 2013, *The Messenger* 151, 29
Buzzoni, A. 2005, *MNRAS*, 361, 725
Cucciati, O., De Lucia, G., Zucca, E., *et al.* 2012, *A&A*, 548, A108
Dahlen, T., Strolger, L.-G., Riess, A. G., *et al.* 2004, *ApJ*, 613, 189
Dahlen, T., Strolger, L.-G., & Riess, A. G. 2008, *ApJ*, 681, 462
Dahlen, T., Strolger, L.-G., Riess, A. G., *et al.* 2012, *ApJ*, 757, 70
Dilday, B., Smith, M., Bassett, B., *et al.* 2010, *ApJ*, 713, 1026
Cappellaro, E., Evans, R., & Turatto, M. 1999, *A&A*, 351, 459
Cappellaro, E., Riello, M., Altavilla, G., *et al.* 2005, *A&A*, 430, 83
Frieman, J. A., Bassett, B., Becker, A., *et al.* 2008, *AJ*, 135, 338
Graur, O., Poznanski, D., Maoz, D., *et al.* 2011, *MNRAS*, 417, 916
Graur, O. & Maoz, D. 2013, *MNRAS*, 430, 1746

Greggio, L. 2005, *A&A*, 441, 1055
Greggio, L. 2010, *MNRAS*, 406, 22
Hardin, D., Afonso, C., Alard, C., *et al.* 2000, *A&A*, 362, 419
Horesh, A., Poznanski, D., Ofek, E. O., & Maoz, D. 2008, *MNRAS*, 389, 1871
Kennicutt, R. C., Jr. 1998, *ARAA*, 36, 189
Kuznetsova, N., Barbary, K., Connolly, B., *et al.* 2008, *ApJ*, 673, 981
Leaman, J., Li, W., Chornock, R., & Filippenko, A. V. 2011, *MNRAS*, 412, 1419
Li, W., Chornock, R., Leaman, J., *et al.* 2011, *MNRAS* 412, 1473
Madau, P., della Valle, M., & Panagia, N. 1998, *MNRAS*, 297, L17
Madgwick, D. S., Hewett, P. C., Mortlock, D. J., & Wang, L. 2003, *ApJ* (Letters), 599, L33
Mannucci, F., Della Valle, M., Panagia, N., *et al.* 2005, *A&A* 433, 807
Mannucci, F., Della Valle, M., & Panagia, N. 2006, *MNRAS*, 370, 773
Mannucci, F., Della Valle, M., & Panagia, N. 2007, *MNRAS*, 377, 1229
Maoz, D., Mannucci, F., & Brandt, T. D. 2012, *MNRAS*, 426, 3282
Mattila, S., Dahlen, T., Efstathiou, A., *et al.* 2012, *ApJ*, 756, 111
Melinder, J., Dahlen, T., Mencía Trinchant, L., *et al.* 2012, *A&A*, 545, 96
Miluzio, M., Cappellaro, E., Botticella, M. T., *et al.* 2013, *Preprint* arXiv:1303.3803
Minkowski, R. 1964, *ARAA*, 2, 247
Neill, J. D., Hudson, M. J., & Conley, A. 2007, *ApJ* (Letters), 661, L123
Pain, R., Fabbro, S., Sullivan, M., *et al.* 2002, *ApJ*, 577, 120
Poznanski, D., Maoz, D., Yasuda, N., *et al.* 2007, *MNRAS*, 382, 1169
Quimby, R. M., Yuan, F., Akerlof, C., Wheeler, J. C., & Warren, M. S. 2012, *AJ*, 144, 177
Rau, A., Kulkarni, S. R., Law, N. M., *et al.* 2009, *PASP*, 121, 1334
Rodney, S. A. & Tonry, J. L. 2010, *ApJ*, 723, 47
Scannapieco, E. & Bildsten, L. 2005, *ApJ*, 629, L85
Smartt, S. J. 2009, *ARAA*, 47, 63
Smith, M., Nichol, R. C., Dilday, B., *et al.* 2012, *ApJ*, 755, 61
Strolger, L.-G., Riess, A. G., Dahlen, T., *et al.* 2004, *ApJ*, 613, 200
Tonry, J. L., Schmidt, B. P., Barris, B., *et al.* 2003, *ApJ*, 594, 1

Discussion

ANDERSON: The discrepancy between IAU and ATEL SN announcements is worrying. Will it be possible to unify these in the future, and how important is this?

CAPPELLARO: There has been a number of discussion to solve this issue, but not definite solution yet. I agree that this is an important issue. At least nearby events (where nearby is to be defined) should be recorded homogeneously.

GABICI: Which is the best estimate for the supernova rate in the milky way?

CAPPELLARO: Estimates taken from 1 in 20 year to 1 in 50 year, the later appears to find more support in recent literature.

Supernova Environmental Impacts
Proceedings IAU Symposium No. 296, 2013
A. Ray & R. A. McCray, eds.

doi:10.1017/S1743921313009216

Binary Effects on Supernovae

Philipp Podsiadlowski

Dept. of Astrophysics, University of Oxford,
Denys Wilkinson Building, Keble Road, Oxford, OX1 3RH, United Kingdom
email: podsi@astro.ox.ac.uk

Abstract. Here we review how binary interactions affect the final pre-supernova structure of massive stars and the resulting supernova explosions. (1) Binary-induced mass loss and mass accretion determine the final envelope structure, the mass, radius and chemical composition, which are mainly responsible for the supernova appearance and supernova (sub-)type. (2) Mass loss can also drastically change the core evolution and hence the final fate of a star; specifically, around $10\,M_{\odot}$, it determines whether a star explodes in a supernova or forms a white dwarf, while for larger masses it can dramatically increase the minimum main-sequence mass above which a star is expected to collapse to a black hole. (3) Mass loss before the supernova directly affects the circumstellar medium (CSM) which can affect the supernova spectrum (e.g. account for the IIn phenomenon), produce powerful radio emission and, in extreme cases, lead to a strong interaction with the supernova ejecta and thus strongly modify the lightcurve shape; it may even be responsible for some of the superluminous supernovae that have recently been discovered.

Keywords. binaries (including multiple): close, stars: mass loss, supernovae: SN 1987A, supernova remnants

---

1. Introduction

It is well established that most stars are members of binary systems. Indeed, in a large fraction of systems, the two stars are close enough to interact directly by Roche-lobe overflow (in 30 – 50 % of systems, depending on the mass of the system). For massive stars, the fraction of interacting binaries is even higher: recently Sana *et al.* (2012) showed that $\sim 75\,\%$ of O stars are members of interacting binaries. As stars generally expand most dramatically after hydrogen core burning, while they spend most of their lifetime on the main sequence, observed systems are more likely to have not yet interacted. On the other hand, supernovae probe the final structure of massive stars; as most of these will have experienced at least one binary interaction at this point, the majority of supernovae will also have been affected by binary interactions. Indeed, the results of Sana *et al.* (2012) show that thinking of supernova progenitors as single stars that have evolved in isolation is not even the correct zeroth-order approximation.

In this review, we discuss the various ways by which binary interactions affect supernovae. In § 2 we summarize how they affect the final envelope structure and resulting supernova appearance, and illustrate this with the important case of SN 1987A in § 3. In § 4 we discuss how they can alter the final fate of a star and in § 5 how they shape the circumstellar environment and the observational signatures this produces.

2. Binary Interactions and Supernova Lightcurves

When a star fills its Roche lobe, there are two main types of main transfer: *stable* and *unstable mass transfer*.

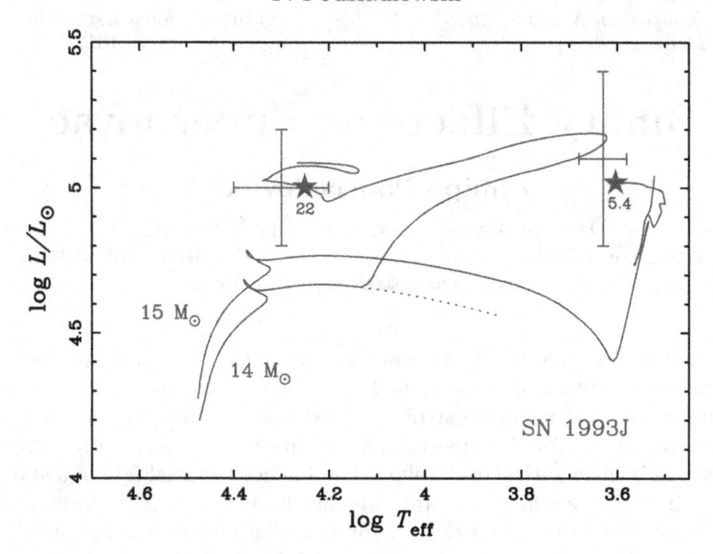

Figure 1. Hertzsprung-Russell diagram showing the evolution of the two components of a massive binary that experiences very late (so-called Case C) mass transfer. At the time of the explosion, the primary is a $5.4\,M_{\odot}$ star with a small $0.3\,M_{\odot}$ H-rich envelope. Because of the accreted mass the secondary never becomes a red supergiant and explodes as a blue supergiant similar to SN 1987A. The locations of the two components in the observed progenitor system of the IIb supernova SN 1993J are indicated by large error bars. (From Maund *et al.* 2004.)

Stable Mass Transfer

In the case of stable mass transfer, mass is transferred from one star, the donor, to the companion star, the accretor. While there may be significant mass loss from the system (in particular, at the highest mass-transfer rates), mass transfer can be largely treated as conservative, i.e. to lowest order both mass and angular momentum of the binary are conserved, and both stars are strongly affected by the mass-transfer phase. This also implies that, at the end of the mass-transfer phase, the system tends to be wider than the initial system. The mass-transfer phase generally ends when the donor star has lost most of its envelope: in the case of a hydrogen-rich donor, most of its hydrogen envelope, producing a helium (or Wolf-Rayet) star, in the case of a donor star with a helium envelope (so-called case BB mass transfer), a star without a hydrogen or helium envelope. It should be noted that generally, at the end of the mass-transfer phase, a hydrogen-rich donor star becomes detached from its Roche lobe when it still has some hydrogen left in its envelope (typically a few $0.1\,M_{\odot}$). As this may show up in the supernova spectrum, this can lead to hybrid supernova types, such as SNe IIb (of which SN 1993J was the proto-type), which first looks like a hydrogen-rich Type II supernova, but later morphs into a SN Ib.

The further evolution of the accretor depends on the evolutionary stage at the time of mass transfer. If it was still on the main sequence, it will generally be rejuvenated, which means that it will subsequently behave like a more massive star but with its clock reset. On the other hand, if it has already left the main sequence, its core structure is now fixed, but because of the increase in the mass of the envelope, its further evolution may be drastically altered; specifically, it may never become a red supergiant, but spend most of its remaining evolution as a blue supergiant where it will explode, producing a supernova similar to SN 1987A (Podsiadlowski & Joss 1989).

Figure 1, which represents a detailed binary evolution model for the system that produced SN 1993J (from Maund *et al.* 2004), illustrates the effects on both the donor and the accretor. At the time of the explosion, the initial primary still has a small hydrogen-rich envelope, leading to a SN IIb, while the secondary, having accreted after having left the main sequence, ends its evolution as a blue supergiant similar to the case of SN 1987A.

Unstable mass transfer

Unstable mass transfer occurs when the companion cannot accrete all of the accreted mass and starts to fill and overfill its Roche lobe: essentially the envelope of the mass donor (often a giant) expands to completely engulf the companion, producing a common-envelope system, consisting of the core of the donor and the companion star, embedded in the envelope of the donor. Because this immersed binary experiences friction with the envelope, the binary orbit shrinks, releasing orbital energy that is directly deposited in the envelope. While the details of common-envelope evolution are still poorly understood, it is clear that, in general, there are two possible outcomes:

(1) If the orbital energy is sufficient to unbind the common envelope, the latter can be ejected, stopping the spiral-in process and leaving a much closer binary, consisting of the core of the donor (again a helium star if the donor was initially hydrogen-rich) and the largely unaffected companion star.

(2) If the orbital energy is not sufficient to eject the envelope, the spiral-in process continues till the two stars have merged completely. In this case, the end product is a single star with possibly some unusual properties. It will, at least initially, be rapidly rotating (possibly having the appearance of a B[e] supergiant), and because of the mixing induced by the merger process may have an unusual chemical profile. In addition, because of the added mass in the envelope (coming from the destroyed companion star), it may end its evolution as a blue supergiant similar to the case of a star that accreted mass (see Podsiadlowski, Joss & Rappaport 1990 and § 3). Also, as this may produce a rapidly rotating core at the time of the explosion, this provides a potential channel for the progenitors of gamma-ray bursts within the framework of the collapsar model (for a review see Fryer *et al.* 2007).

Supernova types

The envelope structure of an exploding star, its mass, radius and composition, is the main factor that determines the appearance of a supernova and hence determines its type and sub-type. For example, a progenitor with a smaller radius (such as the progenitor of SN 1987A; see § 3) will be fainter initially because a larger fraction of the explosion energy is used up in unbinding the more tightly bound envelope. The mass of the envelope critically determines the photon diffusion time in the ejecta, which in turn determines the lightcurve width (and the length of the plateau phase in a SN II-P).

In other words, the sequence of supernova (sub-)types

$$\text{SN II-P} \rightarrow \text{SN II-L} \rightarrow \text{SN IIb} \rightarrow \text{SN Ib} \rightarrow \text{SN Ic}$$

just describes a sequence of increased mass loss, first of the hydrogen-rich envelope, then the helium envelope. As the main effect of binary interactions is to change the envelope structure (see, e.g., Podsiadlowski, Joss & Hsu 1992), one should expect that most of the diversity along this sequence is a direct consequence of binary interactions, rather than just variations in wind mass loss in single stars.

3. Supernova 1987A (SN 1987A)

SN 1987A is a particularly interesting example. While the detected neutrino signal from the collapse confirmed the basic theory of core collapse, it was highly anomalous in many respects: the star that exploded was a blue supergiant instead of a red supergiant, and the progenitor was surrounded by a complex triple-ring nebula (Wampler *et al.* 1990), consisting of material that was ejected by the progenitor system $\sim$ 20000 yr *before* the explosion. The best explanation to-date is that the system initially consisted of a massive binary system that merged 20000 yr before the explosion when part of the progenitor's envelope was ejected and subsequently swept up by the energetic blue-supergiant wind, once the merged object had become a blue supergiant (for more details see Podsiadlowski, Morris & Ivanova 2007; Morris & Podsiadlowski 2007; and Fitzpatrick & Podsiadlowski [these proceedings]).

However, detailed light-echo studies by Sugerman *et al.* (2005) have shown that the circumstellar environment is even more complex, providing an imprint of the whole mass-loss history of the progenitor system. These include: (1) the fast wind in the main-sequence phase, producing a CSM bubble; (2) the slow red-supergiant wind, lasting $10^4 - 10^5$ yr; (3) an (early) stable mass-transfer phase before the actual merger ($\sim 10^3 - 10^4$ yr), in which one expects both a gravitationally flattended red-supergiant wind and a bipolar outflow from the accreting component; (4) the mass loss associated with the merger and finally (5) the energetic blue-supergiant wind.

Fitzpatrick (2012) has shown that the inclusion of all of these, in particular the red-supergiant wind and the early mass-transfer phase, can indeed explain many of the observed structures and light echoes (some of these results are presented in Fitzpatrick & Podsiadlowski [these proceedings]).

4. The Final Fates of Single and Binary Stars

Binary evolution does not only affect the final envelope structure, but can also drastically change the final fate of a star. While it is clear that a star that loses mass early on the main sequence will effectively behave like a less massive star (even a star as massive as 17 $M_\odot$ may end its life as a white dwarf; e.g. Podsiadlowski, Rappaport & Han 2003), the effects can be even more dramatic as illustrated in Figures 2 and 3.

Figure 2 shows how the evolution of a star around 10 $M_\odot$ is affected by the presence (in single stars) or absence of the second dredge-up phase (in the case of stars in binaries that have lost their hydrogen-rich envelopes before the AGB phase). Because the second dredge-up drastically decreases the core mass in single stars, they are more likely to end their evolution as ONeMg white dwarfs, rather than experience core collapse, the more likely fate for the binary case. As the supernovae produced from these objects are likely to be relatively faint electron-capture supernovae, which are not expected to impart large natal kicks to the newborn neutron stars, this may explain why low-kick neutron stars appear to prefer relatively close binary systems (see Podsiadlowski, Langer, *et al.* 2004 for detailed discussions).

Figure 3 illustrates how the loss of the hydrogen-rich envelope before or early during helium core burning (so-called Case B mass transfer) dramatically changes the minimum initial mass where a massive star is expected to become a black hole. As shown by Brown *et al.* (2001), because of the lack of a H-burning shell, the convective core does not grow during helium core burning, and stars end up with much smaller CO and ultimately iron cores. Indeed, because of this, such H-deficient stars formed in Case B binaries are expected to end their evolution as neutron stars rather than as black holes, even for

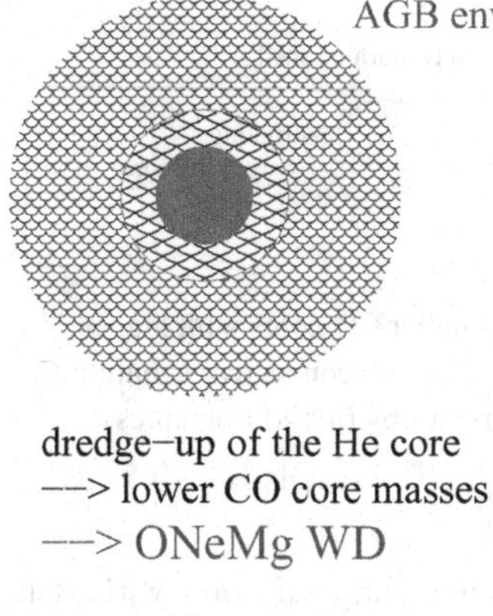

Figure 2. Schematic comparison of the late evolution of stars with an initial mass around $10\,M_\odot$ with and without a H-rich envelope. The former experience a so-called second dredge-up phase at the beginning of the AGB phase reducing the size of the helium core; these stars are therefore more likely to end up as ONeMg white dwarfs. On the other hand, stars without the second dredge-up may eventually collapse and experience an *electron-capture supernova.*

initial masses as high as $\sim 60\,M_\odot$, much larger than the likely estimated initial mass for single stars ($\sim 20 - 25\,M_\odot$).

5. Supernova Environments and Signatures

The complex mass loss from binary systems produces important signatures for binary interactions. The spectacular structures of many planetary nebulae, generally now believed to be largely shaped by various binary interactions, are a testament to this, as is the nebula around SN 1987A, already discussed in § 3.

Figure 4 shows a simulation of how the presence of a binary companion affects the structure of a red giant in a binary (from Mohamed, Booth & Podsiadlowski 2013). The wind is strongly focused towards the orbital plane and has the shape of a double spiral, which itself is very clumpy. These clumps can produce absorption features in high-resolution supernova spectra, which will vary depending on the line of sight. Indeed, the variable Na lines observed in a significant fraction of Type Ia supernovae may provide a direct signature of these wind structures, which lends direct support to the idea that these systems contain red-giant donors in a single-degenerate progenitor scenario (Patat *et al.* 2007). Direct evidence for such spiral wind structures may also have been seen in the periodic radio lightcurves in some radio supernovae, such as SN 1979C (Weiler *et al.* 1991).

LBV Supernovae

There has been mounting evidence in recent years (e.g. Kotak & Vink 2006; Gal-Yam *et al.* 2007) that some supernovae occur in an LBV (luminous-blue-variable) phase. From a stellar evolution point of few this is very surprising as the LBV phase is believed to occur

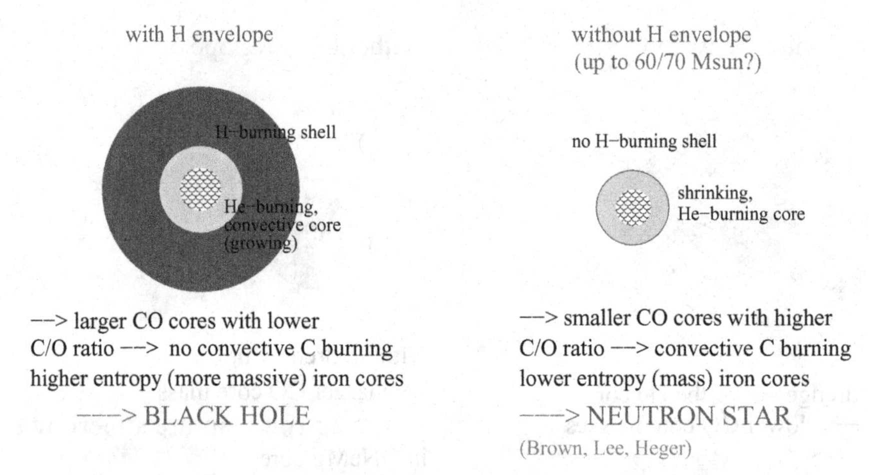

Figure 3. Schematic comparison of the helium core-burning phase of stars with (left) and without a hydrogen envelope (right) for massive stars ($M > 20\,M_\odot$). Because of the lack of a H-burning shell, stars without a hydrogen envelope produce smaller He-exhausted cores with a larger C/O fraction. This makes them more likely to ultimately collapse to a neutron star instead of a black hole.

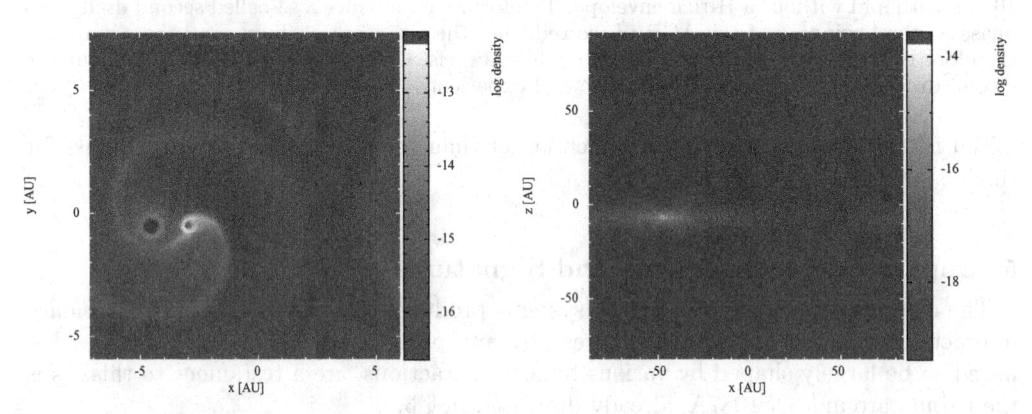

Figure 4. SPH simulations illustrating the mass loss from a symbiotic binary with a red-giant donor (simulating a system like RS Oph). The wind from the donor is assumed to be spherically symmetric, but is gravitationally focused by the companion and shaped by the orbital motion. The panels show the density structure in the orbital plane (left) and in a meridional projection (right). (From Mohamed, Booth & Podsiadlowski 2013.)

soon (or even during) the hydrogen-core-burning phase, where massive stars lose their hydrogen-rich envelopes and become Wolf-Rayet stars, where they spend the last $>10^5$ yr of their evolution. Indeed, if true the theory of single, massive stars would seriously need to be re-written.

Fortunately, the situation is quite different in a binary where both components merge soon after the primary's main-sequence phase: the resulting merged object will spend most of its helium-core-burning phase as a blue supergiant and only enter the LBV phase after helium core burning and may still explode in this phase (Justham, Podsiadlowski & Vink 2013; see Fig. 5). In fact, if both stars are of similar mass and have already developed a helium-rich core, the cores may also merge, producing a much more massive

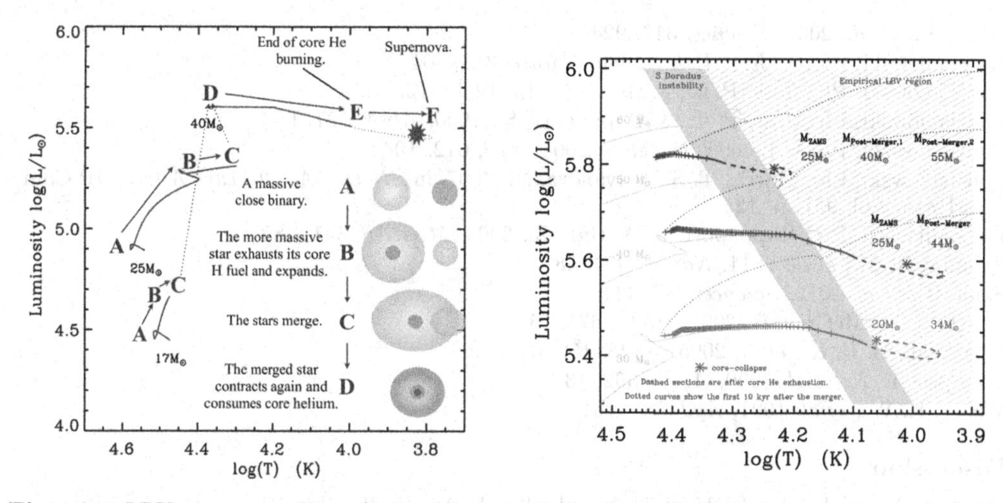

Figure 5. LBV supernovae from binary mergers. *Left:* schematic H-R diagram illustrating the evolution of a massive Case B merger. *Right:* Selected binary models, illustrating the evolution of massive mergers to the LBV phase. The most luminous case simulates a triple merger. (From Justham *et al.* 2013.)

core than a single, massive star could at the same metallicity. Such objects are potential candidates for pair-instability supernovae even at solar metallicity.

Interaction supernovae and superluminous supernovae

The effects of a dense circumstellar medium can be even more dramatic if the mass in the immediate neighborhood of an exploding star is a significant fraction of the ejecta mass. In this case, the interaction of the ejecta with the CSM will slow down the ejecta, efficiently converting its kinetic energy into thermal energy and ultimately radiation. If a CSM shell is at the right distance, this can power the lightcurve and may even be responsible for some of the rare, overluminous supernovae that have been discovered in recent years (Quimby *et al.* 2011). In this case, these supernovae would be overluminous not because they are particularly energetic, but because they are efficient in radiating their kinetic energy (Smith & McCray 2007). One question that still remains to be answered is why does a small subset of progenitors eject a large amount of their envelopes in the last few years/decades before the explosion. One possibility is that these could be ejected common envelopes in binary systems (Chevalier 2012): depending on the precise timing of the ejection this could account for a variety of both IIn and overluminous supernovae.

References

Brown, G. E., *et al.* 2001, *NewA*, 6, 457

Chevalier, R., 2012, *ApJ*, 752, 2

Fitzpatrick, B., 2012, *D. Phil. Thesis (Oxford University)*

Fryer, C. L., *et al.* 2007, *PASP*, 119, 1211

Gal-Yam, A., *et al.* 2007, *ApJ* 656, 372

Justham, S., Posiadlowski, Ph., & Vink, J.-S., 2013, *submitted*

Kotak, R. & Vink, J. S., 2006 *A&A* 460, L5

Maund, J. R., Smartt, S. J., Kudritzki, R. P., Podsiadlowski, Ph., & Gilmore, G. F., 2004, *Nature*, 427, 129

Mohamed, S., Booth, R., & Podsiadlowski, Ph. 2013, *in preparation*

Morris, T., & Podsiadlowski, Ph. 2007, *Science*, 315, 1103

Patat, F., *et al.*, 2007, *Science*, 317, 924
Podsiadlowski, Ph., & Joss, P. C., 1989, *Nature*, 338, 401
Podsiadlowski, Ph., Joss, P. C., & Hsu, J. J. L., 1992, *ApJ*, 391, 246
Podsiadlowski, Ph., Joss, P. C., & Rappaport, S., 1990, *A&A*, 227, L9
Podsiadlowski, Ph. & Langer, N., *et al.* 2004, *ApJ*, 612, 1044
Podsiadlowski, Ph., Morris, T. S., & Ivanova, N., 2007, in *SN 1987A: 20 Years After*, AIP Conf. Proc., Vol. 937, p. 125
Podsiadlowski, Ph., Rappaport, S., & Han, Z., 2003, *MNRAS*, 341, 385
Quimby, R. M., *et al.* 2011, *Nature*, 474, 487
Sana, H., *et al.* 2012, *Science*, 337, 444
Smith, N., & McCray, R. 2007, *ApJL*, 671, L17
Sugerman, B. E. K., *et al.* 2005, *ApJS*, 159, 60
Wampler, E. J., *et al.* 1990, *ApJL*, 362, 13
Weiler, K. W., *et al.* 1991, *ApJ*, 380, 161

Discussion

RAY: 1. Why do the LBV-phase reached only for stellar binaries with a lower mass threshold; what role do the initial separations play in reaching this end stages? 2. Do the LBV states reaching BSG, YSG or RSG's have differences in the stellar wind speeds?

PODSIADLOWSKI: 1. The lower-mass constraint arises from the assumption that you want the post-merger mass to be large enough that the system will experience LBV outbursts later. The initial separation is constrained by the requirement that the system experiences early case of mass transfer. 2. Yes, presumably the wind velocities for the different progenitors will have different wind velocities, but this will depend on the detailed physics of the LBV phenomenon that is not really understood and we did not model.

CHIOTELLIS: Type Ia SNR, look rather spherical symmetric. The CSM of the symbiotic R Ne are rather asymetric. Do you expect the interaction of the SN ejecta with such an AM will lead to a spherical symmetric SNR?

PODSIADLOWSKI: Indeed the CSM is asymetric but the mass contained in the medium is much smaller compared to ejecta mass. So it is not expected substantial effects on the symmetry of the resulting SNR.

Supernova Environmental Impacts
Proceedings IAU Symposium No. 296, 2013
A. Ray & R. A. McCray, eds.

doi:10.1017/S1743921313009228

Recent developments in supernova research with VLBI

Norbert Bartel[1] **and Michael F. Bietenholz**[1,2]

[1]York University, Toronto, M3J 1P3 Canada
email: bartel@yorku.ca

[2]Hartebeesthoek Radio Observatory, PO Box 443, Krugersdrop 1740, South Africa
email: mbieten@yorku.ca

Abstract. Very long baseline interferometry (VLBI) observations during the last 30 years have resolved many supernovae and provided detailed measurements of the expansion velocity and deceleration. Such measurements are useful for estimating the radial density profiles of both the ejecta and the circumstellar medium left over from the progenitor. VLBI measurements are also the most direct way of confirming the relativistic expansion velocities thought to occur in supernovae associated with gamma-ray bursts. Well-resolved images of a few supernovae have been obtained, and the interaction of the ejecta as it expands into the circumstellar medium could be monitored in detail. We discuss recent results, for SN 1979C, SN 1986J, and SN 1993J, and note that updated movies of the latter two of the supernovae from soon after the explosion to the present are available from the first author's personal website.

Keywords. techniques: high angular resolution, radio continuum: stars, supernovae: general, supernovae: individual (SN 1979C, SN 1986J, SN 1993J), stars: mass loss, stars: neutron, galaxies: distances and redshifts.

1. Introduction

Radio emission of supernovae originates from the interaction of the ejecta with the circumstellar medium (CSM) left over from the massive progenitor star. It could also originate from the interior region close to the stellar corpse generated in the explosion. Very long baseline interferometry (VLBI) observations of a supernova can provide us with detailed information about the supernova's size, expansion velocity, deceleration, age, and morphology.

The characteristics of the expansion can be used to estimate density profiles of the ejecta and sometimes bounds on the mass-loss-to-wind-velocity ratio, $\dot{M}/w$, of the progenitor star, independent of radio-lightcurve fitting. When combined with velocities obtained from optical spectra, they can also be used in determining the geometrical distance to the supernova and its host galaxy. Further, they are the most direct way of confirming or disputing relativistic expansion velocities expected in supernovae related to gamma-ray bursts. The morphology gives us information on the clumpiness of the CSM and/or possibly the ejecta. The thickness of the radio-emitting shell region, thought to be between the forward and reverse shocks, is in turn related to the density profiles. Any emission from the central region of the image of the supernova could be related to particularly dense condensations in the shell itself, to an interior shell or to the stellar corpse expected to be left over after the explosion of the massive star. In Fig. 1 we show on the left side part of a radio image of a supernova and on the right side a sketch of a supernova model to facilitate interpretation of observations.

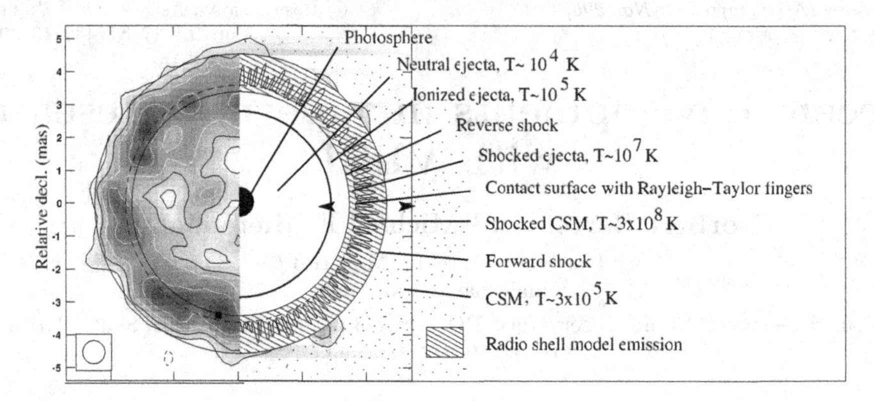

Figure 1. Left side: Composite VLBI image of SN 1993J at 8.4 GHz taken at three epochs between 1998 and 2000. Circles indicate fit inner and outer radii of the shell. Right side: Sketch of the standard model of a core-collapse supernova interacting with the CSM and matched to the radio image. It shows different regions with typical temperatures. Radio emission is expected to emanate between the forward and the reverse shock (taken from Bartel *et al.* 2007).

Since the first VLBI observations of a supernova (SN 1979C) in 1982 (Bartel *et al.* 1985), many supernovae have been detected with VLBI, but only five of them could be imaged in detail. These are SN 1979C in M100 (Bartel & Bietenholz 2008), SN 1986J in NGC 891 (Bietenholz *et al.* 2010a), SN 1987A in the LMC (Ng *et al.* 2011) and SN 1993J in M81 (Bietenholz *et al.* 2003; Bietenholz *et al.* 2010b; Marcaide *et al.* 2009a), all optically identified, and SN 2008iz in the highly optically opaque central region of M82 (Brunthaler *et al.* 2010a). In the remainder we discuss recent results on three of these supernovae.

2. Supernova VLBI

SN 1979C in M100 is the first supernova studied with VLBI. It has also been observed for the longest time and showed almost free expansion for two decades (Bartel & Bietenholz 2003; Marcaide *et al.* 2009b).

That was surprising. With a limit of the kinetic energy of 10^{51} erg for the shocked ejecta and the CSM shells it was suggested that $\dot{M}_w/w \sim 1 \times 10^{-5}\ M_\odot\ \mathrm{yr}^{-1}$ per $w = 10\ \mathrm{km\ s}^{-1}$, an order of magnitude smaller than estimated from radio-lightcurve fitting. However, strong deceleration is expected to start soon. In Fig. 2 (left) we show an image of SN 1979C. This image showed for the first time that the morphology was a shell. Despite the supernova's age, there was no indication of a central source above a limit of 15 times the 5 GHz spectral luminosity of the Crab Nebula (Bartel & Bietenholz 2008).

SN 1986J in NGC 891 is the first supernova seen to have a compact component in its projected center (Bietenholz *et al.* 2004, Fig. 2 right;). The component was first seen in 2003 in images at 8 and 15 GHz, but was not seen at 5 GHz. Before that an inversion in the integrated spectrum above 8 GHz had already been seen in 1999. Currently the compact component dominates images even at 5 GHz, and the spectral inversion has moved to lower frequencies. The component has a small proper motion of $1500 \pm 1500\ \mathrm{km\ s}^{-1}$, is marginally resolved at 22 GHz, and has an optically-thin spectral index indistinguishable from that of the shell. Three interpretations have been suggested. The component could be a pulsar wind nebula, with a spectral luminosity at 5 GHz $\sim$ 15 times that of the Crab Nebula, or emission from close to the black hole left over from the explosion. It could also be part of the shell and be caused by the shock front imparting on a dense condensation

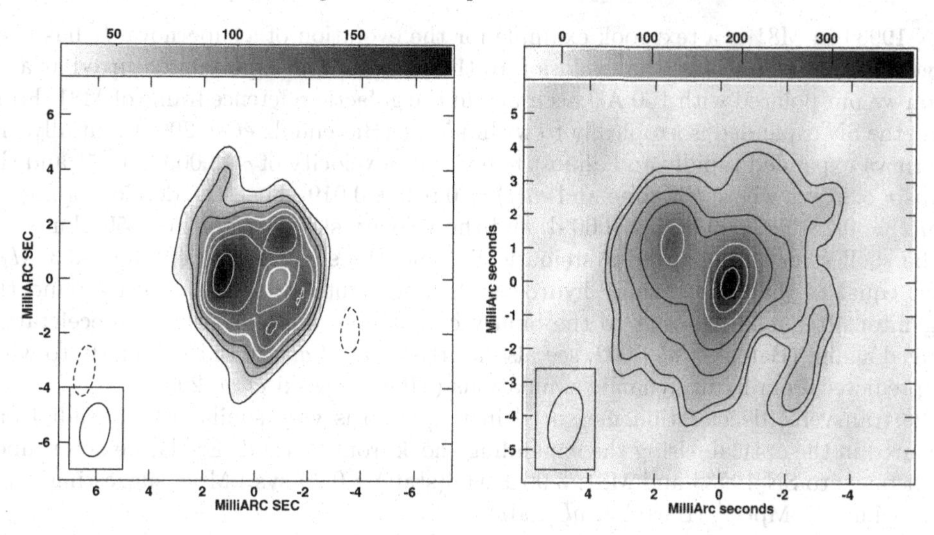

Figure 2. Left: VLBI image of SN 1979C at 5.0 GHz at epoch 2005 February 25. The contour levels are at −17, 17, 30, 40, 50 (first white contour), 60, 70, and 90% of the peak brightness of 186 μJy beam^{-1}. The scale at the top is in μJy beam^{-1}. Here and hereafter the FWHM of the beam is given in the left corner (taken from Bartel & Bietenholz 2008). **Right:** VLBI image of SN 1986J at 5.0 GHz at epoch 2008 October 25. The contour levels are at −7, 7, 10, 20, 30, 40, 50 (first white contour), 70, and 90% of the peak brightness of 388 μJy beam^{-1}. The scale at the top is in μJy beam^{-1} (Bietenholz *et al.* 2010a).

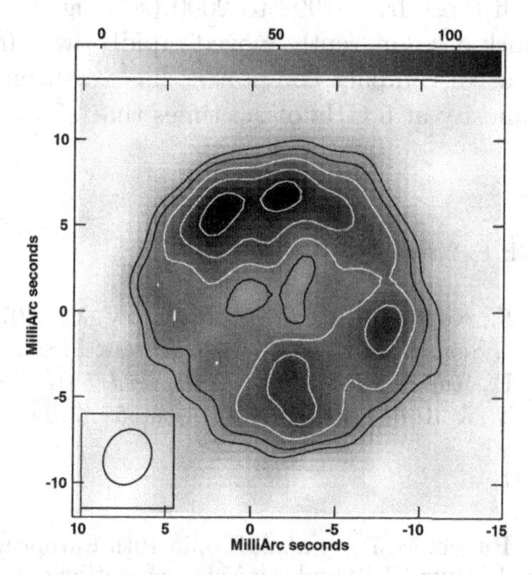

Figure 3. The latest VLBI image of SN 1993J, at 1.7 GHz, at epoch 2010 March 6. The contours are at 25, 35, 50 (first white contour), 70, and 90% of the peak brightness of 117 μJy beam^{-1}. The scale at the top is in μJy beam^{-1}.The coordinate origin is within 64 μas (rms) at the explosion center (taken from Bietenholz *et al.* 2010b).

of the CSM ~1000 times denser than the corresponding average CSM and fortuitously close to the projected center of the shell (Bietenholz *et al.* 2010a). Or the component is associated with the interaction of the shock with the slower parts of a highly anisotropic CSM formed in a binary system where the companion, left over from an earlier supernova explosion, has sprialled into the envelope of the progenitor of the observed supernova (Chevalier 2012). SN 2008iz also showed a compact central component but it has dimmed and is no longer visible in the latest images (Brunthaler *et al.* 2010b).

SN 1993J in M81 is a textbook example for the evolution of a supernova. It has been imaged from a month after the explosion to the present. The explosion occurred at a location we pin-pointed with 160 AU accuracy in the galactic reference frame of M81. From there, the SN expanded isotropically to within 5.5% (Bietenholz *et al.* 2001). Initially, the supernova expanded rapidly and almost freely with a velocity of ~18,000 km s^{-1} and the radius $r \propto t^{m(t)}$, where t is time and $m(t) = 0.919 \pm 0.019$. Then the deceleration grew to $m(t) = 0.781 \pm 0.009$ at $t \sim 1000$ d, and the velocity slowed down to ~8500 km s^{-1}.

The shell structure changed systematically, and the supernova swept up ~0.3 $M_\odot$, about equal to that of the small hydrogen shell thought to have remained around the progenitor after mass transfer to the binary companion. Subsequently the deceleration changed again, (Bartel *et al.* 2002, see also Martí-Vidal *et al.* 2011a,b) similarly to what was predicted from hydrodynamic simulations (Mioduszewski *et al.* 2001).

The transverse deceleration measured in the radio is very similar to the radial one measured in the optical. Using the expanding shock front method (ESM), we determined the distance to SN 1993J and M81: 3.96 ± 0.06(stat.) ± 0.29(sys.) Mpc, suggesting $H_0 = 66 \pm 11$ km s^{-1} Mpc^{-1} (Bartel *et al.* 2007).

Our latest image, from 2010, is displayed in Fig. 3. The supernova shell remains quite circular. The brightness is modulated around the rim, and the pattern of modulation has remained fairly similar over the last few years, and is therefore likely intrinsic rather than due to noise or deconvolution errors. A recent development is that the shell thickness is increasing rapidly (Bietenholz *et al.* 2010b), whereas only a slow increase is predicted by hydrodynamic simulations. This thickening of the shell can be seen when our latest images are compared with those from 1998 to 2000 (see Fig. 1) and earlier. In the last few years, the reverse shock has apparently moved rapidly away from the forward shock, and into the inner ejecta. No compact component has yet been found, with an upper limit on its spectral luminosity at 5 GHz of 0.3 times that of the Crab Nebula.

References

Bartel, N. & Bietenholz, M. F. 2003, *ApJ*, 591, 301
—. 2008, *ApJ*, 682, 1065
Bartel, N., Bietenholz, M. F., Rupen, M. P., & Dwarkadas, V. V. 2007, *ApJ*, 668, 924
Bartel, N., Rogers, A. E. E., Shapiro, I. I., *et al.* 1985, *Nature*, 318, 25
Bartel, N., Bietenholz, M. F., Rupen, M. P., *et al.* 2002, *ApJ*, 581, 404
Bietenholz, M. F., Bartel, N., & Rupen, M. P. 2001, *ApJ*, 557, 770
—. 2003, *ApJ*, 597, 374
—. 2004, *Science*, 304, 1947
—. 2010a, *ApJ*, 712, 1057
Bietenholz, M., Bartel, N., Rupen, M. P., *et al.* 2010b, in 10th European VLBI Network Symposium and EVN Users Meeting: VLBI and the New Generation of Radio Arrays (Proceedings of Science), 57B
Brunthaler, A., Martí-Vidal, I., Menten, K. M., *et al.* 2010a, *A&A*, 516, A27
Brunthaler, A., Martí-Vidal, I., Menten, K. M., *et al.* 2010b, in 10th European VLBI Network Symposium and EVN Users Meeting: VLBI and the New Generation of Radio Arrays (Proceedings of Science), 55B
Chevalier, R. A. 2012, *ApJ* (Letters), 752, L2
Marcaide, J. M., Martí-Vidal, I., Alberdi, A., *et al.* 2009a, *A&A*, 505, 927
Marcaide, J. M., Martí-Vidal, I., Perez-Torres, M. A., *et al.* 2009b, *A&A*, 503, 869
Martí-Vidal, I., Marcaide, J. M., Alberdi, A., *et al.* 2011a, *A&A*, 526, A142
—. 2011b, *A&A*, 526, A143
Mioduszewski, A. J., Dwarkadas, V. V., & Ball, L. 2001, *ApJ*, 562, 869
Ng, C.-Y., Potter, T. M., Staveley-Smith, L., *et al.* 2011, *ApJ* (Letters), 728, L15

Discussion

CAPPELLARO: I remember that 79C was also seen to flaten in optical, few years after explosion. They deduce a mass loss $10^{-4} M_{\odot} yr^{-1}$. Is this consistent?

BARTEL: Not quite, since our estimate is an order of magnitude lower.

RAY: Is it possible to break the degeneracy of n & s (ejecta density profile index & CSM density profile index) from the VLBI data?

BARTEL: It is not possible analytically by assuming the standard model with self-similar expansion. But the changing thickness of the shell and the changing deceleration if that occurs, together with hydrodynamic simulations may give clues as to the density profiles of the ejecta and the CSM.

Supernova Environmental Impacts
Proceedings IAU Symposium No. 296, 2013
A. Ray & R. A. McCray, eds.

doi:10.1017/S174392131300923X

Early Emission of Core-Collapse Supernovae

Melina C. Bersten[1], Omar Benvenuto[2] and Ken'ichi Nomoto[1]

[1]Kavli Institute for the Physics and Mathematics of the Universe, Todai Institutes for Advanced Study, University of Tokyo, 5-1-5 Kashiwanoha, Kashiwa, Chiba 277-8583, Japan
email: melina.bersten@ipmu.jp

[2]Facultad de Ciencias Astronómicas y Geofísicas, Universidad Nacional de La Plata, Paseo del Bosque S/N, B1900FWA La Plata, Argentina
email: obenvenu@fcaglp.unlp.edu.ar

Abstract. We present a study of the early UV/Optical emission of the stripped-envelope supernovae based on a one-dimensional, Lagrangian model that solves the hydrodynamics and radiation transport in an expanding ejecta. The models are compared with observations to constrain the physical properties of the progenitor star, such as radius and mixing of radioactive nickel synthesized during the explosion. In particular, we present models for the early emission of the type IIb SN 2011dh and the Type Ib SN 2008D.

Keywords. supernovae: general, supernovae: individual (SN 2008D), supernovae: individual (SN 2011dh)

1. Introduction

Two main phases can be distinguished in a typical light curve (LC) of stripped-envelope supernovae (SE SNe, i.e. types IIb Ib and Ic): (a) the early UV/optical emission or "cooling phase" powered by the energy deposited by the shock wave and (b) a re-brightening to a "second peak" due to the decay of radioactive material synthesized during the explosion. Global properties of a SNe such as explosion energy (E), ejected mass ($M_{\rm ej}$), and the ^{56}Ni mass can be obtained by modeling the LC around the second peak. On the other hand, observations of the early emission provide unique information about the structure of the star previous to the explosion as well as the mixing process. The short duration of this early phase, specially for compact Wolf-Rayet stars, makes its observation elusive. However, in the last few years the early emission could be detected for a handful of objects and there is an increasing number of early observations, which will allow us to perform detailed studies. Here we present a study of the early SN emission using a one-dimensional, Lagrangian code that solves the hydrodynamics and radiation transport in an expanding ejecta (Bersten *et al.* (2011)). In particular, we analyze the case of the type Ib SN 2008D and the type IIb SN 2011dh which were observe at early times.

2. Early emission models

Early UV/optical emission is expected to occur after the arrival of the shock wave at the surface of the progenitor (shock break-out) and before the re-brightening due to the decay of radioactive material. This emission is a consequence of the nearly adiabatic cooling due to the expansion of the outermost layers of the ejecta. The duration of this early phase depends strongly on the size of the progenitor as shown in Fig. 1, where we present a set of LC models for progenitor radii of 2, 50, 100, 150, 200 and 270 $R_{\odot}$. All these initial configurations have the same He core of 4 $M_{\odot}$ calculated from stellar evolution models (Nomoto & Hashimoto 1988) plus a H-rich envelope smoothly attached to the

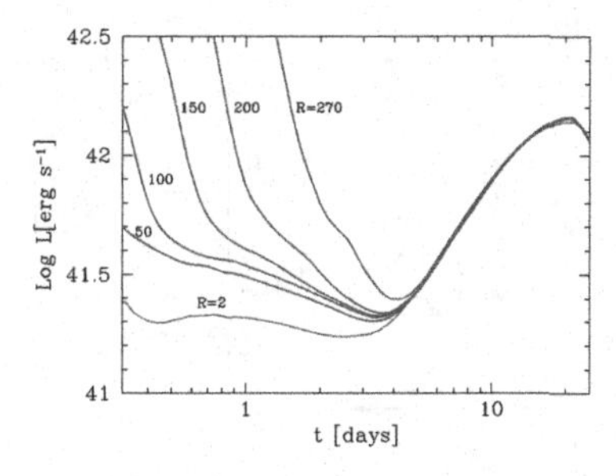

Figure 1. Bolometric LCs for models with different initial radii. The radius variation is accomplished by attaching essentially massless ($\lesssim 0.01\ M_\odot$) envelopes to the He4 model.

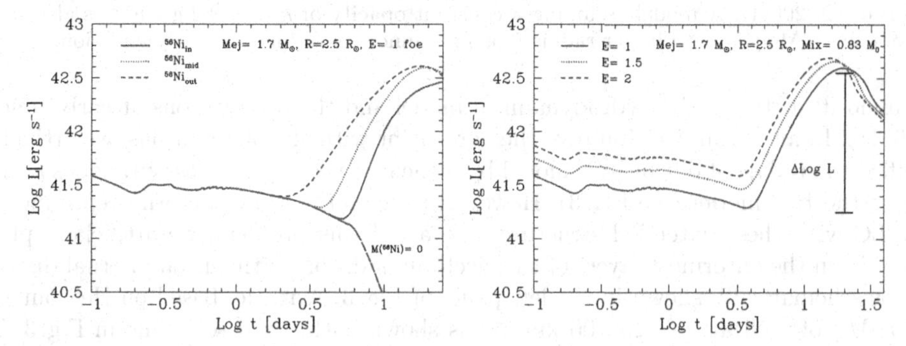

Figure 2. LCs for different ^{56}Ni mixing (left) and explosion energy (right).

the core to modify the progenitor structure. Clearly, the effect of the progenitor radius is only noticeable for times $\lesssim 5$ days (the exact epoch depends on the ^{56}Ni distribution). Observations before this epoch are required to discriminate between different progenitor radii. As previously noted by Dessart *et al.* (2011), models for compact progenitors show an initial plateau in the LC that is produced by He recombination. To test the dependence of the initial plateau on the ^{56}Ni mixing and explosion energy we calculated a set of hydrodynamical models for He core masses of 3 and 4 $M_\odot$. Fig. 2 shows these results for He core mass of 3 $M_\odot$. The models predict show that the initial plateau duration depends on the mixing of ^{56}Ni (left panel) but not on the energy (right panel). The luminosity of the plateau instead seems to depend mainly on the energy. For all the models we calculated the plateau duration (Δt_p) and the difference in magnitude between plateau luminosity and main peak luminosity (ΔM). We found that the models predict (1) $\Delta t_p \approx 4 - 10$ days, and (2) $\Delta M \approx 3$. Such values indicate that the initial plateaus can be detected in ongoing surveys. Note that the models presented here are helium-rich, so they should may be useful only for SNe Ib. We plan to extend our analysis to He-poor SNe.

3. SN 2008D

The type Ib SN 2008D attracted a good deal of attention because of its unusual observational features such as the detection of an X-ray transient (XRT) and an early optical light-curve peak. Here we present the first radiation-hydrodynamical models for the early emission of SN 2008D. We adopt a pre-SN model (from here on He8) with He core of 8 $M_\odot$, $R = 1.4\ R_\odot$, $E = 8.4 \times 10^{51}$ erg s^{-1}, and $M_{\rm Ni} = 0.07\ M_\odot$ similar to that assumed in Tanaka *et al.* (2009); (T09). Fig. 3 (right panel) shows the resulting LC for the He8 model (dashed line) as compared with the observations (Modjaz *et al.* 2009). Clearly, this model cannot explain the early emission shown by the observations. The difference

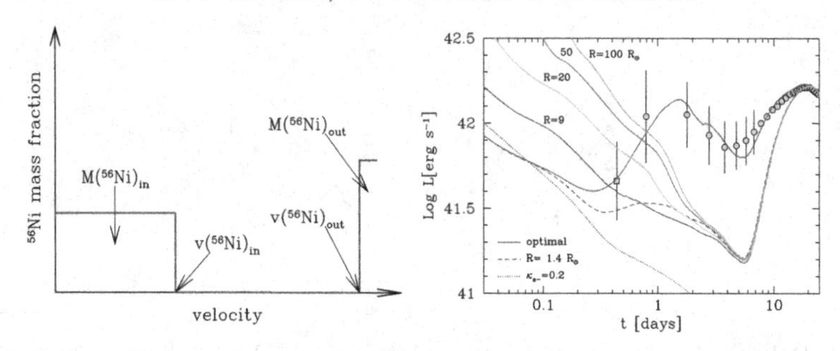

Figure 3. Left: Schematic doubly-peaked ^{56}Ni distribution. **Right**: Bolometric LC for model He8 with (red solid line) and without (red dashed line) external ^{56}Ni compared with the observations of SN 2008D. A model assuming a constant opacity of $\kappa_{e^-} = 0.2$ g cm^{-2} is also shown (red dot line). Models with larger radius (colors) cannot reproduce the early emission.

in luminosity between the hydrodynamical model and the observations at early times is > 0.5 dex. In an attempt to improve the agreement with the observations, we artificially modified the radius of our pre-SN model by attaching nearly mass-less envelopes (< 0.01 $M_\odot$) to the He8 model (see Fig. 3). However, we could not satisfactorily reproduce the early LC with these extended structures. To avoid this problem, we artificially placed some ^{56}Ni in the outermost layers of the ejecta in addition to the usual internal distribution, as schematically shown in the left panel of Fig. 3. A model based on He8 but with $\approx 0.01\,M_\odot$ of ^{56}Ni at $v > 20,000$ km s^{-1} is shown with a red solid line in Fig. 3. The agreement between this model and the observations is excellent.

4. SN 2011dh

The Type IIb SN 2011dh was discovered in the famous nearby galaxy M51. Soon after discovery, a source was identified as the possible progenitor of SN 2011dh in archival, multi-band HST images (Maund *et al.* 2011; Van Dyk *et al.* 2011). Photometry of the source was compatible with a yellow super-giant (YSG) star with $R \sim 270R_\odot$. However, some authors have suggested a more compact progenitor ($R \sim R_\odot$), claiming that the YSG star detected in the pre-SN images may be its binary companion or even an unrelated object (Arcavi *et al.* 2011; Van Dyk *et al.* 2011; Soderberg *et al.* 2011. In Bersten *et al.* (2012) we have used hydrodynamical models to show that an extended progenitor with $R \gtrsim 200\ R_\odot$ in concordance with a YSG star is required to reproduce the early LC (see left panel of Fig 4). Our prediction has recently been confirmed by the reported disapearance of the YSG star (Van Dyk *et al.* 2013). In addition, modeling the LC around the second peak we found that a progenitor with He core mass of $\approx 4M_\odot$, an explosion energy of 8×10^{50} erg, and a ^{56}Ni mass of 0.063 $M_\odot$ reproduces very well the observations. This optimal model (He4) is presented in Fig. 4 and it is consistent with a main-sequence mass of $\approx$13 $M_\odot$. Furthermore, our models rule out progenitors with He core mass >8 $M_\odot$, implying that $M_{\rm ZAMS} < 25M_\odot$. Considering the limitations of single stars of such stellar masses to almost entirely expel the H-rich envelope via stellar winds, as required for SNe IIb, this result is indicative of a binary origin for SN 2011dh. In Benvenuto *et al.* (2013) we performed binary evolution calculations with mass transfer to reproduce the possible progenitor system of SN 2011dh. In Fig. 5 (left panel) we show the H-R digram for our prefered configuration of 16 $M_\odot$ + 10 $M_\odot$ and an initial orbital period of 150 days. Note that the primary star ends its evolution at the right position as compared with the YSG star detected in the pre-SN images (red rectangle). Moreover, the final mass of the primary is $\approx 4M_\odot$, which is consistent with our hydrodynamical modeling, and it includes a thin hydrogen envelope of $\approx 4 \times 10^{-3} M_\odot$ which is required

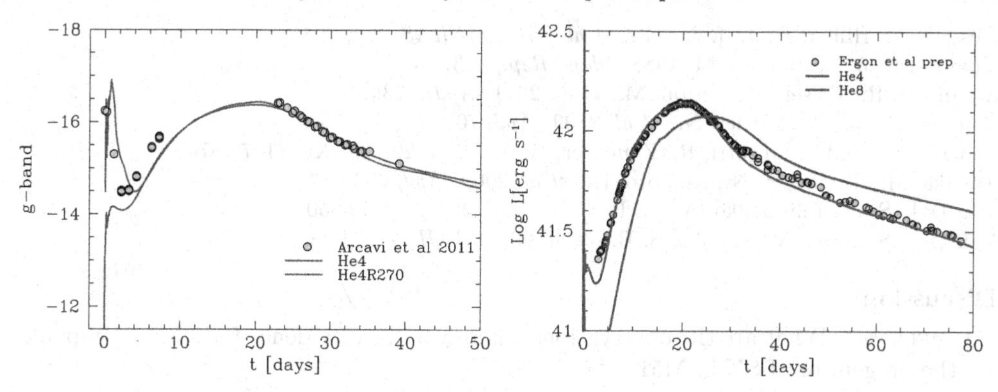

Figure 4. Observed and modeled g'-band LCs **(left panel)** for the compact model He4 and the extended model. Clearly, the progenitor is required to be a supergiant in order to reproduce the observation. (**Right panel**) Bolometric LC for our optimal model, He4, with a He core mass of $\approx 4\ M_\odot$ (red line) and for a model with He core mass of $8\ M_\odot$ (He8; blue line) as compared with the observed bolometric LC of SN 2011dh (cyan dots). Note that the model with large helium core mass is not compatible with the observations.

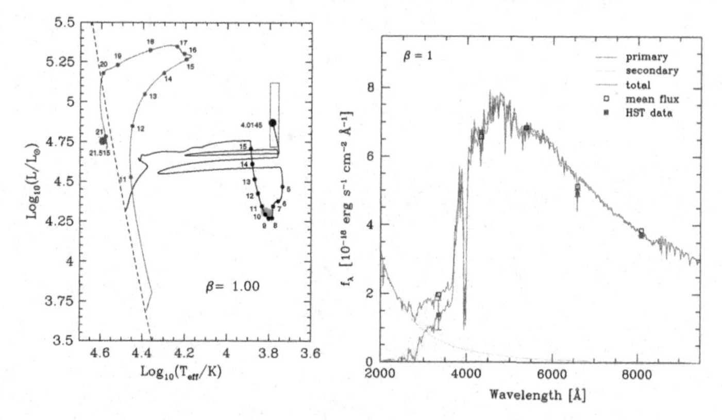

Figure 5. (**Left panel**) Evolutionary tracks in the H-R diagram for both components of a binary system with $M_{\rm ZAMS}$ of 16 $M_\odot$ and 10 $M_\odot$ and an initial period of 150 days. Labeled dots along the tracks indicate the masses of the stars (in solar units) while mass transfer by Roche-Lobe overflow occurs. (**Right panel**) Spectra of the donor and accreting stars (red and blue lines, respectively), and the sum of both (gray line). The mean synthetic (black squares) and observed fluxes (red squares) in each bandpass are shown.

to yield a SN IIb. The binary scenario was further tested by estimating the effect of the putative companion star on the pre-explosion photometry and comparing this with the observations (see Fig. 5, right panel). Because the secondary star is predicted to be an OB type star, the largest effect appears in the blue and UV range. The contribution of the secondary to the flux in the F336W band, however, is marginal, at the 1.5-σ level for a conservative mass-transfer case, and at the 0.6-σ level for a non-conservative case. The existence of the binary companion can be tested in the near future by a search for a blue object at the location of the SN.

References

Arcavi, I., Gal-Yam, A., Yaron, O., *et al.* 2011, *ApJL* , 742, L18

Benvenuto, O. G., Bersten, M. C., & Nomoto, K. 2013, *ApJ*, 762, 74

Bersten, M. C., Benvenuto, O. G., Nomoto, K., *et al.* 2012, *ApJ*, 757, 31

Bersten, M. C., Benvenuto, O., & Hamuy, M. 2011, *ApJ*, 729, 61

Dessart, L., Hillier, D. J., Livne, E., *et al.* 2011, *MNRAS*, 414, 2985
Nomoto, K. & Hashimoto, M. 1988, *Phys. Rep.*, 163, 13
Maund, J. R., Fraser, M., Ergon, M., *et al.* 2011, *ApJL*, 739, L37
Modjaz, M., Li, W., Butler, N., *et al.* 2009, *ApJ*, 702, 226
Soderberg, A. M., Margutti, R., Zauderer, B. A., *et al.* 2011, arXiv:1107.1876
Tanaka, M., Tominaga, N., Nomoto, K., *et al.* 2009, *ApJ*, 692, 1131
Van Dyk, S. D., Filippenko, A. V., Fox, O., *et al.* 2013, ATel 4850
Van Dyk, S. D., Li, W., Cenko, S. B., *et al.* 2011, *ApJL*, 741, L28

Discussion

CAPPELLARO: What are the observations where you have evidence for a blue companion for the progenitor of SN in M51

BERSTEN: I used the model information, T- effective and L for the primary and secondary stars, and calculated the SED of each star using Atmospheric models available in the literature. Then I calculated the composite spectra and the mean fluxes in the same photometric band and then compared with the pre-explosion observations. The blue companion is only marginally detected (1.4 sigma) in the Bluest observation band.

Supernova Environmental Impacts
Proceedings IAU Symposium No. 296, 2013
A. Ray & R. A. McCray, eds.

doi:10.1017/S1743921313009241

CSM-Interacting Stripped-Envelope Supernovae

Massimo Turatto and Andrea Pastorello

Osservatorio Astronomico di Padova - INAF,
vicolo dell'Osservatorio 5, 35122, Padova, Italia
email: massimo.turatto@oapd.inaf.it

Abstract. Type Ibn supernovae (SNIbn) are a small group of core-collapse events whose ejecta interact with a He-rich and H-deprived circumstellar medium. The explosion of the prototype of this family, SN 2006jc, was heralded by a major stellar eruption two years before its core collapse. This unexpected discovery increased significantly the interest of the astronomical community toward these unusual events. A number of other objects have been included to the SNIbn zoo. In these contribution we review our current knowledge on this rare family of interacting, stripped-envelope supernovae (SE-SNe).

Keywords. supernovae: general, supernovae: SN 2006jc, SN 1999cq, SN 2000er, SN 2002ao, SN 2005la, SN 2010al, SN 2011hw, LSQ12btw, PS1-12sk, stars: winds

1. Introduction

Growing evidence has been recently accumulated for the presence of dense circumstellar media around supernova (SN) precursors. This is primarily true for core-collapse SNe descending from massive progenitors M $\geqslant$ 8 $M_{\odot}$, though sometimes controversially claimed also for thermonuclear SNe (Hamuy *et al.* 2003, Benetti *et al.* 2006, Silverman *et al.* 2013). The most massive stars stars lose large fraction of their initial mass via intense winds or episodic giant eruptions that can remove up to tens of solar masses (about 10 $M_{\odot}$ in η Car; Smith *et al.* 2003). Some type IIn SNe, whose observables are explained with the interaction of the SN ejecta with dense H-rich circum-stellar matter (CSM), have been associated to very massive progenitors such as LBV, though several theoretical issues remain unsolved. Also in the $25 - 40$ $M_{\odot}$ initial mass range stars can lose large amounts of their outer envelopes before the final explosions via powerful winds. The SNe associated to progenitors stripped of their envelopes are called of type Ib, if deprived only of the H layer, or of type Ic, if deprived also of He (cfr. Turatto *et al.* 2007 and references therein). At lower masses ($\leqslant$25$M_{\odot}$) the stellar winds are less efficient and the stars reach the collapse of the core still retaining a significant fraction of the H envelope, and the corresponding SNe are called of type II. The amount of H left in the envelope plays a major role in the optical display, because H recombination supports long-lasting plateaus in the light curves (Pumo & Zampieri 2011 and references therein).

The impact of SN ejecta on the slow material close to dying stars is revealed via radio, X-ray and optical emissions. The optical emission originates from the optically thin circumstellar matter ionized by the shock and the SN radiation field. The optical spectra are characterized by blue colours, narrow (a few 1000 km s^{-1}) Balmer emission lines and slow spectral evolution. Objects with such features are usually called as of type IIn (cfr. Turatto *et al.* 2007). Recently, however, a small number of objects have have been discovered with similar blue colours but narrow HeI emissions and no, or very faint, H lines. In close analogy to type IIn these objects have been considered as SE SNe expanding in a dense He-rich CSM, and dubbed as type Ibn SNe(Pastorello *et al.* 2007).

Aim of this paper is to check if such scenario hold to the inclusion of the most recent members.

2. Previous Wisdom

Pastorello *et al.* 2008a analyzed synoptically the observations of type Ibn SNe available at the time, and found that they followed an overall similar photometric and spectroscopic evolution, though still having a number of distinctive peculiarities. The earliest epochs, available only for SN 2000er, showed a blue continuum with HeI lines having narrow ($\sim 800-900$ km s^{-1}) P-Cyg absorptions, forming in the unshocked CSM, and broader (~ 5000 km s^{-1}) red wings, possibly arising in the SN ejecta (a sort of a SNIb expanding in a He-rich CSM).

The subsequent epochs up to late times were best studied in SN 2006jc which surprisingly showed a pre-discovery outbust about two years before the SN explosion (Pastorello *et al.* 2007). Remarkable was a near-UV/blue ($\lambda < 5700$ Å) pseudo-continuum caused by a blend of broad fluorescent lines mainly of Fe (Foley *et al.* 2007). Together with broad (4000–9000 km s^{-1}) lines of intermediate mass elements in the SN ejecta, it showed moderately narrow (2000–3000 km s^{-1}) HeI emissions with narrower absorptions pointing out unshocked CSM at $\sim$ 1000 km s^{-1}, typical of WR stars. At later phases (t > 130d) the blue continuum faded away while narrow H emissions appeared indicating that finally the ejecta entered H-rich material (Pastorello *et al.* 2008a). Two other objects (SNe 1999cq and 2002ao) fitted well the evolution described above.

The light curves were relatively narrow, not dissimilar from those of other SE-SNe. All observations were reasonably well interpreted as the result of the explosion of a WR progenitor exploding in a dense He-rich CSM.

3. Increasing the Sample

After the initial unexpected discoveries, the newly announced SNIbn candidates have been studied in detail.

Probably the best studied object is SN 2010al, observed from the UV to the radio domain (ATEL 2532, ATEL 2513). Initially classified as a SNIIn similar to SN 1998S, because of the clear but non dominant presence of H (Cooke *et al.* 2010), the early blue spectra revealed narrow lines of HeI, HeII, NIII and CIII, typical of WR stars over a blue continuum (Pastorello *et al.* 2011, Pastorello *et al.* 2013a) with P-Cyg absorptions indicative of a CSM expanding at velocity $v \sim 1000$ km s^{-1}. One month later the spectra transformed to match those of SN 2006jc, with lines of intermediate mass elements with FWHM(IME)$\sim$ 5000km s^{-1}.

If SN 2010al displayed weak H even at early stages, implying that the CSM lately ejected by the star had a non-negligible amount of H, a further step in bridging the gap between type Ibn and IIn was made with SN 2005la and SN 2011hw. In the early spectra of the former, H lines had comparable strengths and widths as HeI ($\sim$ 1000 km s^{-1}). These showed also P-Cyg absorptions, while IME lines, interpreted as representative of the ejecta, had $FWHM(IME) \sim 4900$ km s^{-1} (Pastorello *et al.* 2008b). The narrow P-Cyg absorptions disappeared soon and the line widths increased up to $v_{FWHM} = 4200$ km s^{-1}, closely matching the IME velocity. The light curve was definitely broader than other SNIbn, with modulations possibly due to an increased CSM interaction. SN 2011hw

(Smith *et al.* 2012, Pastorello *et al.* 2013b) was similar to SN 2005la with strong, moderately narrow ($v \sim 1900$ km s^{-1}) HeI and H from the very beginning, indicating that the CSM was not completely H-deprived. It also showed unresolved ($v < 250$ km s^{-1}) coronal lines (e.g. [ArX], [FeV], [FeVII]), observed in strongly interacting type IIn, e.g. 1988Z (Turatto *et al.* 1993), indicating that X-ray penetrates the slow, unshocked CSM. Also its light curve was flatter and clearly different from those of other SNIbn. In other words, differently from previously known SNIbn, these two objects revealed the presence of He- and H-rich material ejected short before the explosion.

Puzzling has been the discovery of PS1-12sk in the Brightest Cluster Galaxy CGCG208-042 (Sanders *et al.* 2013). Overall similar to SN 2006jc, but for early evidence of H, it rose the issue of its parent population because of the elliptical morphology of the host. A detailed analysis of the possible scenario involving degenerate progenitors does not explain the observables (e.g. presence of He, H, dense CSM, ejected ^{56}Ni mass, etc.). Though there is no direct evidence of star formation at the site of explosion, the activity of the nucleus and the indication of cooling flows seem to indicate that PS1-12sk is one of the rare (but known) cases of CC-SN exploding in early type galaxies with signs of SF (Hakobyan *et al.* 2008).

The recent LSQ12btw, another victim of the current anarchy in naming SNe, adds little to the story. It showed the typical blue pseudo-continuum due to blends of FeII lines, relatively broad HeI ($v \sim 4000$ km s^{-1}), and a fast evolving light curve (Pastorello *et al.* 2013b).

A few other objects might belong to this class, e.g. OGLE-2012-SN-006, but scanty material has been published so far.

4. The updated Scenario

Figure 1 has been drawn to sketch a scenario for type Ibn events, in the context of interacting SNe. The spectra are displayed according to a decreasing relative intensity of H CSM lines (form top to bottom).

At the top we show the spectrum of SN 2010jl, a SNIIn dominated by Balmer lines with multiple components. Weak He lines are present with comparable velocities. Interestingly, a luminous blue source has been detected at the position of the SN in pre-SN HST imaging. Whether this is really the SN progenitor itself or a young star cluster, the conclusion seems unavoidable that the progenitor had M> 30M$_\odot$ (Smith *et al.* 2011). Consistently, the spectral evolution indicates that the progenitor (possible a LBV) underwent a tremendous mass loss in the last decades before the explosion (Zhang *et al.* 2012). However, the presence of slow preshock winds (v = 28 km s^{-1}, Smith *et al.* 2011) cannot be easy reconciled with the LBV scenario.

Below the SN 2010jl spectrum are reported those of SN 2005la and SN 2011hw. Their H and He lines are comparable in strength and width at all epochs, indicating that they arise from the same layers of shocked CSM located close to the exploding star. Lines of intermediate mass elements, present in the red parts of the spectra, are broader (v $\sim$ 5000 km s^{-1}) suggesting that they are forming in the SN ejecta. The presence of H has been invoked (Smith *et al.* 2012) to explain the existence of a 6000 K black-body which dilutes the typical near-UV/blue continuum, clearly visible in 2006jc-like objects, and reduces the contrast of the spectral features. Such 6000K BB may contribute to sustain the light curves of the two SNe that appear broader than those of other He-rich interacting SNe. We are probably facing stars that ejected short before the explosions consistent amounts of material containing He and H and may be transitioning from the LBV to the WR stages (Smith *et al.* 2012, Pastorello *et al.* 2013a).

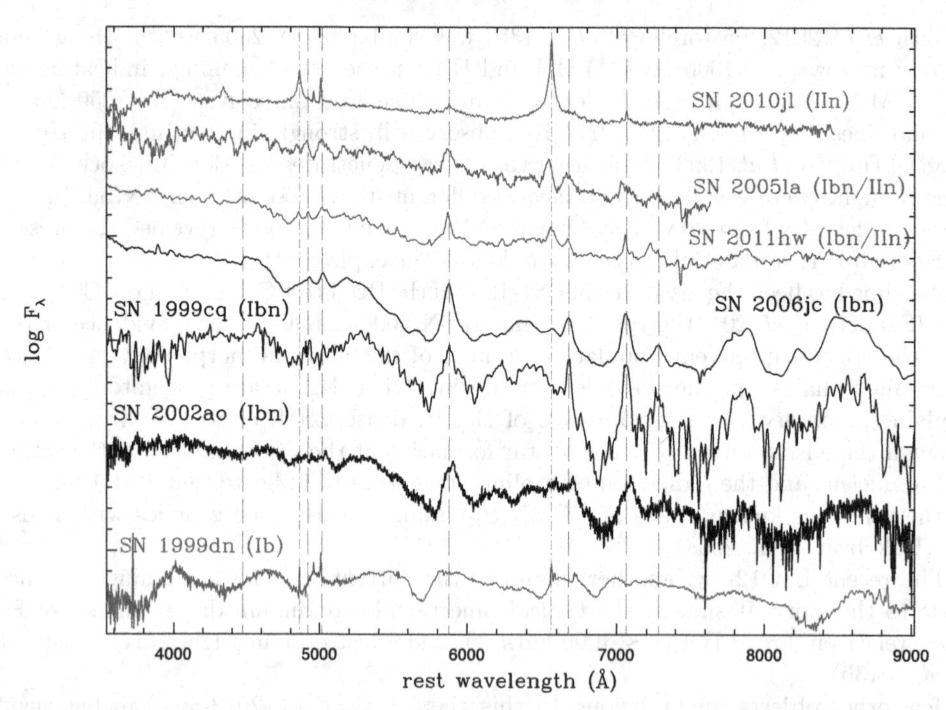

Figure 1. Comparison of spectra of different flavors of CSM interacting SNe, obtained at about 3-4 weeks after the estimated epoch of explosion. Vertical orange dashed lines mark the positions of the main He features, while red dot-dashed lines reveal the rest wavelength positions of Hα and Hβ. At the top we show the spectrum of SN 2010lj, strongly interacting with a dense H-rich CSM (IIn). Below are reported the spectra of SN 2005la and SN 2011hw, whose CSM seems enriched also He (IIn/Ibn). The CSM of the three subsequent SNe, 2006jc, 199cq and 2002ao, does not show H lines (Ibn). At the bottom is reported for comparison the spectrum of the normal SN 1999dn at comparable phase.

The middle-lower part of Fig. 1 shows spectra of three canonical SNIbn, 2006jc, 1999cq and 2002ao (Matheson *et al.* 2000, Pastorello *et al.* 2008a). Their spectra show: 1- prominent He emission lines and weak evidence of H (if any), 2- strong nearUV/blue pseudo-continuum, 3- broad lines of IME, 4- evidence of fast (~1000 km s^{-1}) CSM. The light curves are narrow and resembling those of non-interacting type Ibc SNe. All characteristics are, therefore, well compatible with having WCO progenitors exploded in He-dominated CSM. The overall observed homogeneity of SNIbn is quite surprising for objects dominated by a pre-existing CSM whose geometry and density can be very diverse, and probably is indicative that the role of interaction is not dominant. At the very bottom is reported for comparison the spectrum of the prototypical type Ib SN 1999dn (Benetti *et al.* 2011). The comparison highlights as some of the major ondulations in the continuum of SNIbn are probably associated to an underlying type Ibc SN. All SNIbn exploded in spiral galaxies spanning a broad range of metallicities (Pastorello *et al.* 2013b), with the exception of PS1-12sk exploded in an elliptical with indications of SF.

We conclude, therefore, that the early scenario proposed to explain the observations of intermediate-width He and H features, holds, and that the observed properties of type Ibn are consistent with the explosion of massive WR or transitional LBV/WR precursors that recently enriched their CSM with He-rich material.

References

Anupama, G. C., Sahu, D. K., Gurugubelli, U. K., Prabhu, T. P., Tominaga, N., Tanaka, M., & Nomoto, K. 2009, *MNRAS*, 392, 894

Benetti, S., Cappellaro, E., Turatto, M., Taubenberger, S., Harutyunyan, A., & Valenti, S., 2006, *ApJ* (Letters), 653, L129

Benetti, S., *et al.* 2011, *MNRAS*, 411, 2726

Cooke, J., Ellis, R. S., Nugent, P. E., Howell, D. A., Sullivan, M., & Gal-Yam, A., 2010, *ATEL*, 2491

Foley, R. J., *et al.* 2007, *ApJ*, 657, L105

Hakobyan, A. A., Petrosian, A. R., McLean, B., Kunth, D., Allen, R. J., Turatto, M., & Barbon, R., 2008, *A & A*, 488, 523 (2008).

Hamuy, M., *et al.* 2003, *Nature*, 424, 651

Matheson, T., *et al.* 2000, *ApJ*, 119, 2303

Mattila, S., *et al.* 2008, *MNRAS*, 389, 141

Pastorello, A., *et al.* 2007, *Nature*, 447, 829

Pastorello, A., *et al.* 2008a, *MNRAS*, 389, 113

Pastorello, A., *et al.* 2008b, *MNRAS*, 389, 131

Pastorello, A., Benetti, S., Bufano, F., Kankare, E., Mattila, S., Turatto, M., & Cupani, G. 2011, *AN*, 332, 266

Pastorello, A., *et al.* 2012, in preparation

Pastorello, A., *et al.* 2012, in preparation

Pumo, M. L. & Zampieri, L. 2011, *ApJ*, 741, 41

Sanders, N. E., *et al.* 2013, *ApJ* submitted (eprint arXiv:1303.1818)

Silverman, J. M., *et al.* 2013, *ApJ*, submitted (eprint arXiv:1304.0763)

Smith, N., *et al.* 2003, *AJ*, 125, 1458

Smith, N., *et al.* 2011, *ApJ*, 732, 63

Smith, N., *et al.* 2012, *MNRAS*, 426, 1905

Turatto, M., Cappellaro, E., Danziger, I. J., Benetti, S., Gouiffes, C., & Della Valle, M., 1993, *MNRAS*, 262, 128

Turatto, M., Benetti, S., & Pastorello, A., 2007, in: S. Immler & K. Weiler (eds.), *AIP Conference Proceedings 937*, p. 187

Zhang, T., *et al.* *ApJ*, 144, 131

Discussion

CHEVALIER: The light curves are bright; is this due to interaction or high 56Ni?

TURATTO: The light curves of SNIbn are moderately bright ($L_{bol} \sim 10^{43}$ erg/s) Ni masses have been estimated only in few objects because light curves are in most cases not sufficiently extended. For SN 2006jc the estimated Ni mass is $M_{Ni} = 0.250.40$ (Pastorello 2008). Most probably the interaction acts flattening the light curve.

MAEDA: Do you see any signature of the dust formation in SNe other than SN 2006jc? For example, when you have late-time spectra (e.g. $\geqslant$ 2 months), do you see the blue-shift in lines?

TURTTO: The objects identified and studied so far are relatively distant. Spectral observations, therefore, are limited in most cases to the first 2 months best in SN 2006jc, in which dust evidence emerged just at about day 50. I am not aware of the results of the observations of 2010 al at late time.

Supernovae Environmental Impacts
Proceedings IAU Symposium No. 296, 2013
A. Ray & R. A. McCray, eds.

doi:10.1017/S1743921313009253

Superluminous Supernovae

Robert M. Quimby
Kavli Institute for the Physics and Mathematics of the Universe (WPI)
Todai Institutes for Advanced Study, University of Tokyo
5-1-5 Kashiwa-no-Ha, Kashiwa City, Chiba 277-8583, Japan
email: `robert.quimby@ipmu.jp`

Abstract. Not long ago the sample of well studied supernovae, which were gathered mostly through targeted surveys, was populated exclusively by events with absolute peak magnitudes fainter than about −20. Modern searches that select supernovae not just from massive hosts but from dwarfs as well have produced a new census with a surprising difference: a significant percentage of supernovae found in these flux limited surveys peak at −21 magnitude or brighter. The energy emitted by these superluminous supernovae in optical light alone rivals the total explosion energy available to typical core collapse supernovae ($> 10^{51}$ erg). This makes superluminous supernovae difficult to explain through standard models. Adding further complexity to this picture are the distinct observational properties of various superluminous supernovae. Some may be powered in part by interactions with a hydrogen-rich, circumstellar material but others appear to lack hydrogen altogether. Some could be powered by large stores of radioactive material, while others fade quickly and have stringent limits on 56-Ni production. In this talk I will discuss the current observational constrains on superluminous supernova and the prospects for revealing their origins.

Keywords. supernovae: general

In the year 2000, after decades of surveys for time variable objects, the highest luminosity supernova published was SN 1999cy at $M_V < -20.1$ (Turatto *et al.* 2000; excluding light attributable to the optical afterglows of gamma-ray bursts). At that time, there were a handful of candidates for even more luminous supernovae, but these were still–and largely remain–unvetted by the refereeing process. The true luminosities of these objects are uncertain. Most are based simply on the discovery data presented in IAU Circulars and often include "quick and dirty" magnitude estimates divined from photographic plates in fields lacking reliable comparison stars†.

In 1995, the Supernova Cosmology Project (SCP) announced eleven high redshift supernovae from their ongoing, CCD based, batch discovery search program (Perlmutter *et al.* 1995). Ten of these proved to be Type Ia supernovae, and several of these would be used to reveal the accelerating expansion of our universe (Perlmutter *et al.* 1999; see also Riess *et al.* 1998). The eleventh discovery, the probable Type IIn SN 1995av, is perhaps the first well observed, high-luminosity supernova ($M_R < -20.8$). While a mere footnote to the larger discovery of our accelerating universe, SN 1995av gives the first indication that non-targeted, flux limited searches are sensitive to rare, high luminosity events that could easily have been passed over previously. It also suggested that such discoveries could prove to be a non-negligible contributor to such surveys.

The high redshift oriented SCP spawned a low redshift counterpart in the Spring of 1999 (and later the SNFactory, Aldering *et al.* 2002), tasked with selecting nearby

† For example, for its distance, the reported magnitude of SN 1988O would correspond to an absolute magnitude of about −22, but this source was later classified as a subluminous SN Ia. Given this classification, the magnitude reported in the announcement was likely incorrect (J. Mueller, priv. comm. 2009).

supernovae in a manner equivalent to the distant search. This survey quietly netted two discoveries of unprecedented luminosities: SNe 1999as and 1999bd (Knop *et al.* 1999, Nugent *et al.* 1999). Deng *et al.* (2001) show that the Type Ic SN 1999as reached a peak luminosity of at least $M < -21.5$, and the discovery report for the Type IIn SN 1999bd implies a peak of $M < -21.6$. For comparison, Li *et al.* (2011a) have presented a volume limited sample of supernova discoveries from their targeted LOSS program, and the brightest of these 179 events is the Type Ia supernova SN 2006lf at $M_R = -19.55 \pm 0.12$.

Following in the footsteps of the SCP and SNFactory, the Texas Supernova Search (TSS; Quimby 2006) began patrolling the skies in 2004 with a modified version of the same image subtraction code. The program also had a similar objective of discovering supernovae with as little bias to host environment as possible. The TSS was relatively modest in its scope (the main survey instrument was a 0.45 m telescope, ROTSE-IIIb), but this was not without its advantages. As few discoveries were made and these were all rather bright ($m \gtrsim 18$ mag), spectroscopic confirmation could be carried out for all new supernova candidates selected. The main facilitator for this follow-up was the 9.2 m Hobby-Eberly Telescope, which is queue scheduled and permits low resolution spectroscopic observations to be triggered as needed, even in bright time when most observatories switch to high resolution or NIR instruments. The TSS began finding significant outliers from the established supernova population almost immediately. The fourth supernova discovered by the TSS, SN 2005ap, would prove to be ten times more luminous than most Type Ia supernovae, and at $M = -22.2$, it remains one of the most luminous supernovae ever found.

Since this discovery, several new optical transient searches have come on line including the Catalina Real-Time Transient Survey (CRTS), the Palomar Transient Factory (PTF; now iPTF), Pan-STARRS, La Silla-QUEST (LSQ), Skymapper, and the Dark Energy Survey. Many of these searches have reported supernovae with peak luminosities brighter than $M = -21$.

To be considered "superluminous," a supernova should be both brighter than the brightest normal (non-interacting) SNIa and significantly brighter than peak magnitude distributions of normal thermonuclear and core-collapse events. Figure 1 shows the pseudo-absolute magnitude distributions of core-collapse and Type Ia supernovae from the LOSS volume limited sample (Li et al 2011a). Pseudo-absolute magnitudes are the observed magnitudes corrected for distance and Galactic extinction, but not corrected for any absorption by the host environment. We have scaled the distributions to the volumetric rates derived from the LOSS search (Li *et al.* 2011b) to give the cumulative rate of events fainter than a given magnitude (top panel), and the rates per half magnitude bin (lower panel). The distribution of core collapse supernovae is reasonably well approximated by a Gaussian distribution with $M_{\rm peak} = -16.4$ and $\sigma = 1.0$ mag (dotted line in the figure), ignoring the surplus of events between $-13.5 < M_{\rm peak} < -14.0$. The resemblance to a Gaussian is surprising given that the sample is not corrected for host absorption (we would expect an intrinsically Gaussian distribution to be skewed to fainter magnitudes by dust).

The most luminous SN Ia published so far, SN 2007if, reached a peak optical magnitude of about $M = -20.4$ (Scalzo *et al.* 2010, Yuan *et al.* 2010). Yasuda & Fukugita (2010) have calculated the intrinsic luminosity function from the SN Ia discovered by the SDSS-II. They find a Gaussian distribution with $M_B = -19.423$ (roughly $M_R = -19.5$ assuming a normal spectrum) and $\sigma = 0.237$.

We define superluminous supernovae to be events with peak absolute magnitudes in the optical that are brighter than any known, non-interacting Type Ia supernova, or $M_{\rm peak} \leqslant -20.5$. The most luminous interacting SN Ia published so far, SN 2005gj (Aldering *et al.*

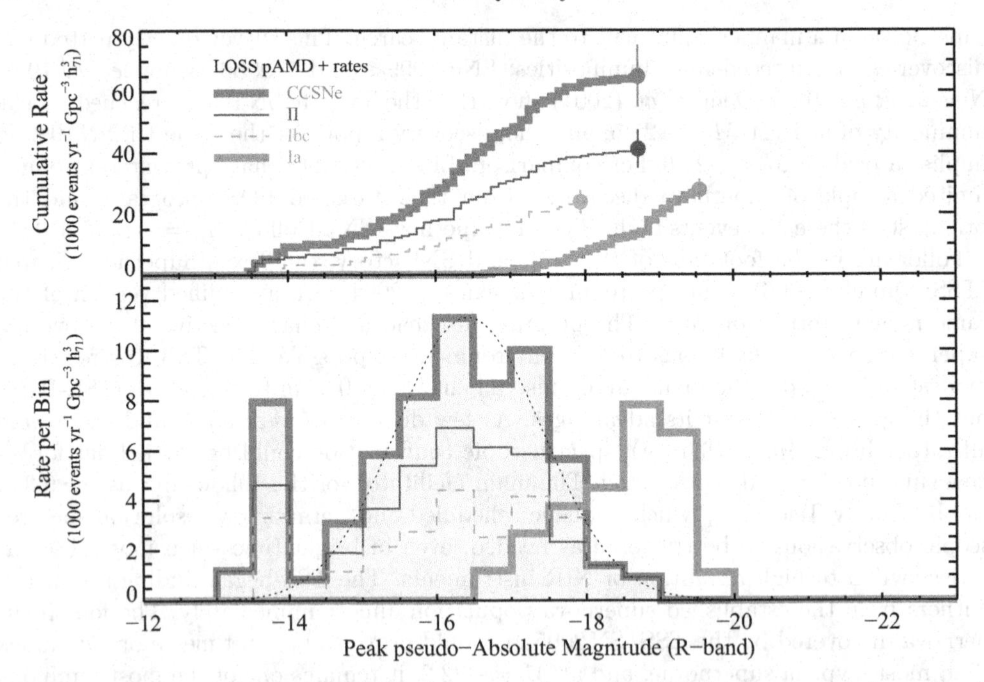

Figure 1. Peak pseudo-absolute R-band magnitude distribution for supernovae in an local, volume limited sample demonstrating that normal supernovae do not get brighter than about $M = -20$ mag. Based on data from the Lick Observatory Supernova Search Li *et al.* 2011a.

2006), reached roughly this limit, and it is possible that some higher luminosity Type IIn events may have a similar physical origin (cf. Dilday *et al.* 2012; Silverman *et al.* in prep.). This cutoff is about 4σ brighter than either the CCSNe distribution from LOSS or the SN Ia distribution from the SDSS-II sample (Li *et al.* 2011a, Yasuda & Fukugita 2010).

We have searched the literature for all known supernovae with peak magnitudes brighter than $M_{\rm peak} \leqslant -20.5$. Table 1 lists the 22 supernovae brighter than this cut-off that have so far been published, and Table 2 lists 28 further events that have been announced. Note that we exclude from this list objects with lower intrinsic luminosities that have been magnified via gravitational lensing even when this raises the effective luminosity above our defined threshold (e.g. the normal Type Ia supernova, PS1-10afx; Chornock *et al.* 2013, Quimby *et al.* 2013b).

Although the sample is relatively small, some events share certain characteristics that distinguish them from the others. The most basic division is that the spectra some SLSNe do not show obvious hydrogen features (SLSN-I), but others do (SLSN-II). Figure 2 shows representative spectra of SLSN-I and SLSN-II compared to examples of the more familiar SNIa and SNII classes. The light curves of the published SLSN sample are shown in figure 3. It is evident from this sample that there is significant dispersion in both rise and decay time scales. These differences could indicate some diversity in the progenitors of SLSNe.

Some SLSNe, like SN 2007bi for example, may be powered mainly by radioactive decay ^{56}Ni, but others (e.g. PTF09cnd, SN2010gx) reach peak luminosities too great to be explained exclusively by the ^{56}Ni production allowed by their late time photometric limits (Pastorello *et al.* 2010, Quimby *et al.* 2011, Chen *et al.* 2013). Some SLSNe, like SN 2006gy for example, may be powered mainly by interactions of the SN ejecta with pre-SN winds (e.g. Smith *et al.* 2007), but others (e.g. SN 2008es; Miller *et al.* 2009, Gezari *et al.* 2009) show no obvious signs of such ongoing interactions. It may therefore

Table 1. Published SLSNe

Name	RA	Dec	z	Type	Peak Mag[a]	Reference
SN 2003ma	05:31:01.9	−70:04:15.9	0.289	IIn	$M = -21.6$	Rest *et al.* 2011
SN 2005ap	13:01:14.8	+27:43:31.4	0.283	Ic	$M = -22.2$	Quimby *et al.* 2007
SN 2005gj	03:01:12.0	+00:33:14	0.0616	IIna	$M = -20.5$	Aldering *et al.* 2006
SN2213-1745	22:13:39.970	−17:45:24.486	2.0458	?	$M_{UV} \sim -21.2$	Cooke *et al.* 2012
SCP 06F6	14:32:27.4	+33:32:24.8	1.189	Ic	$M = -22.1^b$	Barbary *et al.* 2009
SN1000+0216	10:00:05.872	+02:16:23.621	3.8993	?	$M_{UV} \sim -21.4$	Cooke *et al.* 2012
SN 2006gy	03:17:27.1	+41:24:19.5	0.019	IIn	$M = -20.7$	Smith *et al.* 2007
SN 2006oz	22:08:53.6	+00:53:50.4	0.376	Ic	$M = -21.7?$	Leloudas *et al.* 2012
SN 2006tf	12:46:15.8	+11:25:56.3	0.074	IIn	$M = -20.5$	Smith *et al.* 2008
SN 2007bi	13:19:20.2	+08:55:44.3	0.1279	Ic	$M = -21.0$	Galyam *et al.* 2009
SN 2007va	14:26:23.24	+35:35:29.1	0.1907	II?	$M_{IR} = -24.2$	Kozlowski *et al.* 2010
SN 2008am	12:28:36.2	+15:34:49.1	0.2338	IIn	$M = -21.8$	Chatzopoulos *et al.* 2011
SN 2008es	11:56:49.1	+54:27:25.7	0.205	II	$M = -22.0$	Miller *et al.* 2009, Gezari *et al.* 2009
SN 2008fz	23:16:16.6	+11:42:47.5	0.133	IIn	$M = -21.9$	Drake *et al.* 2010
PTF09atu	16:30:24.5	+23:38:25.0	0.501	Ic	$M = -21.6$	Quimby *et al.* 2011
SN 2009jh	14:49:10.1	+29:25:10.4	0.349	Ic	$M = -21.7$	Quimby *et al.* 2011
PTF09cnd	16:12:08.9	+51:29:16.2	0.258	Ic	$M = -21.9$	Quimby *et al.* 2011
CSS100217	10:29:13.0	+40:42:20.0	0.147	IIn?	$M = -22.8$	Drake *et al.* 2011
SN 2010gx	11:25:46.7	−08:49:41.4	0.230	Ic	$M = -21.5$	Pastorello *et al.* 2010, Quimby *et al.* 2011
PS1-10ky	22:13:37.8	+01:14:23.6	0.956	Ic	$M = -21.9$	Chomiuk *et al.* 2011
PS1-10awh	22:14:29.8	−00:04:03.6	0.908	Ic	$M = -21.9$	Chomiuk *et al.* 2011
PS1-11bam	08:41:14.192	+44:01:56.95	1.566	Ic	$M_{UV} \sim -22.3$	Berger *et al.* 2012

[a] Peak magnitudes are in the rest-frame, unfiltered ROTSE-IIIb system except for SN2213-1745, SN1000+0216, and PS1-11bam, which are in the rest frame UV, and SN 2007va, which is in the rest frame IR.

[b] Observed with filters that do not overlap well with the rest frame ROTSE-IIIb system.

Table 2. Announced SLSNe

Name	RA	Dec	z	Type	Peak Mag[a]	Reference
SN1995av	02:01:36.75	+03:38:55.2	0.30	II?	$M_R < -20.8$	IAUC 6270
SN1999as	09:16:30.86	+13:39:02.2	0.1270	Ic	$M_V < -21.5$	IAUC 7128; Deng *et al.* 2001
SN1999bd	09:30:29.17	+16:26:07.8	0.151	IIn	$M < -21.6$	IAUC 7133
SN 2000ei	04:17:07.2	+05:45:53	0.60	II?	$M_R < -19.9$	IAUC 7516
SN 2007bt	14:27:47.73	+12:48:47.1	0.04	IIn	$M < -20.8$	CBET 941
SN 2007bw	17:11:01.99	+24:30:36.4	0.14	IIn	$M < -21.8$	CBET 941
2007-Y-155	01:07:56.083	+00:17:41.51	0.797	Ic	$M_R < -21.3$	Garnavich *et al.* 2010
SN 2009ca	21:26:22.20	−40:51:48.6	0.090	Ic	$M < -20.9$	CBET 1787
SN 2009nm	10:05:24.54	+51:16:38.7	0.21	IIn	$M < -21.3$	CBET 2106
PTF10heh	12:48:52.0	+13:26:24.5	0.338	IIn	$M_R < -21.1$	ATel 2634
PTF10nmn	15:50:02.79	−07:24:42.1	0.123	Ic	$M_R \sim -20.8$	Gal-Yam 2012
PTF10vqv	03:03:06.8	−01:32:34.9	0.452	Ic	$M_R < -21.8$	ATel 2979
SN 2010hy	18:59:32.89	+19:24:25.9	0.19	Ic	$M < -20.7$	CBET 2461, 2476
SN 2010jk	01:12:35.63	+15:28:28.5	0.28	IIn	$M < -20.6$	CBET 2534
SN 2010kd	12:08:01.1	+49:13:32.8	0.101	Ic	$M = -21.1$	CBET 2556; Vinko *et al.* 2012
SN 2011af	02:25:54.36	+10:23:11.1	0.064	IIn	$M < -20.6$	CBET 2659
SN 2011cp	07:52:33	+21:53:30	0.39	IIn	$M_V = -21.7$	ATel 3340, CBET 2733
CSS110406	13:50:57.77	+26:16:42.8	0.143	Ic	$M_R < -21.4$	ATel 3343, 3344, 3351
PTF11dsf	16:11:33.55	+40:18:03.5	0.385	IIn	$M_R < -22.2$	ATel 3465
SN 2011ep	17:03:41.78	+32:45:52.6	0.28	Ic	$M < -21.8$	CBET 2787
PTF11rks	01:39:45.51	+29:55:27.0	0.19	Ic	$M_R < -20.7$	ATel 3841
CSS120121	09:46:13	+19:50:28	0.175	Ic	$M_g = -21.1$	ATel 3873, 3918
CSS111230	14:36:58	+16:30:57	0.245	Ic	$M_V < -21.5$	ATel 3883
LSQ12byu	12:16:05.88	+09:38:07.1	0.34	I?	$M \sim -20.5$	ATel 4063
PTF12dam	14:24:46.20	+46:13:48.3	0.107	Ic	$M_R < -21.2$	ATel 4121
LSQ12dlf	01:50:29.78	−21:48:45.4	0.23?	Ic	$M_V < -21.5$	ATel 4299, 4329
SSS120810	23:18:01.82	−56:09:25.7	0.18	Ic	$M_V < -21.6$	ATel 4313, 4329
CSS121015	00:42:44	+13:28:27	0.286	Ic?	$M_V < -22.5$	ATel 4498, 4512
MLS121104	02:16:43	+20:40:09	0.14	Ic	$M_V < -21.3$	ATel 4599

[a] Peak absolute magnitudes are estimated from the observed magnitudes and redshift only; K-corrections are not applied.

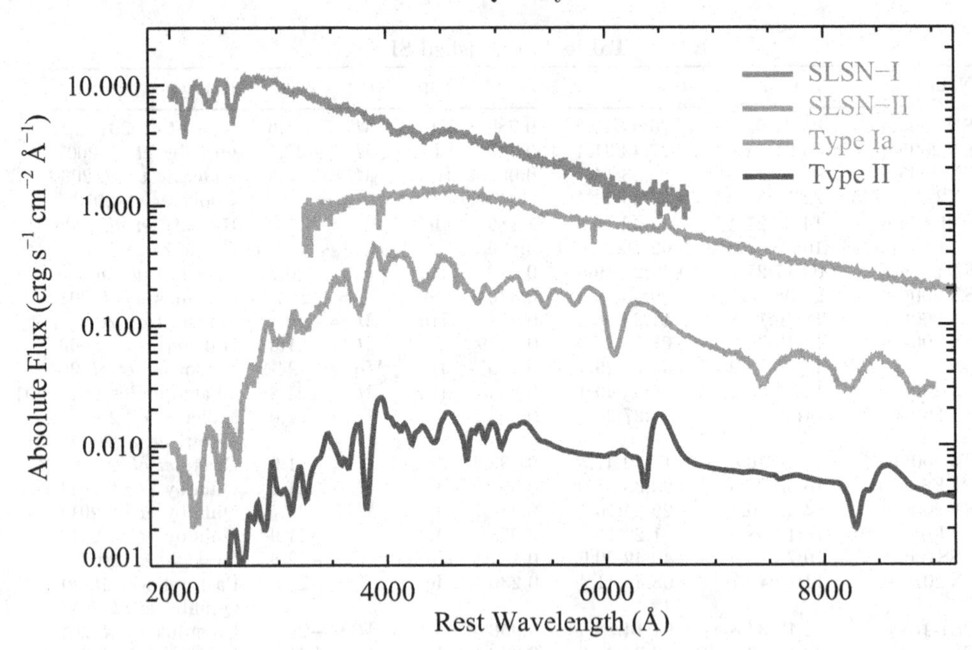

Figure 2. Spectra of normal and superluminous supernovae taken near peak (optical) brightness. The SLSN-I spectrum is a composite of SCP 06F6 (Barbary *et al.* 2009), PTF09cnd (Quimby *et al.* 2011), and SN 2005ap (Quimby *et al.* 2007), the SLSN-II is SN 2006gy from Smith *et al.* 2007, the Type Ia is a combination of SN 1992A (Kirshner *et al.* 1993) and SN 2003hv (Leloudas *et al.* 2009), and the Type II is a Nugent template (see supernova.lbl.gov/~nugent/nugent_templates.html). Flux values have been scaled to typical values for each class. SLSNe are about 10 times brighter than typical Type Ia supernovae in the optical, but in the UV, they can be a thousand times more luminous.

be that there are fundamentally different engines powering these observationally distinct events. On the other hand, the principle sources of power may yet be related; stochastic differences in the final years of the progenitors may simply color the observations.

A possible process connecting the engines powering at least some SLSNe is the conversion of kinetic energy in the supernova ejecta into radiant energy via an interaction with slower moving material. This is most clearly evident in SLSNe-II such as SN 2006gy, where the narrow emission features seen in the spectra require slow-moving material in the vicinity of the explosion (fast-moving material would give rise to broad, not narrow emission features). It is possible that events like SN 2008es also derive some of their power from ejecta/CSM interactions, but in this case the distribution of CSM must be truncated such that the slow moving material has mostly been overtaken by the SN ejecta by the time the spectroscopic observations begin (Moriya & Tominaga 2012). Extending this model further, if the CSM was depleted of its hydrogen (for example, if the progenitor was striped of its hydrogen long before the SN explosion), the ejecta/CSM interaction could in principle provide a similar transfer of kinetic energy into photons. A possible source for such hydrogen poor CSM may be material cast off by instabilities in the cores of very massive stars in their final years (e.g. Woosley *et al.* 2007; Umeda & Nomoto 2008.

Another possibility is that the high luminosities are achieved by thermalization of energy deposited into an expanding SN envelope by a compact remnant that formed as a result of the core-collapse. In the magnetar model (Kasen & Bildsten 2010; Woosley 2010), rotational energy from the nascent neutron star is transferred (by an unspecified process)

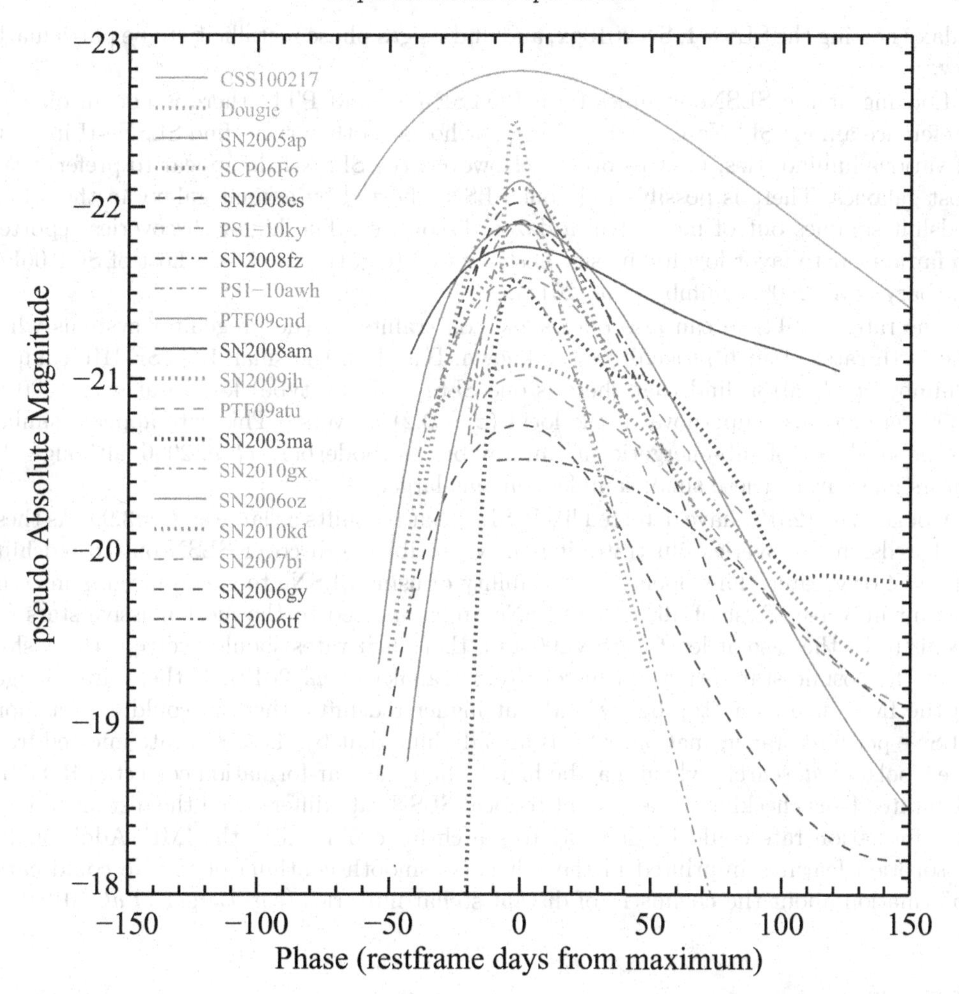

Figure 3. Approximate light curves in the rest frame optical band (ROTSE-IIIb unfiltered system) for a collection of published SLSNe. Adapted from Quimby *et al.* (2013a), which includes data from Quimby *et al.* (2007), Smith *et al.* (2007), Smith *et al.* (2008), Miller *et al.* (2009), Gezari *et al.* (2009), Barbary *et al.* (2009), Gal-Yam *et al.* (2009), Drake *et al.* (2010), Pastorello *et al.* (2010), Drake *et al.* (2011), Quimby *et al.* (2011), Chomiuk *et al.* (2011), Rest *et al.* (2011), and Leloudas *et al.* (2012).

to the ejecta mass. Kasen & Bildsten (2010) show that such models can reproduce at least the light curves of events like SN 2008es and even SN 2007bi with plausible initial rotation periods and magnetic field strengths. In this case, the progenitors could be of more modest initial masses. More exotic compact remnants may similarly inject additional energy into the SN ejecta (Ouyed *et al.* 2012).

We can get some insights into the progenitors by studying the broader environments in which these SLSNe explode (e.g. their hosts). Neill *et al.* (2011) have studied the NUV-Optical color vs. Optical magnitude distribution of a number of high luminosity supernovae and they find a preference for fainter, bluer hosts when compared to the broader population of GALEX to SDSS matched galaxies. However, the sample studied is still consistent with the giant to dwarf host distribution of normal luminosity core-collapse supernovae from PTF (Arcavi *et al.* 2010). Chen *et al.* (2013) have studied the galaxy hosting the SLSN-I, SN 2010gx, and find its gas phase metallicity to be particularly low.

Looking at the SLSNe samples from ROTSE-IIIb and PTF, there is not an obvious preference among SLSNe-II for dwarf or giant hosts (both surveys find SLSNe-II in hosts of various luminosities, faint to bright). However, the SLSN-I do appear to prefer dwarf host galaxies. There is possibly only one SLSN-I hosted by a giant galaxy in these low-redshift samples out of more than a dozen discoveries. The high-z discoveries reported so far appear to favor low luminosity hosts as well (e.g. the $M > -18$ host of SCP 06F6; Barbary *et al.* 2009, Quimby *et al.* 2011).

The rates of SLSNe can also offer some constraints on the progenitor systems when the birth rates of such progenitors are known. Based on the small ROTSE-IIIb sample, Quimby *et al.* 2013a find that there is one SLSN (of any type) for about every 400 to 1300 core-collapse supernova in the local ($z \sim 0.2$) universe. This rate appears similar to the local rate of sub-energetic gamma-ray bursts (Soderberg *et al.* 2006, although the errors inherent to these small samples remains large.

Cooke *et al.* (2012) have detected likely SLSNe at redshifts as large as $z = 3.90$. As these high redshift discoveries illustrate, it is now possible to discover SLSN out to redshifts of $z = 4$ or greater. This opens the possibility of using SLSNe to gleam insights into the distant universe. First of all, if the SLSNe are connected to the most massive stars (as seems to be the case at least for SN 2006gy), then their rates should evolve with redshift with the cosmic star formation history (e.g. Tanaka *et al.* 2012). If there are changes in the IMF such as a "top-heavy" IMF at higher redshifts, then we could expect more SLSNe per unit star formation. This is already hinted at by the $z \sim 4$ rate inferred from the Cooke *et al.* search, which may be higher than the star-formation corrected ROTSE-IIIb rate. Thus checking if the distant to local SLSN rate differs from the distant to local star-formation rate could be one way to search for evolution in the IMF. Additionally, absorption features imprinted in the otherwise smooth continua of SLSNe could carry information about the chemistry of distant stellar nurseries (e.g. Berger *et al.* 2012).

References

Aldering, G., Adam, G., Antilogus, P., *et al.* 2002, in Society of Photo-Optical Instrumentation Engineers (SPIE) Conference Series, Vol. 4836, Society of Photo-Optical Instrumentation Engineers (SPIE) Conference Series, ed. J. A. Tyson & S. Wolff, 61–72

Aldering, G., Antilogus, P., Bailey, S., *et al.* 2006, *ApJ*, 650, 510

Arcavi, I., Gal-Yam, A., Kasliwal, M. M., *et al.* 2010, *ApJ*, 721, 777

Barbary, K., Dawson, K. S., Tokita, K., *et al.* 2009, *ApJ*, 690, 1358

Berger, E., Chornock, R., Lunnan, R., *et al.* 2012, *ApJ* (Letters), 755, L29

Chatzopoulos, E., Wheeler, J. C., Vinko, J., *et al.* 2011, *ApJ*, 729, 143

Chen, T.-W., Smartt, S. J., Bresolin, F., *et al.* 2013, *ApJ* (Letters), 763, L28

Chomiuk, L., Chornock, R., Soderberg, A. M., *et al.* 2011, *ApJ*, 743, 114

Chornock, R., Berger, E., Rest, A., *et al.* 2013, *ApJ*, 767, 162

Cooke, J., Sullivan, M., Gal-Yam, A., *et al.* 2012, *Nature*, 491, 228

Deng, J. S., Hatano, K., Nakamura, T., *et al.* 2001, in Astronomical Society of the Pacific Conference Series, Vol. 251, New Century of X-ray Astronomy, ed. H. Inoue & H. Kunieda, 238

Dilday, B., Howell, D. A., Cenko, S. B., *et al.* 2012, *Science*, 337, 942

Drake, A. J., Djorgovski, S. G., Mahabal, A., *et al.* 2011, *ApJ*, 735, 106

Drake, A. J., Djorgovski, S. G., Prieto, J. L., *et al.* 2010, *ApJ* (Letters), 718, L127

Gal-Yam, A. 2012, *in IAU Symposium, Vol. 279,* IAU Symposium, 253–260

Gal-Yam, A., Mazzali, P., Ofek, E. O., *et al.* 2009, *Nature*, 462, 624

Garnavich, P. M., Aguilera, C., Becker, A., *et al.* 2010, in Bulletin of the American Astronomical Society, Vol. 42, American Astronomical Society Meeting Abstracts #215
Gezari, S., Halpern, J. P., Grupe, D., *et al.* 2009, *ApJ*, 690, 1313
Kasen, D. & Bildsten, L. 2010, *ApJ*, 717, 245
Kirshner, R. P., Jeffery, D. J., Leibundgut, B., *et al.* 1993, *ApJ*, 415, 589
Knop, R., Aldering, G., Deustua, S., *et al.* 1999, IAU circ., 7128, 1
Kozłowski, S., Kochanek, C. S., Stern, D., *et al.* 2010, *ApJ*, 722, 1624
Leloudas, G., Chatzopoulos, E., Dilday, B., *et al.* 2012, *A&A*, 541, A129
Leloudas, G., Stritzinger, M. D., Sollerman, J., *et al.* 2009, *A&A*, 505, 265
Li, W., Chornock, R., Leaman, J., *et al.* 2011a, *MNRAS*, 412, 1473
Li, W., Leaman, J., Chornock, R., *et al.* 2011b, *MNRAS*, 412, 1441
Miller, A. A., Chornock, R., Perley, D. A., *et al.* 2009, *ApJ*, 690, 1303
Moriya, T. J. & Tominaga, N. 2012, *ApJ*, 747, 118
Neill, J. D., Sullivan, M., Gal-Yam, A., *et al.* 2011, *ApJ*, 727, 15
Nugent, P., Aldering, G., Phillips, M. M., *et al.* 1999, IAU Circ., 7133, 1
Ouyed, R., Kostka, M., Koning, N., Leahy, D. A., & Steffen, W. 2012, *MNRAS*, 423, 1652
Pastorello, A., Smartt, S. J., Botticella, M. T., *et al.* 2010, *ApJ* (Letters), 724, L16
Perlmutter, S., Aldering, G., Goldhaber, G., *et al.* 1999, *ApJ*, 517, 565
Perlmutter, S., Pennypacker, C. R., Goldhaber, G., *et al.* 1995, *ApJ* (Letters), 440, L41
Quimby, R. M. 2006, PhD thesis, The University of Texas at Austin
Quimby, R. M., Aldering, G., Wheeler, J. C., *et al.* 2007, *ApJ* (Letters), 668, L99
Quimby, R. M., Kulkarni, S. R., Kasliwal, M. M., *et al.* 2011, *Nature*, 474, 487
Quimby, R. M., Yuan, F., Akerlof, C., & Wheeler, J. C. 2013a, *MNRAS*, 431, 912
Quimby, R. M., Werner, M. C., Oguri, M., *et al.* 2013b, *ApJ* (Letters), 768, L20
Rest, A., Foley, R. J., Gezari, S., *et al.* 2011, *ApJ*, 729, 88
Riess, A. G., Filippenko, A. V., Challis, P., *et al.* 1998, *AJ*, 116, 1009
Scalzo, R. A., Aldering, G., Antilogus, P., *et al.* 2010, *ApJ*, 713, 1073
Smith, N., Chornock, R., Li, W., *et al.* 2008, *ApJ*, 686, 467
Smith, N., Li, W., Foley, R. J., *et al.* 2007, *ApJ*, 666, 1116
Soderberg, A. M., Kulkarni, S. R., Nakar, E., *et al.* 2006, *Nature*, 442, 1014
Tanaka, M., Moriya, T. J., Yoshida, N., & Nomoto, K. 2012, *MNRAS*, 422, 2675
Turatto, M., Suzuki, T., Mazzali, P. A., *et al.* 2000, *ApJ* (Letters), 534, L57
Umeda, H. & Nomoto, K. 2008, *ApJ*, 673, 1014
Vinko, J., Zheng, W., Pandey, S. B., *et al.* 2012, in American Astronomical Society Meeting Abstracts, Vol. 219, American Astronomical Society Meeting Abstracts #219
Woosley, S. E. 2010, *ApJ* (Letters), 719, L204
Woosley, S. E., Blinnikov, S., & Heger, A. 2007, *Nature*, 450, 390
Yasuda, N. & Fukugita, M. 2010, *AJ*, 139, 39
Yuan, F., Quimby, R. M., Wheeler, J. C., *et al.* 2010, *ApJ*, 715, 1338

Discussion

KAMBLE: 1. Are there any early features that can be used to distinguish SLSN from other SN? 2. Based on the proposed models would you expect them to take place in the local universe?

QUIMBY: The fastest way to identify SLSNe is through spectral observations. SLSN-I in particular show spectral features that are unique among SNe, such as the broad OII features (although some normal luminosity type Ic briefly show this soon after explosion). More generally, the spectra may provide a distance (redshift) that combined with the photometry could indicate a SLSN. A slowly rising light curve can also signal a SLSN candidate, but this requires time to collect. 2. From the older models, there was no expectation to find SLSNe in the local universe, but one of the first discoveries, SN

2006gy, was found just 73 Mpc away. Some of the more modern SLSN models now allow for such local events.

MILISAVLJEVIC: Can you comment on the late-time optical emission from SLSN? I am unfamiliar with such spectra outside of SN 2007bi.

QUIMBY: SN 2007bi had a slow-evolving light curve that could be monitored for ~ 1 year. This permitted late-time optical spectra to be obtained that exhibited strong iron features consistent with large nickel production. Other SLSN, however, have generally exhibited faster-fading light curves, preventing late-time observations. Currently, there is a dearth of late-time spectra for these objects.

Supernova Environmental Impacts
Proceedings IAU Symposium No. 296, 2013
A. Ray & R. A. McCray, eds.

doi:10.1017/S1743921313009265

Supernova Optical Observations and Theory

Keiichi Maeda, Melina C. Bersten, Takashi J. Moriya, Gaston Folatelli, and Ken'ichi Nomoto

Kavli Institute for the Physics and Mathematics of the Universe, University of Tokyo,
5-1-5 Kashiwanoha, Kashiwa, Chiba 277-8583, Japan
email: keiichi.maeda@ipmu.jp

Abstract. We review emission processes within the supernova (SN) ejecta. Examples of the application of the theory to observational data are presented. The emission processes and thermal condition within the SN ejecta change as a function of time, and multi-epoch observations are important to obtain comprehensive views. Through the analyses, we can constrain the progenitor radius, compositions as a function of depth, ejecta properties, explosion asymmetry and so on. Multi-frequency follow-up is also important, including radio synchrotron emissions and the inverse Compton effect, γ-ray emissions from radioactive decay of newly synthesized materials. The optical data are essential to make the best use of the multi-frequency data.

Keywords. supernovae: general, radiation mechanisms: general.

---

1. Introduction

Supernovae (SNe) are classified into several types by characteristics in their spectra around the maximum light (e.g., Filippenko 1997). The spectral typing has been linked to the progenitor scenarios through the amount of the envelope present at the time of explosion (e.g., Nomoto *et al.* 1995). Stars that retain their H envelope produce SNe with H-rich spectra, classified as SNe II. SNe II are divided into a few sub categories. If there are strong hydrogen (narrow) emission lines, an SN is classified as SN IIn. Stars that have lost at least a large fraction of the H envelope produce He-rich SNe IIb, H-deficient SNe Ib, and both H and He-deficient SNe Ic, in order of increasing degree of the envelope stripping (called stripped-envelope SNe; SE-SNe). They are believed to be a core-collapse (CC) SNe from a massive star. SNe Ia show neither H nor He, and they are characterized by a strong Si absorption. This matches well the standard scenario that SNe Ia are a result of a thermonuclear explosion of a C+O white dwarf, reaching (nearly) the Chandrasekhar limiting mass (e.g., Nomoto *et al.* 1984; Woosley & Weaver 1986).

2. Optical Emission Processes and Model Examples

Generally, an SN is described as a metal-rich fireball, homologously expanding into a space with $\sim 10,000$ km s^{-1}.Thus the optical depth decreases with time. Photosphere moves back toward the center as time goes by, and so does the line forming region until the photosphere disappears. This property allows one to extract the chemical and structure information within the ejecta as a function of the depth (Stehle *et al.* 2006).

SNe of different types are also associated with different energy sources. In the beginning, the energy source could be the thermal energy deposited at the explosion. The thermal energy is more important for SNe with a more extended envelope. This is a dominant source in SNe IIp for the first ~ 100 days, while it lasts only for a few days in SE SNe. SNe create a huge amount of radioactive ^{56}Ni, and its decay is another major

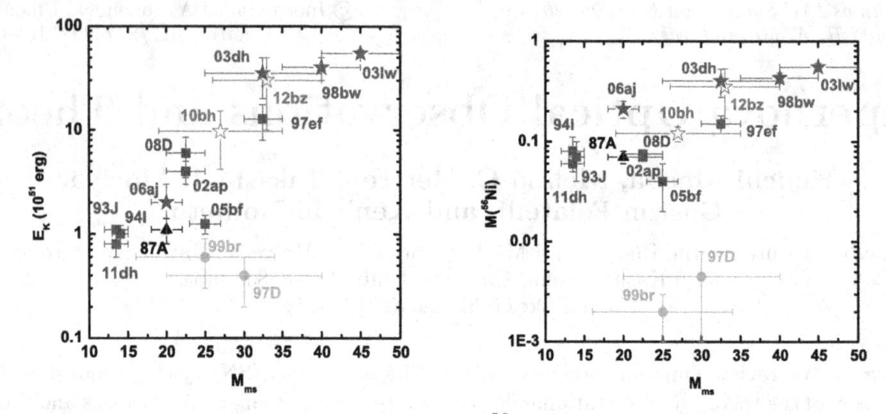

Figure 1. Relations in $M_{\rm ms} - E_{\rm K}$ and $M_{\rm ms} - M(^{56}{\rm Ni})$. Filled symbols are for those derived with numerical radiation transfer modeling, while open symbols are those estimated with simple scaling relations. Shown here are GRB-SNe Ic (red/magenta stars), SE SNe (blue squares), faint SNe IIP (green circles), and SN 1987A (black triangle).

source. If the CSM density is high, the energy input from the ejecta-CSM interaction could dominate, and this is believed to be a power source of SNe IIn.

2.1. *Maximum-Light Analysis*

Details in analyzing the observed properties around the maximum brightness are different for different types of SNe, and here we show a specific example of SE SNe. In these ^{56}Ni-powered SNe with little or no H-envelope, the peak luminosity (L), light curve time scale (Δt), and the expansion velocity from absorption features (V), are connected to the mass of ^{56}Ni [$M(^{56}{\rm Ni})$], ejecta mass ($M_{\rm ej}$), the kinetic energy of the explosion ($E_{\rm K}$), as follows: $L \propto M(^{56}{\rm Ni})$, $\Delta t \propto E^{-1/4} M^{3/4}$, $V \propto E_{\rm K}^{1/2} M_{\rm ej}^{-1/2}$. The main-sequence mass ($M_{\rm ms}$) can be estimated from $M_{\rm ej}$. Results from this kind of analyses are shown in Figure 1.

There is a trend that a more massive progenitor results in a more energetic explosion with a large amount of ^{56}Ni. Some SNe Ic show broad absorption features, indicating a larger energy than other SNe ($E_{\rm K} \sim 10^{51}$ erg) (e.g., Iwamoto *et al.* 1998). The most energetic broad-line SNe Ic reach $E_{\rm K} \gtrsim 10^{52}$ erg, and some of them are associated with long-soft Gamma-Ray Bursts (hereafter GRBs) (Galama *et al.* 1998; Hjorth *et al.* 2003).

2.2. *Early-Phase Emission*

There are an increasing number of SNe discovered just after the explosion (e.g., Nugent *et al.* 2011). Such an early emission is sensitive to the outer envelope structure of the progenitor. An SN from a more extended progenitor is predicted to show a higher luminosity persisting for a longer time scale. A specific example here is shown for SN IIb 2011dh in M51. The SN properties around maximum are very similar to a prototypical SN IIb 1993J. Figure 2 (left) shows an example of the maximum light modeling (§2.1). The He core mass was derived to be $\sim 4 M_\odot$, the explosion energy to be $\sim 0.8 \times 10^{51}$ erg, and $M(^{56}{\rm Ni}) \sim 0..06 M_\odot$ (Bersten *et al.* 2012). Thus SN 2011dh is almost a twin of SN1993J. The main sequence mass must have been $\sim 12 - 15 M_\odot$. This suggests that the binary evolution has to be invoked to have the hydrogen envelope stripped away.

Pre-explosion HST images show that unexpectedly there was an yellow supergiant (YSG) at the SN position (Maund *et al.* 2011; Van Dyk *et al.* 2011). While there had been discussion about whether this was a progenitor or not, later images showed that the YSG has disappeared–it was a progenitor (Van Dyk *et al.* 2013). Before this confirmation,

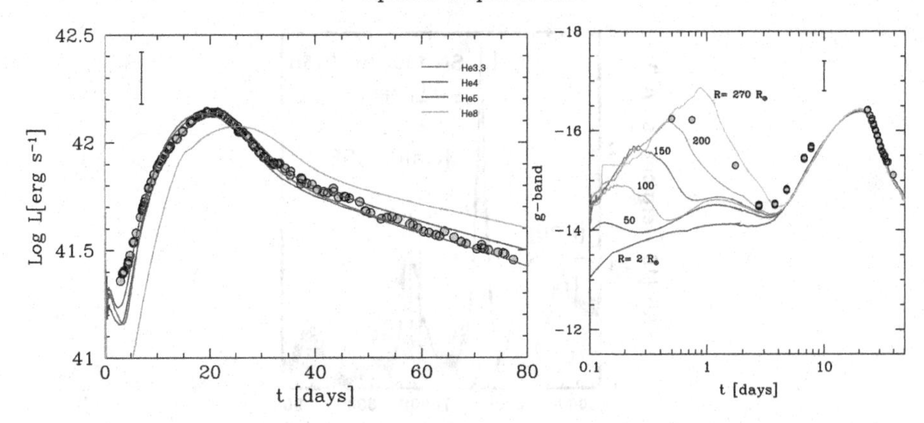

Figure 2. Radiation hydrodynamic simulations for SN 2011dh (Bersten *et al.* 2012). Left panel shows the maximum-light analysis. Right panel shows the early-phase analysis.

Bersten *et al.* (2012) analyzed the early emission to address the issue. Figure 2 (right) shows the early-phase light curve compared to the models with different progenitor radii. There is the initial emission from the shock-deposited thermal energy clearly detected, which requires the progenitor as large as $\sim 200R_{\odot}$, consistent with the YSG.

The very early phase spectra also contain the abundance information at the progenitor surface. Recent highlights include a discovery of carbon lines in many SNe Ia (Parrent *et al.* 2011; Folatelli *et al.* 2012). The carbon content is a key in clarifying progenitor and explosion scenarios (Maeda *et al.* 2010b; Tanaka *et al.* 2011).

Another example is about Super Luminous SNe (SLSNe). While the leading model for its power is an SN-CSM interaction within a very dense CSM (Moriya *et al.* 2013), a few other scenarios such as a magnetor input model have not been rejected (Kasen *et al.* 2010; see also Maeda *et al.* 2007a). A new approach to discriminate the interaction model from the others has been proposed by Moriya & Maeda (2012), which predicts a brief 'dip' before the LC rise as was indicated for SNSN 2006oz (Leloudas *et al.* 2012).

2.3. *Late-Phase Emission*

Late-time observations also have unique advantage. The SN ejecta become fully transparent in about a year. At this phase the emission from the innermost region is directly viewed by an observer. It thus provides a unique way to probe the abundance and structure of the innermost region, which are closely related to the nature of the explosion. Deriving the explosion geometry through a late-time spectrum is a unique and powerful way to tackle to the explosion physics, and this will be separately discussed in §3.

As an example, Figure 3 shows a late-time spectrum of SN 2006aj (Maeda *et al.* 2007b), which was discovered in association with an X-Ray Flash (XRF) 060218 (Pian *et al.* 2006). Unlike typical SE SNe, the feature at 7400Å is identified as [Ni II], not [Ca II]. It was then derived that the amount of stable ^{58}Ni is $\sim 0.05M_{\odot}$. This large amount of neutron-rich materials may indicate that there was a strong wind from a newly-formed neutron star, and such a strong activity might well be linked to a birth of the XRF.

SN Ia explosion physics is another application. Maeda *et al.* (2010a) examined late-time spectra of SNe Ia and concluded that one can associate different thermonuclear burning products to different emission lines: [Fe III] blend at $\sim$ 4700Å comes from a low-density ^{56}Ni-rich region, while [Fe II]λ7155 and [Ni II]λ7388 are from a high-density, stabile Fe/Ni–rich region. The existence of stable Fe-peaks could be a smoking gun to distinguish several popular progenitor and explosion models (Maeda *et al.* 2010b).

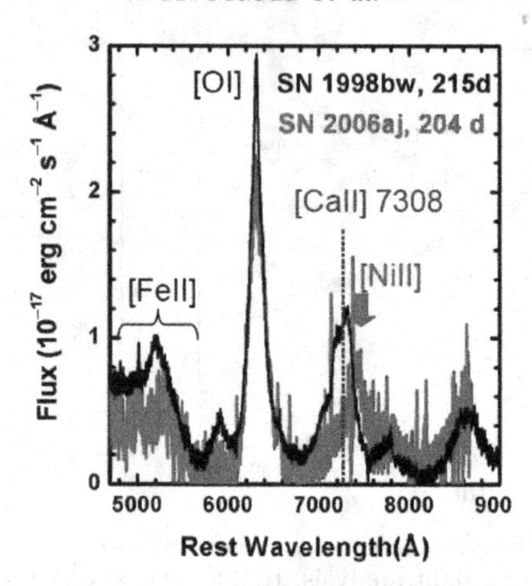

Figure 3. A late-time spectrum of SN 2006aj (Maeda *et al.* 2007b).

Late-time observations have led to new discoveries as well. SN Ib 2005cz was peculiar, as this was discovered in an elliptical galaxy. Similar objects are also found in elliptical or outskirts. Except for being faint [$M(^{56}\mathrm{Ni}) \sim 0.01 M_{\odot}$], the early-phase spectral evolution was not peculiar as SE SNe. However, their late-time spectra are characterized by a lack of [OI] and by strong [Ca II] (Kawabata *et al.* 2010). The origin of these 'Ca-rich' SNe is still under debate (Parets *et al.* 2010), including suggestion that it is an explosion of a star at the lower boundary to become an SN ($M_{\mathrm{ms}} \sim 8 - 10 M_{\odot}$) (Kawabata *et al.* 2010).

3. Explosion asymmetry

Explosion mechanisms of SNe have not been clarified. The geometry is probably a key. Different scenarios predict different types of explosion geometry, thus *observationally* deriving the geometry of the innermost ejecta is of highest importance. The ejecta become transparent at about 100 or 200 days after the explosion. Thanks to the homologous expansion, the Doppler shift indicates where the photon was emitted. A photon emitted from the near/far side of the ejecta is blueshifted/redshifted. Thus, the line profile can be used as a direct tracer of the distribution of the emitting materials.

In SE SNe, [O I] $\lambda\lambda 6300, 6363$ is the strongest in their late-time spectra. Maeda *et al.* (2002) predicted that a bipolar explosion should produce double peaks in the [O I] if viewed from the equatorial direction. Figure 4 (left) shows the [O I] line profiles of 18 SE SNe we have obtained by *Subaru* and *VLT* (Maeda *et al.* 2008). We found that the doubly-peaked profile is not rare, suggesting that the asphericity is a generic feature in SE SNe, and a bipolar-type geometry is a straightforward interpretation.

We have also investigated explosion asymmetry in SNe Ia through late-time spectra, based on the idea that the distribution of different burning products can be traced by line profiles of different lines (§2.3). We discovered that [Fe II]$\lambda 7155$ and [Ni II]$\lambda 7378$ do show variations in its central wavelength for different SNe Ia; some SNe show redshift and others blueshift. The above observations indicate that an asymmetric, offset ignition is a generic feature of SNe Ia (Maeda *et al.* 2010a; Maeda *et al.* 2011).

The geometry derived here could be a key to understanding various diverse properties of SNe. For example, it has been pointed out that normal SNe can be categorized into two groups based on the Si absorption feature, i.e., the high velocity group and low velocity group (Benetti *et al.* 2005). In addition, SNe Ia show a variation in their peak colors

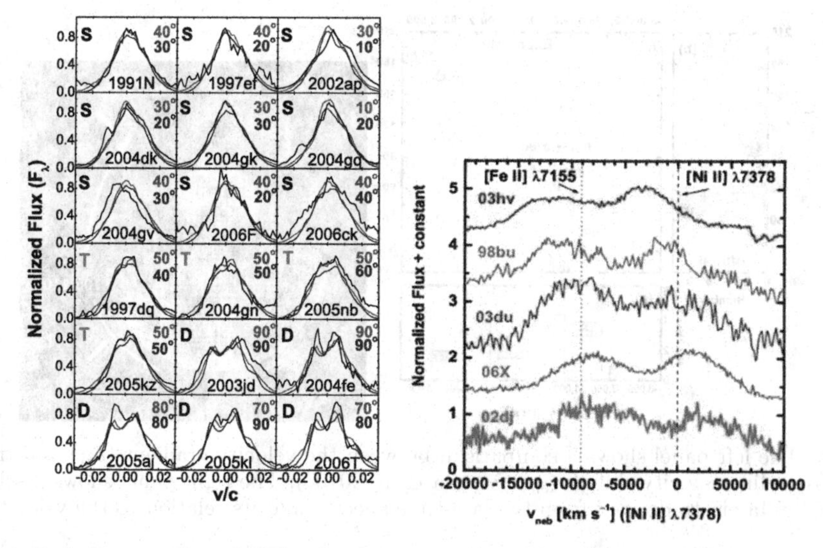

Figure 4. Left: line profiles of [OI] in late-time spectra of SE SNe (black) (Maeda *et al.* 2008). Also shown are the bipolar explosion model predictions (Maeda *et al.* 2002; Maeda *et al.* 2006). Right: Late-time spectra of some selected SNe Ia (Maeda *et al.* 2010a; Maeda *et al.* 2011). .

(e.g., Folatelli *et al.* 2010). The origin of these diversities has not yet been clarified, and solving these could have a high impact in the SN cosmology.

We discovered that the Si velocity in the early phase is correlated to the velocity shift in the nebular lines in the late-phases (Figure 5: Maeda *et al.* 2010c). The correlation indicates that there is no intrinsic difference between high velocity SNe and (at least a part of) low velocity SNe, but the different appearance is a consequence of different viewing directions. At the same time, there has been accumulating evidence that a part of low-velocity SNe belong to a population intrinsically different than high velocity SNe (Wang *et al.* 2013). Combining all these results suggests that there are likely (at least) two populations in normal SNe Ia, one including both high velocity and low velocity SNe, while the other including only low velocity SNe Ia. The late-time spectroscopic study suggests that there is not intrinsic difference in the high velocity and low velocity SNe in the former group, and may well be explained by the viewing angle effect.

SNe Ia show a variation in their peak colors. This is another major issue in SN cosmology since the estimate of extinction (thus distance) relies on the intrinsic color. By comparing the 'viewing direction' and the color, we have shown that that the intrinsic color variation is at the level of $B - V \sim 0.2$ mag, and the variation within this level can be attributed to the intrinsic variation (Maeda *et al.* 2011; Cartier *et al.* 2011).

4. Optical Study in Multi-wavelength Context

Looking at the same object in different wavelengths provides a comprehensive view of astrophysical objects, and it is also the case in SNe. Radio synchrotron emissions from SNe (so far detection limited for CC SNe) provide not only the CSM density/distribution but also the ejecta mass and the kinetic energy (Maeda 2012a; Maeda 2013a). Figure 6 shows a relation between the peak date and peak luminosity at 5 GHz. SNe with the same set of the ejecta mass and the kinetic energy should follow a line in this diagram. So far the observational data indicate that SE SNe have quite uniform ejecta properties,

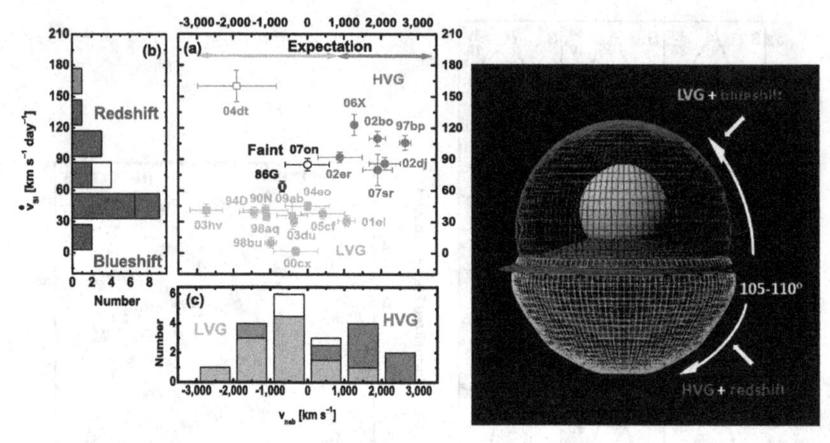

Figure 5. The left panel shows a comparison between the velocity gradient ($\dot{v}_{\rm Si}$) and the late-time emission line velocity shift ($v_{\rm neb}$) (Maeda *et al.* 2010c). The right panel shows a schematic picture of SN Ia ejecta derived from the late-time spectra and its relation to the velocity.

with the ejecta mass seemingly clustered into $\sim 1 - 5M_{\odot}$. This may indicate that the binary path could be important. This can be compared to the insight from the optical data provided through the maximum light analysis (§2.1). For example, in case of SN IIb 2011dh, the radio multi-band light curves can be explained quite well by adopting the ejecta properties obtained through the optical analysis (Maeda 2012a).

Combining the optical data with other wavelengths is critical in understanding the shock wave physics. The optical data directly provide the thermal photon density at the shock wave. Depending on the density of relativistic electrons there, one expects to see the inverse Compton cooling effect in the radio emission. By this way, one can constrain the relativistic electron population at the shock wave, as suggested by Maeda (2013a) (see also Maeda 2012a). Analyzing a few SNe, the efficiency of the electron acceleration is found to be less than 1% (i.e., less than 1% of the shock heated thermal energy goes into the relativistic electrons). This is an important input in constraining the electron energy distribution at the SN shock through multi-wavelength study (Maeda 2013b).

Another example is the high energy emission from radioactive decay. Nearby SNe Ia are within reach by NuStar and Astro-H in hard X and soft γ-rays (Fig. 7: Maeda *et al.* 2012b), hopefully leading to detection of radioactive decay chain ^{56}Ni $\rightarrow$ ^{56}Co $\rightarrow$ ^{56}Fe in the near future. This will be direct evidence of the syntheses of ^{56}Ni and its amount. This will provide a test for our analysis of optical data (e.g., §2.1) through a totally independent way, and this could bring us a better understanding of the nature of SNe.

5. Summary

We have reviewed emission processes within SN ejecta. Various examples of applying the theory to observational data have been presented. The emission processes and thermal condition of the SN ejecta change as a function of time, and thus multi-epoch follow-up is important to obtain comprehensive views. The very early phase observations can be used to estimate the progenitor radius and surface compositions, both are closely related to the evolution toward an SN. Such an early follow-up may also shed light on still unresolved nature of SLSNe. Maximum-light properties have been leading to relations between various explosion/progenitor parameters. Late-phase observations are critical to

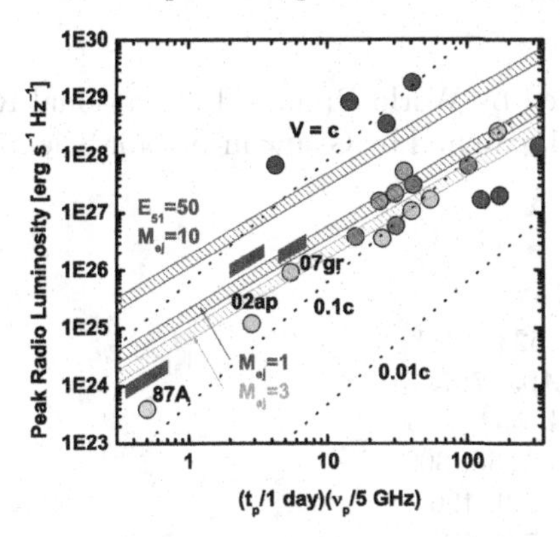

Figure 6. Radio properties of SE SNe. The peak luminosity is plotted as a function of the peak date. See Maeda (2013a) for details. The expected constant lines for a given set of ejecta properties are shown by the shaded area (blue, red, and green, with E_{51} and $M_{\rm ej}$ given in labels).

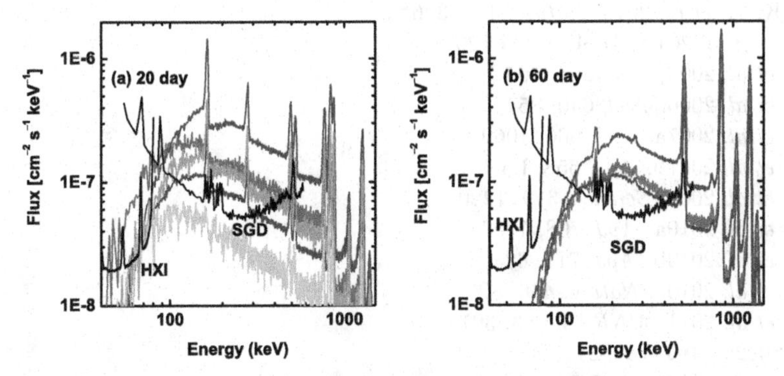

Figure 7. Exampels of synthetic high-energy spectra at (a) 20 days and (b) 60 days after the SN Ia explosion (at 10 Mpc). Sensitivity curves (10^6 seconds) of *HXI* and *SGD* on board *Astro-H* are shown by black lines. See Maeda *et al.* 2012b for details.

tackle to the still unresolved nature of SN explosions, as the late-phase emission is a probe to materials deep in the ejecta, which directly reflect the nature of the explosion,

Application of the late-time data is highlighted by deriving the geometry of the explosion, We have shown that both CC SNe and SNe Ia are asymmetric, but a way how it deviates from spherically symmetry is different. This interpretation offers the asymmetry and viewing direction as a possible origin of some observational diversities of SNe.

Not only multi-epoch, but also multi-frequency follow-up is important. Ejecta properties as estimated from the optical data analysis can be tested with radio data in an independent way. Further, considering the cooling effect in radio has been suggested to place a strong constraint on the efficiency of relativistic electron acceleration at the SN shock wave, and the analysis requires the optical data as an input. In the near future, we also expect to detect and analyze hard X and soft γ-rays from radioactive isotopes produced at the SN explosions, and a combined analysis of such data and the optical data will lead to a comprehensive understanding of the SN progenitors and explosions.

Acknowledgements

This research is supported by World Premier International Research Center Initiative (WPI Initiative), MEXT, Japan and by Grant-in-Aid for Scientific Research (23740141).

References

Benetti, S., *et al.* 2005, *ApJ*, 623, 1011
Bersten, M. C., *et al.* 2012, *ApJ*, 757, 31
Cartier, R., *et al.* 2011, *A&A*, 534, L15
Filippenko, A. V. 1997, *ARAA*, 35, 309
Folatelli, G., *et al.* 2010, *AJ*, 139, 120
Folatelli, G., *et al.* 2012, *ApJ*,745, 74
Galama, T. J., *et al.* 1998, *Nature*, 395, 670
Hjorth, J., *et al.* 2003, *Nature*, 423, 847
Iwamoto, K., *et al.* 1998, *Nature*, 395, 672
Kasen, D. & Bildsten, I. 2010, *ApJ*, 717, 245
Kawabata, K. S., *et al.* 2010, *Nature*, 465, 326
Leloudas, G., *et al.* 2012, *A&A*, 541, L129
Maeda, K., *et al.* 2002, *ApJ*, 565, 405
Maeda, K., *et al.* 2006, *ApJ*, 640, 854
Maeda, K., *et al.* 2007a, *ApJ*, 666, 1069
Maeda, K., *et al.* 2007b, *ApJ*, 658, L5
Maeda, K., *et al.* 2008, *Science*, 319, 1220
Maeda, K., *et al.* 2010a, *ApJ*, 708, 1703
Maeda, K., *et al.* 2010b, *ApJ*, 712, 624
Maeda, K., *et al.* 2010c, *Nature*, 466, 82
Maeda, K., *et al.* 2011, *MNRAS*, 413, 3075
Maeda, K. 2012a, *ApJ*, 758, 81
Maeda, K., *et al.* 2012b, *ApJ*, 760, 54
Maeda, K. 2013a, *ApJ*, 762, 14
Maeda, K. 2013b, *ApJ*, 762, L24
Maund, J. R., *et al.* 2011, *ApJ*, 739, L37
Moriya, T. J. & Maeda, K. 2012, *ApJ*, 756, L22
Moriya, T. J., *et al.* 2013, *MNRAS*, 428, 1020
Nomoto, K., Thielemann, F.-K., & Yokoi, K. 1984, *ApJ*, 286, 644
Nomoto, K., Iwamoto, K., & Suzuki, T. 1995, *Phys. Rep.*, 256, 173
Nugent. P., *et al.* 2011, *Nature*, 480, 344
Parets. H. B., *et al.* 2010, *Nature*, 465, 322
Parrent. J. T., *et al.* 2011, *ApJ*, 732, 30
Pian, E., *et al.* 2006, *Nature*, 442, 1011
Stehle, M., *et al.* 2006, *MNRAS*, 360, 1231
Tanaka, M., *et al.* 2011, *MNRAS*, 410, 1725
Van Dyk, S. D., *et al.* 2011, *ApJ*, 741, L28
Van Dyk, S. D., *et al.* 2013, *ATEL*, 4850
Wang, X., *et al.* 2013, *Science*, 340, 170
Woosley, S. E. & Weaver, T. A. 1986, *ARAA*, 24, 205

Discussion

TURATTO: Is it correct that the best spectral range for investigating the assymmetries in SNR is the NIR?

MAEDA: Yes, the NIR can be a better place to see than the optical. The line blend is still there, but solving the blend is easier is the NIR than in the optical. Observationally the NIR is tough, but we are running NIR late-time observations to increase the late-time NIR spectra.

Supernova Environmental Impacts
Proceedings IAU Symposium No. 296, 2013
A. Ray & R. A. McCray, eds.

doi:10.1017/S1743921313009277

Light Curve Modeling of Superluminous Supernovae

Takashi Moriya[1,2,3], Sergei I. Blinnikov[4,5,6], Nozomu Tominaga[7,1], Naoki Yoshida[8,1], Masaomi Tanaka[9], Keiichi Maeda[1], and Ken'ichi Nomoto[1]

[1]Kavli IPMU, University of Tokyo, Kashiwanoha 5-1-5, Kashiwa, Chiba 277-8583, Japan
email: takashi.moriya@ipmu.jp

[2]Dept. of Astronomy, University of Tokyo, Hongo 7-3-1, Bunkyo-ku, Tokyo 113-0033, Japan

[3]RESCEU, University of Tokyo, Hongo 7-3-1, Bunkyo-ku, Tokyo 113-0033, Japan

[4]ITEP, Bolshaya Cheremushkinskaya 25, 117218 Moscow, Russia

[5]Novosibirsk State University, Novosibirsk 630090, Russia

[6]SAI, Moscow University, Universitetski pr. 13, 119992 Moscow, Russia

[7]Dept. of Physics, Konan University, 8-9-1 Okamoto, Kobe, Hyogo 658-8501, Japan

[8]Dept. of Physics, University of Tokyo, Hongo 7-3-1, Bunkyo-ku, Tokyo 113-0033, Japan

[9]NAOJ, 2-21-1 Ohsawa, Mitaka, Tokyo 181-8588, Japan

Abstract. Origins of superluminous supernovae (SLSNe) discovered by recent SN surveys are still not known well. One idea to explain the huge luminosity is the collision of dense CSM and SN ejecta. If SN ejecta is surrounded by dense CSM, the kinetic energy of SN ejecta is efficiently converted to radiation energy, making them very bright. To see how well this idea works quantitatively, we performed numerical simulations of collisions of SN ejecta and dense CSM by using one-dimensional radiation hydrodynamics code STELLA and obtained light curves (LCs) resulting from the collision. First, we show the results of our LC modeling of SLSN 2006gy. We find that physical parameters of dense CSM estimated by using the idea of shock breakout in dense CSM (e.g., Chevalier & Irwin 2011, Moriya & Tominaga 2012) can explain the LC properties of SN 2006gy well. The dense CSM's radius is about 10^{16} cm and its mass about 15 $M_{\odot}$. It should be ejected within a few decades before the explosion of the progenitor. We also discuss how LCs change with different CSM and SN ejecta properties and origins of the diversity of H-rich SLSNe. This can potentially be a probe to see diversities in mass-loss properties of the progenitors. Finally, we also discuss a possible signature of SN ejecta-CSM interaction which can be found in H-poor SLSN.

Keywords. supernovae: general, supernovae: individual (SN 2006gy, SN 2006oz), stars: mass loss, circumstellar matter

---

1. Introduction

There are several suggested ways to explain the huge luminosity of superluminous supernovae (SLSNe), e.g., large production of ^{56}Ni (Gal-Yam *et al.* 2009, Moriya *et al.* 2010), interaction between dense circumstellar medium (CSM) and SN ejecta (or ejecta from stellar surface) (e.g., Woosley *et al.* 2007), magnetar spin-down (e.g., Kasen & Bildsten 2010), fallback (e.g., Dexter & Kasen 2012), etc. Most of SLSNe II (see, e.g., Gal-Yam 2012 for a review of the SLSN classification) show narrow lines which are expected to appear from dense CSM and it is natural to think they are coming from the interaction.

2. Shock Breakout for Superluminous Supernovae

Here, we investigate an SLSN model in which a SN explosion occurred in dense CSM and the SN gets superluminous because of the deceleration by the dense CSM. To explain the huge luminosities of a SLSN in this model, the dense CSM needs to be so dense that we need to take the effect of the shock breakout into account (Chevalier & Irwin 2011, Moriya & Tominaga 2012). From the shock breakout model, we can estimate two observable timescales,

$$t_d \simeq \begin{cases} \frac{R_o}{v_s}\left[\left(\frac{c/v_s + x^{1-w}}{c/v_s+1}\right)^{\frac{1}{1-w}} - x\right] & (w \neq 1), \\ \frac{R_o}{v_s}\left(x^{\frac{1}{1+c/v_s}} - x\right) & (w = 1). \end{cases} \tag{2.1}$$

$$t_s = \frac{R_o - xR_o}{v_s}, \tag{2.2}$$

where t_d is the diffusion timescale in the dense CSM after the shock breakout which corresponds to the SLSN LC rising time, t_s is the time required for the shock wave to go through the entire CSM, R_o is the CSM radius, v_s is the shock velocity, c is the speed of light, xR_o is the radius of the shock breakout, and w is the CSM density slope ($\rho \propto r^{-w}$). For the best observed SLSN 2006gy, $t_d \simeq 70$ days and $t_s \simeq 200$ days (Smith *et al.* 2010) and we can estimate CSM properties for given w and v_s. Estimated CSM structures reproduces SN 2006gy properties very well, as reported in Moriya *et al.* (2013b). What is particularly interesting is that the steady mass loss model could not reproduce the SN 2006gy LC. We also note that the shell shocked diffusion model suggested by Smith & McCray (2007) does not work for SN 2006gy (Moriya *et al.* 2013a).

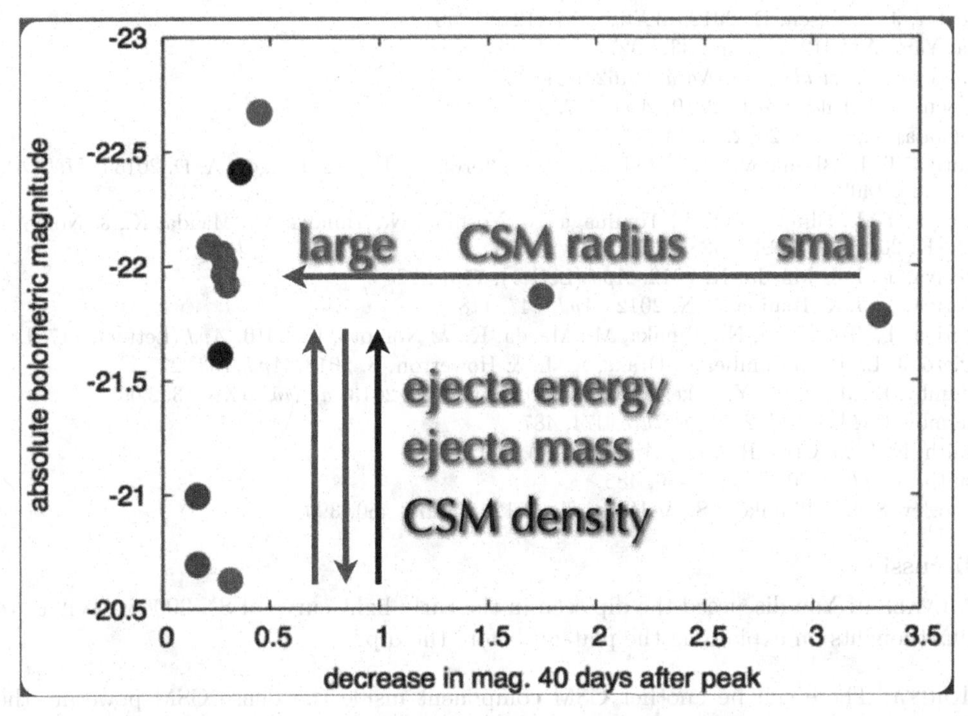

Figure 1. Direction of changes in decline rate-peak luminosity plane.

3. Diversity of Superluminous Supernovae

Quimby *et al.* (2013) reported the distribution of SLSNe in a decline rate-luminosity plane. Based on the interaction model of SN 2006gy, we can change SN ejecta and CSM properties to see the origin of the diversity. In Figure 1, we show in which directions the SLSN properties move when we change SN ejecta and CSM properties obtained by the preliminary numerical LC calculations. More detailed study will show us the origin of the diversity of SLSNe and the mass loss of the progenitors.

4. 'Dip' in Superluminous Supernova Light Curve

Currently, SLSNe I (e.g., Quimby *et al.* 2011) does not have clear signatures of the physical processes making them superluminous. Leloudas *et al.* (2012) reported the LC observations of a SLSN-I 2006oz. They detected a precursor of the SN and there existed a 'dip' in the LC between the main LC and the precursor. This may indicate the existence of dense C+O CSM around the SN ejecta and SLSN-I may also be related to the SN ejecta-dense CSM interaction (Moriya & Maeda 2012). This 'dip' is also observed in Type IIn SN 2009ip (e.g., Prieto *et al.* 2013), in which the existence of the dense CSM is clear because of the SN type.

Acknowledgement

This research was supported by a grant from the Hayakawa Satio Fund awarded by the Astronomical Society of Japan.

References

Chevalier, R. A. & Irwin, C. M. 2011, *ApJ* (Letters), 729, L6
Dexter, J. & Kasen, D. 2012, *arXiv*, arXiv:1210.7240
Gal-Yam, A. 2012, *Science*, 337, 927
Gal-Yam, A., *et al.* 2009, *Nature*, 462, 624
Kasen, D. & Bildsten, L. 2010, *ApJ*, 717, 245
Leloudas, G., *et al.* 2012, *A&A*, 541, 192
Moriya, T. J., Blinnikov, S. I., Baklanov, P. V., Sorokina, E. I., & Dolgov, A. D. 2013a, *MNRAS*, 430, 1402
Moriya, T. J., Blinnikov, S. I., Tominaga, N., Yoshida, N., Tanaka, M., Maeda, K., & Nomoto, K. 2013b, *MNRAS*, 428, 1020
Moriya, T. J. & Maeda, K. 2012, *ApJ* (Letter), 756, L22
Moriya, T. J. & Tominaga, N. 2012, *ApJ*, 747, 118
Moriya, T., Tominaga, N., Tanaka, M., Maeda, K., & Nomoto, K. 2010, *ApJ* (Letter), 717, L83
Prieto, J. L., Brimacombe, J., Drake, A. J., & Howerton, S. 2013, *ApJ*, 763, 27
Quimby, R. M., Fang, Y., Akerlof, C., & Wheeler, J. C. 2013, *arXiv*, arXiv:1302.0911
Quimby, R. M., *et al.* 2011, *Nature*, 474, 487
Smith, N. & McCray, R. 2007, *ApJ* (Letter), 671, L17
Smith, N., *et al.* 2010, *ApJ*, 686, 485
Woosley, S. E., Blinnikov, S., & Heger, A. 2012, *Nature*, 450, 390

Discussion

CHEVALIER: You discussed the dip seen in the early light curve of SN 2006oz. What are your thoughts on explaining the plateau before the dip?

MORIYA: There can be another CSM component inside the dense CSM powering the main LC and that component may power the plateau before the dip.

SAHA, L.: 1. Is there any limit on the mass of the ejecta so that SN could be superluminous or not? 2. How does density gradient help for this kind of SLSN scenario or what are the implications of CSM density gradient for SLSN?

MORIYA: 1. Ejecta mass should be similar to or less than the CSM mass. This is because ejecta must be decelerated by CSM to make SLSNe bright. 2. From the LC point of view, the density gradient can affect the LC decline after the peak. The density gradient also affects spectra and I refer Moriya & Tominaga (2012) for details.

Supernova Environmental Impacts
Proceedings IAU Symposium No. 296, 2013
A. Ray & R. A. McCray, eds.

doi:10.1017/S1743921313009289

Distance determination to six nearby galaxies using type IIP supernovae.

Subhash Bose and Brijesh Kumar

Aryabhatta Research Institute of Observational Sciences, Nainital, India
email: bose@aries.res.in; email@subhashbose.com

Abstract. We use early optical photometric and spectroscopic data of six Type-IIP SNe to derive distances to their host galaxies using the expanding photosphere method (EPM). Our sample consists of luminous to sub-luminous SNe 1999gi, 2004et, 2005cs, 2008in, 2009md and 2012aw; having absolute V-magnitudes from -17 to -15 and host galaxy distances from 5 to 22 Mpc. The SN 2008in is peculiar in nature showing dual behavior of a luminous as well as sub-luminous event. The EPM distances for four of the events in our sample are derived for the first time. We take utmost care in minimizing the errors arising from photospheric velocity determination and the broadband filter responses, hence leaving out uncertainty in dilution factor models as the only major source of error. Our preliminary results indicate that EPM-derived distances using Dessart model is found to be consistent with the distances quoted in the literature. We find that EPM method is applicable only to the early (<50 d) photometric data of supernovae and dense spectroscopic data is necessary to estimate accurate distances.

Keywords. galaxies: distances and redshifts, (stars:) supernovae: general,(stars:) supernovae: individual (2012aw, 2004et, 2005cs, 2008in, 2009md, 1999gi)

---

1. Introduction

Supernovae owing to their high luminosity, are the objects of interest for extragalactic distance estimation. Type II supernovae are not used traditionally as standard candles for distance measurement. However, in case of type II-P SNe, assuming spherically symmetric expansion of their ejecta and radiating isotropically as a blackbody at a well defined temperature, we can apply the Expanding Photosphere Method (EPM; Kirshner & Kwan 1974) a variant of Baade-Wesselink method to determine distances to their host galaxies.

1.1. *Expanding Photosphere Method*

The EPM is fundamentally a geometrical technique, in which we compare the linear radii determined from the expansion velocity and angular radii of the supernova by fitting blackbody with the observed fluxes at different epochs.

Assuming homologous expansion, we may relate photospheric velocity v_{phot}, angular radius θ and distance D at time t and t_0 be the explosion epoch, we may write

$$\theta = \frac{v_{phot}(t - t_0)}{D}$$
$$t = D\left(\frac{\theta}{v_{phot}}\right) + t_0 \tag{1.1}$$

Hence we get a linear equation, whose slope yields the distance to SN (D) and the $y-$intercept as the explosion epoch (t_0).

To determine θ we assume SN radiating isotropically as blackbody and accounting for the conservation of radiative energy we may write,

$$4\pi R^2 \cdot \pi B_\lambda(T_c) = 4\pi D^2 f_\lambda^{dered} \tag{1.2}$$

where $B_\lambda(T_c)$ is Planck Blackbody function at color temperature T_c and f_λ^{dered} is the extinction corrected (de-reddened) observed flux. Absorbing the R and D into θ we may write,

$$f_\lambda^{dered} = \theta^2 \pi B_\lambda(T_c) \tag{1.3}$$

Now introducing the wavelength dependent extinction A_λ and writing in terms of observed flux f_λ,

$$f_\lambda = \theta^2 \pi B_\lambda(T_c) 10^{-0.4A_\lambda} \tag{1.4}$$

On minimizing the above equation with two or more known observed fluxes f_λ^{obs} and the Planckian blackbody model function we can determine the angular radius θ and the color temperature T_c both together. Thus to derive distance by EPM, all we need are observed flux f_λ and photospheric velocity v_{phot}.

One of the most important assumption that goes into EPM, is to consider the expanding photosphere radiating as blackbody. There might be a significant departure from a blackbody atmosphere and this will be directly reflected into the distance estimated by EPM. The thermalization layer from which the thermal photons are generated is significantly deeper than photospheric layer from which photons start to flow freely without any further scattering (i.e. the surface of last scattering at $\tau = \frac{2}{3}$), to take care of this discrepancy, we use "dilution factor" ξ as

$$\xi = \frac{R_{therm}}{R_{phot}} \tag{1.5}$$

and rewrite the equation (1.4) as,

$$f_\lambda = \xi^2 \theta^2 \pi B_\lambda(T_c) 10^{-0.4A_\lambda} \tag{1.6}$$

In principle this factor should depend upon many physical properties including chemical properties, density profile of the SNe etc. However, studies have shown (Eastman *et al.* 1996) that ξ more or less behaves as one-dimensional function of color temperature, T_c only. The computation of ξ requires realistic SN atmosphere models. Till date, two prescription for dilution factors are available, Hamuy *et al.* (2001) (hereafter H01) which is improved estimate of ξ over Eastman *et al.* (1996) used 63 stellar atmospheric models, whereas the other Dessart & Hillier (2005) (hereafter D05) which uses CMFGEN models for SN to determine ξ for different filter pass-band combinations.

In order to improve the accuracy of our study, we used `SYNOW` (Branch *et al.* 2001, Branch *et al.* 2001, Elmhamdi *et al.* 2006) to model the spectra for each of the event to determine their photospheric velocities v_{phot} at different phases. Further to remove the effect of filter response which is intrinsically embedded in observed flux, we convolve the response function $\Re_\lambda(\lambda')$ for each pass-band filter with the blackbody model to obtain the synthetic model flux. be the normalized response function of a particular filter whose effective wavelength is λ, then the convolved synthetic flux b_λ is,

$$b_\lambda(T_c) = \int_0^\infty \Re_\lambda(\lambda') \pi B(\lambda', T_c) d\lambda' \tag{1.7}$$

Hence in the equation 1.6, the blackbody flux is replaced with convolved blackbody flux b_λ for each filter and rewritten as,

$$f_\lambda = \xi^2 \theta^2 b_\lambda(T_c) 10^{-0.4A_\lambda} \tag{1.8}$$

Table 1. Adopted parameters

SN	E(B-V)	Reference epoch (JD)	Recession Velocity ($km\ s^{-1}$)
SN1999gi	0.21	2451518.3 ± 3.1	552
SN2004et	0.41	2453270.5 ± 0.9	45
SN2005cs	0.05	2453549.0 ± 1.0	463
SN2008in	0.098	2454825.6 ± 2.0	1567
SN2009md	0.1	2455162.0 ± 8.0	1308
SN2012aw	0.075	2456002.6 ± 0.8	778

Table 2. Table of derived EPM distances to host galaxies and comparison with NED result.

Host galaxy	SN event	EPM Distance[a] Mpc	Distance Modulus	NED Distance[b] Mpc
NGC 3351/M95	SN 2012aw	9.83 ± 0.41	29.96 ± 0.09	10.11 ± 0.98
NGC 6946	SN 2004et	5.86 ± 0.76	28.84 ± 0.28	5.96 ± 1.97
NGC 3184	SN 1999gi	11.62 ± 0.29	30.33 ± 0.05	11.95 ± 2.71
NGC 4303/M61	SN 2008in	14.51 ± 1.38	30.81 ± 0.21	16.46 ± 10.84
NGC 5194/M51a	SN 2005cs	8.01 ± 0.62	29.52 ± 0.17	7.91 ± 0.87
NGC 3389	SN 2009md	23.29 ± 1.96	31.84 ± 0.18	21.29 ± 2.21

[a] EPM distances are using D05 prescription.
[b] NED (`http://ned.ipac.caltech.edu/`) distance is mean value of all redshift independent distances listed in NED for the galaxy. Errors are the STD DEV of the listed distances.

2. Sample selection and data

For the EPM study we select a sample of six recent II-P SNe viz., 1999gi (Leonard *et al.* 2002), 2004et (Sahu *et al.* 2006), 2005cs (Pastorello *et al.* 2006, Pastorello *et al.* 2009), 2008in (Roy *et al.* 2011), 2009md (Fraser *at al.* 2011) and a very recent event 2012aw (in preparation Bose *et al.*); which comprises of normal to sub-luminous type events having good photometric and spectroscopic follow-up including early plateau phase. Among these, 1999gi, 2004et and 2012aw are normal events; 2005cs and 2009md are sub-luminous whereas 2008in lies in between normal and sub-luminous II-P events. Th adopted parameters from literature are tabulated in Table 1, $E(B-V)$are used to de-redden corresponding photometric and spectroscopic data, *Recession velocity* are used to doppler correct spectra and *reference epoch* are the explosion epochs adopted in corresponding literature, in reference to these epochs we independently estimate explosion epochs from EPM.

3. Preliminary Results

EPM is applied to each of these six events and corresponding distance and explosion epoch are determined for each set of dilution factor models viz., D05 and H01 with each of three filters subsets *BV, BVI* and *VI.* For the SNe 2012aw, 2005cs and 2004et, EPM is also applied by fixing explosion epochs to their observationally constrained explosion epochs having accuracy less than a day and thus keeping Distance as the only free parameter for EPM fit. Figure 1 is the EPM fit for SN 2012aw with both explosion epoch and distance, while the Figure 2 is the EPM fit with fixed explosion epoch having distance as only free parameter. Similar EPM fits are done for all other events and

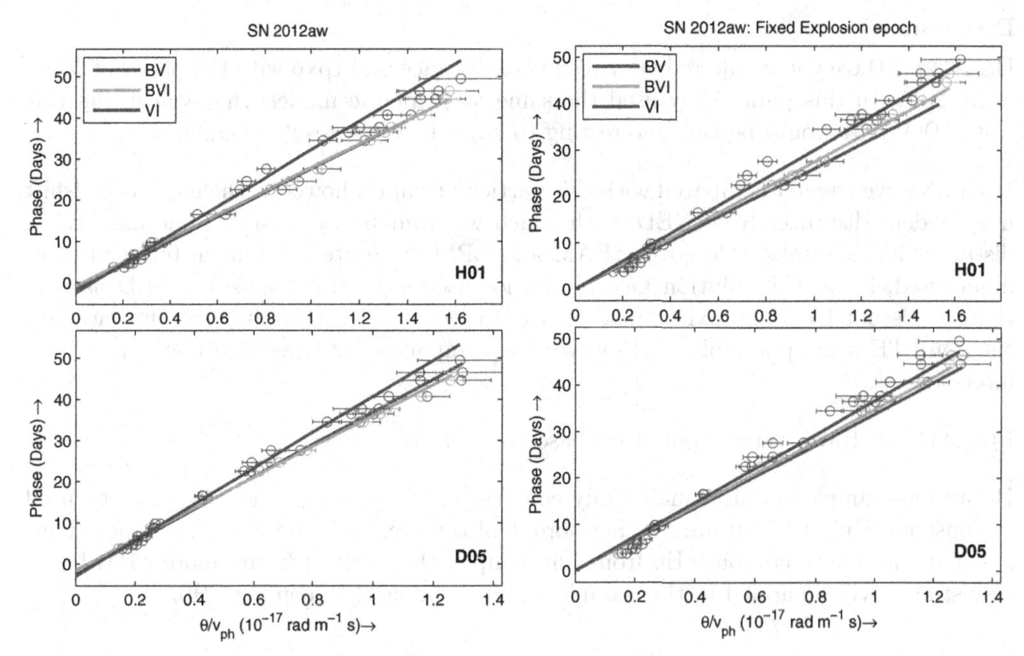

Figure 1. EPM fitting for SN 2012aw using both D05 and H01 prescriptions.

Figure 2. EPM fitting for SN 2012aw using both D05 and H01 prescriptions with fixed explosion epoch and Distance as the only free parameter.

EPM determined distances using D05 prescription are tabulated in Table 2 (a detailed analysis is in preparation, Bose *et al.*). The EPM distances using D05 are found to be more consistent than those by H01 with NED listed distances for corresponding galaxies.

4. Conclusion

The D05 prescription of dilution factors are found to be more suitable and accurate for EPM analysis. Also the SYNOW derived photospheric velocities significantly improve the accuracy of distance determination.

References

Branch, D., Baron, E., & Jeffery, D. J. 2001, arXiv:astro-ph/0111573

Branch, D., Benetti, S., Kasen, D., *et al.* 2002, *ApJ*, 566, 1005

Dessart, L. & Hillier, D. J. 2005, *A&A*, 439, 671

Eastman, R. G., Schmidt, B. P., & Kirshner, R. 1996, *ApJ*, 466, 911

Elmhamdi, A., Danziger, I. J., Branch, D., *et al.* 2006, *A&A*, 450, 305

Fraser, M., Ergon, M., Eldridge, J. J., *et al.* 2011, *MNRAS*, 417, 1417

Hamuy, M., Pinto, P. A., Maza, J., *et al.* 2001, *ApJ*, 558, 615

Kirshner, R. P. & Kwan, J. 1974, *ApJ*, 193, 27

Leonard, D. C., Filippenko, A. V., Li, W., *et al.* 2002, *AJ*, 124, 2490

Pastorello, A., Sauer, D., Taubenberger, S., *et al.* 2006, *MNRAS*, 370, 1752

Pastorello, A., Valenti, S., Zampieri, L., *et al.* 2009, *MNRAS*, 394, 2266

Roy, R., Kumar, B., Benetti, S., *et al.* 2011, *ApJ*, 736, 76

Sahu, D. K., Anupama, G. C., Srividya, S., & Muneer, S. 2006, *MNRAS*, 372, 1315

Discussion

BERSTEN: Have you compared your results of distance and tpxp with the paper of Jones *et al.* 2010. In this paper they used the same atmospheric models that you use in this work (DO5)? It would be very interesting to know how the results compare.

BOSE: No, we haven't compared with this particular paper, however, the mean of redshift independent distances from NED, with which we compare our result, some has SEAM distances included also. Moreover SEAM and EPM are quite different in terms of technique used, in SEAM, dilution factors are not used separately, since the SED used in this case is not LTE approximation, the SED used in SEAM, itself taken into account the non-LTE atmosphere of SN. However the comparison of these two methods will be interesting.

FOLLATELLI: Have you computed Ho based on your distances?

BOSE: Our sample is quite small. Only comprising of 6 events, hence it cannot be used to construct Hubble diagram. Furthermore, 5 of the events are nearby (<15 Mpc). Thus it is not suitable to compute H0 from this sample. However, in future more EPM based new studies will be added to this sample and may be used to compute H0.

Supernova Environmental Impacts
Proceedings IAU Symposium No. 296, 2013
A. Ray & R. A. McCray, eds.

doi:10.1017/S1743921313009290

Supernova interaction with dense mass loss

Roger A. Chevalier

Dept. of Astronomy, University of Virginia
P.O. Box 400325, Charlottesville, VA 22903, USA
email: rac5x@virginia.edu

Abstract. Supernovae of Type IIn (narrow line) appear to be explosions that had strong mass loss before the event, so that the optical luminosity is powered by the circumstellar interaction. If the mass loss region has an optical depth $> c/v_s$, where v_s is the shock velocity, the shock breakout occurs in the mass loss region and a significant fraction of the explosion energy can be radiated. The emission from the superluminous SN 2006gy and the normal luminosity SN 2011ht can plausibly be attributed to shock breakout in a wind, with SN 2011ht being a low energy event. Superluminous supernovae of Type I may derive their luminosity from interaction with a mass loss region of limited extent. However, the distinctive temperature increase to maximum luminosity has not been clearly observed in Type I events. Suggested mechanisms for the strong mass loss include pulsational pair instability, gravity-waves generated by instabilities in late burning phases, and binary effects.

Keywords. (stars:) supernovae: general

1. Introduction

Once the radiation from the supernova shock wave emerges from the stellar surface, the interaction with the surrounding medium commences. In the case of core collapse supernovae, the initial interaction is with mass loss from the progenitor star. At low circumstellar densities, the emission from the interaction is a small part of the supernova power and typically involves X-ray and radio radiation. This is the case for Type IIP supernovae, where the inferred circumstellar density is roughly consistent with that expected from the red supergiant progenitors of the these events (Chevalier *et al.* 2006). However, there are supernovae, of Type IIn, that are thought to have most of their peak optical power come from interaction with a circumstellar medium (e.g., Chugai & Danziger 1994). Here I discuss cases where the circumstellar interaction is an important part of the optical emission. The shock breakout phase is treated in Section 2 and the initial viscous shock propagation in Section 3. Speculations on the formation of the dense circumstellar medium are discussed in Section 4 and the final comments in Section 5.

2. Shock Breakout

A useful way of considering the enhanced supernova luminosity that can result from very dense circumstellar interaction is in terms shock breakout in the dense gas (Chevalier & Irwin 2011, Balberg & Loeb 2011). While a shock front is inside the supernova progenitor, the shock front is mediated by radiation for standard supernova conditions. The optical depth across the shock transition for a radiation dominated shock is $\sim c/v_s$, where v_s is the shock velocity and c is the speed of light. This result is determined by the condition that the hydrodynamic time for the shock wave to cross the shock thickness is approximately equal to the diffusion time for the radiation. When the shock front reaches a position that is at an optical depth of $\sim c/v_s$ from the surface, radiation can diffuse out to optically thin layers and shock breakout occurs. For low circumstellar densities, the timescale for the breakout event is typically quite brief and depends on the radius of the progenitor star, R. For a red supergiant progenitor, the most extended normal progenitor, the supernova shock takes about a day to traverse the star and the breakout phase lasts about half an hour, determined by the light travel time across the star, R/c.

A low density, optically thin circumstellar medium is not expected to have much influence on the optical light near maximum, although it can be very important for radio and X-ray emission. At higher circumstellar densities, the medium becomes optically thick and the breakout radiation must diffuse out through the circumstellar medium. The result is that the radiated energy in the breakout is not changed, but the radiation is emitted over a longer period of time, lowering the luminosity.

When the circumstellar optical depth, τ_w, becomes of order c/v_s or more, the radiation dominated shock can propagate into the circumstellar region. In this case, substantial supernova energy can be present as internal energy at a larger radius than would otherwise be the case, so that the effects of adiabatic expansion are lessened and the luminosity rises. Chevalier & Irwin (2011) suggested that this model can describe the initial rise of 10's of days of superluminous supernovae like SN 2006gy. When the circumstellar interaction occurs at a large radius where diffusion of radiation can occur rapidly and the circumstellar medium is sufficiently massive to thermalize the supernova kinetic energy, a large fraction of the supernova energy can be radiated. The physical parameters, a circumstellar medium of $\sim 10\ M_\odot$ extending out to $\sim 10^{16}$ cm, are similar to those in the diffusion model of Smith & McCray (2007). Chevalier & Irwin (2011) used an analytic model for the emission, which has limitations; in particular, the initial interaction was described by a self-similar solution, but that solution breaks down as the reverse shock wave propagates into the supernova ejecta. Ginzburg & Balberg (2012) undertook numerical simulations that did not have this problem and obtained comparable parameters for SN 2006gy. In their simulations, Moriya *et al.* (2013) found that a density law $\rho \propto r^{-2}$ did not give a good fit to light curve; their favored explosion parameters are mass loss as high as $15\ M_\odot$ and energy no more than 4×10^{51} ergs.

A general expectation of the shock breakout model is that the initial rise to maximum luminosity should be primarily due to heating of the photosphere as the shock wave begins to affect the stellar emission. This effect is not clearly observed in the data on SN 2006gy, although there is some sign of an increasing temperature in the spectra of Smith *et al.* (2010).

A Type IIn event that apparently does show the increasing temperature in the rise to maximum is SN 2011ht. Early Swift observations show a clear rise in temperature to maximum (Roming *et al.* 2012), which is characteristic of shock breakout in the circumstellar gas. The rise time to maximum, 40 days, is an indicator of the density of the circumstellar gas. The supernova radiation is expected to be efficiently radiated, so that the radiated energy is a significant fraction of the supernova energy if the energy is thermalized by the interaction.. Interestingly, the light curve is similar in shape to that of SN 2006gy, but the absolute magnitude is fainter by 5 magnitudes, with a radiated energy $\sim 2.5 \times 10^{49}$ ergs. The implication is that the explosion energy is smaller by a factor up to $\sim 10^2$. The possibility that the event was not a supernova was noted by Roming *et al.* (2012). However, there are supernovae that occur in this low energy range and Mauerhan *et al.* (2013) suggest that the late emission from SN 2011ht is due to a small amount of ^{56}Ni. In the end, it is not possible to definitively answer the question of whether SN 2011ht was a supernova.

For luminous supernovae that are of Type IIn, it is very plausible that the high optical luminosity is produced by circumstellar interaction (see next section). There are also luminous supernovae that are not of the "n" type. SN 2008es is a luminous Type II (Gezari *et al.* 2009, Miller *et al.* 2009) and SN 2010gx is an example of a luminous Type Ib/c (Pastorello *et al.* 2010). The Type Ib/c objects outnumber those of Type II. Chevalier & Irwin (2011) suggested that the optical light from these objects is powered by circumstellar interaction in which the mass loss region ends before the diffusion radius is reached by the shock front. Ginzburg & Balberg (2012) carried out numerical simulations of this situation and successfully modeled the optical light curve from SN 2010gx. As described above, an expectation in this model is that the temperature should increase in the rise to maximum luminosity. In the case of SN 2006oz, there are observations on the rise to maximum and there is no evidence for an increasing temperature; the temperature remains roughly constant (Leloudas *et al.* 2012). Another mechanism, such as magnetar power (Kasen & Bildsten 2010), may be indicated.

Another superluminous event of interest is PS1-10afx (Chornock *et al.* 2013). In this case, the luminosity and temperature at maximum implied a photospheric radius of 5×10^{15} cm, while the observed lines indicated an expansion velocity of 11,000 km s^{-1}. The implied expansion timescale is 50 days, but the rise time for the light curve is $\sim 10 - 15$ days. This evolution is naturally explained in the shock breakout view (Chevalier & Irwin 2011); the shock front must traverse the optically thick star before shock breakout occurs. However, the rise to luminosity maximum did not show the increasing temperature that might be expected in the breakout scenario and there is no spectral evidence for circumstellar interaction, so the nature of the event is still in doubt.

Overall, the shock breakout scenario is attractive for the events that show narrow line spectra because there is evidence for slowly expanding circumstellar matter ahead of the shock front. When these lines are not present, there is no clear evidence for dense circumstellar matter. There may be more than one mechanism that gives rise to very luminous supernovae.

3. Viscous Shock Wave

Once the forward shock wave gets into a regime where the optical depth to the surface $\tau < c/v_s$, a radiation dominated shock can no longer be maintained and there is a transition to a viscous shock front. There may initially be radiative acceleration of the unshocked gas, so that a viscous shock does not initially form, but, if the circumstellar medium is extended, the formation of a viscous shock in inevitable. The maximum optical depth at which it can form is somewhat less than c/v_s.

At low circumstellar densities the emission from the reverse shock dominates that from the forward shock because the density is higher and the shock velocity is lower, ~ 1000 km s^{-1} or less. However, as the circumstellar density rises, the reverse shock becomes radiative first, which leads a lower increase in luminosity with increasing density and to a dense shell that can absorb radiation from the reverse shock, thus affecting the X-ray luminosity.

The forward shock, with a velocity of 1000's of km s^{-1} or more, can heat gas to a temperature of $100v_4^2$ KeV, where v_4^2 is the shock velocity in units of 10^4 km s^{-1}. Katz *et al.* (2011) found that inverse Compton cooling can balance the shock heating at a temperature of 60 keV if the optical depth is c/v_s (breakout conditions) and the shock velocity is 10^4 km s^{-1}. This estimate assumed that the electrons are heated only by Coulomb collisions with the ions. If there is additional collisionless heating, the gas temperature is raised. Chevalier & Irwin (2012) estimated regions of the shock velocity vs circumstellar density plane where the cooling time is short compared to the age and where inverse Compton cooling is larger than bremsstralung cooling. The regime where the cooling is rapid roughly overlaps the regime where the electron scattering optical depth is >1. If inverse Compton losses are larger than bremsstrahlung, the X-ray emission from the hot shocked gas is reduced as a fraction of the shock power.

The X-rays that are emitted by the hot gas have to propagate through the preshock circumstellar medium. One effect of the surrounding medium is Compton recoil (Chevalier & Irwin 2012, Svirski *et al.* 2012), which reduces the energies of the high energy photons to a maximum of $\sim 511/\tau_{es}^2$ keV, where τ_{es} is the electron scattering optical depth and is assumed > 1. It can be seen that $\tau_{es} > 8$ is necessary to have an effect in the energy range around 10 keV that is accessible to X-ray telescopes like *Chandra*. *NuSTAR*, with its sensitivity to 80 keV, has a better chance of detecting this effect.

Another effect is photoabsorption by the preshock medium (Chevalier & Irwin 2012), which has a larger effect at low X-ray energies. An important aspect of photoabsorption is that its occurrence requires that the absorbing medium not be completely ionized. For the X-ray emission that is expected from the hot gas, there is the possibility of complete ionization (Chevalier & Irwin 2012). Complete ionization is expected at higher shock velocities $\sim 10^4$ km s^{-1}, but not at lower velocities, <5000 km s^{-1}. The result also depends on clumping and asymmetries in the circumstellar gas. If the gas is not fully ionized, the column density corresponding to an electron scattering optical depth τ_{es} is $3 \times 10^{23} \tau_{es} (Z/Z_\odot)^{-1}$ cm^{-2}, where Z is the supernova metallicity and $Z_\odot$ is the solar metallicity. Even for $\tau_{es} = 1$, the column density is orders of magnitude larger than a typical interstellar column density and would have a dramatic effect on X-ray emission.

A supernova that provides some test of these models is the Type IIn (narrow line) SN 2010jl. The optical spectrum showed narrow, presumably circumstellar, emission lines on top of broad wings that could be attributed to electron scattering (Smith *et al.* 2010). The scattering requires an electron scattering optical depth $\gtrsim 1$. An X-ray spectrum with the *Chandra* observatory at 2 months required a column density $\sim 10^{24}$ cm^{-2}, (Chandra *et al.* 2012). At an age of 12 months, the column dropped to $\sim 3 \times 10^{23}$ cm^{-2}, confirming that the absorption is connected to circumstellar gas. The observations with *Chandra* showed a hot thermal component with temperature $\gtrsim 10$ keV, corresponding to a shock velocity $\gtrsim 3000$ km s^{-1}. The emission is presumably from the forward shock front in SN 2010jl.

The presence of the large column density in SN 2010jl shows that it is possible for a large fraction of the X-ray emission to be extinguished. The absorption in SN 2010jl corresponds to $\tau_{es} \sim 1$; larger values of τ_{es} are expected close to the time of shock breakout in a wind. The expected absorption optical depth at 10 keV is $\sim 1.5\tau_{es}(Z/Z_\odot)$ if the circumstellar region is not fully ionized. Chevalier & Irwin (2012) noted that the observed low X-ray luminosity of SN 2006gy, $\lesssim 10^{40}$ ergs s^{-1} (Smith *et al.* 2007; Ofek *et al.* 2007) compared to an optical luminosity of 3×10^{44} ergs s^{-1} at a time close to maximum light (Smith *et al.* 2010), is consistent with breakout in a wind. The case of SN 2011ht was also mentioned above as a likely case of a wind breakout. An X-ray luminosity of 10^{39} ergs s^{-1} was initially reported for this object based on *Swift* observations (Roming *et al.* 2012), but higher spatial observations with *Chandra* showed that the source is not coincident with SN 2011ht (Pooley 2012). Taking a conservative upper limit to the X-ray luminosity of 10^{39} ergs s^{-1}, the X-ray emission is again a small fraction of the optical luminosity, which was 3×10^{42} ergs s^{-1}. The strong suppression of the X-ray emission in SN 2006gy and SN 2011ht is consistent with both of these objects being shock breakouts in a wind, although the explosion energy is probably much smaller in SN 2011ht.

Dwarkadas & Gruszko (2012) make the point that the evolution of X-ray emission from supernovae is typically not what is expected for a steady wind, so that estimates of mass loss rate can be misleading. However, the preshock density at any time can be estimated from $\rho_0 = L/(2\pi\eta R^2 v_s^3)$, where η is the efficiency of conversion of the shock power into the luminosity L. If observations are made over a period of time, an estimate of the mass required for the luminosity can be made and if the velocity of the mass loss can be estimated from narrow lines, the timescale for the emission can be obtained. The mass and timescale estimates provide useful constraints on possible mechanisms for the mass loss.

An interesting aspect of the viscous shock front in dense mass loss is that the shock wave is expected to be collisionless, so that diffusive shock acceleration of particles to relativistic energies can occur (Katz *et al.* 2011, Murase *et al.* 2011). These studies show that an unusually nearby dense interaction supernova would have to occur to have a chance of detecting high energy gamma-ray emission from the supernova.

4. Origin of the Dense Mass Loss

Many of the characteristics of Type IIn supernovae require an optical depth to electron scattering $\geqslant 1$ in the circumstellar medium, which can be expected to produce wings to narrow line profiles and obscure the inner high velocities related to the supernova (if present). Typical parameters for Type IIn events involve masses of $0.01 - 10\ M_\odot$ on scales of $10^{15} - 10^{16}$ cm from the supernova. As noted above, a circumstellar mass as high as $10 - 20\ M_\odot$ has been estimated for SN 2006gy. These events have been linked to luminous blue variables (LBVs) because they are some of the only objects known to have such extreme mass loss (Kiewe *et al.* 2012). In addition, the Type IIn SN 2005gl was found to have a probable LBV progenitor in pre-explosion images (Gal-Yam & Leonard 2009). Although this link has been made, there is still the issue that the reason for the extreme mass loss from LBVs is not understood. In addition, stellar evolution models do not predict supernova explosions at the time of the LBV phase. The timing of the explosion close in time to the mass loss is a remarkable feature of the Type IIn events.

One explanation for SN 2006gy was pulsational pair instability of massive stars (Woosley *et al.* 2007). In this model, there is no terminal explosion, but the emission results from the interaction of pulsationally driven shells. The shells can have a mass of several $M_\odot$ and the radiated energy from their interaction can be 10^{50} ergs.

An interesting proposal by Quataert & Shiode (2012) is that super-Eddington fusion reactions at the end of the life of a massive star can generate gravity waves that deposit their energy/momentum in the outer parts of the star, driving strong mass loss. The prime burning phases for this are Ne and O burning, which occur in the last year of evolution before core collapse. However, only a fraction of massive stars show evidence for the late mass loss and, in this hypothesis, it is unclear what determines the fraction that show the mass loss. A variant of this hypothesis is that the burning instability leads to expansion of the star instead of mass loss, and the mass loss is driven by binary interaction (Soker 2013).

Ofek *et al.* (2013) found emission from SN 2010mc in the 40 days leading up to the explosion and suggested that the observations support the gravity wave driven mass loss hypothesis. The reasons are that the circumstellar velocities are relatively high, ~ 2000 km s^{-1} and that the brightening was observed 40 days before explosion, when the late unstable burning phases are expected. However, Smith *et al.* (2013) show that the light curve of SN 2010mc is very much like that of SN 2009ip during the period June - Sept 2012, indicating that they involve a similar physical situation. SN 2009ip showed eruptive behavior over the period 2009 to 2012, somewhat longer than might be expected for the Ne and O burning phases. Also, high velocities observed in spectral lines are possibly the result of electron scattering. In any case, Fraser *et al.* (2013) note that there is some question whether SN 2009ip was in fact a supernova in Sept 2012, a question that can also be raised for SN 2010mc.

Another possibility for the strong mass loss is that it is driven by close binary interaction. Common envelope evolution is expected in close binaries, accompanied by strong mass loss. This has the advantage that there can be considerable variety in the events. Also, recent observations have shown that massive stars are in close binaries more frequently than had been thought (Sana *et al.* 2012). The problem is to have an explosion close to the time that the mass loss occurs. Chevalier (2012) speculated that the event involves the spiral in of a neutron star due to common envelope evolution. In cases where the interaction occurs at a relatively small separation, the neutron star spirals into the core of its companion because of the steep density gradient. In cases where there is spiral in, but the separation is somewhat greater, the companion star can become a red supergiant with a flat density profile, which causes spiral in to stop. The result is a neutron star – helium star binary. When the He star evolves, there is again the possibility of mass loss accompanied by the spiral in of the neutron star to the core.

In one view, the spiral in of the neutron star to the core gives rise to a red supergiant with a neutron star core, a Thorne-Żytkow star (Thorne & Żytkow 1977). However, neutrino losses of the matter near the neutron star may lead to strong accretion onto the neutron star, especially when the neutron star is in the core of the companion star (Chevalier 1996). The rapid accretion may lead to black hole formation and an explosion, as in the model for gamma-ray bursts (GRBs) of Fryer & Woosley (1998). A variant of this model is that the rapid accretion is accompanied by strong magnetic field amplification of the neutron star, leading to an explosion (Barkov & Komissarov 2011).

In this scenario for explosions preceded by strong mass loss, the explosion mechanism is not the neutrino mechanism, which is generally favored for most core collapse supernovae, but is related to rapid rotation and accretion on a central compact object. Spiral in of a compact object is an efficient way of achieving rapid rotation in the central region. The mechanism is believed to operate for GRBs, which have been associated with energetic ($>10^{52}$ ergs) supernovae. Some Type IIn supernovae also appear to have had unusually large energies; for example, SN 2003ma had an integrated bolometric luminosity of 4×10^{51} ergs over 4.7 years and an estimated explosion energy $> 10^{52}$ ergs (Rest *et al.* 2011). Studies of the neutrino mechanism for core collapse supernova explosions indicate that the maximum energy that can be attained is $\sim 2 \times 10^{51}$ ergs (Janka 2012). Both GRBs and some Type IIn supernovae may require an explosion mechanism involving rotation and magnetic fields in order to produce high energy explosions.

5. Discussion

Although circumstellar interaction at low densities can be successfully described in terms of spherical models, there are signs that at the higher densities considered here, where the interaction contributes to the optical luminosity, the situation is more complex. Galactic LBVs show a complex circumstellar structure, binary evolution can lead to aspherical strcuture (as apparently occurred in SN 1987A), and Type IIn supernovae often show significant polarization in their optical light. In addition to these complexities, the Type IIn events appear to include a wide range of stellar masses and explosion energies, indicating a variety of evolutionary paths. We are still at the early stages of understanding these events.

Acknowledgement

I am grateful to my collaborators for their participation in this research, and to NASA grant NNX12AF90G and NSF grant AST-0807727 for support.

References

Balberg, S. & Loeb, A. 2011, *MNRAS*, 414, 1715
Barkov, M. V. & Komissarov, S. S. 2011, *MNRAS*, 415, 944
Chandra, P., Chevalier, R. A., Irwin, C. M., *et al.* 2012, *ApJ*(Letters), 750, L2
Chevalier, R. A. 1996, *ApJ*, 459, 322
Chevalier, R. A. 2012, *ApJ*(Letters), 752, L2
Chevalier, R. A. & Irwin, C. M. 2011, *ApJ*(Letters), 729, L6
Chevalier, R. A. & Irwin, C. M. 2012, *ApJ*(Letters), 747, L17
Chevalier, R. A., Fransson, C., & Nymark, T. K. 2006, *ApJ*, 641, 1029
Chornock, R., Berger, E., Rest, A., *et al.* 2013, preprint, arXiv:1302.0009
Chugai, N. N. & Danziger, I. J. 1994, *MNRAS*, 268, 173
Dwarkadas, V. V. & Gruszko, J. 2012, *MNRAS*, 419, 1515
Fraser, M., Inserra, C., Jerkstrand, A., *et al.* 2013, *MNRAS*, submitted, arXiv:1303.3453
Fryer, C. L. & Woosley, S. E. 1998, *ApJ*(Letters), 502, L9
Gal-Yam, A. & Leonard, D. C. 2009, *Nature*, 458, 865
Gezari, S., Halpern, J. P., Grupe, D., *et al.* 2009, *ApJ*, 690, 1313
Ginzburg, S. & Balberg, S. 2012, *ApJ*, 757, 178
Janka, H.-T. 2012, *ARNPS*, 62, 407
Kasen, D. & Bildsten, L. 2010, *ApJ*, 717, 245
Katz, B., Sapir, N., & Waxman, E. 2011, preprint, arXiv:1106.1898
Kiewe, M., Gal-Yam, A., Arcavi, I., *et al.* 2012, *ApJ*, 744, 10
Leloudas, G., Chatzopoulos, E., Dilday, B., *et al.* 2012, *A&A*, 541, A129
Mauerhan, J. C., Smith, N., Silverman, J. M., *et al.* 2013, *MNRAS*, 1053
Miller, A. A., Chornock, R., Perley, D. A., *et al.* 2009, *ApJ*, 690, 1303
Moriya, T. J., Blinnikov, S. I., Tominaga, N., *et al.* 2013, *MNRAS*, 428, 1020
Murase, K., Thompson, T. A., Lacki, B. C., & Beacom, J. F. 2011, *Phys. Rev. D*, 84, 043003
Ofek, E. O., Cameron, P. B., Kasliwal, M. M., *et al.* 2013, *ApJ*(Letters), 659, L13
Ofek, E. O., Sullivan, M., Cenko, S. B., *et al.* 2013, *Nature*, 494, 65
Pastorello, A., Smartt, S. J., Botticella, M. T., *et al.* 2010, *ApJ*(Letters), 724, L16
Pooley, D. 2012, The Astro. Tel., 4062, 1
Quataert, E. & Shiode, J. 2012, *MNRAS*, 423, L92
Rest, A., Foley, R. J., Gezari, S., *et al.* 2011, *ApJ*, 729, 88
Roming, P. W. A., Pritchard, T. A., Prieto, J. L., *et al.* 2012, *ApJ*, 751, 92
Sana, H., de Mink, S. E., de Koter, A., *et al.* 2012, *Science*, 337, 444
Smith, N. & McCray, R. 2007, *ApJ*(Letters), 671, L17
Smith, N., Li, W., Foley, R. J., *et al.* 2007, *ApJ*, 666, 1116
Smith, N., Chornock, R., Silverman, J. M., Filippenko, A. V., & Foley, R. J. 2010, *ApJ*, 709, 856
Smith, N., Mauerhan, J. C., Kasliwal, M. M., & Burgasser, A. J. 2013, preprint, arXiv:1303.0304
Soker, N. 2013, preprint, arXiv:1302.5037
Svirski, G., Nakar, E., & Sari, R. 2012, *ApJ*, 759, 108
&Thorne, K. S., Żytkow, A. N. 1977, *ApJ*, 212, 832
Woosley, S. E., Blinnikov, S., & Heger, A. 2007, *Nature*, 450, 390

Discussion

SUTARIA: Concerning SN2011ht there is a controversy about the nature of this event; is it a true SN IIn or is it a SN imposter. Could you please comment?

CHEVALIER: In the shock breakout view of this event, the circumstellar density is high, but the shock velocity, and thus the energy, are low. The model does not clearly answer the SN vs. imposter question.

NOMOTO: What is your model for Type I Superluminous Supernovae?

CHEVALIER: In the common envelope (CE) view, there is an initial CE episode with the neutron star spiraling in the 1f envelope. Depending on parameters, the outcome may be a neutron star He star binary. Eventually the He star expands and there is another spiraling phase for the compact object. This can lead to a Type I SLSN.

SURNIS: What are the chances of detecting the central compact object in X-rays and radio for Galactic SNRs having dense CSM? What if the optical density of the CSM is low?

CHEVALIER: The dense surroundings might make it more difficult to see a compact object because of obscuration, or continuing interaction could compete with central emission. The effect of the region immediately around the center depends on the supernova properties, which are not well understood.

Supernova Environmental Impacts
Proceedings IAU Symposium No. 296, 2013
A. Ray & R. A. McCray, eds.

doi:10.1017/S1743921313009307

X-rays from Core-collapse Supernovae

David Pooley

Department of Physics, Sam Houston State University, Huntsville, Texas 77341
and Eureka Scientific, Inc.
email: dave@shsu.edu

Abstract. Core-collapse supernovae can produce X-rays through a variety of mechanisms, which are briefly reviewed. Through a combination of targeted searches of specific supernovae and archival searches for serendipitous coverage of supernovae, the number of known X-ray supernovae has grown by a factor of five in the past 13 years, when the *Chandra X-ray Observatory* and *XMM-Newton* were launched. The *Swift* satellite has contributed greatly to the discovery of X-ray emitted supernovae, but care must taken with all *Swift* detections given its spatial resolution and the number of X-ray binaries typically seen in external galaxies. About half of the reported *Swift* detections of X-ray emission from supernovae are in fact not due to the supernovae but from unrelated nearby sources in the host galaxies.

Keywords. supernovae: general; stars: mass loss

1. Introduction

Over thirty years ago, X-rays were discovered in the direction of supernova (SN) 1980K (Canizares *et al.* 1982) in the near aftermath (years) of its discovery, marking it the first supernova detected in X-rays at such a young age. The number of additional young supernovae detected in X-rays grew very slowly in the ensuing decade and a half, totaling nine known X-ray emitting supernovae before the launch of the *Chandra X-ray Observatory* and *XMM-Newton* in 1999.

In the past 13 years, the number of X-ray supernovae has grown by more than a factor of five to over 50 because of the combined efforts of *Chandra*, *XMM*, and the *Swift* satellite, which was launched in 2004. Figure 1 shows the growth of the known X-ray supernova population as a function of time. Even with the current generation of very powerful X-ray satellites, only a handful of new X-ray supernova detections are made each year. The sensitivity of a typical X-ray observation may be on the order of 10^{-14} erg cm^{-2} s^{-1}, and the typical luminosity of an X-ray emitting supernova is 10^{38} - 10^{39} erg/s, constraining this field of study to the very local universe (Figure 1).

2. X-ray Production Mechanisms

Core-collapse supernovae start with a burst of X-rays as the shock breaks out from the dying star, seen for the first time in SN 2008D (Soderberg *et al.* 2008; Modjaz *et al.* 2009). The timescale of this event is difficult to explain in a spherically symmetric explosion, but aspherical, jet-driven supernova simulations reproduce both the spectrum and lightcurve of SN 2008D reasonable well (Couch *et al.* 2011). After the shock breakout, strong X-ray emission is seen again at days to months after the explosion, and in some cases persists for years and even decades.

The X-ray emission of type II supernovae is convincingly explained as thermal radiation ($kT \lesssim 10$ keV) from the "reverse shock" region that forms within the expanding SN ejecta as it interacts with the dense stellar wind of the progenitor star. The interaction of a

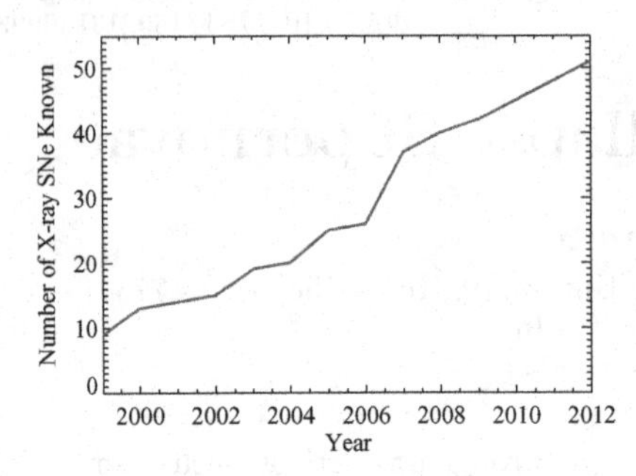

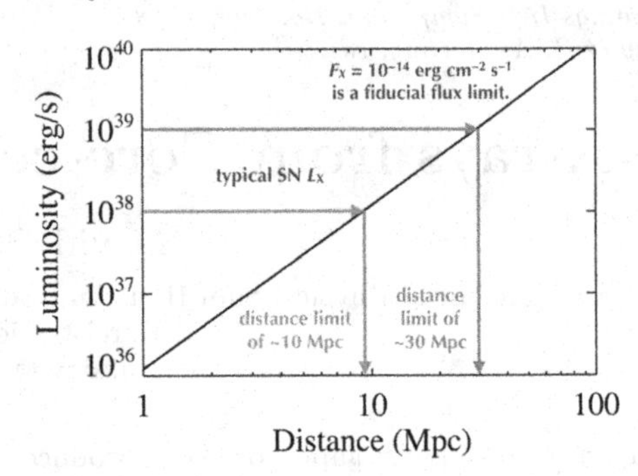

Figure 1. *Left:* The number of young supernovae known to emit X-rays as a function of time since the launch of *Chandra*, *XMM*, and *Swift*. *Right:* The typical X-ray luminosity of a supernova is between 10^{38} – 10^{39} erg/s, and the typical sensitivity of an X-ray observation is 10^{-14} erg cm^{-2} s^{-1}, limiting most observations to within 30 Mpc.

spherically symmetric SN shock and a smooth CSM has been calculated in detail (Chevalier 1982a,b; Chevalier & Fransson 1994; Suzuki & Nomoto 1995; Fransson *et al.* 1996). As the supernova shock emerges from the star, its characteristic velocity is $\sim 10^4$ km/s, and the density distribution in the outer parts of the ejecta can be approximated by a power-law in radius, $\rho \propto r^{-n}$, with $7 \lesssim n \lesssim 20$. The outgoing shock propagates into the dense circumstellar material (CSM) formed by the pre-supernova stellar wind. For red giant progenitors, this wind is slow ($v_w \sim 10$ km/s) and results from a high mass loss rate ($\dot{M} \sim 10^{-4} - 10^{-6}\ M_\odot$/yr). The density for such a wind follows $\rho = \dot{M}/4\pi r^2 v_w$. The collision between supernova ejecta and CSM also produces a "reverse" shock, which travels outward at $\sim 10^3$ km/s slower than the fastest ejecta. Interaction between the outgoing shock and the CSM produces a hot shell ($\sim 10^9$ K), while the reverse shock produces a denser, cooler shell ($\sim 10^7$ K) with much higher emission measure from which most of the observable X-ray emission arises. Within this framework of CSM interaction, Chevalier *et al.* (2006) have shown how X-ray and radio measurements of type IIP SNe

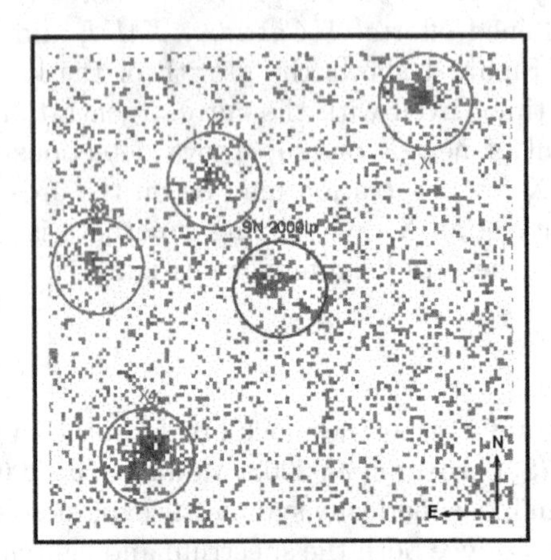

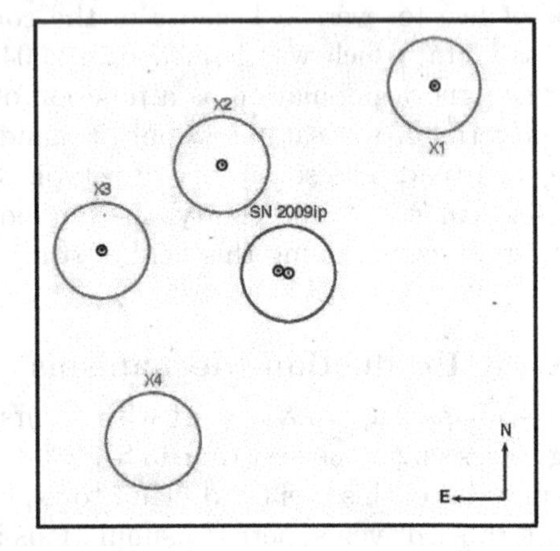

Figure 2. *Left: Swift* XRT image of the field around SN 2009ip, which is indicated with a blue circle. Nearby sources are indicated by red circles. All circles in this image are 1′ in diameter. *Right: Chandra* image of the same field, revealing an unrelated point source about 6″ from 2009ip which is unresolved from it in the *Swift* XRT image. Black circles indicate *Chandra* sources and are 6″ in diameter. Note the extreme variability of source X4 between the *Swift* and *Chandra* observations; such behavior is typical of X-ray binaries, the types of sources expected to be present around the sites of core-collapse supernovae.

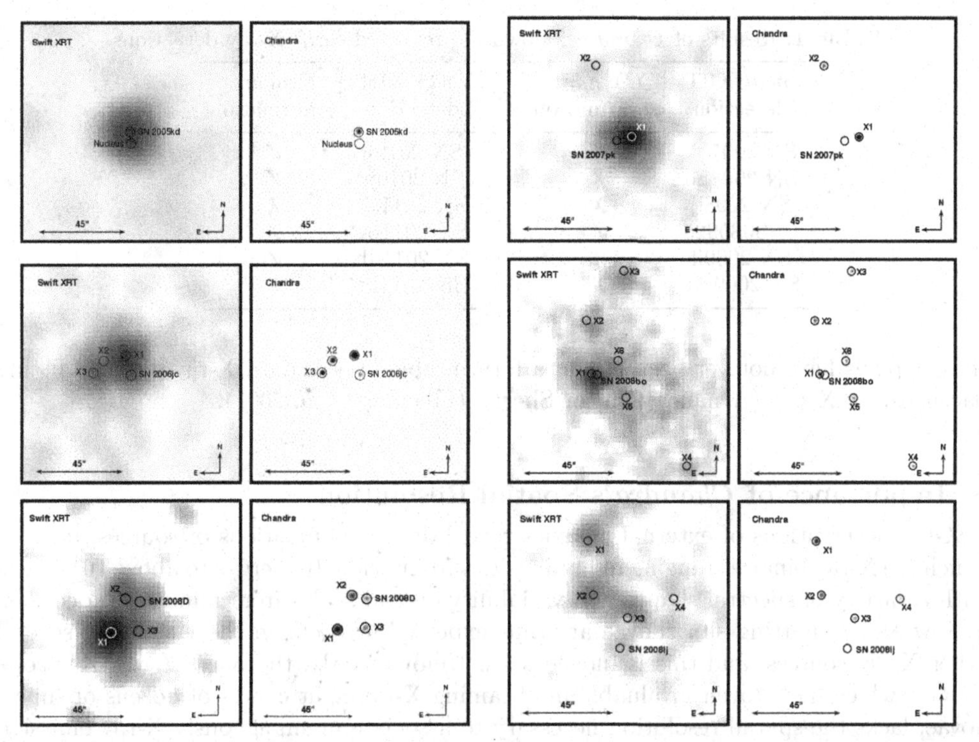

Figure 3. *Swift* XRT and *Chandra* images of 2005kd, 2006jc, and 2008D (left) and 2007pk, 2008bo, and 2008ij (right). Each image is ~1.8′ on a side. In all cases, there is strong X-ray emission around the site of the supernova as seen with the *Swift* XRT. In the supernovae on the left, some amount of the flux is attributable to the SN (although contaminated by the flux from nearby sources). However, in the cases on the right, none of the supernovae were detected with *Chandra*, and all of the *Swift* XRT flux is due to nearby sources.

are excellent probes of the mass loss of the progenitor star. Dwarkadas & Gruszko (2012) give an excellent review of CSM interaction and discuss the expected temporal evolution of the X-ray luminosity.

The origin of X-rays from Ib/c supernovae is less clear; the interaction of the supernova shock with the low-density CSM around Ib/c progenitors results in a reverse shock which is too weak to explain the observed X-rays. Chevalier & Fransson (2006) have suggested an inverse-Compton and synchotron mechanism for the production of X-rays from these supernovae.

The type IIn SNe are perhaps the least understood but can be the most X-ray luminous subtype. The narrow optical lines that characterize the IIn subclass are clear evidence of dense circumstellar gas; they probably arise from reprocessing of X-ray emission. The X-ray emission could result from the shocked ejecta, as in the case of the normal type II SNe, or it could originate from shocked clumps of gas in the CSM (e.g., Chugai 1993). These two scenarios predict vastly different widths for X-ray emission lines, but we have not yet obtained an X-ray spectrum of sufficient quality to make the distinction. For the case where the emission comes from the shocked ejecta, Nymark *et al.* (2006) have shown the complexity of the resulting X-ray spectrum (e.g., see their Fig. 11) and the dangers of using single-temperature spectral models. Their calculations reveal the rich emission-line spectrum and temperature profile of the radiative shocks. Unfortunately,

Table 1. Results of *Chandra* followup of reported *Swift* X-ray detections.

Swift XRT detection	*Chandra* confirmation	*Swift* XRT detection	*Chandra* confirmation
SN 2007pk	✗	SN 2010jl	✓
SN 2008bo	✗	SN 2010jr	✓
SN 2008ij	✗	SN 2011ay	✗
CSS080928	✓	SN 2011by	✗
SN 2009if	✓	SN 2011dh	✓
SN 2009mk	✗	SN 2011ht	✗

these models have not yet been put in a form usable by standard X-ray spectral fitting packages like Xspec (Arnaud 1996) or Sherpa (Freeman *et al.* 2001).

3. Importance of *Chandra*'s Spatial Resolution

X-ray observations of external galaxies reveal diverse populations of sources, most of which are X-ray binaries ranging in luminosities from below 10^{36} erg/s to above 10^{41} erg/s with a variety of spectral shapes and variability on timescales from minutes to decades.

New X-ray emitting supernovae are thus expected to occur in the vicinity of several other X-ray sources, and this is indeed seen. Unfortunately, the *Swift* X-ray Telescope (XRT), which has proven invaluable in obtaining X-ray light curves of dozens of supernovae, lacks the spatial resolution necessary to associate unambiguously X-ray emission in the direction of a young supernova with the supernova itself. For this, *Chandra* is necessary, with its spatial resolution over 30 times better than the *Swift* XRT. A recent example of the necessity of such resolution is SN 2009ip, in which the X-rays seen by *Swift* XRT are due to both 2009ip and an unrelated source about $6''$ away (Figure 2).

For the past five years, my colleagues and I have been awarded *Chandra* observations of new *Swift* XRT supernova detections with the aim of determining whether the X-ray emission is due to the supernova or to nearby sources (or both). We have followed up two or three supernovae a year, and Table 1 lists results. A checkmark indicates that at least some of the X-ray emission seen by *Swift* XRT is due to the supernova, and an x-mark indicates that all of the X-rays seen by the XRT are due to unrelated sources. Less than half of the *Swift* XRT detections are in fact X-ray emitting supernovae.

References

Arnaud, K. A. 1996, *Astronomical Data Analysis Software and Systems V*, 101, 17
Canizares, C. R., Kriss, G. A., & Feigelson, E. D. 1982, *ApJ*, 253, L17
Chevalier, R. A. 1982a, *ApJ*, 258, 790
Chevalier, R. A. 1982b, *ApJ*, 259, 302
Chevalier, R. A. & Fransson, C. 1994, *ApJ*, 420, 268
Chevalier, R. A., Fransson, C., & Nymark, T. K. 2006, *ApJ*, 641, 1029
Chevalier, R. A. & Fransson, C. 2006, *ApJ*, 651, 381
Chugai, N. N. 1993, *ApJ*, 414, L101
Couch, S. M., Pooley, D., Wheeler, J. C., & Milosavljević, M. 2011, *ApJ*, 727, 104
Dwarkadas, V. V. & Gruszko, J. 2012, *MNRAS*, 419, 1515
Fransson, C., Lundqvist, P., & Chevalier, R. A. 1996, *ApJ*, 461, 993
Freeman, P., Doe, S., & Siemiginowska, A. 2001, *Proc. SPIE*, 4477, 76
Modjaz, M., Li, W., Butler, N., *et al.* 2009, *ApJ*, 702, 226
Nymark, T. K., Fransson, C., & Kozma, C. 2006, *A&A*, 449, 171

Pooley, D. 2007, *Supernova 1987A: 20 Years After*, 937, 381
Sell, P. H., Pooley, D., Zezas, A., *et al.* 2011, *ApJ*, 735, 26
Soderberg, A. M., Berger, E., Page, K. L., *et al.* 2008, *Nature*, 453, 469
Suzuki, T. & Nomoto, K. 1995, *ApJ*, 455, 658

Discussion

VINK: Has the inverse Compton emission ever been used in conjunction with the radio to estimate the magnetic fields?

POOLEY: Yes, see Sayan Chakraborti's paper and the next talk by Alak Ray.

CHANDRA: How far away is the x-ray contaminating source in SN 2009ip field?

POOLEY: About 6-7"

MORIYA: Are X-ray faint SNe IIn faint even in optical? Are there any correlations between X-ray luminosity and optical luminosity in SNe IIn.

POOLEY: I have not yet tried to correlate X-ray luminosity with optical luminosity for IIn SNe, but that is something I plan to do in the future.

Supernova Environmental Impacts
Proceedings IAU Symposium No. 296, 2013
A. Ray & R. A. McCray, eds.

doi:10.1017/S1743921313009319

A tale of two shocks in SN 2004dj

Alak Ray[1], Sayan Chakraborti[2], Naveen Yadav[1], Randall Smith[2], Poonam Chandra[3] and David Pooley[4]

[1]Tata Institute of Fundamental Research, Mumbai 400005, India

[2]Harvard-Smithsonian Center for Astrophysics, Cambridge, MA 02138, USA

[3]National Centre for Radio Astrophysics, TIFR, Pune 411007, India

[4]Sam Houston State University, Huntsville, TX 77341, USA

Abstract. Type IIP SNe constitute a major fraction of all core-collapse supernovae and arise from massive stars that end their lives close to Red Supergiants. The blastwave from the SN interacting with the progenitor's circumstellar matter produces a hot region bounded by a forward and a reverse shock from which most of the X-ray emission originates. Analysis of archival Chandra observations of SN 2004dj, one of the nearest supernovae since SN 1987A, together with published data from radio and optical bands determines the pre-explosion mass-loss rate, blastwave speed, electron acceleration and magnetic field amplification efficiencies. X-ray emission arises from both inverse Compton scattering by non-thermal electrons accelerated in the forward shock and from thermal emission from the supernova ejecta hit by the reverse shock. Determination of the properties of the radiating plasma based on the separation of thermal and non-thermal radiation differentiates different types of supernovae and their environments.

Keywords. supernovae, X-rays: general, radio continuum: general, circumstellar matter

---

1. Introduction

An important aspect of understanding of supernovae (SNe) relates to the kind of progenitor star they arise from and the circumstellar medium they explode into. The most common type of supernovae in a volume limited sample is Type IIP SNe (Smith *et al.* 2011). They are shown to arise from red supergiants (RSG) (Smartt *et al.* 2009 and references therein) with large Hydrogen envelope and a main sequence mass $\sim 8-18M_{\odot}$. When their pre-explosion images are compared to those several years post-explosion they reveal the disappearance of the original RSG progenitor star (Maund and Smartt 2009). However the location of the progenitor stars on the H-R diagram, in some cases, shows that these stars are not at the predicted end points for single star tracks computed with evolutionary codes, and IIP SN 2008cn, for example, exploded as a yellow supergiant (YSG) rather than a RSG (Maund *et al.* 2011). Thus both progenitor and its environment in these explosions are complex and require probes, such as when the shock launched from the explosion interacts with the circumstellar medium (CSM) and radio and X-ray emission generated in the hot, shocked region. The shock accelerates particles to high energies which provide us through radio and X-ray measurements such tools (see e.g. Chevalier *et al.* 2006). Usually, radio and X-ray emission from IIP SNe are rather faint, but if detailed spectral data could be obtained for a few X-ray bright SNe, they could provide a discrimination of thermal and nonthermal parts and constrain physical nature of the shock interaction. Due to the long lasting, bright (optical) light during the plateau stage, shock accelerated electrons that emit radio waves are likely to be subject to energy loss that affects the radio light curves and at the same time generate signals as nonthermal X-rays via Inverse Compton scattering (IC). SN 2004dj was one such nearby

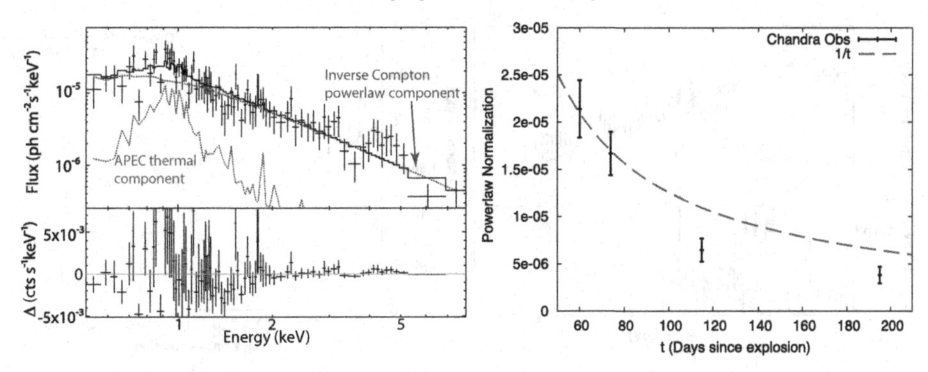

Figure 1. a) X-ray spectrum of SN 2004dj on 2004 August 9. Bars are counts from Chandra, dotted line is the power-law model for IC, dashed line is APEC model for the thermal plasma, while the solid line is the full model. The full model is dominated by the non-thermal flux even at $\sim$ 1 keV; b) Time evolution of the normalization of the non-thermal IC component obtained from simultaneous fits to Chandra spectra of SN 2004dj. The time interval is measured with respect to the explosion date of June 11 from Zhang *et al.* 2006. Line is best fit $\propto t^{-1}$ function.

SN ($d = 3.06$ Mpc) which was extensively followed up in optical, radio and X-ray bands and here we discuss the X-ray lightcurve and spectral evolution. Together with radio data (arising from power law electrons with index p with $N(E) \propto E^{-p}$) they constrain the properties of the shocked plasma responsible for thermal and nonthermal emission, and determine the pre-SN mass loss rate (Chakraborti *et al.* 2012).

2. SN 2004dj: X-ray and radio observations

SN 2004dj was observed as a Target of Opportunity and detected by Chandra X-ray Observatory on 2004 August 09 (Pooley and Lewin 2004). Chandra observed it for another three epochs during 2004 with ACIS-S (without gratings) for 50 ks each time. The X-ray flux (0.5-8 keV) ranged from 8.8×10^{-14} erg cm^{-2} s^{-1} to 2.0×10^{-14} erg cm^{-2} s^{-1} between Aug 9 and Dec 22 (Chakraborti *et al.* 2012). Similarly, extensive observations of this SN took place with GMRT, MERLIN and the VLA (see Chandra *et al.* 2013).

The SN spectra was extracted into XSPEC 12.7.1 for further analysis. Data from each epoch of observation were jointly fitted with a combination of power law spectrum and collisionally ionized diffuse gas using APEC models (Smith *et al.* 2001) that is subject to photoelectric absorption column (wabs), see Fig 1a. We kept the same column density for wabs, plasma temperature for APEC, and photon index for power law at all epochs. The emission measure for APEC and the normalization of the power law were determined separately for each epoch (best fit $N_H = (1.7 \pm 0.5) \times 10^{21}$ cm^{-2}).

Chandra observations of SN 2004dj indicate that the IC component of the X-ray flux falls with time. The spectrum of X-rays also softens with time. The quality of the SN 2004dj spectra allows a quantitative explanation for the first time. Although both thermal and non-thermal components decrease eventually, their proportion changes with time. At early times the spectra is dominated by the inverse Compton flux and is therefore harder. As the supernova ages, the source of seed optical photons to be scattered turns off. So at late times, the spectra is dominated by emission from the reverse-shocked thermal plasma and is softer.

The blastwave shock is the site of acceleration of particles, including electrons, to relativistic speeds. These electrons emit radio radiation via synchrotron emission in the magnetic field which is in turn amplified by the shock. The simplest model by Chevalier 1982, is to assume that a fraction ϵ_e or ϵ_B of the thermal energy is used to accelerate electrons

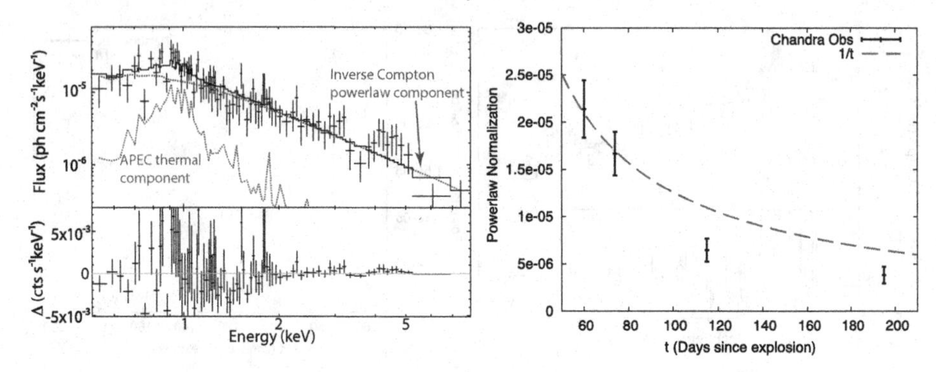

Figure 2. a) Normalizations of thermal (APEC model) and non-thermal (power-law model) fluxes for SN 2004dj on 2004Aug09; b) Chandra data breaks the degeneracy between the two components and determine their temperature and the photon index.

and amplify magnetic fields respectively. The radio emission from a supernova is dependent upon these fractions. According to Chevalier *et al.* 2006, the radio emission measures $S_\star = A_\star \epsilon_{B-1} \alpha^{8/19} = 1.0(f/0.5)^{-8/19}(F_{\rm p}/({\rm mJy}))^{-4/19}({\rm D}/({\rm Mpc}))^{-8/19}(\nu/5\ {\rm GHz})^{-4/19} {\rm t}_{10}^2$ where $F_{\rm p}$ is the peak flux at peak frequency ν at $10 \times t_{10}$ days from the explosion and $\epsilon_{B-1} = \epsilon_B/0.1$. The equipartition factor is defined as $\alpha \equiv \epsilon_e/\epsilon_B$. Using the 4.99 GHz, radio light curve from Beswick *et al.* 2005 for SN 2004dj, we have $S_\star = 5.1$. These relativistic electrons also contribute to the X-ray flux through IC scattering of optical photons. For an electron index p = 3 this is expected to generate a power law in X-rays with photon index 2, consistent with the observations. The normalization of the inverse Compton flux obtained for the first epoch ($t_{10} \sim 6$) of Chandra observations, is found to be $(3.8 \pm 0.5) \times 10^{37}$ ergs s^{-1} (see Fig 2a). This together with $S_\star = 5.1$ as found in this work, $V_4 = 0.9$ as implied by the temperature of the reverse-shocked plasma (Fig 2b), $L_{\rm bol} = 0.89 \times 10^{42}$ erg s^{-1} from Zhang *et al.* 2006, we get $\alpha \sim 23 \times \gamma_{\rm min}^{-19/11}$. We take $\gamma_{\rm min} \sim 2.5$ as used by Chevalier and Fransson 2006 for SN 2002ap, which then implies $\alpha \sim 4.8$. Since we know $S_\star$ from radio synchrotron, $A_\star$ from thermal X-rays and α from inverse Compton, we can use the definition of $S_\star$ to get $\epsilon_B \equiv 0.1 \times \epsilon_{B-1} \sim 0.082$, which in turn implies $\epsilon_e \equiv \alpha \times \epsilon_B \sim 0.39$. Hence both ϵ_e and ϵ_B are determined independently.

3. Implications

The time variation of the IC flux is due to the expansion of the blastwave and the fading away of the supernova's supply of the seed photons. The normalization of the IC flux scales as (Chevalier *et al.* 2006): $E \times dL_{\rm IC}/dE \propto L_{\rm bol}(t)/t$. Therefore during the plateau phase of the optical light curve of the SN where the bolometric luminosity L_{bol} is nearly constant, the IC flux is expected to fall off as $\propto t^{-1}$. At late times, since L_{bol} itself would be decreasing, the IC flux would deviate from the t^{-1} dependence. In Fig 1b this prediction (shown by the dashed line) is consistent with the observed X-ray flux components of SN 2004dj. The emission measure from the thermal part of the X-ray flux (from the reverse shocked ejecta) can be related to the mass loss parameter $\dot{M}/v_W$ if we can determine the blastwave shock radius from the velocity (e.g. with the postshock temperature or from radio broadband spectra assumed to be a synchrotron self absorbed spectrum). The APEC fit, along with distance of 3.06 Mpc and the emission measure gives: $\dot{M} = (3.2 \pm 1.1) \times 10^{-7} (v_{\rm w}/10\ {\rm km/s})\ {\rm M}_\odot\ {\rm yr}^{-1}$ This mass loss rate is however substantially less than that predicted empirically for RGB branch following de Jager *et al.* 1988, for an estimated progenitor mass of $\sim 15 M_\odot$ (Maiz-Applellaniz *et al.* 2004).

Around a third of the energy thermalized by the collision of the ejecta with circumstellar matter is processed by the shock to accelerate electrons to relativistic velocities. Another tenth of the thermal energy available is used in the turbulent amplification of magnetic fields. Electrons can "cool" via Inverse Compton cooling against low-energy photons from the SN photosphere. The cooling break frequency in the radio bands can be obtained by demanding that the IC loss time scale is equal to the age of the SN. For the parameters relevant to SN 2004dj, the predicted $\nu_{\rm IC}$ is within the coverage of the VLA band for about 180 days which should be seen in its radio spectral index variations.

4. Conclusions

For bright IIP SN like 2004dj we can separate the thermal component and the IC components and determine the pre-explosion mass-loss rate, electron acceleration and magnetic field amplification efficiencies. Another IIP SN 2011ja X-ray brightened between two Chandra observations (Chakraborti *et al.* 2013). This may imply a complex circumstellar medium, e.g. an explosion inside a rarefied cavity surrounded by a denser medium characteristic of an older and slower Red Supergiant wind, like SN 1987A. Progenitor environments of IIP SNe may be more complex than what has been interpreted so far. Radio observations of another IIP SN 2012aw (Yadav *et al.* 2013) show evidence of electron cooling during the (optical light curve) plateau phase.

References

Beswick, R. J., *et al.* 2005, *ApJ Letters*, 623, L21
Chakraborti, S., *et al.* 2013, *these Proceedings & arXiv:* 1302.7067
Chakraborti, S., Yadav, N., & Ray, A., *et al.* 2012, *ApJ*, 761, 100
Chandra, P., *et al.* 2013, *In preparation*
Chevalier, R. A. 1982, *ApJ*, 258, 790
Chevalier, R. A., Fransson, C., & Nymark, T. K. 2006, *ApJ*, 641, 1029
Chevalier, R. A. & Fransson, C. 2006, *ApJ*, 651, 381
de Jager, C., Nieuwenhuijzen, H., & van der Hucht, K. A. 1988, *Astron. Astroph. Suppl.*, 72, 259
Maiz-Apellaniz, J., *et al.* 2004, *ApJ Letters*, 615, L113
Maund, J., *et al.* 2011, *ApJ Letters* 739, L37
Maund, J. & Smartt, S. 2009, *Science* 324, 486
Pooley, D. & Lewin, W. H. G. 2004, *IAUC*, 8390, 1
Smartt, S., *et al.* 2009, *MNRAS* 395, 1409
Smith, N., *et al.* 2011, *MNRAS*, 412, 1522
Smith, R. K., *et al.* 2001, *ApJ Letters*, 556, L91
Yadav, N., *et al.* 2013, *these Proceedings*
Zhang, T., *et al.* 2006, *AJ* 131, 2245

Discussion

SAHA, LAB: Is there any time variability in flux for SN2004dj. Due to presence of reverse shocks, there could be turbulence in the SN region. So there could be magnetic amplification. Did you see any flux enhancement due to the reason mentioned above?

RAY: The X-ray flux (0.5-8 keV band) decreases with time. Radio emission is the characteristic indicator of synchrotron radiation in a magnetic field amplified near the forward shock although electrons accelerated at the reverse shock may contribute a minor fraction of the total flux density. The flux density at 5 GHz decreases initially.

Supernova Environmental Effects
Proceedings IAU Symposium No. 296, 2013
A. Ray & R. A. McCray, eds.

doi:10.1017/S1743921313009320

Radio Observations Of A Nearby Type IIP SN 2012aw

Naveen Yadav[1], Alak Ray[1], Sayan Chakraborti[2], Christopher Stockdale[3], Poonam Chandra[4], Randall Smith[2], Rupak Roy[5], Vikram Dwarkadas[6], Firoza Sutaria[7], Dave Pooley[8], Brijesh Kumar[5] and Subhash Bose[5]

[1]Tata Institute of Fundamental Research,
Mumbai, India
email: naveen.phys@gmail.com

[2]Harvard-Smithsonian CfA, USA

[3]Marquette University, USA

[4]National Center for Radio Astronomy-TIFR, India

[5]Aryabhhata Research Institute of Observational Sciences, India

[6]University of Chicago, USA

[7]Indian Institute of Astrophysics, India

[8]Sam Houston State University, Huntsville, USA

Abstract. SN 2012AW is a type-IIP supernova which exploded in M95. In this paper we discuss the radio observations of this supernova and model them to determine the important parameters relevant to the explosion and the evolution of blast wave. We also determine the dominant cooling process important to this source.

Keywords: Radio continuum: general, Supernovae: individual (SN 2012aw), Methods: data analysis, Techniques: image processing, interferometric

1. Introduction

SN 2012aw is a bright type II-P supernova which exploded in the galaxy M95 (d 10 Mpc). Spectra taken 4 – 5 days after discovery showed it to be a type IIP explosion (Fagotti *et al.* (2012)). Fraser *et al.*identified a candidate progenitor in archival HST images. Fraser *et al.* (2012) and have inferred a progenitor mass in range 14 – 26 $M_\odot$, whereas Van Dyk *et al.* (2012) inferred a progenitor mass in range 17 – 18 $M_\odot$. Its progenitor seems to be a faint red supergiant. SN 2012aw is being extensively studied from optical to X-ray bands. We followed the object at radio wavelengths using JVLA and GMRT, targeting it at *L,C,S,X,K & Ka* bands at multiple epochs.

2. Radio Observations, Reduction & Modeling

SN 2012aw was first detected in radio JVLA-*K* band $\sim$ 10 days by Stockdale *et al.* (2012). We conducted *EVLA* and *GMRT* observations of 2012aw at various epochs extending up to 187 days after the explosion. These observations have been reduced using Astronomical Image Processing Software (*AIPS*) standard techniques. Interferometric visibilities have been calibrated using standard calibrators. The single source data has been extracted using *AIPS* task *SPLIT* and imaged using *IMAGR*.

The radio emission can be modeled as synchrotron emission with a combination of absorption and cooling processes. It has an optically thin component and an optically thick

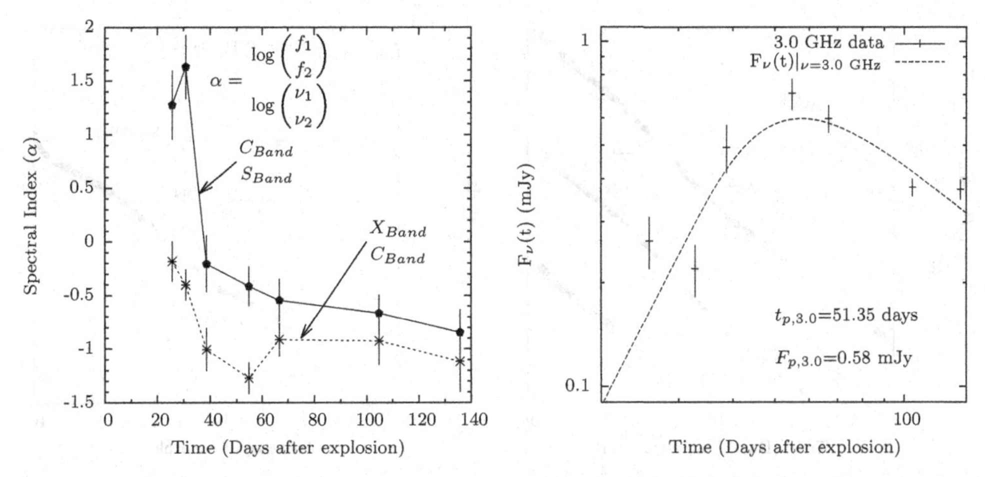

Figure 1. Left: Spectral index curves made using the S, C & X band data. Note the dip in the spectral index curve marked as X_{Band}/C_{Band}, is a signature of a cooling mechanism which was present at early time and is slowly turning off as the index approaches , -1, its value in the optically thin regime. **Right:** An SSA fit to the 3.0 GHz radio lightcurve using $\gamma = 3.1$. It gives value of $t_{p,3.0} = 51.35$ days & $F_{p,3.0} = 0.58$ mJy.

component which can be modeled as a combination of synchrotron self absorption (SSA) & free-free absorption (FFA). We use Chevalier model-I (Chevalier (1996)) to study this emission. In this model the radius of interaction shell increases as, $R \propto t^m$ and energy density in relativistic electrons and magnetic fields also follows the hydrodynamical evolution with $u_e, u_B \propto t^{-2}$. In the analysis we assume that, $m \sim 1.0$. Electron index can be obtained by fitting a power law to the optically thin component. The equation for the radio flux evolution in such case of is given in Chevalier (1998) for the case of a supernova expanding in to a circumstellar medium set up by a uniform wind. In case of a source dominated by SSA the spectral index approaches value, $-(\gamma - 1)/2$, as the source enters the optically thin regime. If we try to model this object by simply using *SSA+FFA*, the fit gives m greater than 1, implying an accelerated blast wave, which is incorrect as the blast wave decelerates due to its interaction with the circumstellar matter. Physically this can be explained as follows: the radio light curves of a source which is affected by a cooling mechanism (which slowly turns off) will be similar to a source which expands slowly at early times and the rate of expansion increases with time. This indicates the need to include a cooling mechanism which is dominant at early times and turns off at later times. In Figure 1(Left) the spectral index curve shows a sign of a cooling mechanism. We need to compare the cooling timescales for various mechanisms. The cooling timescales for an electron of energy E can be written using formulas for energy loss from Pacholczyk (1970) as $t^{-1}_{Compton} = 3.97 \times 10^{-2} u_{rad} E$ and $t^{-1}_{sync} = 5.95 \times 10^{-2} u_B E$.

To get the synchrotron cooling timescale we need an estimate of magnetic field. In Chevalier model-I the magnetic field evolves as t^{-1}. If we know the magnetic field at one epoch it can simply be scaled to get the field at any other epoch using, $B(t) = B_0 \left(t/t_0\right)^{-1}$. To get the value of magnetic field we can either use a late time radio spectrum or a low frequency radio lightcurve which are free from the electron cooling effects. To get an estimate of the FFA we use $\dot{M}_{-5}/v_{w1}$ determined from time of X-ray detection. This object was first detected in X-ray (0.2 – 10 KeV band) by Immler *et al.* (2012) approximately 4 days after the explosion. This can be used to get an upper limit on the quantity $\dot{M}_{-5}/v_{w1}$ which describes mass loss by a uniform wind. Using $t_X = 4$ days, $v_{s4} \sim 1.0$ and $E_{KeV} = 1.0$ and adopting the value $C_5 = 2.6 \times 10^6$in to the Equation 2.17

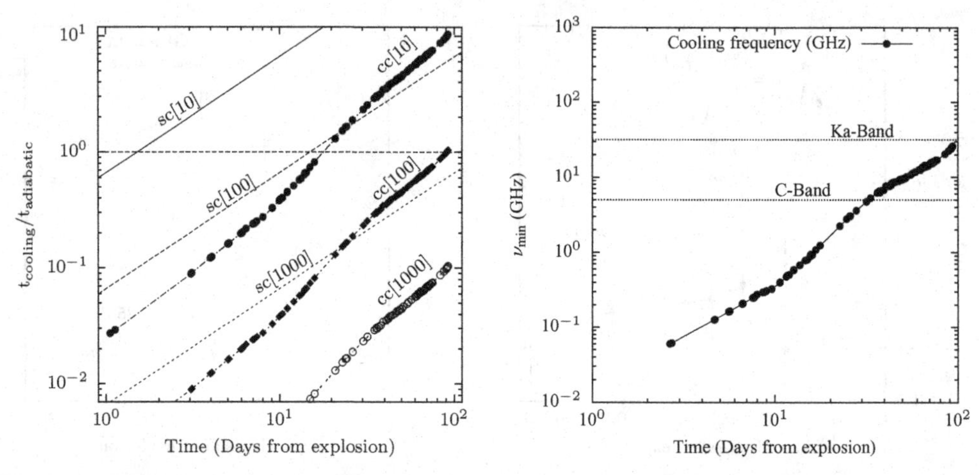

Figure 2. Left: Cooling timescales for electrons calculated using the computed bolometric light curve for various values of Lorentz factors as marked in the diagram by sc[γ_i] for synchrotron Cooling & cc[γ_i] for Compton cooling respectively. Note that Compton cooling process is dominant over the synchrotron cooling process at any given Lorentz factor γ_i. **Right:** The minimum frequency affected by cooling plotted as a function of the age of SN 2012aw. Note that the dotted lines at 32.0 GHz and 5.0 GHz show that at early time all the VLA bands will have some effect of electron cooling. For the C-Band the cooling phase lasts till $\sim$ 30 days.

from Chevalier *et al.* (1994), we get, $\dot{M}_{-5}/v_{w1} < 8.64\times10^4 t_X v_{s4} C_5^{-1} E_{KeV}^{8/3} < 0.13$. where $\dot{M}_{-5}$ is mass loss rate in units of 10^{-5} M$_\odot$, v_{w1} is wind velocity in units of 10 Km/s, v_{s4} is outer shock velocity in units of 10^4 Km/s and t_X is the time at which the medium becomes optically thin to X-rays of energy E_{KeV} and C_5 is a constant. This value can be used to get an upper limit on the time for which free-free absorption dominates. Using Equation 4 from Chevalier *et al.* (2006) we get, $t_{ff} \approx 6\left(\dot{M}_{-6}/v_{w1}\right)^{2/3} T_{cs5}^{-1/2} v_{s4}^{-1} (\nu/8.46\ \text{GHz})^{-2/3}$. which gives $t_{ff} \leqslant 16.0$ days at 3.0 GHz and $t_{ff} \leqslant 11.0$ days at 5.0 GHz for $T_{cs5} = 1.0$. This shows that this object is not dominated by free-free absorption in 3.0 GHz at early times. The 3.0 GHz light curve can be fitted by a simple SSA model with a value of $m \sim 1.0$ as shown in Figure 1 (right). This can be used to derive the value of radius and magnetic field strength. Using the Equation 11 & 12 from Chevalier (1998) gives $B_0 \sim 0.48$ Gauss and $R_0 \sim 3.9\times10^{15}$ cm on $\sim$ 51 days assuming equipartition. We require the bolometric luminosity to determine the Compton cooling timescale. We construct a bolometric lightcurve using published photometric data from Bayless *et al.* (2013) & Munari *et al.* (2013).

The calculated cooling timescales are shown in the Figure 2 (left). It is evident that for electrons of any given Lorentz factor γ_i Compton cooling dominates over the synchrotron cooling mechanism. Therefore in order to model the radio spectrum at early epochs and at high frequency we need to consider the effect of cooling mechanism on emission. Assuming that the an electron emits synchrotron radiation at its characteristic frequency we can get an estimate of frequencies which are affected at a given age. Using $\nu_c \sim c_1 B E^2$, where c_1 is a constant, the minimum frequency above which there are significant effects due to Compton cooling can be written as, $\nu_{min} > 0.78\,(t/10\ \text{days})\left(L_{Bol}/10^{42}\right)^{-2}$ GHz. The minimum frequency which is affected by cooling is shown in Figure 2 (right). It shows that at early times all the radio band is affected by cooling, but as the supernova fades ν_{min} goes to larger and larger values because of a decrease in bolometric luminosity and an increase in the radius of forward shock. As a result the cooling is important only for the highest frequencies at late times.

3. Conclusions

We discuss the radio observations of SN 2012aw. We study the rate at which the relativistic electrons at the radiosphere lose energy due to inverse Compton process and synchrotron process. We find that in the case of SN 2012aw Compton cooling dominates over the synchrotron cooling process. Therefore we note that Compton cooling effects need to be considered at early times in order to do a consistent modeling of the high frequency radio emission. This may help us to probe the particle acceleration process at the forward shock in a young radio bright supernova. In order to to gain a better understanding of the importance of cooling mechanisms we need to follow bright type II-P supernova as quickly as possible after type classification in radio and X-ray wavelengths.

4. Acknowledgements

The National Radio Astronomy Observatory is a facility of the National Science Foundation operated under cooperative agreement by Associated Universities, Inc. We would like to thank the Director *EVLA* for granting us DDT time on this object. One of the author Naveen Yadav wishes to acknowledge the support of CSIR-SPM fellowship.

References

Chevalier, R. & Fransson, C. 1994, *ApJ*, 420, 268
Chevalier, R. 1998, *ASP-CS*, 93, 125
Chevalier, R. 1998, *ApJ*, 499, 810
Chevalier, R. & Fransson, C. 2006, *ApJ*, 641, 1029
Fraser, M., Maund, J. R., Smartt, S. J., *et al.* 2012, *Ap. Lett.*, 759, 13
Van Dyk, S. D., Cenko, S. B., Poznanski, D., *et al.* 2012, *ApJ*, 756, 131
Stockdale, C. J., Ryder, S. D., Van Dyk, S. D., *et al.* 2012, *ATEL*, 4012, 1
Yadav, N., Chakraborti, S., & Ray, A. 2012, *ATEL*, 4010, 1
Fagotti, P., Dimai, A., Quadri, U., Strabla, L., *et al.* 2012, *CBET*, 3054, 1
Immler, S. & Brown, P. J. 2012, *ATEL*, 3995, 1
Pacholczyk, A. G. 1970, *Radio Astrophysics, San Francisco: Freeman, 1970*
Munari, U., Henden, A., Belligoli, R., *et al.* 2013, *New Astron.*, 20, 30
Bayless, A. J., Pritchard, T. A., *et al.* 2013, *Ap. Lett.*, 764, L13

Discussion

POONAM CHANDRA: The formula for free-free absorption is steeper than a power law, but you seem to have used a power law form for FFA?

NAVEEN YADAV: The actual formula used in the fitting has the exponential dependence (Eqn.9 from Chevalier (1998)).

ROGER CHEVALIER: Your best fit has an $m > 1.0$, which is not possible in the CSM interaction model?

NAVEEN YADAV: I agree that $m > 1$ is unphysical, and have addressed the issue above.

Supernova Environmental Impacts
Proceedings IAU Symposium No. 296, 2013
A. Ray & R. A. McCray, eds.

doi:10.1017/S1743921313009332

The optical photometric and spectroscopic investigation of Type IIP supernova 2012A

Rupak Roy[1], Firoza Sutaria[2], Subhash Bose[1], Sean Johnson[3], Vikram Dwarkadas[3], Brian York[4], Brijesh Kumar[1], Brajesh Kumar[1], Vijay K. Bhatt[1], Sayan Chakraborti[5], Don York[3], Adam Ritchey[6], Gabrielle Saurage[7] and Mary Beth Kaiser[8]

[1]Aryabhatta Research Institute of Observational Sciences, Nainital, India
[2]Indian Institute of Astrophysics, Bangalore, India
[3]University of Chicago, Chicago, USA
[4]Space Telescope Science Institute, Baltimore, USA
[5]Institute for Theory and Computation, Harvard, USA
[6]University of Washington, Seattle, USA
[7]Apache Point Observatory, Sacramento Mountains, Sunspot, USA
[8]Johns Hopkins University, Baltimore, Maryland, USA

Abstract. Supernova 2012A was discovered on 7.39UT, January, 2012 in the nearby galaxy NGC 3239 at an unfiltered magnitude of 14.6 and classified spectroscopically as a Type IIP event. Here, we present the optical photometric and spectroscopic follow-up of the event during 14d to 130d post explosion.

Keywords. (Stars:) Supernovae: individual (SN 2012A); techniques: photometric, spectroscopic

---

1. Introduction

Given the considerable development in automated sky survey programs, the detection rate of supernovae (SNe) has increased enormously (Lennarz *et al.* 2012) in the last few years. In a volumetric study it was found that more than 50% of all SNe are Hydrogen rich (Type II), of which about 70% show prolonged plateau (IIP) in their light curves (Li *et al.* 2011). Type II SNe show diversities in their light curves and spectra. Indeed there are distinct classes – normal and subluminous events - which show different properties, along with several peculiar events like SN 1987A, which resulted from explosion of a blue supergiant star. Following the discovery of SN 2008in (Roy *et al.* 2011a), a different class of SNe has been suggested with characteristics in-between normal and subluminous Type IIP. SN 2009js is a new entry in this category (Gandhi *et al.* 2013). The nearby event SN 2012A also showed similar behaviour, and is the prime target of our study.

SN 2012A was discovered on 7.39UT, January, 2012 in the nearby ($\sim$ 9 Mpc) galaxy NGC 3239 at an unfiltered magnitude of 14.6 (Moore *et al.* 2012). It was classified as a Type IIP event, with a spectrum similar to that of SN 2004et at about 14d post explosion, confirming its identity as a young Type IIP (Stanishev and Pursimo 2012; Roy and Chakraborti 2012).

2. Observations and data reductions

The ground based optical photometric observations were carried out at the 104-cm Sampurnanand Telescope (ST) using Johnson UBV and Cousins RI filters, and also from

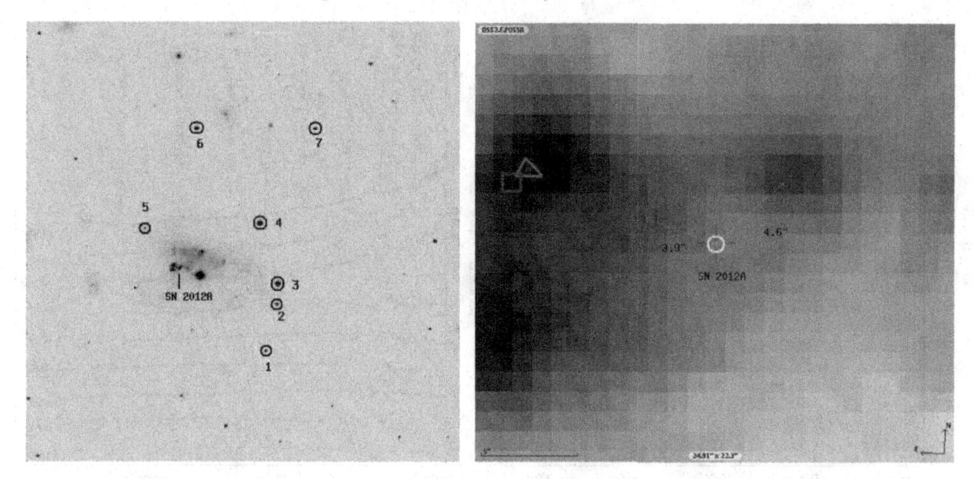

Figure 1. Left Panel: Identification chart of the field of SN 2012A in NGC 3239. The image is about $13' \times 13'$ taken in B-band with the 104-cm ST at ARIES, Nainital. The SN location is marked along with the secondary stars used for calibration. North is up and East is to the left. **Right Panel:** $24.''91 \times 22.''3$ region around SN 2012A in pre-SN DSS image. The SN location is at the center and marked with the yellow circle. The red triangle is an extended source, whereas the blue rectangles are point sources tabulated in 2MASS catalog. The angular separation between SN and two nearby sources are respectively $3.''9$ and $4.''6$. These are the prime sources of contamination of SN flux.

the 130-cm Devasthal Fast Optical Telescope (DFOT) using BVR filters. Photometric observations were conducted at 20 epochs between 14d and 350d. The field of SN 2012A is calibrated using Landolt (1992) standard stars of the fields of SA98. Left panel of Fig. 1 shows the SN position and location of 7 local standards in the field. The detailed methodology for data analysis is discussed elsewhere (Roy *et al.* 2011a,b). A closer view of the transient location obtained from a pre-SN DSS image is presented in the right panel of Fig. 1. Several star-forming knots, cataloged in the 2MASS survey and marked around the SN position, are prime sources of contamination of SN flux. In order to calculate the correct SN flux, a 'template-subtraction technique' is required, but due to absence of pre-SN $UBVRI$ images of this field, in the present work the instrumental magnitudes of the SN are derived by the profile (PSF) fitting method. SDSS magnitudes of the nearest star forming knot have been used for a rough estimation of background flux. Therefore the measurements are expected to represent closely the true SN magnitude when it is bright, i.e., in the early plateau phase, while in later epochs the background flux is substantial, and hence the estimated parameters will be marginally affected.

The long-slit low-resolution spectroscopy in the optical range ($0.40 - 0.95\mu$m) was performed at eight epochs during 14d and 134d: three epochs from the 2-m IUCAA Girawali Observatory (IGO), one epoch from the 3.5-m Astrophysical Research Consortium (ARC) telescope and at five epochs from 2-m Himalayan Chandra Telescope (HCT). We also report on the high resolution spectroscopy from 3.5-m ARC at the early plateau phase of the transient. The methodology for spectroscopic data analysis is same as described in Roy *et al.* 2011b.

3. Results

The calibrated light curves of SN 2012A in $UBVRI$ bands are presented in the left panel of Fig. 2, and its comparison with that of the normal Type IIP SN 2004et (Sahu

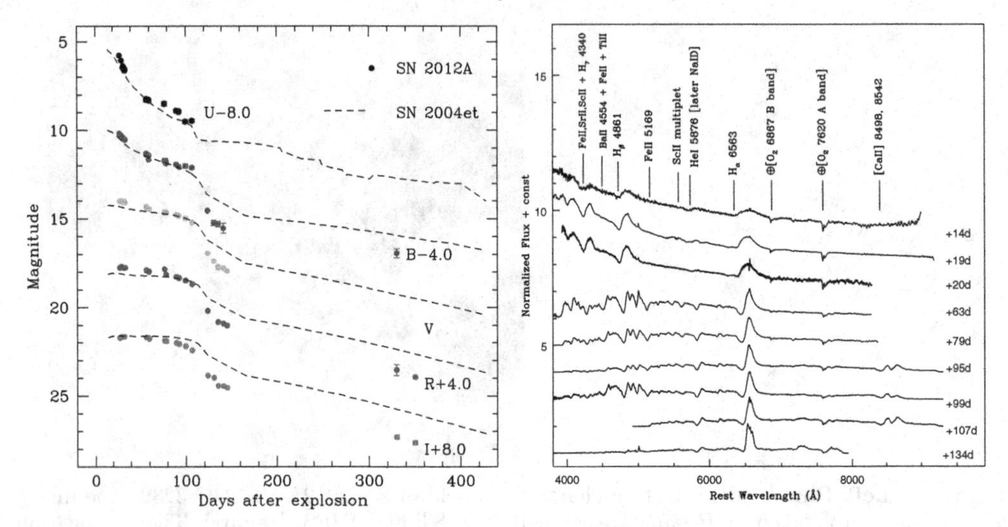

Figure 2. Left Panel: Light curves of SN 2012A in *UBVRI* bands. The light curves are shifted for clarity, while for SN 2004et they are scaled in magnitude and time to match with SN 2012A. **Right Panel:** Doppler-corrected (recession velocity ~752 km.s^{-1}) flux spectra of SN 2012A from the plateau (14d) to the nebular phase (134d). Prominent hydrogen and metal lines are marked.

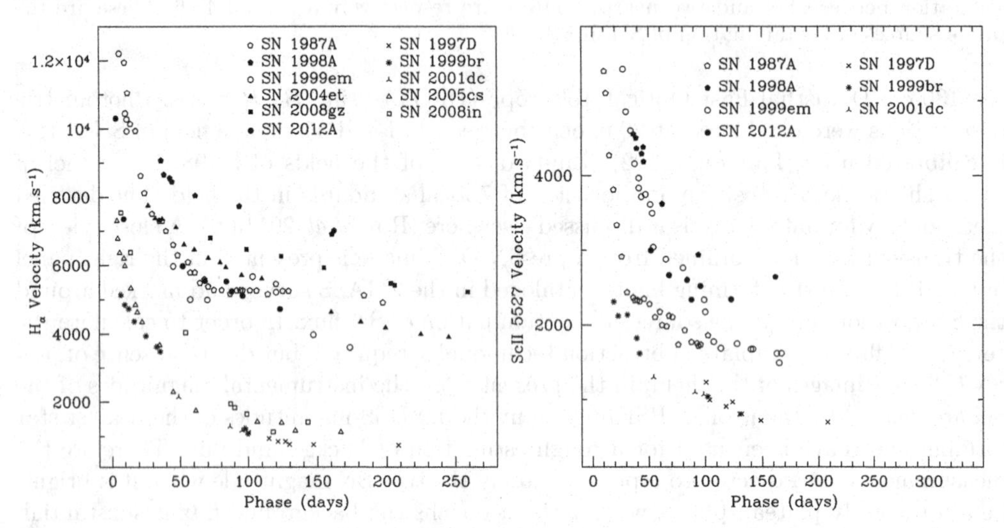

Figure 3. Comparison of velocity profile of SN 2012A with other Type IIP SNe. **Left Panel:** H$_\alpha$ velocity, showing the velocity near outer ejecta. **Right Panel:** ScII 5527 velocity, showing the velocity near the photosphere. *Ref:* Elmhamdi *et al.* (2003a), Pastorello *et al.* (2004, 2005, 2009), Roy *et al.* (2011a, b), Sahu *et al.* (2006).

et al. 2006) shows that the plateau luminosity of both events decreases in similar fashion, though plateau to nebular conversion is more rapid for SN 2012A than SN 2004et. In nebular phase SN 2012A becomes much fainter than SN 2004et.

The right panel of Fig. 2 shows the evolution of rest-frame spectra of SN 2012A, where the prominent lines of Hydrogen, Helium and metal lines along with several telluric absorption lines have been marked. The line identification has been done using Leonard *et al.* (2002). The continuum dominated early spectra (14d,19d and 20d) demonstrate the high photospheric temperature during the initial epochs. The spectral evolution is similar to normal Type IIP SNe.

Fig. 3 depicts a comparison of outer ejecta and photospheric velocities of SN 2012A with other Type IIP SNe. H_α profiles, representing the velocity of the outer ejecta, show a clear bimodal distribution: low luminosity events asymptotically attain a lower velocity than normal events, though they are much more systematic than normal Type IIP. SN 2012A is located near the lower edge of the velocity distribution of normal events. Photospheric velocity measured from the ScII 5527 line is almost similar in both cases, and SN 2012A preserves the characteristic features of normal Type IIP SNe.

Galactic reddening along the line of sight, $E(B-V) = 0.32 \pm 0.0005$ mag (Schlegel *et al.* 1998) and the mean distance of the host is 8.6±0.5 Mpc. In our analysis of the high-resolution 3.5m ARC spectra, no sign of NaID at host galaxy velocity is found, although there is weak CaII H and K absorption. The upper limit on the column density ratio of NaID/CaII at the host galaxy velocity is very low (NaID/CaII < 0.04), indicating very little dust grain depletion of Ca. Therefore we presume that extinction is mainly dominated by the Milky-Way.

From the estimated mid-plateau absolute V-band magnitude (~ -15.1 mag), photospheric velocity ($\sim 2600\ \mathrm{km\,s^{-1}}$) and plateau duration ($\sim$110 days), the basic parameters of the progenitors have been calculated using the prescriptions of Elmhamdi et al (2003b) and Litvinova and Nadezhin (1985). The amount of synthesized ^{56}Ni is ~ 0.012 $M_\odot$, ejected mass is $\sim 26\ M_\odot$, explosion energy is $\sim 7 \times 10^{50}$ erg, and the radius of the pre-SN progenitor is about 86 $R_\odot$. Assuming the mass of the compact remnant $\sim 2\ M_\odot$, main sequence mass of the progenitor can be constrained to be around 29 $M_\odot$. This is certainly a crude estimation and more detailed modelling is essential.

Acknowledgements

We thank George Wallerstein for providing us one of the APO echelle spectra.

References

Elmhamdi, *et al.* 2003a, *MNRAS*, 338, 939
Elmhamdi, *et al.* 2003b, *A&A*, 404, 1077
Gandhi, P., *et al.* 2013, *ArXiv*, 1303.1565
Landolt, A. U., 1992, *AJ*, 104, 340
Lennarz, D., Altmann, D., & Wiebusch, C., 2012, *A&A*, 538, A120
Leonard, D. C., *et al.* 2002, *PASP*, 114, 35
Litvinova, I. Y. & Nadezhin, D. K., 1985, *Soviet Astronomy Letters*, 11, 145
Li, W., *et al.* 2011, *MNRAS*, 412, 1441
Moore, B., Newton, J., & Puckett, T., 2012, *CBET*, 2974, 1
Pastorello, A., *et al.* 2004, *MNRAS*, 347, 74
Pastorello, A., *et al.* 2005, *MNRAS*, 360, 950
Pastorello, A., *et al.* 2009, *MNRAS*, 394, 2266
Roy, R., *et al.* 2011a, *ApJ*, 736, 76
Roy, R., *et al.* 2011b, *MNRAS*, 414, 167
Roy, R. & Chakraborti, S., 2012, *CBET*, 2975, 2
Sahu, D. K., Anupama, G. C., Srividya, S., & Muneer, S., 2006, *MNRAS*, 372, 1315
Schlegel, D. J., Finkbeiner, D. P., & Davis, M., 1998, *ApJ*, 500, 525
Stanishev, V. & Pursimo, T. 2012, *CBET*, 2974, 3

Discussion

BERSTEN: The physical parameters found for supernova 2012A seems somewhat inconsistent. With a ejecta mass of 24 $M_\odot$ and the very small energy found, the plateau should be extremely long.

ROY R.: This work is preliminary. We have applied LN85 to calculate the burst parameters. Detailed calculations with more improved theoretical models are in progress.

NOMOTO: A comment: The relation between the H velocity and the luminosity you showed may be interpreted as follows: The H-alpha velocity is a good indicator of mixing, thus suggesting the progenitor mass as follows. The lower H velocity means a larger scale of mixing of H down to the deeper layer. Such a deep mixing occurs in smaller mass progenitor because more massive envelope (due to less mass loss) causes a stronger Rayleigh-Taylor instability. The smaller mass progenitor tends to be fainter. The higher H velocity suggests more massive one.

Supernova environmental impacts
Proceedings IAU Symposium No. 296, 2013
A. Ray & R. A. McCray, eds.

doi:10.1017/S1743921313009344

Supernova progenitor mass constraints through spatial correlations with host galaxy star formation

Joseph P Anderson

Departamento de Astronomía, Universidad de Chile, Casilla 36-D, Santiago, Chile
email: anderson@das.uchile.cl

Abstract. We present progenitor mass constraints on supernovae (SNe), from correlations with star-forming regions within host galaxies. Through a pixel statistics method used together with H-alpha imaging of host galaxies, we present a progenitor mass sequence running from supernovae type Ia (SNIa) arising from the lowest, through SNII, SNIb, and finally SNIc arising from the highest mass progenitors, implied from an increasing association of their explosion sites with star formation (SF). We also present constraints on the various core-collapse (CC) sub-types, finding the perhaps surprising result that SNIIn show the lowest degree of association with SF of any CC type, implying relatively low-mass progenitors. Finally, we compare the SNIIn environment distribution to that of SNIa, posing the provocative question that additional SNIIn may be linked to the SNIa phenomenon where the latter's spectra are hidden beneath that of circumstellar material (CSM) interaction.

Keywords. (stars:) supernovae: general, (ISM:) HII regions

1. Introduction

Constraining SN progenitor characteristics through observing their immediate environments within host galaxies has become a strong area of research in recent years. This approach, consisting of investigating the nature of stellar populations in close proximity to the explosion sites of SNe enables one to build statistically significant samples of events, while also using the characteristics of stellar populations most representative of that of their progenitors for obtaining constraints.

To date these investigations have involved two distinct approaches. The first involves using imaging of nearby SN host galaxies and has often used some form of pixel statistics to measure the degree of association of different SN types with different types of (usually young) stellar populations (see e.g. Leloudas *et al.* 2010; Kelly & Kirshner 2012). This has led to various constraints on differences in the ages of environments of CC SNe, and hence progenitor mass constraints. The second approach consists of obtaining spectra of host HII regions of SNe within galaxies, which can be used to investigate the environment metallicities of different SN types. This approach has led to a series of CC SN progenitor metallicity constraints, in particular constraints on differences between the different SN types which may explain differences in their transient evolution (Modjaz *et al.* 2008; Anderson *et al.* 2010; Leloudas *et al.* 2011; Modjaz *et al.* 2011; Sanders *et al.* 2012; Stoll *et al.* 2012).

In these proceedings we summarise the most recent results to arise from an investigation of the association of different SN types with SF within galaxies, using SN host galaxy imaging and pixel statistics.

2. Host galaxy pixel statistics

Over the last decade we have obtained a large sample of Hα imaging of SN host galaxies. Hα line emission within galaxies traces on-going SF, as the emission is a result of the recombination of ISM hydrogen ionised by the UV flux of massive stars. Therefore one can use this narrow-band emission as a tracer of the distribution of massive stars within galaxies, and hence use this to probe differences in the massive star environments of different types of SNe. Within our current sample (published in Anderson *et al.* 2012) we have a sample of host galaxy imaging for: 163.5 SNII; which can be separated into 58 IIP, 13 IIL, 13.5 IIb, 19 IIn and 12 'impostors', plus 48 with no sub-type classification, and 96.5 SNIbc; 39.5 Ib, 52 Ic and 5 with no sub-type classification. In addition we have a sample of 98 SNIa which we are analysing, comparing environment characteristics with light-curve properties (Anderson *et al.* in prep.).

To analyse these data we use a pixel statistics method first outlined in James & Anderson (2006), then further described in Anderson & James (2008). This method produces a value between 0 and 1 for every pixel (an 'NCR' value) of the continuum-subtracted Hα image, where 0 indicates a pixel consistent with sky values or zero emission flux, while

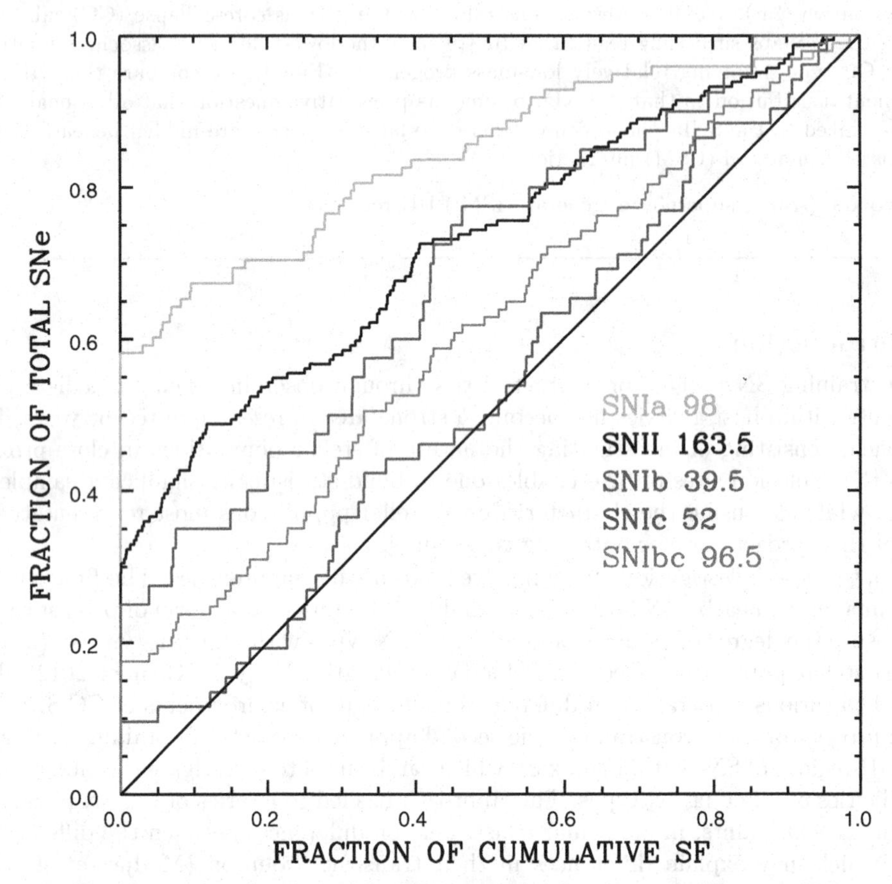

Figure 1. Cumulative distributions of the NCR statistics for the main SN types: SNIa in green, SNII in black, SNIb in red, SNIc in blue, and the overall SNIbc population in magenta. The diagonal black line shows a hypothetical population which shows a one-to-one correlation with the Hα emission within host galaxies. As a distribution moves away from this population to the upper left of the plot, the SN population is showing a lower degree of association to the emission.

1 indicates the pixel has the highest count of any pixel within the image. Using this statistic, distributions for different SN types can be built. The statistic is formalised in such a manner that if the Hα count within a pixel is directly proportional to the amount of SF occurring within that region of the galaxy, and a SN population directly traces the SF within their host galaxies, then we expect a flat distribution of NCR values for that population, with a mean NCR value of 0.5. Following this, these distributions are analysed and used to provide progenitor constraints: if a population has a lower mean NCR value then it is assumed that this relates to *on average* longer stellar lifetimes, and lower progenitor masses, implied from the lower association to the on-going SF.

3. Progenitor constraints

In figure 1 the NCR distributions for different SN types are presented. These are plotted cumulatively: as distributions move away to the top left of the plot from the diagonal black line, they are displaying a progressively lower degree of association with the on-going SF as traced by Hα, which implies longer stellar lifetimes and lower progenitor masses. In fig. 1 we see a clear implied sequence of progenitor masses, starting with the SNIa showing the lowest correlation with SF, and therefore having the lowest progenitor masses (as expected), through the SNII, the SNIb and finally the SNIc showing the highest correlation with host galaxy HII regions, therefore arising from the highest mass progenitors. While the SNII and SNIb distributions are quite similar, with the SNIb showing a slightly higher HII region association than the SNII, the SNIc are statistically significantly separated from SNIb: this implies a difference in the mean lifetimes and hence masses between these two types.

In fig. 2 the distributions of various SN sub-types are presented. The most surprising result to arise from these statistics would appear to be that the SNIIn population shows one of the lowest degrees of association with the SF, indeed showing a similar (but slightly lower) correlation with host galaxy HII regions as the SNIIP. This is surprising because the progenitors of SNIIn are often claimed to be Luminous Blue Variable stars (LBVs) (e.g. Smith 2008). However, LBVs are thought to be very massive stars which one would expect to be associated with on-going SF. Hence, in contrast our results imply

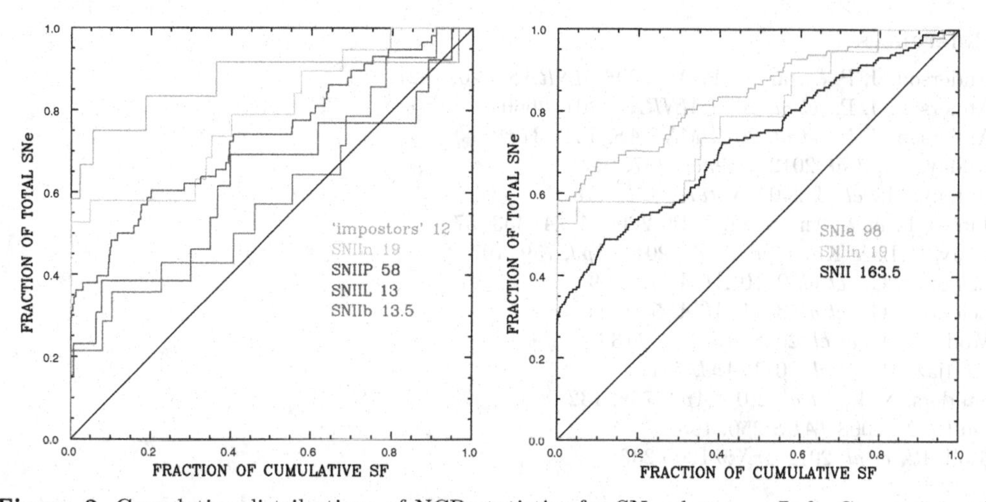

Figure 2. Cumulative distributions of NCR statistics for SN sub-types. ***Left***: Comparison of SNII sub-types: 'impostors' in green, SNIIn in cyan, SNIIP in blue, SNIIL in red, and SNIIb in magenta. ***Right***: Comparison of the SNIa (green), SNIIn (cyan) and overall SNII (black) distributions.

that at least the majority of SNIIn arise from relatively low mass progenitors. This is also consistent with Kelly & Kirshner (2012), who found similar environment properties for SNIIn and SNIIP. It is also noted that in the left panel of fig. 2 there appears to be some evidence that SNIIL and SNIIb are more highly correlated with the line emission than SNIIP, possibly suggesting higher mass progenitors for the former.
Finally, in the right panel of fig. 2 a comparison is made between the SNIIn, SNIa and SNII NCR distributions. The fact that the SNIIn distribution lies between the other two populations is intriguing. A number of SNIa have been reported to explode within a high density CSM environment (see Hamuy *et al.* 2003; Dilday *et al.* 2012). Therefore, one may speculate that there are additional SNIa where the CSM interaction is sufficiently strong to hide characteristic SNIa features, hence these are classified as SNIIn. If this speculation is true then this would naturally explain the low degree of association of SNIIn with on-going SF.

4. Conclusions

SN environmental studies can provide useful constraints on progenitor differences between different SN types. Using pixel statistics and Hα pixel statistics a mass sequence has been proposed for the main SN types running SNIa-SNII-SNIb-SNIc, in terms of increasing progenitor mass implied from an increasing association to host galaxy HII regions. In addition, it has been found that the SNIIn show a relatively low degree of association with the line emission, a result which is inconsistent with the majority of their progenitors being LBV stars, but which could be consistent with the hypothesis that additional SNIIn have SNIa-like progenitors, where the spectra are dominated by features of CSM interaction rather than those from the thermonuclear runaway.

Acknowledgments: J.A. acknowledges support from FONDECYT grant 3110142, and grant ICM P10-064-F (Millennium Center for Supernova Science), with input from 'Fondo de Innovacin para la Competitividad, del Ministerio de Economa, Fomento y Turismo de Chile'.

References

Anderson, J. P. & James, P. A., 2008 *MNRAS*, 390, 1527
Anderson, J. P., *et al.* 2010 *MNRAS*, 407, 2660
Anderson, J. P., *et al.* 2012 *MNRAS*, 424 1372
Dilday, B., *et al.* 2012 *Science*, 337, 942
Hamuy, M., *et al.* 2003 *Nature*, 424, 651
James, P. A. & Anderson, J. P., 2006 *A&A*, 453, 57
Kelly, P. L. & Kirshner, R. P., 2011 *ApJ*, 759, 107
Leloudas, G., *et al.* 2010 *A&A*, 518, 29
Leloudas, G., *et al.* 2011 *A&A*, 530, 95
Modjaz, M., *et al.* 2008 *AJ*, 136, 1136
Modjaz, M., *et al.* 2011 *ApJ*, 731, 4
Sanders, N. E., *et al.* 2012 *ApJ*, 758, 132
Smith, N. 2008 *IAUS* 250, 193
Stoll, R., *et al.* 2012 *arXiv* 1205.2338

Discussion

CAPPELLARO: Are you worried of a possible bias for classification of type IIn in particular when they occur in the middle of bright HII regions?

ANDERSON: Indeed this is a worry. We are currently compiling all available data on the SNIIn in our sample to look at these possible selection effects.

KOO: I wonder if you can discriminate whether SNe that you observe belong to the current generation of stars or to the previous generation of stars.

ANDERSON: There is no way of knowing which episode of SF the SN is related to. However, as stated in my talk, our method is completely statistical in nature: we would not claim that any individual SN is young or old dependent on an association. We only claim differences between SNe in a statistical sense.

WANG: Is it possible to establish the association on a more probability base?

ANDERSON: Yes, such approach has been adopted for association of SNe with galaxy types. It may also been done for HII regions based on the separation of a SN from an H alpha-emitting region.

Supernova Environmental Impacts
Proceedings IAU Symposium No. 296, 2013
A. Ray & R. A. McCray, eds.

doi:10.1017/S1743921313009356

Light Echoes of Historic Transients

Armin Rest[1], B. Sinnott[2], D. L. Welch[2], J. L. Prieto[3] and F. Bianco[4]

[1]STScI, 3700 San Martin Dr., Baltimore, MD 21218, USA
email: arest@stsci.edu

[2]Department of Physics and Astronomy, McMaster University,
Hamilton, Ontario L8S 4M1, Canada

[3]Department of Astrophysical Sciences, Princeton University,
4 Ivy Lane, Princeton, NJ 08544, USA

[4]Center for Cosmology and Particle Physics, New York University,
4 Washington Place, New York, NY 10003, USA

Abstract. Light echoes, light from a variable source scattered off dust, have been observed for over a century. The recent discovery of light echoes around centuries-old supernovae in the Milky Way and the Large Magellanic Cloud have allowed the spectroscopic characterization of these events, even without contemporaneous photometry and spectroscopy using modern instrumentation. Here we review the recent scientific advances using light echoes of ancient and historic transients, and focus on our latest work on SN 1987A's and Eta Carinae's light echoes.

Keywords. reflection nebulae, supernovae: general, supernovae: individual (SN 1987A), stars: individual (η Car), stars: variables: other

---

1. Introduction

Light echoes (LEs) arise when light from a transient or variable source is scattered off circumstellar or interstellar dust, reaching the observer after a time delay resulting from the longer path length (e.g., Couderc 1939; Chevalier 1986; Schaefer 1987; Xu *et al.* 1994; Sugerman 2003; Patat 2005). Over a century ago, in 1901, the first scattered LEs were discovered around Nova Persei (Ritchey 1901a,b, 1902). They were recognized as such shortly thereafter by Kapteyn (1902) and Perrine (1903). Since then, LEs have been observed around a wide variety of objects: the Galactic Nova Sagittarii 1936 (Swope 1940), the eruptive variable V838 Monocerotis (Bond *et al.* 2003), the Cepheid RS Puppis (Westerlund 1961; Havlen 1972), the T Tauri star S CrA (Ortiz *et al.* 2010), and the Herbig Ae/Be star R CrA (Ortiz *et al.* 2010). Echoes have also been observed from extragalactic SNe, with SN 1987A being the most famous case (Crotts 1988; Suntzeff *et al.* 1988b), but also including SNe 1980K (Sugerman *et al.* 2012), 1991T (Schmidt *et al.* 1994; Sparks *et al.* 1999), 1993J (Sugerman & Crotts 2002; Liu *et al.* 2003), 1995E (Quinn *et al.* 2006), 1998bu (Garnavich *et al.* 2001; Cappellaro *et al.* 2001), 2002hh (Welch *et al.* 2007; Otsuka *et al.* 2012), 2003gd (Sugerman 2005; Van Dyk *et al.* 2006; Otsuka *et al.* 2012), 2004et (Otsuka *et al.* 2012), 2006X (Wang *et al.* 2008; Crotts & Yourdon 2008), 2006bc (Gallagher *et al.* 2011; Otsuka *et al.* 2012), 2006gy (Miller *et al.* 2010), 2007it (Andrews *et al.* 2011), and 2008bk (Van Dyk 2013). All of the aforementioned LEs had the common selection criterion that they were found serendipitously while the transient source was still bright.

Early on in the last century, Zwicky (1940) had the idea that it might be possible to learn more about historical SNe by studying their scattered LEs. However, the few dedicated surveys trying to implement this idea for historic SNe (van den Bergh 1965a,b, 1966; Boffi *et al.* 1999) and novae (van den Bergh 1977; Schaefer 1988) were not successful. With the emergence of CCDs as astronomical detectors in combination with the advancement in telescope technology that allowed to image larger field-of-views, the wide-field time-domain surveys at the beginning of this century significantly improved in depth and area. These improvements led to the first discoveries of LEs at angular distances from the ancient transients too large to suggest an immediate association. The i400-900 year-old LEs from three LMC supernovae were found by Rest *et al.* (2005b) as part of the SuperMACHO survey (Rest *et al.* 2005a). Subsequent targeted searches in our Galaxy found LEs of Tycho's SN (Rest *et al.* 2007, 2008b), Cas A (Rest *et al.* 2007, 2008b; Krause *et al.* 2008a), and η Carinae (Rest *et al.* 2012a).

2. Light Echo Spectroscopy

Spectroscopy of LEs allows the transient to be studied long after it has already faded. The first LE spectrum was a 35 hour exposure of one of Nova Persei 1901 LEs (Perrine 1903), confirming that the nebulous moving features seen around Nova Persei were indeed its echoes. LE spectra of SN 1987A were most similar to those of the SN near maximum light (Gouiffes *et al.* 1988; Suntzeff *et al.* 1988b). Serendipitously, Schmidt *et al.* (1994) found the spectrum of SN 1991T taken 750 days after maximum again similar to the one at peak, indicating that at that time echoes from the SN at peak dominated the spectrum.

Spectroscopy of ancient SN LEs discovered in the LMC (Rest *et al.* 2005b) resulted in the first opportunity to classify ancient transients long after their direct light had encountered Earth (Rest *et al.* 2008a). The subsequent discovery and spectroscopy of LEs of Cas A and Tycho (Rest *et al.* 2007, 2008b; Krause *et al.* 2008a,b) allowed their spectroscopic classification as a SN IIb and normal SN Ia, respectively.

At first, it was widely believed that an observed LE spectrum would be the lightcurve-weighted integration of the transients' individual epochs (e.g., Patat *et al.* 2006; Rest *et al.* 2008a; Krause *et al.* 2008a). Such an integration is equivalent to assuming that the dust filament is thick, which is not always the case(Rest *et al.* 2011b, 2012b). Spectral features which persist for long periods of time during the evolution of an outburst will be weighted much more strongly when scattered by a thick filament than a thin filament.

In addition to spectroscopic classification, LEs also offer two more exciting scientific opportunities: "3D spectroscopy and "spectroscopic time series" of transients. Examples of realization of these techniques are provide in the following two sections using SN 1987A and η Carinae.

2.1. *3D Spectroscopy*

LEs scattered by different dust structures offer an opportunity that is unique in astronomy: probe the same object directly from different directions. Fig. 1 shows the light paths of seven SN 1987A LEs. By observing these LEs, SN 1987A can be analyzed as if the observers are at different line-of-sights, allowing to directly compare different hemispheres of one and the same object. This technique was first applied to observe η Carinae central star from different directions using spectra of the reflection nebula (Boumis *et al.* 1998; Smith *et al.* 2003).

Rest *et al.* (2011a) obtained spectra of three different Cas A LEs, viewing the Cas A SN from very different lines-of-sight. After accounting for the effects of the scattering dust, they found that the He I λ5876 and Hα features of one LE are blue-shifted by an additional $\sim$4000 km s^{-1} relative to the other two LE spectra. X-ray and optical data of the Cas A remnant also show a Fe-rich outflow in the same direction (Burrows *et al.* 2005; Wheeler *et al.* 2008; DeLaney *et al.* 2010). This indicates that Cas A was an intrinsically

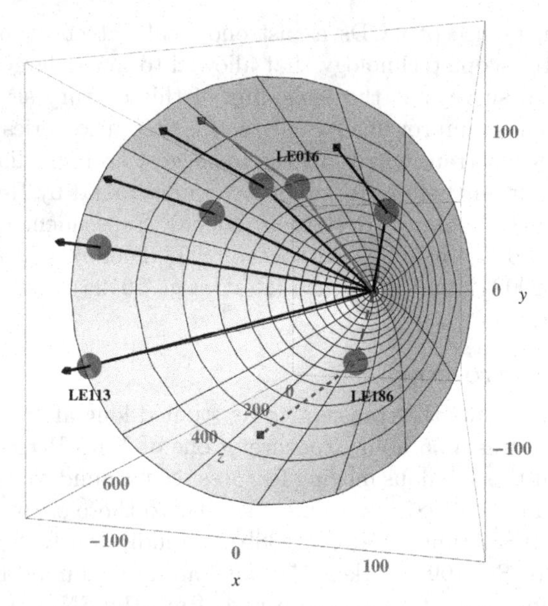

Figure 1. Light paths and 3D scattering dust locations for seven LEs of SN 1987A spectroscopically observed by Sinnott *et al.* (2013). Solid red and dashed blue lines highlight the extreme north and south viewing angles corresponding to LE016 and LE186 (see Figure 2). North is towards the positive y axis, east is towards the negative x axis, and z is the distance in front of the SN. All units are in light years. This figure is courtesy of Sinnott *et al.* (2013).

asymmetric SN. The blue-shifted SN ejecta is in the direction approximately opposite the motion of the resulting neutron star, suggesting that the explosion mechanism that gave the neutron star its kick affected the outer layers of the SN. This appears to be the first instance where the structure of the SN remnant can be directly associated with asymmetry observed in the explosion itself.

SN 1987A has a very rich set of LEs scattering off circumstellar (Crotts & Kunkel 1991; Crotts *et al.* 1995; Sugerman *et al.* 2005a,b) and interstellar dust (Crotts 1988; Suntzeff *et al.* 1988b; Gouiffes *et al.* 1988; Couch *et al.* 1990; Xu *et al.* 1994, 1995). The LEs shown in Figure 1 were spectroscopically observed by Sinnott *et al.* (2013). After correcting for the effects of the scattering dust, these spectra can be directly compared to a LE spectrum constructed from the spectro-photometric library of SN 1987A (Menzies *et al.* 1987; Catchpole *et al.* 1987, 1988; Hamuy *et al.* 1988; Suntzeff *et al.* 1988a; Whitelock *et al.* 1988; Catchpole *et al.* 1989; Phillips *et al.* 1988, 1990). Sinnott *et al.* (2013) find an excess in redshifted Hα emission and a blueshifted knee for the LE LE016 at position angle PA=16° (see red line in Figure 2). Both asymmetry signatures disappear as the PA increases, and then reappear in the form of an excess in blueshifted Hα emission and a redshifted knee at the opposite PA=186° LE, LE186 (see blue line in Figure 2).

In Figure 3, the light paths of the LE186 and LE016 echoes are illustrated and compared to the structure of the SN 1987A ejecta (Kjær *et al.* 2010). Even though the opening angle of the two LEs is only ~40°, the differences in the Hα lines are quite striking. Sinnott *et al.* (2013) argues that these differences are caused by a two-sided and asymmetric ^{56}Ni distribution in the outer H envelope. The symmetry axis defined by the 16°/186° viewing angles is in excellent agreement with the PA of the symmetry axis of the elongated ejecta that was measured to be ~15° (Wang *et al.* 2002; Kjær *et al.* 2010). Kjær *et al.* (2010) also found the present-day ejecta to be blueshifted in the north and redshifted in the south, inclined out of the plane of the sky by ~25°. The two-sided ^{56}Ni

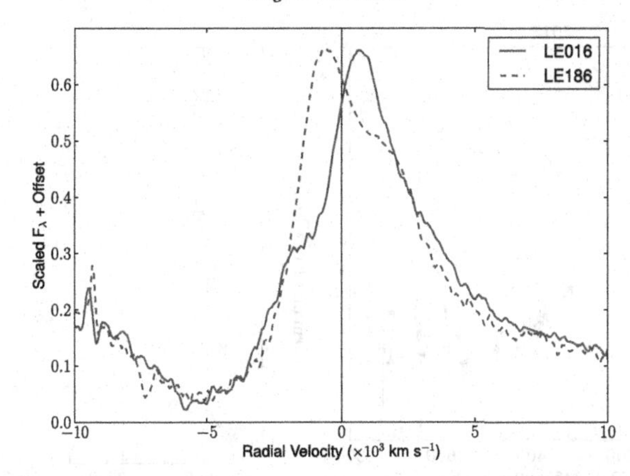

Figure 2. Observed Hα lines from LE016 and LE186. Emission peaks have been interpolated with high-order polynomials. Spectra are scaled and offset for comparison purposes, as well as smoothed with a boxcar of 3 pixels. Although this plot does not take into account the important differences in LE time-integrations between the spectra, it highlights the overall difference in fine-structure in the two LE spectra from opposite PAs. Observing Hα profiles with opposite asymmetry structure at opposite PAs is surprising considering the opening angle between the two LEs is $< 40°$. This figure and caption is courtesy of Sinnott *et al.* (2013).

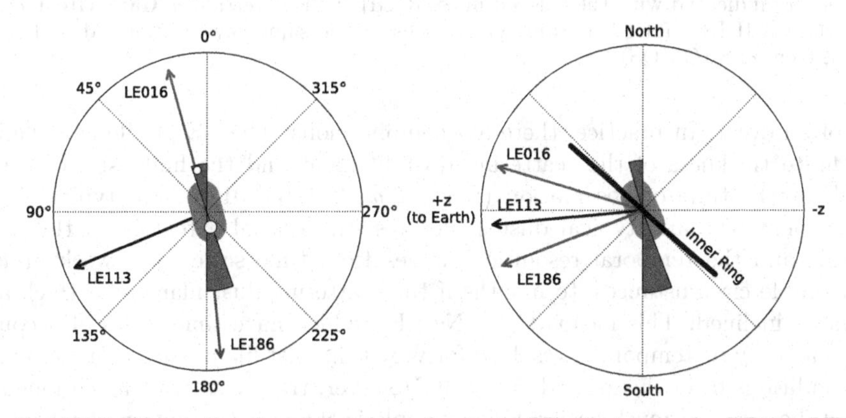

Figure 3. LEFT: PAs on the sky of the three dominant LE viewing angles LE016, LE113 and LE186. The central grey region denotes the orientation on the sky of the elongated remnant ejecta (PA = 16°) from Kjær *et al.* (2010). The green wedges illustrate the proposed two-sided distribution of ^{56}Ni, most dominant in the southern hemisphere. White circles illustrate the locations of the two mystery spots as identified by Nisenson & Papaliolios (1999), with radius proportional to relative brightness in magnitudes of the two sources. Only the relative distance from the center of the SNR to the mystery spots is to scale in the image. RIGHT: Schematic with viewing angle perpendicular to Earths line of sight. The inclination of the inner circumstellar ring is shown along with the proposed two-sided distribution of ^{56}Ni in green. Note that the green wedges are to highlight the proposed geometry (not absolute velocity) of the ^{56}Ni asymmetry probed by the LE spectra, illustrating that the southern overabundance is most dominant. This figure and caption is courtesy of Sinnott *et al.* (2013).

distribution proposed by Sinnott *et al.* (2013) as an explanation for the asymmetry seen in the LE observations is therefore roughly aligned with the ~25-year-old emerging SN remnant both in PA and inclination out of the sky.

2.2. *Spectroscopic Time Series*

If the scattering dust filament is infinitely thin, then the LE is just the projected light curve of the transient. Therefore it is, in theory, possible to obtain spectra from different

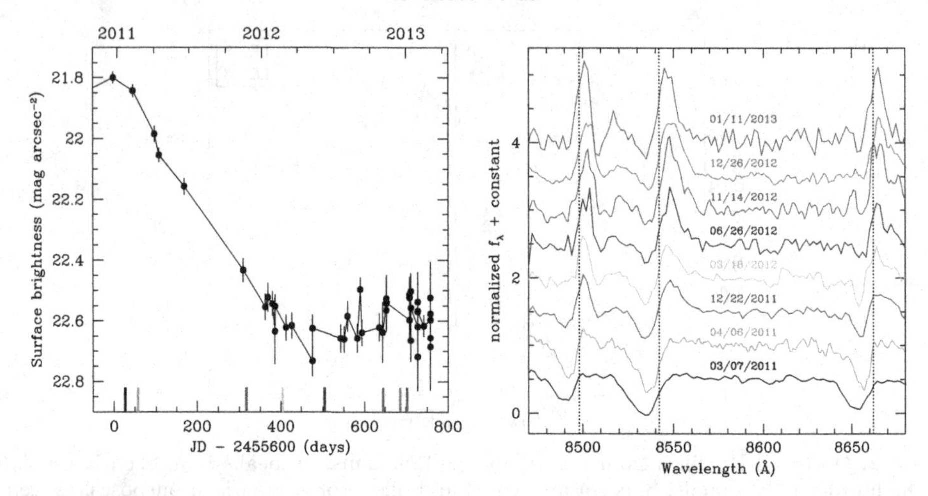

Figure 4. LEFT: Light curve of part of η Car's Great Eruption, derived from its LEs. Shown is the surface brightness in i band from the same sky position at different epochs (Blanco 4m (MOSAIC II, DECam), Swope (Direct CCD), FTS (Spectral), SOAR (SOI)). The LE is the projected light curve of the source transient. Since the LE has an apparent motion, the light curve of the source event "moves" through a given position on the sky. The epochs spectra were taken are indicated with the colored lines. RIGHT: LE spectra of η Car's Great Eruption showing the Ca II IR triplet from different epochs at the same sky location (Magellan Baade (IMACS), Gemini-S (GMOS)).

epochs of an event. In practice, there are complications: The LE profile gets convolved by the finite thickness of the scattering dust filament and the finite size of their slit and PSF at the time of observation (Rest *et al.* 2011b, 2012b). For typical Galactic LEs of ancient SNe and typical dust structures, the spatial extent is on the order of arcseconds and the temporal resolution ranges from time scales of a week under the most favorable circumstances, to months if the scattering dust filament is thick and/or unfavorably inclined. This means that SNe LEs, which have time-scales of a couple of months, can only be temporally resolved for very thin dust filaments and under excellent seeing conditions from the ground or space. However, transients with much longer time scales can be resolved much easier. One example is the Great Eruption of η Car, which lasted two decades and showed temporal variability on time-scales of months. In this case, LE spectra from different epochs are only marginally affected by dust width, slit and PSF size.

The left panel of Figure 4 shows the LE flux of η Car's Great Eruption at a given RA and Dec for various epochs. The light curve "moves" through a given position, and the shown flux is the light curve of some part of the Great Eruption convolved with the scattering dust thickness. The colored vertical lines indicate when we took a spectrum of the LE. The right panel of Figure 4 shows the corresponding spectrum in the same colors for the wavelength range covering the Ca II IR triplet. The spectrum taken close to the peak in the light curve correlates best with spectra of G2-to-G5 supergiants, a later range of types than predicted by standard opaque wind models (Rest *et al.* 2012a). The Ca II IR triplet is in absorption with an average blueshift of 200 km s^{-1}, and the lines are asymmetric extending up to blueshifts of 800 km s^{-1}(Rest *et al.* 2012a). Spectra taken at later epochs during the declining part of the light curve show a transition of the Ca II IR triplet from pure absorption through a P-Cygni profile to a nearly pure emission line spectrum. In addition, strong CN molecular bands develop (see Figure 5). This is unlike other LBV outbursts which move back to the earlier stellar types toward

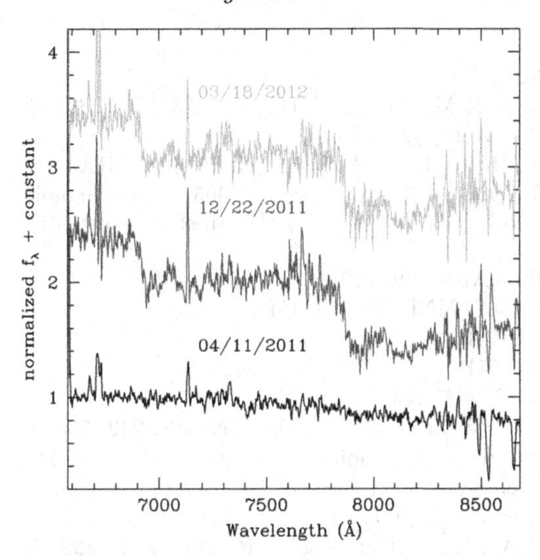

Figure 5. LE spectra of η Car's Great Eruption showing how the CN molecular bands develop for later epochs.

the end of the eruptions. The LEs of η Car indicate that the Great Eruption was not a typical LBV giant eruption. A paper detailing these observations is in preparation.

3. Summary

In the last decade, LE spectroscopy has emerged as powerful tool to spectroscopically classify ancient and historic SNe, for which no contemporary observations with modern instrumentation were possible. More recently, the technique of LE spectroscopy has been refined and improved, and it is now possible to utilize it to directly probe the asymmetries of the transient sources of the LEs. Furthermore, a spectroscopic time series can be obtained for favorable scattering dust filament structure or if the source transient is an event with a long time-scale.

4. Acknowledgments

We thank all the observers that have contributed to the monitoring of η Car's LEs, especially E. Hsiao (and the Carnegie Supernova Project II) and T. Matheson. Based on observations of program GS-2012B-Q-57 obtained at the Gemini Observatory, which is operated by the Association of Universities for Research in Astronomy, Inc., under a cooperative agreement with the NSF on behalf of the Gemini partnership: the National Science Foundation (United States), the National Research Council (Canada), CONICYT (Chile), the Australian Research Council (Australia), Ministério da Ciência, Tecnologia e Inovação (Brazil) and Ministerio de Ciencia, Tecnología e Innovación Productiva (Argentina). This paper includes data gathered with the 6.5 meter Magellan Telescopes located at Las Campanas Observatory, Chile. Based on observations at the Cerro Tololo Inter-American Observatory, National Optical Astronomy Observatory, which are operated by the Association of Universities for Research in Astronomy, under contract with the National Science Foundation. The SOAR Telescope is a joint project of: Conselho Nacional de Pesquisas Cientificas e Tecnologicas CNPq-Brazil, The University of North Carolina at Chapel Hill, Michigan State University, and the National Optical Astronomy Observatory.

References

Andrews, J. E., *et al.* 2011, ApJ, 731, 47

Boffi, F. R., Sparks, W. B., & Macchetto, F. D. 1999, A&AS, 138, 253

Bond, H. E., *et al.* 2003, Nature, 422, 405

Boumis, P., Meaburn, J., Bryce, M., & Lopez, J. A. 1998, MNRAS, 294, 61

Burrows, A., Walder, R., Ott, C. D., & Livne, E. 2005, in Astronomical Society of the Pacific Conference Series, Vol. 332, The Fate of the Most Massive Stars, ed. R. Humphreys & K. Stanek, 350–+

Cappellaro, E., *et al.* 2001, ApJ, 549, L215

Catchpole, R. M., *et al.* 1987, MNRAS, 229, 15P

——. 1988, MNRAS, 231, 75P

——. 1989, MNRAS, 237, 55P

Chevalier, R. A. 1986, ApJ, 308, 225

Couch, W. J., Allen, D. A., & Malin, D. F. 1990, MNRAS, 242, 555

Couderc, P. 1939, Annales d'Astrophysique, 2, 271

Crotts, A. 1988, IAU Circ., 4561, 4

Crotts, A. P. S. & Kunkel, W. E. 1991, ApJ, 366, L73

Crotts, A. P. S., Kunkel, W. E., & Heathcote, S. R. 1995, ApJ, 438, 724

Crotts, A. P. S. & Yourdon, D. 2008, ApJ, 689, 1186

DeLaney, T., *et al.* 2010, ApJ, 725, 2038

Gallagher, J. S., *et al.* 2011, in Bulletin of the American Astronomical Society, Vol. 43, American Astronomical Society Meeting Abstracts #217, 337.22–+

Garnavich, P. M., *et al.* 2001, in Bulletin of the American Astronomical Society, Vol. 33, American Astronomical Society Meeting Abstracts, 1370

Gouiffes, C., *et al.* 1988, A&A, 198, L9

Hamuy, M., Suntzeff, N. B., Gonzalez, R., & Martin, G. 1988, AJ, 95, 63

Havlen, R. J. 1972, A&A, 16, 252

Kapteyn, J. C. 1902, Astronomische Nachrichten, 157, 201

Kjær, K., Leibundgut, B., Fransson, C., Jerkstrand, A., & Spyromilio, J. 2010, A&A, 517, A51

Krause, O., Birkmann, S. M., Usuda, T., Hattori, T., Goto, M., Rieke, G. H., & Misselt, K. A. 2008a, Science, 320, 1195

Krause, O., Tanaka, M., Usuda, T., Hattori, T., Goto, M., Birkmann, S., & Nomoto, K. 2008b, Nature, 456, 617

Liu, J.-F., Bregman, J. N., & Seitzer, P. 2003, ApJ, 582, 919

Menzies, J. W., *et al.* 1987, MNRAS, 227, 39P

Miller, A. A., Smith, N., Li, W., Bloom, J. S., Chornock, R., Filippenko, A. V., & Prochaska, J. X. 2010, AJ, 139, 2218

Nisenson, P. & Papaliolios, C. 1999, ApJ, 518, L29

Ortiz, J. L., Sugerman, B. E. K., de La Cueva, I., Santos-Sanz, P., Duffard, R., Gil-Hutton, R., Melita, M., & Morales, N. 2010, A&A, 519, A7+

Otsuka, M., *et al.* 2012, ApJ, 744, 26

Patat, F. 2005, MNRAS, 357, 1161

Patat, F., Benetti, S., Cappellaro, E., & Turatto, M. 2006, MNRAS, 369, 1949

Perrine, C. D. 1903, ApJ, 17, 310

Phillips, M. M., Hamuy, M., Heathcote, S. R., Suntzeff, N. B., & Kirhakos, S. 1990, AJ, 99, 1133

Phillips, M. M., Heathcote, S. R., Hamuy, M., & Navarrete, M. 1988, AJ, 95, 1087

Quinn, J. L., Garnavich, P. M., Li, W., Panagia, N., Riess, A., Schmidt, B. P., & Della Valle, M. 2006, ApJ, 652, 512

Rest, A., *et al.* 2007, in Bulletin of the American Astronomical Society, Vol. 38, Bulletin of the American Astronomical Society, 935

Rest, A., *et al.* 2011a, ApJ, 732, 3

——. 2008a, ApJ, 680, 1137

——. 2012a, Nature, 482, 375

Rest, A., Sinnott, B., & Welch, D. L. 2012b, PASA, 29, 466

Rest, A., Sinnott, B., Welch, D. L., Foley, R. J., Narayan, G., Mandel, K., Huber, M. E., & Blondin, S. 2011b, ApJ, 732, 2
Rest, A., *et al.* 2005a, ApJ, 634, 1103
——. 2005b, Nature, 438, 1132
——. 2008b, ApJ, 681, L81
Ritchey, G. W. 1901a, ApJ, 14, 293
——. 1901b, ApJ, 14, 167
——. 1902, ApJ, 15, 129
Schaefer, B. E. 1987, ApJ, 323, L47
——. 1988, ApJ, 327, 347
Schmidt, B. P., Kirshner, R. P., Leibundgut, B., Wells, L. A., Porter, A. C., Ruiz-Lapuente, P., Challis, P., & Filippenko, A. V. 1994, ApJ, 434, L19
Sinnott, B., Welch, D. L., Rest, A., Sutherland, P. G., & Bergmann, M. 2013, ApJ, 767, 45
Smith, N., Davidson, K., Gull, T. R., Ishibashi, K., & Hillier, D. J. 2003, ApJ, 586, 432
Sparks, W. B., Macchetto, F., Panagia, N., Boffi, F. R., Branch, D., Hazen, M. L., & della Valle, M. 1999, ApJ, 523, 585
Sugerman, B. E. K. 2003, AJ, 126, 1939
——. 2005, ApJ, 632, L17
Sugerman, B. E. K., *et al.* 2012, ApJ, 749, 170
Sugerman, B. E. K. & Crotts, A. P. S. 2002, ApJ, 581, L97
Sugerman, B. E. K., Crotts, A. P. S., Kunkel, W. E., Heathcote, S. R., & Lawrence, S. S. 2005a, ApJ, 627, 888
——. 2005b, ApJS, 159, 60
Suntzeff, N. B., Hamuy, M., Martin, G., Gomez, A., & Gonzalez, R. 1988a, AJ, 96, 1864
Suntzeff, N. B., Heathcote, S., Weller, W. G., Caldwell, N., & Huchra, J. P. 1988b, Nature, 334, 135
Swope, H. H. 1940, Harvard College Observatory Bulletin, 913, 11
van den Bergh, S. 1965a, AJ, 70, 667
——. 1965b, PASP, 77, 269
——. 1966, PASP, 78, 74
——. 1977, PASP, 89, 637
Van Dyk, S. D. 2013, ArXiv e-prints, 1305.6639
Van Dyk, S. D., Li, W., & Filippenko, A. V. 2006, PASP, 118, 351
Wang, L., *et al.* 2002, ApJ, 579, 671
Wang, X., Li, W., Filippenko, A. V., Foley, R. J., Smith, N., & Wang, L. 2008, ApJ, 677, 1060
Welch, D. L., Clayton, G. C., Campbell, A., Barlow, M. J., Sugerman, B. E. K., Meixner, M., & Bank, S. H. R. 2007, ApJ, 669, 525
Westerlund, B. 1961, PASP, 73, 72
Wheeler, J. C., Maund, J. R., & Couch, S. M. 2008, ApJ, 677, 1091
Whitelock, P. A., *et al.* 1988, MNRAS, 234, 5P
Xu, J., Crotts, A. P. S., & Kunkel, W. E. 1994, ApJ, 435, 274
——. 1995, ApJ, 451, 806
Zwicky, F. 1940, Reviews of Modern Physics, 12, 66

Discussion

ZANARDO: Can the light echoes from SN 1987A shed some 'light' into the nature of the explosion and the likely asymmetric explosion?

REST: Yes, with the data in hand, we already see significant asymmetry in H-alpha. We still have to put this more carefully into the context of the asymmetry in the ejecta and original CS dust structure.

PODSIADLOWSKI: Is there a way your light-echo analysis could shed light on the issue of the mystery spot that has been reported in SN87A?

Rest: It depends on how bright the mystery spot is compared to the main explosion. If it is a significant fraction, then yes.

Anderson: How many more events/transients do you think we will be able to apply this method to?

Rest: Probably another 5 10 SN / eruptions in MW galaxy, and then also in some of the other Local Group galaxies like SMC and M31

Milisavljevic: Have you compared your results with the recent measurements of the 3D properties of SN 1987A's ejecta?

Rest: We have just started to look more closely into this.

Sankrit: With several years of experience now, is there some way of identifying Dust-sheets that would be likely places for observing light echoes

Rest: Spitzer 8 micron images seem to correlate best with the light echoes. However, having an 8 micron structure doesn't mean that there are light echoes, since also the distance to the dust sheet needs to be just right.

Supernova Environmental Impacts
Proceedings IAU Symposium No. 296, 2013
A. Ray & R. A. McCray, eds.

doi:10.1017/S1743921313009368

Circumstellar interaction in Type IIn supernovae

Poonam Chandra[1], Roger A. Chevalier[2], Nikolai Chugai[3], Alicia M. Soderberg[4], and Claes Fransson[5]

[1]National Centre for Radio Astrophysics, Pune University Campus, Ganeshkhind, Pune-411 007, INDIA
email: poonam@ncra.tifr.res.in

[2]Dept. of Astronomy, University of Virginia, P.O. Box 400325, Charlottesville, VA 22904, USA

[3]Institute of Astronomy of Russian Academy of Sciences, Pyatnitskaya Street 48, 109017 Moscow, Russia

[4]Smithsonian Astrophysical Observatory, 60 Garden Street, Cambridge, MA 02138, USA

[5]Department of Astronomy, Stockholm University, AlbaNova, SE-106 91 Stockholm, Sweden

Abstract. Type IIn supernovae have bright optical emission and high bolometric luminosities. Due to their high mass loss, their are expected to have dense circumstellar interaction, thus produce bright radio and X-ray emission. We aim to carry out systematic study to understand their circumstellar interaction, mass loss properties. Here, I provide specific examples of two Type IIn supernovae, 2006jd and 2010jl.

Keywords. dense matter, radiation mechanisms: general, circumstellar matter, stars: mass loss, supernovae: individual (2006jd, 2010jl)

1. Introduction

Type IIn supernovae (hereafter SNe IIn) usually have bright optical emission and high bolometric luminosities. They show characteristic narrow emission lines of hydrogen and helium (Schlegel 1990, Filippenko 1997), which are indicative of dense circumstellar medium (CSM). This class of supernovae (SNe) generally show significant heterogeneity in terms of their emission line profiles, luminosity and mass loss rates. In SNe IIn radio, X-ray and late time optical emission comes from circumstellar interaction of the ejecta with the CSM. When the ejecta plow into the CSM with supersonic speeds, a forward shock moves into the circumstellar wind generating $\sim 100v_4^2$ keV plasma, and a reverse shock in the freely expanding SN ejecta generating $\sim 1v_4^2$ keV plasma, where v_4 is the forward shock velocity in units of 10^4 km s^{-1}. Radio emission is expected to be non-thermal synchrotron emission from the forward shock, while X-ray emission is likely to have a thermal origin arising in both reverse and forward shock Chevalier 1982, Chevalier & Fransson 2003. Because of the high density of the CSM, the emission due to the CSM interaction is expected to be high. SNe IIn are relatively X-ray luminous, exceeding 10^{40} erg s^{-1}. However, they have remain elusive at radio wavelengths. Out of couple of hundred total SNe IIn, 81 have been looked in radio bands with only 10 detected. This is merely 5 % of total SNe IIn and 12 % of observed.

To understand the progenitor, evolution and mass loss history of SNe IIn, we are carrying out a campaign to study them in multiwavebands with first observations initiated in radio bands. The detected ones are followed up with *Chandra*, *XMM*-Newton or *Swift*-XRT in X-ray bands. So far we have observed a total of 43 SNe IIn with the Very

Large Array (VLA)†. We have been able to detect radio emission from 4 of them, which are SN 2005kd, 2006jd, 2008iy and 2009ip. In X-ray we have detected SN 2006jd (with *Chandra* and *XMM*-Newton), 2010jl (with *Chandra* at 3 epochs) and SN 2009ip (with *XMM*-Newton). Here we describe SN 2006jd and 2010jl.

SN 2006jd was discovered on 2006 October 12, with an apparent magnitude of 17.2, in the galaxy UGC 4179 at a redshift of $z = 0.0186$ (Prasad & Li 2006). The initial spectrum of SN 2006jd showed features of a Type IIb SN, similar to SN 1993J (Blondin *et al.* 2006). However, based on Keck spectra at late epochs, it was reclassified as a Type IIn SN (A. Filippenko, 2006, private communication). On the basis of a spectrum on 2006 October 17.51 UT, we assume its explosion date to be 2006 October 06.5 UT (Blondin *et al.* 2006). Immler *et al.* (2007) detected X-ray emission from SN 2006jd on 2007 November 16 with the Swift X-ray Telescope (XRT) in a 2.3 ks exposure. The net count rate was $(6.3 \pm 2.0) \times 10^{-3}$ counts s^{-1}. The first radio emission was detected from the SN on 2007 November 21.28 UT in the 5 GHz band, on 2007 November 26.36 UT in the 8.5 GHz band, and on 2007 November 26.38 UT in the 22.5 GHz band Chandra & Soderberg 2007.

SN 2010jl was discovered on 2010 November 3.5 (UT) at a magnitude of 13.5 (Newton & Puckett 2010), and brightened to magnitude 12.9 over the next day. Pre-discovery observations indicate an explosion date in early 2010 October (Stoll *et al.* 2011). Spectra on 2010 November 5 showed it to be a Type IIn event (Benetti *et al.* 2010). The apparent magnitude is brightest for a Type IIn SN since SN 1998S. SN 2010jl is associated with the galaxy UGC 5189A at a distance of 50 Mpc ($z = 0.011$), implying that SN 2010jl reached $M_V \sim -20$ (Stoll *et al.* 2011) and placing it among the more luminous Type IIn events (Kiewe *et al.* 2012). Hubble Space Telescope (HST) images of the site of the SN taken a decade before the SN indicate that the progenitor star had an initial mass $\geqslant 30 M_\odot$ (Smith *et al.* 2011). The *Swift* on board X-ray Telescope (XRT) detected X-rays from SN 2010jl on 2010 November 5.0–5.8 (Immler *et al.* 2010).

2. Observations

2.1. *SN 2006jd Observations*

SN 2006jd was observed in radio bands with the VLA from 2007 Nov 21 onwards and until 2012 Apr 7 in the 22.5 GHz (K band), 8.5 GHz (X), 5 GHz (C), and 1.4 GHz (L) bands at various epochs, along with a 44 GHz band observation at one epoch. For C and X bands, the data were taken in the interferometric mode for an average of 30 minutes (including calibrator time), whereas for L and K bands the data were collected for 1 hr. The data was analysed using standard AIPS routines. We also observed SN 2006jd with the Giant Metrewave Radio Telescope (GMRT) on three occasions between 2009 October to 2010 April in the 1.3 GHz and 0.61 GHz bands and detected in the 1.3 GHz band at both occasions.

SN 2006jd was observed with the Swift on board XRT at 18 epochs between 2007 November and 2011 March in Photon counting mode. All the Swift observations were for exposures less than 10 ks. We also observed SN 2006jd once with the *Chandra* and once with the *XMM*-Newton X-ray observatories. We carried out the *XMM*-Newton observations starting 2009 April 7 at 4:36:46 UT with the EPIC-PN and EPIC-MOS cameras in full frame with thin filter mode. The exposures for the EPIC-MOS1 and EPIC-MOS2 were 42.667 ks and 42.672 ks, respectively, and for the EPIC-PN, the exposure time

† The National Radio Astronomy Observatory is a facility of the National Science Foundation operated under cooperative agreement by Associated Universities, Inc.

was 41.032 ks. We used the EPIC-PN observations to carry out a detailed spectroscopic analysis and 1963 counts were obtained in the 0.2–10 keV range Chandra *et al.* 2012b. We also observed SN 2006jd with *Chandra* ACIS-S in VFAINT mode. The observations started on 2009 September 14 at 00:03:21 UT for a total exposure of 37.24 ks. A total of 888 counts were obtained in the 0.2–10 keV range Chandra *et al.* 2012b.

2.2. *SN 2010jl Observations*

We started radio observations of the SN 2010jl with the VLA starting 2010 Nov 06. The last observation was taken on 2013 Jan 19. The observations have been made in K, C and X bands. So far we have not detected the SN in radio bands. We are continuing to follow it in radio bands.

The Swift detection of SN 2010jl allowed us to trigger our approved *Chandra* Cycle 11 program in 2010 December. We again observed SN 2010jl in 2011 October and then 2012 June under Cycle 13 of *Chandra.* The observations were taken with ACIS-S without grating in a VFAINT mode and the duration of the observations at each epoch was ~ 40 ks. In all cases we detected the SN with absorbed count rates of 1.13×10^{-2} cts s^{-1}, 3.29×10^{-2} cts s^{-1} and 3.68×10^{-2} cts s^{-1}, respectively (see Chandra *et al.* 2012a for first two epochs).

3. Results and Interpretation

3.1. *Supernova 2006jd*

SN 2006jd remained bright in all the radio bands over the four-year span of observations, except for two early upper limits in L band and occasional upper limits in the K band at later epochs. The radio observations of SN 2006jd show an evolution from a somewhat positive spectral index to a negative index. This type of evolution is commonly observed in radio SNe (Weiler *et al.* 2002) and is attributed to the transition from optically thick to optically thin radiation. However, the external free-free and synchrotron absorption models are not able to fit the radio data well, which appear to give more of a power-law rise. This is also observed in other SNe IIn where the rise of the radio flux was followed: SN 1986J (Weiler *et al.* 1990) and SN 1988Z (van Dyk *et al.* 1993, Williams *et al.* 2002). Weiler *et al.* (1990) proposed an internal free-free absorption (FFA) model in which thermal absorbing gas is mixed into the synchrotron emitting gas, so that the flux takes the form

$$
\begin{aligned}
F(\nu,t) &= K_1 \left(\frac{\nu}{5\,\mathrm{GHz}}\right)^{\alpha} \left(\frac{t}{1000\,\mathrm{days}}\right)^{\beta} \left(\frac{1-\exp(-\tau_{\mathrm{intFFA}})}{\tau_{\mathrm{intFFA}}}\right), \\
\tau_{\mathrm{intFFA}} &= K_3 \left(\frac{\nu}{5\,\mathrm{GHz}}\right)^{-2.1} \left(\frac{t}{1000\,\mathrm{days}}\right)^{\delta'}, \qquad (3.1)
\end{aligned}
$$

where K_1 and K_3 are the flux and absorption normalizations, respectively. Here α is the optically thin frequency spectral index, which relates to the electron energy index p ($N(E) \propto E^{-p}$) as $p = 1-2\alpha$. The assumption that the energy density in the particles and the fields is proportional to the postshock energy density leads to $\beta = 3m-(3-\alpha)(ms+2-2m)/2$. Here m is the expansion parameter ($R \propto t^m$, R is the shock radius) and s is the CSM density power-law index ($\rho \propto r^{-s}$). This model seems to fit the data very well for $s = 1.6$ (Fig. 1). The above equation reduces to $F(\nu) \propto \nu^{2.1+\alpha}$ in optically thick limit. The value of α is determined by the late optically thin evolution and is $\alpha = -1.04$ in our case. The fact that the optically thick spectrum is approximately reproduced (Fig. 1)

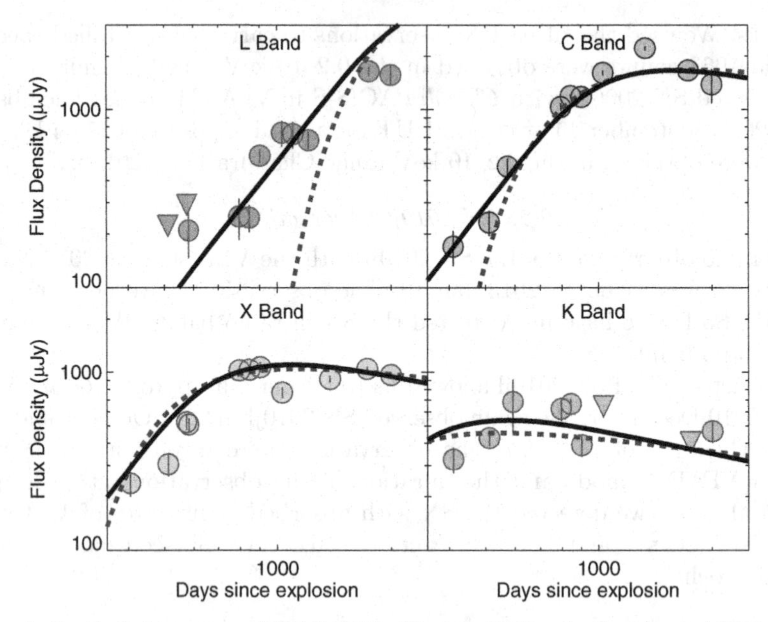

Figure 1. Radio light curves of SN 2006jd. Black solid line is the internal FFA model, whereas, the dashed line is the external FFA model.

provides support for the internal FFA model. We find that external absorption played a negligible role in the best-fit case. We also estimate that the mass of the cool gas required to do internal FFA is $M_a \approx 2 \times 10^{-8} T_4^{5/2} M_\odot$, showing that a modest amount of cool gas mixed into the emitting region can give rise to the needed absorption (Chandra *et al.* 2012b). The source of the cool gas is likely to radiative cooling of dense gas in the shocked region.

The XMM-Newton and Chandra X-ray spectra can be fit by either an optically thin thermal spectrum, which leads to an electron temperature $T_e \geqslant 20$ keV, or a nonthermal spectrum with photon index $\Gamma \approx 1.2$. In the nonthermal interpretation, flux $\propto \nu^{0.2}$ (as $\alpha = 1 - \Gamma$) would be a surprisingly hard spectrum for either synchrotron or inverse Compton emission. Thus, we discard the nonthermal origin of X-rays. The X-ray data were best fit with a thermal plasma model at a temperature above 10 keV, i.e., in a range where the 0.2–10 keV spectral shape is not sensitive to temperature. The spectrum also indicates a column density of 1.3×10^{21} cm^{-2}, which is significantly larger than the Galactic absorption column density of 4.5×10^{20} cm^{-2} in that direction. We have a clear detection of 6.9 keV Fe line, which is due to Fe xxvi. A Mekal model with five times the solar abundance does fit the spectrum well and reproduces the iron line at the correct energy but with a narrower width. To look for the possibility of nonequilibrium ionization (NEI), we also fit the XSPEC NEI model to our data. The fit is quite good and it reproduces the iron line at the correct energy. However, the best-fit ionization timescale in the NEI model is $\tau = 6.28 \times 10^{11}$ s cm^{-3}. Since $\tau = nt$ (n is the number density and t is the age) and the XMM spectrum was taken when the SN age was 908 days (or 7.84×10^7 s), this implies a number density of $n = 7.7 \times 10^3$ cm^{-3}. This seems low and makes NEI model unlikely.

The Chandra spectrum is best fit with a high-temperature thermal plasma and a column density of 1.6×10^{21} cm^{-2}, which agree with our fits to the XMM-Newton spectrum very well. Even though the Chandra spectrum is not as detailed as that of the XMM-Newton, we clearly detect the 6.9 keV Fe line. For Swift observations, we converted the

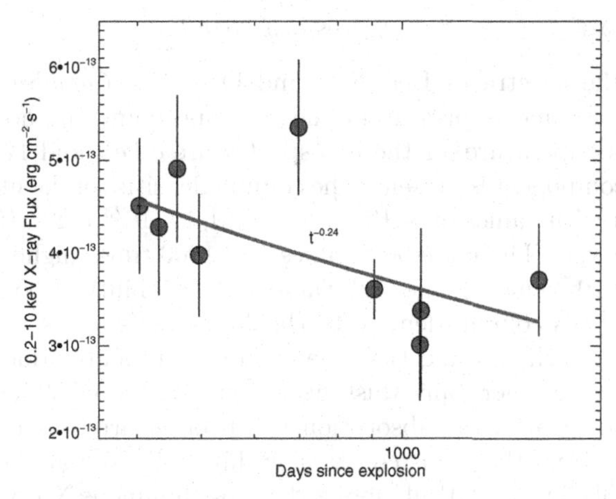

Figure 2. X-ray 0.3–10 keV light curve of SN 2006jd.

count rates to 0.2–10 keV fluxes using a thermal plasma model with a temperature of 60 keV and column density of 1.3×10^{21} cm^{-2}. This gives us X-ray light curve spanning 4 years (Fig. 2). The light curve luminosity L_ν decays as $L_\nu \propto t^{-0.24}$. For bremsstrahlung emission $L_\nu \propto t^{2m(1-s)+1}$. This implies $s = 1.7$ for $m = 0.9$. The radio and X-ray observations thus give $s \approx 1.6$.

In the thermal interpretation, the temperature T is assumed to be produced by shock heating, with the shock velocity $v_{sh} = [16kT/(3\mu m_p)]^{1/2} = 6700(kT/60\text{keV})^{1/2}$ km s^{-1}, where k is Boltzmann's constant, μ is the mean molecular weight, and m_p is the proton mass. The shock velocity is closer to expectations for the forward shock than the reverse shock wave at this age, although X-ray emission from young SNe is typically attributed to the reverse shock wave (Chevalier & Fransson 2003). Using results from Fransson *et al.* (1996) for L_ν, the mass-loss rate $\dot{M}$, normalized to $R = 10^{15}$ cm, is $\dot{M}_{-3}/v_{w2} \approx 5v_4^{0.6}$, where $\dot{M}_{-3} = \dot{M}/(10^{-3}\, M_\odot \text{ yr}^{-1})$, v_{w2} is the preshock wind velocity in units of 100 km s^{-1}, and v_4 is the average velocity in units of 10^4 km s^{-1} at 10^3 days. At 10^3 days, the preshock density is $2 \times 10^{18} v_4^{-1}$ g cm^{-3}. The corresponding density in the shocked gas is $n_H \sim 3 \times 10^6$ cm^{-3}.

A robust result of fitting the X-ray spectrum is the hydrogen column density N_H to the source. Of the $N_H = 1.3 \times 10^{21}$ cm^{-2} that is observed, 4.5^{20} cm^{-2} can be attributed to the Galaxy, so there is 8.5×10^{20} cm^{-2} left for the SN host galaxy interstellar medium and the SN CSM. This column density estimate assumes that the absorbing medium has solar abundances, and is cool and not highly ionized. In the model for the X-ray emission discussed above, the expected column density in the unshocked CSM is 410^{22} cm^{-2}, a factor of 50 larger than that inferred from the observations. A possible reason is that the CSM is fully ionized by the X-ray radiation. However, the ionization parameter for this case is $\xi = L/nr^2 = 50$, which indicates that there may be some ionization of CNO elements, but not full ionization of the gas Verner & Ferland 1996. To check on this result, we ran the CLOUDY code Ferland *et al.* 1998 for the parameters discussed above. As expected, we found some ionization of CNO elements, but little change in the absorption of the X-ray spectrum. The radio data also suggests that the FFA along the line of sight to the X-ray emission fails to produce the observed radio absorption. This is consistent with our analysis of the radio emission, which showed that external FFA does not lead to good model fits to the data.

3.2. *Supernova 2010jl*

In Fig. 3, we plot the spectra of Dec 2010 and Oct 2011 *Chandra* observations. The fluorescent 6.4 keV Fe line is present in the first spectrum but not the second. The lower limits on the temperature for the two spectra are 8 keV and 12 keV, respectively, showing that a hot component is present. The column densities of the main X-ray emission component are high with values of $\sim 10^{24}$ cm^{-2} and 3×10^{23} cm^{-2} (for a metallicity of $Z \sim 0.3 Z_{\odot}$), respectively. These are 3000 times and 1000 times higher than the Galactic column density (3×10^{20} cm^{-2}). The high value and variability of N_H point to an origin in the CSM. The excess column density to the X-ray emission is not accompanied by high extinction to the SN, showing that the column is probably due to mass loss near the forward shock wave where any dust has been evaporated. This is the first time that external circumstellar X-ray absorption has been clearly observed in an SN. The unabsorbed emission from the SN is constant within 20%–30% at the two epochs, with $\sim 7 \times 10^{41}$ erg s^{-1}, placing SN 2010jl among the most luminous X-ray SNe yet observed. In the thermal interpretation, a lower limit of 10 keV for the temperature puts a lower limit of the 2700 km s^{-1} on the shock speed. In comparing the observed luminosity to a thermal emission model to find the physical parameters, we note that our measurements give the spectral luminosity, not the total luminosity. We use Equation (3.11) of Fransson *et al.* (1996) for the luminosity. These expressions allow for a variation of the pre-shock density $\propto r^{-s}$, where s is a constant. For the plausible value $m = 0.8$, we find that $s = 1.6$ gives a reasonable representation of the luminosity and N_H evolution. The implied value of the mass loss rate $\dot{M}$, normalized to $R = 10^{15}$ cm, is $\dot{M}_{-3}/v_{w2} \approx 8 v_4^{0.6}$, where $\dot{M}_{-3} = \dot{M}/(10^{-3}\, M_{\odot}\, \mathrm{yr}^{-1})$, $v_w 2$ is the pre-shock wind velocity in units of 100 km s^{-1}, and v_4 is the shock velocity in units of 10^4 km s^{-1} at the first epoch. The high temperature implies that we are observing the forward shock region. In modeling the X-ray absorption in SN 2010jl we have assumed that the absorbing gas is not fully ionized. If the circumstellar gas is photoionized by the X-ray emission, the absorption is reduced (e.g., Fransson 1982).

The 2010 December spectrum shows a 6.4 keV feature, which is identified with the narrow Kα iron line. Since the 6.4 keV Fe line arises from neutral or low ionized iron (Fe I to Fe XI), it supports our finding that the radiation field is not able to completely ionize the circumstellar gas. At the second epoch, N_H is smaller by a factor of three, so the strength of the Fe line should be correspondingly smaller; this is consistent with the nondetection of the line. The problem with this picture is that it assumes Fe is in the low ionization stages that produce the Kα line; this requires an ionization parameter $\xi \leqslant 5$ (Kallman et al. 2004), which is below the inferred value. One possibility is that the circumstellar gas is clumped, with a density $\geqslant 40$ times the average; another is that the Kα line emission is from dense gas that is not along the line of sight.

A thermal fit to the low-temperature component (Fig. 3) implies an absorbing column density of $(1.37 \pm 8.44)10^{20}$ cm^{-2}, much less than the column to the hot component and consistent with the Galactic column density within the errors. This rules out the possibility that the cooler X-rays come from slow cloud shocks in the clumpy CSM or from the reverse shocks. The component is also present in the second epoch. It could arise from a pre-SN mass loss event or from an unrelated source in the direction of the SN. The components are best fit with either a thermal component ($T \sim 1-2$ keV) or a power law with $\Gamma = 1.6 - 1.7$. The luminosities of this component in the 2010 December and 2011 October spectra are 3.5×10^{39} erg s^{-1} and 5×10^{39} erg s^{-1}, respectively. The luminosity range and the power-law index are compatible with a background ultraluminous X-ray source (ULX), which can typically be described by an absorbed power-law spectrum

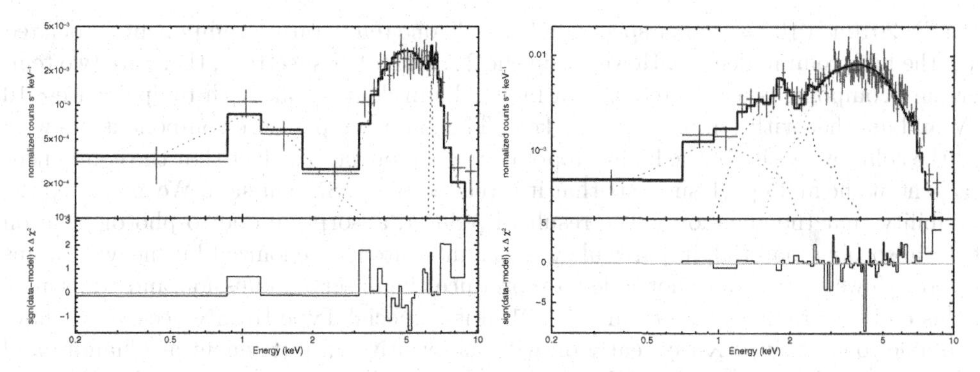

Figure 3. Chandra spectra of SN 2010jl taken in Dec 2010 (left) and in Oct 2011 (right). The presence of Fe 6.4 keV line in the first spectrum is evident. There is also an indicating of an extra component (around 1 keV) in the Oct 2011 spectrum.

Swartz *et al.* 2004. Since the error in the flux determination is between 20%–30%, a factor of 1.4 change in the luminosity at the two epochs is consistent with a constant flux. Thus we attribute this component to a background source, most likely a ULX, which is associated with the blue excess emission region seen in the pre-SN HST images Smith *et al.* 2011. There is an additional low temperature component present in the second epoch which was not present in the Dec 2010 spectrum. Origin of this component is not clear.

4. Discussion and Conclusions

The basic interpretation of the X-ray and radio data in terms of standard spherical models led to inconsistencies in SN 2006jd. One is that the column density of matter to the X-ray emission is about a factor 50 smaller than that needed to produce the X-ray luminosity. The straightforward explanation is that there is a global asymmetry in the distribution of the circumstellar gas that allows a low column density in one direction, while dense interaction is taking place over much of the rest of the solid angle (as viewed from the SN). Chugai & Danziger (1994) had suggested a scenario with equatorial mass loss to explain the presence of fast- and intermediate-velocity shock fronts in the Type IIn SN 1988Z. Polarization observations of SNe IIn have shown evidence for large-scale asymmetry, e.g., SN 1997eg Hoffman *et al.* 2008, SN 1998S Leonard *et al.* 2000, and SN 2010jl Patat *et al.* 2011. In this view, the column density to the radio emission would be small, as for the X-ray emission, because they are both from the same region. The external FFA model for the radio absorption thus fails, and an absorption mechanism internal to the emission is indicated. We have argued that thermal absorption internal to the emitting region is a plausible mechanism, as has previously been proposed for other SNe IIn Weiler *et al.* 1990. There is evidence that the X-ray properties we have found for SN 2006jd are not unusual for a Type IIn event. Chandra observations of SN 2001em and SN 2005dk have shown hard emission Pooley & Lewin 2004, Pooley *et al.* 2007, implying a high temperature. Our observations at both radio and X-ray wavelengths imply that the circumstellar density profile ($\rho \propto r^{-2}$) is flatter than the $s = 2$ case that would be expected for a steady wind. We find $s = (1.5 - 1.6)$; deviation from the steady case is plausible for a SN IIn because the mass loss may be due to an eruptive event. In addition, the optical luminosity evolution of SN 2006jd shows a second peak at $\sim$ 500 days Stritzinger *et al.* 2012 that is indicative of a nonstandard density distribution.

In SN 2010jl 2010 December spectrum has only one temperature component associated with the high column density. However, in the 2011 October spectrum, there are two temperature components associated with a high column density, one with temperature ~10 keV and another with temperature 1.1 keV. The lower temperature component fits with 1/4 the column density of the high-temperature component. The fact that the component is absent at the first epoch suggests that it is related to the SN emission. We examined the possibility that the emission is the result of reduced absorption due to photoionization of the absorbing material, in particular, that lighter atoms are ionized but heavier atoms are not. However, we were not able to reproduce the observed emission and the source of this emission remains uncertain. SN 2010jl is a special Type IIn SN because we have been able to catch it in X-rays early on with as sensitive an instrument as Chandra and trace the early X-ray evolution. We observe dramatic changes over two epochs separated by 10 months. For the first time we see clear evidence of external CSM absorption in an SN. We also find that the CSM is not fully photoionized by the SN emission, the SN is very luminous in X-rays, and the temperature of the emitting gas is $\geqslant$10 keV.

References

Benetti, S., Bufano, F., Vinko, J., *et al.* 2010, *CBET*, 2536, 1

Blondin, S., Modjaz, M., Kirshner, R., *et al.* 2006, *CBET*, 679, 1

Chandra, P., Chevalier, R. A., Irwin, C. M., *et al.* 2012, *ApJ*, 750, L2

Chandra, P., Chevalier, R. A., Chugai, N., *et al.* 2012, *ApJ*, 755, 110

Chandra, P. & Soderberg, A. 2007, *ATel*, 1297, 1.

Chevalier, R. A. & Fransson, C. 2003, in: K. Weiler (eds.), *Supernovae and Gamma-Ray Bursters: Lecture Notes in Physics* (Berlin: Springer), vol. 598, p. 171

Chevalier, R. A. 1982, *ApJ*, 259, 302

Chugai, N. N. & Danziger, I. J. 1994, *MNRAS*, 268, 173

Ferland, G. J., Korista, K. T., Verner, D. A., *et al.* 1998, *PASP*, 110, 761

Filippenko, A. V. 1997, *ARAA*, 35, 309

Fransson, C., Lundqvist, P., & Chevalier, R. A. 1996, *ApJ*, 461, 993

Fransson, C. 1982, *A&A*, 111. 140

Hoffman, J. L., Leonard, D. C., Chornock, R., *et al.* 2008, *ApJ*, 688, 1186

Immler, S., Milne, P., & Pooley, D. 2010, *ATel*, 3012, 1

Immler, S., Brown, P. J., Filippenko, A. V., & Pooley, D. 2007, *ATel*, 1290, 1

Kallman, T. R., Palmeri, P., Bautista, M. A., Mendoza, C., & Krolik, J. H. 2004, *ApJS*, 155, 675

Kiewe, M., Gal-Yam, A., Arcavi, I., *et al.* 2012, *ApJ*, 744, 10

Leonard, D. C., Filippenko, A. V., Barth, A. J., & Matheson, T. 2000, *ApJ*, 536, 239

Newton, J. & Puckett, T. 2010, *CBET*, 2532, 1

Patat, F., Taubenberger, S., Benetti, S., Pastorello, A., & Harutyunyan, A. 2011, *A&A*, 527, L6

Pooley, D., Immler, S., & Filippenko, A. V. 2007, *ATel*, 1023, 1

Pooley, D. & Lewin, W. H. G. 2004, *IAU Circ.*, 8323, 2

Prasad, R. R. & Li, W. 2006, *CBET*, 673, 1

Schlegel, E. M. 1990, *MNRAS*, 244, 269

Smith, N., Li, W., Miller, A. A., *et al.* 2011, *ApJ*, 732, 63

Stoll, R., Prieto, J. L., Stanek, K. Z., *et al.* 2011, *ApJ*, 730, 34

Stritzinger, M., Taddia, F., Fransson, C., *et al.* 2012, *ApJ* 756, 173

Swartz, D. A., Ghosh, K. K., Tennant, A. F., & Wu, K. 2004, *ApJS*, 154, 519

van Dyk, S. D., Weiler, K. W., Sramek, R. A., & Panagia, N. 1993, *ApJ*, 419, L69

Verner, D. A. & Ferland, G. J. 1996, *ApJS*, 103, 467

Weiler, K. W., Panagia, N., Montes, M. J., & Sramek, R. A. 2002, *ARAA*, 40, 387

Weiler, K. W., Panagia, N., & Sramek, R. A. 1990, *ApJ*, 364, 611

Williams, C. L., Panagia, N., Van Dyk, S. D., *et al.* 2002, *ApJ*, 581, 396

Discussion

PODSIADLOWSKI: Two questions: 1. Your modeling of the CSM assumes spherical symmetry. In many models for the mass ejection (and there is probably also observational evidence for this), the medium is very anisotropic. Is there a way to model the radio emission for that case? 2. In your modeling of SN 2006jd you said that the Metal fits required a metallicity of 5 times solar, which seems very extreme. Is there another way to explain the data?

CHANDRA: 1: In radio as of now we do assume spherical symmetry to model the radio emission. It may not be a bad assumption because in VLBI of some supernovae, such as SN 1993J, the emission was quite spherical. One can model the circumstellar inhomogeneities but I am not aware of any radio emission model which takes care of the global asymmetry in the explosion. However, in Type IIn supernovae, circumstellar density clumps can be models in radio emission.

2: We did try various models including non-equilibrium ionization models. The only way the energy and width of the strong iron line would fit would be by assuming 5 times solar metallicity.

CHAKRABORTI: Is the gap between the two observations larger than the cooling time of the forward shocked gas? Why does the Fe line vanish?

CHANDRA: 1. SN 2006jd was classified as Type IIn supernova only after 2 years post discovery. And thats when we started following it in the radio bands regularly. The early post discovery observations were taken only once where it was not detected. So we obviously missed the cooling timescale window. 2. The Iron line vanishes in SN 2010jl two spectra because the column density is 3 times smaller in the second spectrum, which means strength of Iron line should be 3 times smaller than the previous detection, which is consistent with the non-detection.

POOLEY: Concerning the ULX component of the spectrum, why does it fall off so quickly? (Note: I thought you said it had a power-law index of 1.7)

CHANDRA: ULX components usually have power law of this order. Our results are indeed consistent with the ULX interpretation.

WANG: One needs to treat the parameters from the fits of the assumed model with grain of salt, since the model could not really be tested. For example, a combination of a thermal + Power law would give quite different parameters.

CHANDRA: Yes, the combination would have given different fit but would have increased number of free parameters. Yes, fits should be taken with grain of salt but we are trying to fit with a simplistic model which can best represent the data.

Supernova Environmental Impacts
Proceedings IAU Symposium No. 296, 2013
A. Ray & R. A. McCray, eds.

doi:10.1017/S174392131300937X

The dusty debate: core-collapse supernovae and dust

Rubina Kotak

Astrophysics Research Centre, Queen's University, Belfast, BT7 1NN, Northern Ireland
email: r.kotak@qub.ac.uk

Abstract. Dust plays an important role in our understanding of the near and distant Universe. The enormous amounts ($\gtrsim 10^8\, M_\odot$) of dust observed at high redshifts have forced us to revisit the commonly-invoked sites of dust production. Although core-collapse supernovae are the prime candidates for cosmic dust production, their actual contribution to the dust budget has been the subject of much debate in recent years. Here, I will discuss results from several vigorous observational campaigns aimed at quantifying the amount of dust produced by core-collapse supernovae. Although sample sizes are still modest, I will attempt to put the role of supernovae as dust producers into perspective.

Keywords. core-collapse supernovae, dust, echoes

---

1. Introduction

Numerous studies in recent years have emphasised the important role that dust plays in our understanding of the near and distant Universe. Dust formation in the interstellar medium (ISM) has been shown to be extremely inefficient, so the preferred site for dust formation is in the atmospheres of evolved, low-mass ($M \lesssim 8\, M_\odot$) stars from where it is transported into the ISM via stellar winds. This mechanism however, fails to explain the presence of dust at high redshifts as the evolutionary time-scales of these low-mass stars (up to 1 Gyr) begin to become comparable to the age of the Universe. Furthermore, the IR luminosities of $z > 6$ quasars e.g. Bertoldi *et al.* (2003), Dwek *et al.* (2007) imply enormous dust masses ($10^8\, M_\odot$). The short time-scales required for dust enrichment make core-collapse supernovae rather natural candidates for dust producers in the early Universe.

It has long been hypothesized (Cernushi *et al.* 1967, Hoyle & Wickramasinghe, 1970, Tielens *et al.* 1990) that the physical conditions in the ejecta core-collapse supernovae may lead to the condensation of large amounts of dust. A combination of factors form the basis for this presumption: (i) core-collapse supernova ejecta contain large amounts of refractory elements from which dust grains could form; (ii) cooling of the ejecta occurs by adiabatic expansion augmented, in some cases, by molecular emission; (iii) dynamical instabilities in the ejecta results in regions of enhanced density which may further aid the process of grain-growth self-shielding.

Early attempts at modelling dust condensation in supernova ejecta (e.g. Tielens *et al.*, 1990) were easily able to generate substantial amounts (0.1-1 $M_\odot$) of dust. However, see e.g. Cherchneff *et al.* (these proceedings) who point out severe deficiencies in the treatement of grain condensation and growth in earlier work.

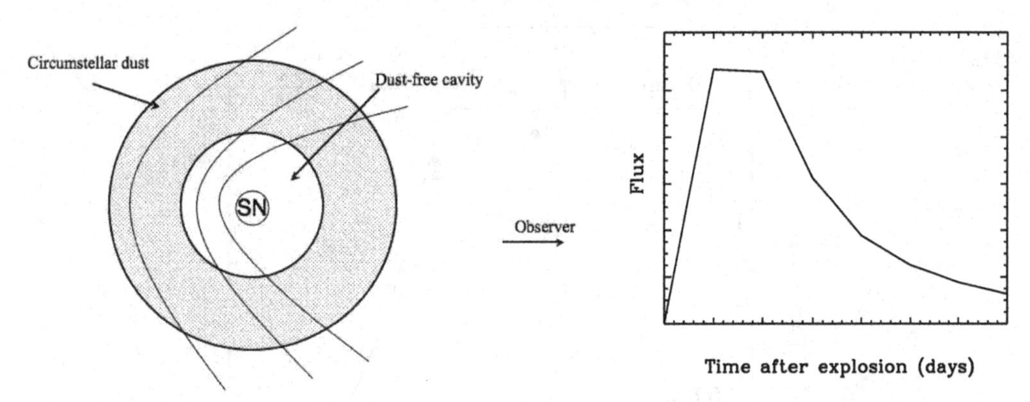

Figure 1. Left: Schematic illustration of an echo arising due to the explosion of a supernova in a dusty circumstellar medium. The series of paraboloids delineate the emitting volume which changes as a function of time. Adapted from Dwek (1983). **Right:** Example of a light curve resulting from a configuration such as that shown in the left-hand panel. Note the characteristic flat-top of the resulting light curve.

2. Observations

The observational support for the hypothesis that grains condense in substantial amounts in core-collapse supernovae is remarkably meagre. Two of the most compelling ways of detecting dust are: (i) the attenuation of spectral lines at optical/near-IR wavelengths in the nebular phase (see e.g. Fig. 16 in Meikle *et al.*, 2011); (ii) thermal emission from dust grains. Until recently, the strongest evidence for dust formation in supernova came from SN 1987A which showed a strong mid-IR excess that was accompanied by a decrease in optical emission and a blueward shift of emission line profiles (Lucy *et al.*, 1989, Danziger *et al.* (1989), Wooden *et al.*, 1991). However, even for this very well-studied albeit peculiar object, recent claims of large ejecta dust masses are controversial and model-dependent.

While the attenuation of spectral lines at late times is a relatively unambiguous signature of the presence of dust, it is difficult to derive quantitative measures of the amount and nature of the dust. As warm grains emit most strongly in the mid-IR, this is the ideal wavelength range for following dust condensation in real time. However, ground-based mid-IR observations are challenging – if not unfeasible – for the vast majority of supernovae. Even for SN 1987A (at only ~50 kpc), most of the mid-IR data came from the *Kuiper Airborne Observatory* (Wooden *et al.*, 1991). Since the launch of the *Spitzer Space Telescope* , with vastly superior sensitivity and spatial resolution compared to previous instrumentation, this situation has been changing dramatically.

2.1. *Near-infrared echoes*

When studying the thermal emission from dust, it is important to bear in mind that even if a near- or mid-IR "excess" is detected, it might not necessarily be due to new dust that has condensed in the ejecta. Thermal emission may arise from pre-existing dust in the circumstellar medium e.g. due to a dusty wind from the progenitor star which has been heated by the flash from the supernova, resulting in an infrared echo (Bode & Evans, 1979, Dwek, 1983). Shock heating due to ejecta-circumstellar matter interaction may be another mechanism which gives rise to an echo. A schematic diagram of a configuration that would give rise to an echo is shown in Fig. 1.

Clearly, an infrared echo could potentially mask any signature of newly condensing dust, given that the magnitude of this effect. However, in general, emission due echoes

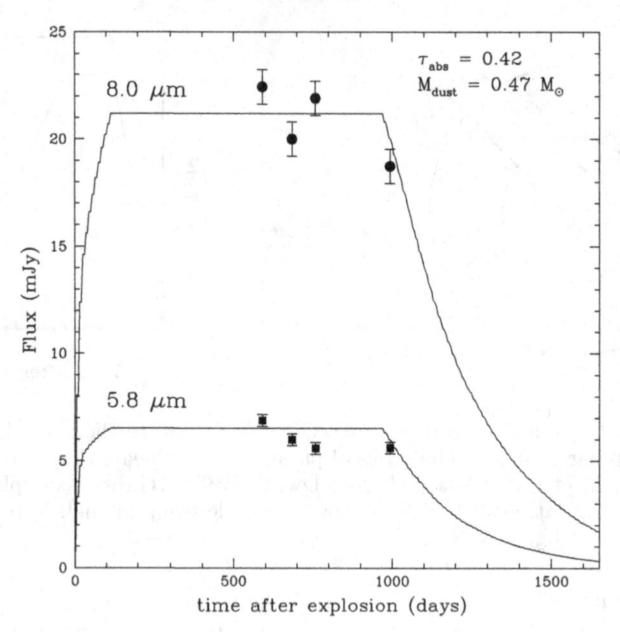

Figure 2. Dereddened 5.8 and 8.0μm photometry of SN 2002hh compared with synthetic light curves generated by an infrared echo model with a dust mass of 0.47 $M_\odot$. The dust temperature at the inner boundary was 345 K during the IR echo plateau phase. Taken from Meikle *et al.* (2006).

tends to appear earlier in the evolution of a supernova, compared to dust formation, which tends to occur at epochs of several hundred days. A caveat, of course, is the geometry and spatial extent of the circumstellar matter. Given the characteristic light curves that arise from echoes, one way of potentially distinguishing between pre-existing and new dust is to monitor the light curve to determine its shape (see Fig. 1). In situations where there is contribution to the infrared luminosity from both an echo, and new dust, the situation is complicated, and detailed modelling is required. Meikle *et al.* (2006) report the first detection of an IR echo in the most common of supernova types, the type-IIP supernova SN 2002hh (see Fig. 2). However, given the complexity of the field around the supernova, it was not possible to conclusively establish whether the echo was due to dusty pre-existing circumstellar matter, or a dusty molecular cloud.

2.2. *The case of SN 2004et*

Extensive mid-IR observations such as those of SN 2004et (Kotak *et al.*, 2009) allow us to witness – in real time – the formation of new ejecta dust (see Fig. 3). Very late-time observations reveal the likely effect of an interstellar infrared echo, which may well dominate the emission longward of $\sim 25\mu$m for this supernova.

The evolution of the spectral energy distributions (Fig. 3) point to infrared emission from dust. Consideration of the (evolution of) the model parameters further allows one to rule out an early-time echo as the source of the mid-infrared luminosity.

The models shown in Fig. 3 are based on a spherical, uniform sphere of isothermal dust grains, following the escape probability treatment of Lucy *et al.* (1989). A typical grain size distribution is used for a mix of refractory materials as predicted by models of dust condensation in supernovae.

In order to estimate the amount of dust, the mass of dust is increased until an adequate match to the spectrum is obtained. In order to be conservative, we model the spectra for

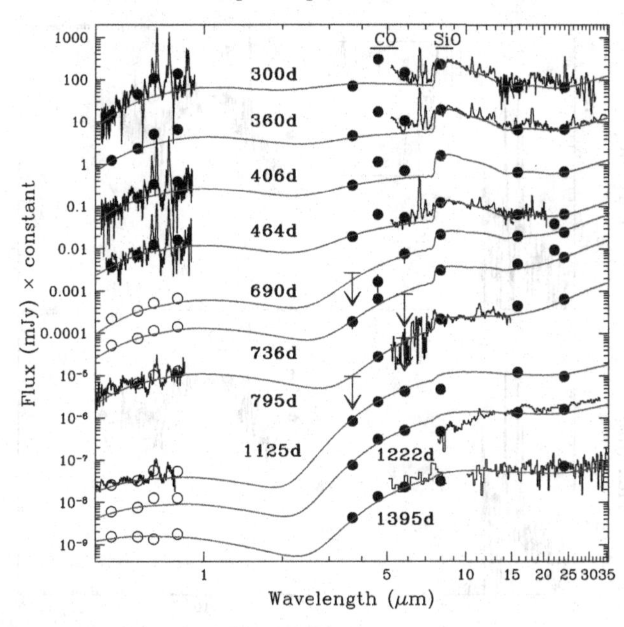

Figure 3. Spectral energy distributions of SN 2004et. The smooth (blue) lines show the model fits, including a hot blackbody component, and a cold (interstellar infrared echo) component to account for the early, and late-time behaviours, respectively. The optical spectra are from Sahu *et al.* (2006). The open circles indicate estimated optical fluxes obtained by the interpolation or extrapolation of the light curves. Adapted from Kotak *et al.* (2009).

the most optically thin case that will still provide an adequate fit to the spectra. This approach was first tested on SN 1987A, and yielded dust masses consistent with other studies.

Interestingly, for SN 2004et, the success of the model (Fig. 4) in reproducing the 8-14 μm feature not only lends further support to the newly-formed silicate dust scenario, but also provides an additional constraint on the model at each epoch in that the optical depth had to be adjusted to match the visibility of this feature. Thus, in spite of the high optical depth, the derived dust masses for SN 2004et (few $\times 10^{-4} M_{\odot}$) are actual values, rather than lower limits.

As with SN 1987A (Spyromilio *et al.* 1988, Wooden *et al.*, 1993), emission due to SiO and CO is clearly evident in SN 2004dj, Kotak *et al.* (2005), and SN 2005af Kotak *et al.* (2006). Also, a cool dust continuum provides a good match to the spectra. The same pattern holds for SN 2005af, with strong molecular emission at earlier epochs (~200 d), which is replaced by a strong cool continuum at later times (~600 d). Our dust mass estimate for SN 2005af comes to $\sim 4 \times 10^{-4} M_{\odot}$. In the latter spectrum, there is a hint of an even cooler component, which might increase this estimate somewhat.

The estimates obtained from the method described above represent the amount of directly detected dust. It is currently difficult to determine how much more dust may be present in optically-thick clumps. This problem was already identified in the context of SN 1987A (Lucy *et al.* 1989, Wooden *et al.*, 1993). The problems persists even at wavelengths as long as (24 μm, the extent of most of our data). However, for most – if not all – of our *Spitzer* targets, current indications are that the clumps are optically-thick in the mid-IR regime before significant dust condensation occurs.

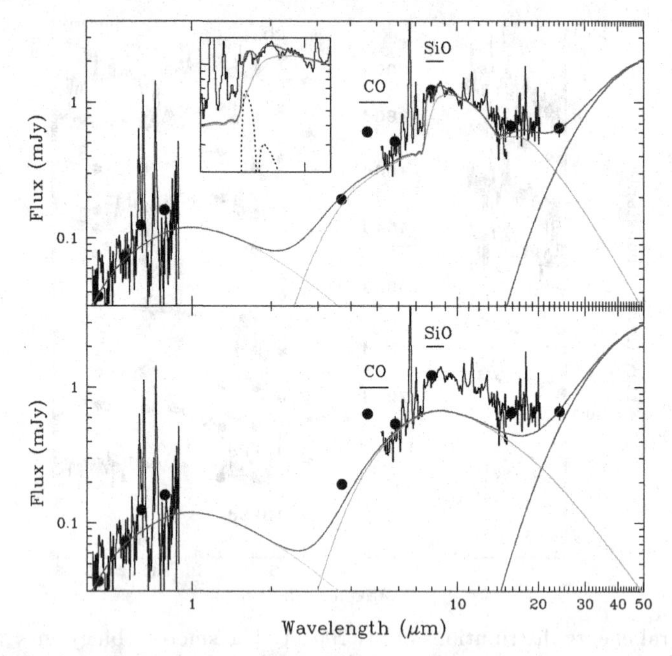

Figure 4. Day 464 observations (black) of SN 2004et compared with models. The upper (main) and lower panels show the isothermal dust models: green for silicate and amorphous carbon grains, respectively. The total model spectrum (blue) also comprises hot (blackbody: cyan) and cold (interstellar IR echo: red) components. The upper panel model also contains a contribution from the SiO fundamental. The inset shows the separate SiO contribution (dotted line). It can be seen that a superior match to the spectrum is achieved with the combined silicate dust and SiO model, as compared with the amorphous carbon dust model. Taken from Kotak *et al.* (2009).

3. Summary

From a sample of well-observed type II-P supernovae in the mid-IR, we find that all objects formed some dust. Less than a decade ago, there was only scant evidence for dust condensation in the ejecta of the most common type of core-collapse supernova.

Grain formation models predict that carbon, silicate, and magnetite grains should be present in substantial quantities, with the silicate grains probably dominating. Our sample of core-collapse supernovae all show evidence of strong emission due to CO, or SiO, or both at epochs as early as ~100 d. Thus, although our sample size remains small, all of the supernovae that showed evidence for dust condensation, also showed evidence of strong molecular emission at earlier epochs. Thus, molecular emission may well be the harbinger of dust formation.

Current estimates of the amount of dust remain small for type II-plateau supernovae, the most common local type of core-collapse SN with ejecta dust masses in the range of 10^{-3} to $10^{-5} M_{\odot}$. This is 10–100 times lower than needed to account for the dust seen at high redshifts. In most – but not all – cases, the estimates are lower limits, and some dust may well exist in optically-thick clumps. Although low-mass AGB stars may account for some fraction of the deficit, it is unlikely that they could account for the entire cosmic dust budget. However, this scenario depends heavily on the assumed initial mass function, star formation history, and dust formation efficiency, all of which are difficult to constrain observationally. The recently-revived proposition (e.g. Draine 2009) that dust grains might be able to survive and grow in the ISM may go some way towards allieviating the problems outlined here. Nevertheless, much work remains to be done in assessing the dust production in core-collapse supernovae which is currently limited mainly by mid-IR facilities.

References

Bertoldi, F. *et al.* , 2003, *A&A*, (Letters) 406, 55
Bode, M. & Evans, A., 1979, *A&A*, 73, 113
Cernushi, F. *et al.* , 1967, *Ann. d'Astrophys*, 30, 1039
Cherchneff, I. *et al.*, these proceedings.
Danziger, J. *et al.*, 1989, *IAUC*, 4746, 1
Draine, B., 2009 *ASPC*, 414, 453
Dwek, E., 1983, *ApJ*, 274, 175
Dwek, E., Galliano, F., & Jones, 2007, *ApJ*, 662, 927
Hoyle, F. & Wickramasinghe, N. C., 1970, *Nature*, 226, 62
Kotak, R., *et al.* 2005, *ApJ*, (Letters) 628, 123
Kotak, R., *et al.* 2006, *ApJ*, (Letters) 651, 117
Kotak, R., *et al.* 2009, *ApJ*, 704, 306
Liu, W. & Dalgarno, A., 1994, *ApJ*, 428, L769
Lucy, L. B. *et al.* , 1989, *Structure and dynamics of the ISM, ed. Springer*, 164
Meikle, W. P. S., *et al.* , 2006, *ApJ*, 649, 332
Meikle, W. P. S., *et al.* , 2011, *ApJ*, 732, 109
Sahu, D. K. *et al.*, 2006, *MNRAS*, 372, 1315
Spyromilio, J., *et al.* 1988, *Nature*, 334, 327
Tielens, A. G. G. M., *et al.* 1990, *NASA Conf. Publ.*, 3061, 59

Discussion

VINK: What about dust in supernova remnants?

KOTAK: The picture is more confusing when remnants are considered. For example, for the young SNR 1E 0102.27219, analyses of mid-IR data by different groups result in dust mass estimates that vary greatly: Stanimirovic *et al.*(2005) find no more than $8 \times 10^4 M_\odot$ at $\sim$120 K associated with the remnant, while Sandstrom *et al.* (2009) derive dust masses of a few $\times 10^3 M_\odot$. Direct evidence for large quantities of dust in SNRs is weak. On the other hand, there are many SNR studies that have emphasised the role of SNRs in the destruction of dust grains.

BOUCHET: It would be good to have an idea where in the ejecta the dust is forming, i.e. Within what velocity of outflow? For instance, for SN87A we got a velocity from the line profiles < 1870 km / sec. Have you this kind of information for other SNe

KOTAK: Yes, we generally do. For the cases where dust is forming in the ejecta of SNE 2004dj (Meikle *et al.* 2011), 2004et (Kotak *et al.* 2009, Sahu *et al.*. The dust forms in slow-moving ejecta $< 2000 km\ s^{-1}$.

SUTARIA: A Spitzer survey made an estimate of dust from type-IIn SNe, using (mainly) archival data. Can you please comment on the veracity of that result?

KOTAK: There have certainly been mid-SR detections based on archival data. The number of available epochs of data per SN is limited it is therefore difficult to determine contributions due to IR echoes.

VINK: A comment: The high dust mass in Cas A reported by Dunne *et al.* was based on an error. One of their lower limits on flux was a flux point and hence there was not much cold dust. For Kepler's SNR the dust seems to originate from the AGB star that formed a companion of the progenitor (Williams *et al.*.)

RAY: Is there an estimate of the Pre-SN dust in various cases of SNe?

KOTAK: Yes, in "pure" echo cases, the parameters required to fit the observed fluxes are available, including the dust mass.

Supernova Environmental Impacts
Proceedings IAU Symposium No. 296, 2013
A. Ray & R. A. McCray, eds.

doi:10.1017/S1743921313009381

Molecules and dust in the ejecta of Type II-P supernovae

I. Cherchneff and A. Sarangi

Department Physik, Universität Basel, Klingelbergstrasse 82, CH-4056 Basel, Switzerland

Abstract. We study the formation of molecules and dust clusters in the ejecta of solar metallicity, Type II-P supernovae using a chemical kinetic approach and follow the evolution of molecules and small dust cluster masses from day 100 to day 1500 after explosion. We predict that large masses of molecules including CO, SiO, SiS, O_2, and SO form in the ejecta. We show that the non-equilibrium chemistry results in a gradual build up of the dust mass from small ($\sim 10^{-5}$ $M_{\odot}$) to large values ($\sim 5 \times 10^{-2}$ $M_{\odot}$) over a five-year period after explosion. This result provides a natural explanation to the discrepancy between the small dust masses detected at infrared wavelengths some 500 days post-explosion and the larger amounts of dust recently detected with the Herschel telescope in supernova remnants.

Keywords. supernovae: general, molecular processes, dust, circumstellar matter, astrochemistry

1. Introduction

Massive stars (8 $M_{\odot}$ < $M_{\star}$ < 30 $M_{\odot}$) end their life as Core-Collapse supernovae (hereafter SNe) of Type II-P. Despite the huge amount of energy released by the explosion ($\sim 1 \times 10^{51}$ erg), and the harsh physical conditions that characterise the ejected stellar gas, dust and molecules have been detected in many SNe, including SN1987A, some hundred days after explosion (e.g., Spyromilio *et al.* 1988, Lucy *et al.* 1989, Kotak *et al.* 2009, Sugerman *et al.* 2006). The fundamental band of SiO has been detected in several SNe, e.g., SN2004et, and the fading of the transition over time was ascribed to the depletion of SiO in the condensation process of silicates at ~400 days post-outburst (Kotak *et al.* 2009). Most important are the small masses of warm dust derived with values that range from 1×10^{-5} $M_{\odot}$ to 1×10^{-3} $M_{\odot}$. The latest data on SN remnants obtained with the submilimetre (submm) Herschel telescope have cast a new light on the dust released by SN events. A large mass of cold ejecta dust amounting to 0.08 $M_{\odot}$ was derived in the 330-year-old SNR Cas A (Barlow *et al.* 2010). In the Crab Nebula, cool dust was recently detected in the filaments and the derived dust masses amount to 0.1 – 0.24 $M_{\odot}$, depending on the type of dust assumed (Gomez *et al.* 2012). Finally, 0.4 to 0.7 $M_{\odot}$ of cool, ejecta dust have been inferred from submm flux data in the young remnant SN1987A (Matsuura *et al.* 2011). The dust mass produced by SNe is thus uncertain. Here we present the results of physico-chemical models of dust synthesis in Type II-P SNe.

2. Ejecta physical and chemical model

A stellar progenitor of mass 15 $M_{\odot}$, typical of Type II-P SNe is considered. The stratified ejecta is described by the mass zones of the progenitor core given by explosion models, and we assume that the gas within each zone is fully-microscopically mixed. No gas leakage between different zones is considered. The initial composition of the ejecta in the form of the elemental mass yields are taken from Rauscher *et al.* (2002). The basic model parameters are similar to those characterising the 20 $M_{\odot}$ progenitor of Cherchneff

& Dwek (2009, 2010), i.e., an explosion energy of 1×10^{51} ergs, an effective γ-ray optical depth at 100 days $\tau(100) = 23.8$, and a ^{56}Ni mass of 0.075 $M_{\odot}$. The temperature and density dependance as a function of time are given by Sarangi & Cherchneff (2013).

All possible atoms, molecules, and ions that form in the SN ejecta are considered in our chemical scheme, and we model the ejecta chemistry considering all possible types of chemical reactions relevant to hot and dense environments. All chemical pathways that lead to the formation of linear molecules, carbon chains and rings, and small dust clusters include neutral-neutral processes such as termolecular, bimolecular, radiative association, and charge exchange reactions, whereas the destruction is described by thermal fragmentation, neutral-neutral processes (i.e., oxidation reactions of carbon chains and all reverse processes of the formation reactions), ion-molecule recombination processes and charge exchange reactions. Furthermore, the radioactive decay of ^{56}Ni to ^{56}Co, and ^{56}Co to ^{56}Fe creates γ-rays that degrade to X rays and UV photons through collision with thermal electrons and triggers the creation of a population of Compton electrons in the gas. These fast electrons ionise atoms and destroy chemical species in the ejecta. The description of the growth pathways of small silicate clusters, namely forsterite dimer $(Mg_2SiO_4)_2$ and enstatite dimer $(MgSiO_3)_2$, is based on the work by Goumans & Bromley (2012), where the growth of enstatite and forsterite clusters from the SiO dimer is described by one oxygen-addition step followed by one Mg-inclusion as a recurrent growth scenario.

3. Results

The chemistry is followed from day 100 until day 1500, a time span which covers the initial formation of molecules at early times until the dust cluster synthesis is fully completed some 4 years after outburst. We find that the zones of the He-core are efficient at forming large amounts of molecules. Because the ejecta is assumed to be hydrogen-free, the number of chemical species formed is limited, and this poor chemistry typical of SN ejecta is well exemplified by the detection of only two molecules, CO and SiO, in several SNe (e.g., Lucy *et al.* 1989, Kotak *et al.* 2009), and CO in SN remnants (Rho *et al.* 2012). The mass of silicon monoxide, SiO, is shown in Fig. 1 as a function of post-explosion time for all the various zones in the ejecta. Masses derived from infrared (IR) observations of a few Type II-P SNe are also shown. The SiO mass follows a rapid increase at day 200 in zones 1B, 2, and 3, when the formation of SiO is delayed to 400 days in zones 4A and 4B. The prevalent formation processes are radiative association and reaction of atomic Si with dioxygen, O_2, while destruction is mainly due to reactions with Ar^+ and Ne^+. The sharp decrease after day 200 is due to the conversion of SiO into SiO dimers and silica and silicate cluster growth.

The evolution of CO mass with post-explosion time for the He-core zones is shown in Fig. 2. CO masses derived from available observational data for SN1987A are also plotted for early times (Liu & Dalgarno 1995). In zones 4A, 4B, 2 and 3, CO forms as early as 200 days to reach masses ranging from 10^{-4} to 10^{-2} $M_{\odot}$ at day 1500. The total ejected CO mass at day 1500 is as large as ~ 0.1 $M_{\odot}$, that is, much larger than the masses derived from IR data before day 600 in SN1987A. These large amounts of CO primarily form in zones 4A and 4B but do not trace efficient carbon dust formation in these two zones. These zones indeed form no carbon dust because of the quick destruction of carbon chains by reaction with O_2 to form CO.

Three other molecules are present in the ejecta, namely SiS, O_2 and SO. The gas-phase chemistries of all molecular species and dust clusters are entangled and the ejecta is made of a molecular component equivalent to ~ 29 % of the 2.5 $M_{\odot}$ of ejected material.

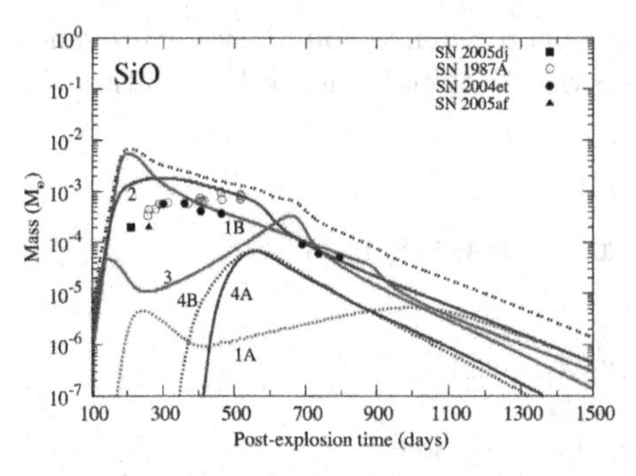

Figure 1. SiO masse variation with post-explosion time for the 15 $M_\odot$ progenitor as a function of ejecta zones (taken from Sarangi & Cherchneff 2013). The masses derived from observational data of several SNe are also shown.

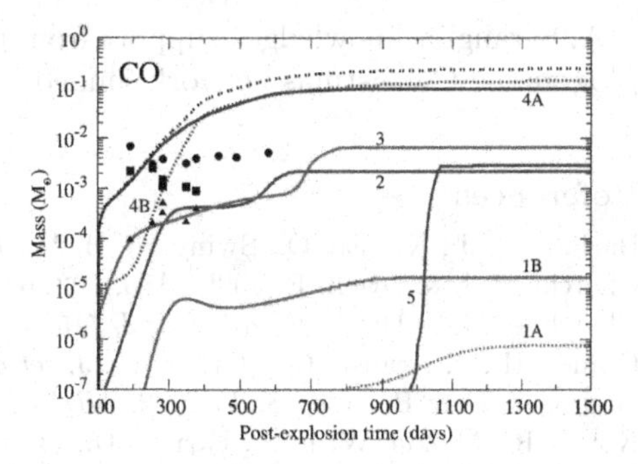

Figure 2. CO mass variation with post-explosion time for the 15 $M_\odot$ progenitor as a function of ejecta zones (from Sarangi & Cherchneff 2013). CO masses derived from observations of SN1987A are also shown.

The variation of dust cluster masses summed over all ejecta zones with time is illustrated in Fig. 3. The timing of dust production highly depends on the local chemistry characterising the zones, as exemplified by the formation of silicate and carbon clusters. Forsterite first nucleates at 300 days in zone 1B, and gradually grows to reach its maximum value at $\sim$ 600 days in zone 2. These two nucleation phases are seen as two SiO depletion events in Figure 1. The gradual growth of the forsterite total mass results from the chemistry of SiO formation and the growth of silicate clusters. The scenario is different for carbon clusters. Carbon chains grow in significant amounts in the only carbon-rich zone of the ejecta, zone 5. This zone is helium-rich and He atoms are ionised by Compton electrons. The produced ions are destroyed by recombination to He and the decrease of Compton electrons with time. He^+ is detrimental to the formation and survival of molecules in zone 5 (Cherchneff & Dwek 2009), as the ion quickly destroys molecules and postpones the time for carbon chain synthesis.

This sequence of epochs at which dust of various kinds condenses in the ejecta results in a gradual dust mass growth from $\sim 10^{-4}$ $M_\odot$ at 400 days to 5 $\times 10^{-2}$ $M_\odot$ at day 1500. This growth sequence provides a possible explanation to the discrepancy between the mass of warm dust derived at early time from IR data and the mass of cool dust inferred from submm fluxes in SN remnants. Because dust cluster masses represent upper limits to the total dust mass formed in the ejecta, we conclude that SNe are efficient but moderate dust suppliers to galaxies.

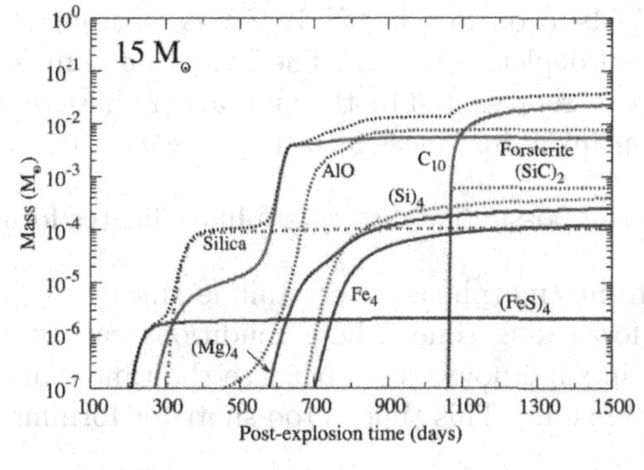

Figure 3. Dust cluster mass variation with post-explosion time for the 15 $M_\odot$ progenitor summed over all ejecta zones (from Sarangi & Cherchneff 2013).

A. Sarangi acknowledges support from the European Science Foundation Eurogenesis programme CoDustmas network funded by the Swiss National Science Foundation.

References

Barlow, M. J., Krause, O., Swinyard, B. M., *et al.* 2010, *A&A*, 518, L138
Cherchneff, I. & Dwek, E. 2009, *ApJ*, 703, 642
Cherchneff, I. & Dwek, E. 2010, *ApJ*, 715, 1
Gomez, H. L., Krause, O., Barlow, M. J., *et al.* 2012, *ApJ*, 760, 96
Goumans, F. & Bromley S. T. 2012, *MNRAS*, 420, 3344
Kotak, R., Meikle, W. P. S., Farrah, D., *et al.* 2009, *ApJ*, 704, 306
Liu, W. & Dalgarno, A. 1995, *ApJ*, 454, 472
Lucy, L. B., Danziger, I. J., Gouiffes, C., & Bouchet, P. 1989, *in G. Tenorio-Tagle, M. Moles & J. Melnick (eds.), IAU Colloq. 120 'Structure and Dynamics of the Interstellar Medium'*, p. 164
Matsuura, M., Dwek, E., Meixner, M., *et al.* 2011 *Science*, 333, 1258
Rauscher, T., Heger, A., Hofmann, R. D., & Woosley, S. E. 2002, *ApJ*, 576, 323
Rho, J., Onaka, T., Cami, J., & Reach, W. 2012, *ApJ*, 747, 6
Sarangi, A. & Cherchneff, I., 2013, *ApJ*, submitted
Spyromilio. J, Meikle, W. P. S., Learner, R. C. M., & Allen, D. A. 1988, *Nature*, 334, 327
Sugerman, B. E. K., Ercolano, B., Barlow, M. J., *et al.* 2006, *Science*, 313, 196

Discussion

RAY: Is there an anticorrelation seen between the dust mass and the Ni56 mass?

CHERCHNEFF: The final dust mass found is not truly correlated to the 56 Ni mass, but the time at which dust forms is. The larger the 56 Ni mass, the later dust will form in the ejecta.

BOUCHET: 1) What is the limit on the mass of dust? Does it depend of the yields in your model? 2) The original metallicity of the progenitor could be important. If so, which LMC metallicity have you considered for SN87A?

CHERCHNEFF: The dust mass depends on the initial elemental yields but this dependence is not linear, because of the gas phase chemistry and the formation of molecules. We have not yet modeled SN1987A but we would assume a solar metallicity divided by 3.

UNIDENTIFIED: So what is wrong with the claim of 0.7 $M_{\odot}$ of dust in SN 1987A?

CHERCHNEFF: The 0.7 Msun of dust in SN 1987A is an upper limit based on the assumption of 100% element depletion in dust. The formation of molecules is not taken into account and the bottleneck represented by the nucleation phase of dust is not considered. This number is by no means a dust mass formed in the ejecta.

KOO: I wonder if there is a possibility that crystalline silicates form in cooling SN ejecta

CHERCHNEFF: To go from amorphous to crystalline silicates, you need to maintain a high gas temperature for a long time. These conditions can be found in discs. These conditions are not met in supernova ejecta because the time scale for dust formation is of the order of a few years only. This time is too short for forming crystalline silicates.

Supernova Environmental Impacts
Proceedings IAU Symposium No. 296, 2013
A. Ray & R. A. McCray, eds.

doi:10.1017/S1743921313009393

Autopsy of the Supernova Remnant Cassiopeia A

Dan Milisavljevic[1] and Robert A. Fesen[2]

[1]Harvard-Smithsonian Center for Astrophysics, 60 Garden Street, Cambridge, MA, 02138
email: `dmilisav@cfa.harvard.edu`

[2]6127 Wilder Lab, Dept. of Physics & Astronomy, Dartmouth College, Hanover, NH 03755
email: `robert.fesen@dartmouth.edu`

Abstract. Three-dimensional kinematic reconstructions of optically emitting ejecta in the young Galactic supernova remnant Cassiopeia A (Cas A) are discussed. The reconstructions encompass the remnant's faint outlying ejecta knots, including the exceptionally high-velocity NE and SW streams of debris often referred to as 'jets'. The bulk of Cas A's ejecta are arranged in several circular rings with diameters between approximately 30″ (0.5 pc) and 2′ (2 pc). We suggest that similar large-scale ejecta rings may be a common phenomenon of young core-collapse remnants and may explain lumpy emission line profile substructure sometimes observed in spectra of extragalactic core-collapse supernovae years after explosion. A likely origin for these large ejecta rings is post-explosion input of energy from plumes of radioactive ^{56}Ni-rich ejecta that rise, expand, and compress non-radioactive material to form bubble-like structures.

Keywords. supernova remnant

1. Introduction

Massive stars ($\geqslant 8$ $M_\odot$) explode as core-collapse supernovae (SNe; Types II, IIb, Ib,c) and are a primary source of heavy elements in the universe. The specific engine that drives these explosions is uncertain, but gravity along with aspherical influences from uneven neutrino heating, rotation, and magnetic fields have been investigated (see Janka 2012 and references therein). Debate over the dynamics and relative strength of these influences stems in large part from a dearth of observational constraints.

Study of the class of young oxygen-rich supernova remnants (SNRs) believed to be the end products of SNe provide opportunities to directly test models of massive stellar explosions. Nearby SNRs permit investigations of explosion kinematics and elemental abundances at spatial scales that are impossible to achieve from extragalactic SN observations, and thus yield unique information about the core collapse explosion mechanism, nucleosynthesis yields, and the nature of their central compact remnants.

The young Galactic remnant Cassiopeia A (Cas A) provides perhaps the clearest look at the explosion dynamics of a high mass SN. With an explosion date most likely around 1681 ± 19, Cas A is the youngest Galactic core-collapse SNR known (Thorstensen *et al.* 2001; Fesen *et al.* 2006), and at an estimated distance of 3.4 kpc (Reed *et al.* 1995), it is also among the closest.

Cas A is the only historical core-collapse SNR with a secure SN subtype classification. The detection of light echoes of the supernova outburst (Rest *et al.* 2011; Besel *et al.* 2012) has enabled follow-up observations which indicate that the original supernova associated with Cas A exhibited an optical spectrum at maximum light similar to those seen for the Type IIb events SN 1993J and SN 2003bg (Krause *et al.* 2008; Rest *et al.* 2011).

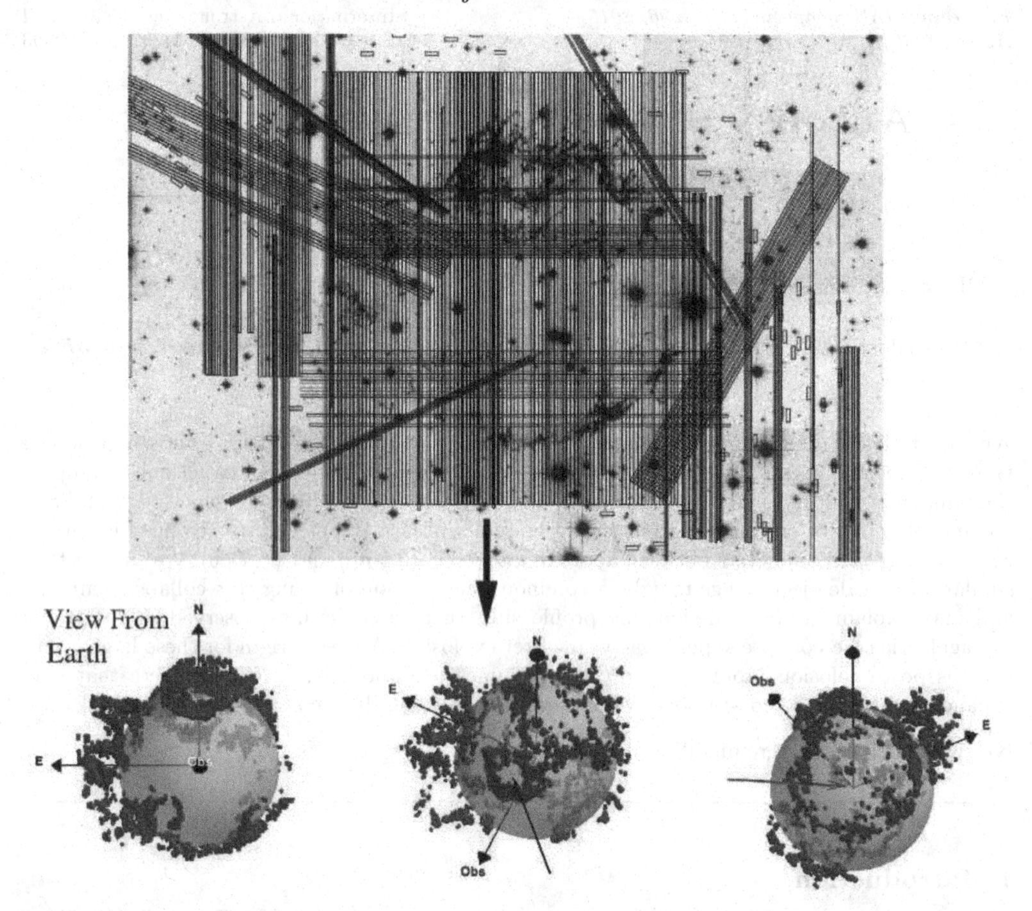

Figure 1. Data from Milisavljevic & Fesen (2013) showing the kinematic map constructed for Cas A. Top: Locations of over 200 slit positions used to develop a 3D reconstruction. Bottom: Various perspectives of the resulting 3D map. Blue arrow points to a region coincident with X-ray emitting Fe-rich material.

2. 3D Reconstruction of Cas A

We recently created a detailed 3D kinematic reconstruction of Cas A. Its optical-emitting ejecta were mapped from a spectroscopic survey involving hundreds of long slit spectra taken over three years (Milisavljevic & Fesen 2013; Figure 1). It is the most complete kinematic map of an SNR to date.

Our data show that Cas A's main shell ejecta are arranged in several well-defined and nearly circular rings with diameters between approximately 30″ (0.5 pc) and 2′ (2 pc). In Figure 2, a Mercator projection of the main shell knots is shown to illustrate the relative scale and distribution of the rings. Some rings form complete circles, while others appear as partial circles or ellipses. Three of these rings encircle X-ray-emitting Fe-rich ejecta that has been mapped by DeLaney *et al.* (2010).

Motivation to undertake a deep reconnaissance of the entire Cas A was driven in part from an interest in finally grasping the kinematic properties of the NE and SW jets. Previous studies of Cas A's NE and SW regions of exceptionally high-velocity knots could not address the question of whether they form a true bipolar structure. Our data indicate that this is indeed the case. The NE and SW jets with expansion velocities approaching $15,000$ km s^{-1} appear to be directed in nearly opposite directions suggestive of opposing

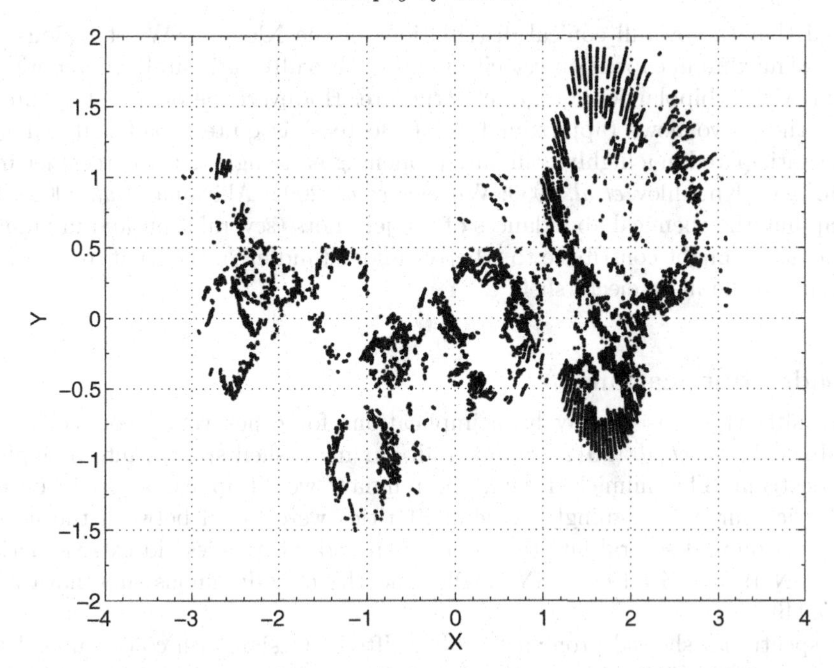

Figure 2. The main shell of Cas A's optically-emitting ejecta as represented in a Mercator projection. The linear scale is equal in all directions around any point and conformal, but the cylindrical map projection distorts the size and shape of large objects, especially towards the poles. From Milisavljevic & Fesen (2013).

flows of SN debris originating from the Si-S-Ca-Ar layer of the progenitor star with opening half-angles of approximately 40 degrees.

3. Discussion

It is clear from both our survey and previous studies that the distribution of Cas A's ejecta is not random. At least half a dozen large and coherent ejecta rings are observed. The sizes and arrangement of these rings may be informing us about important properties of the explosion dynamics and subsequent evolution of the expanding debris. We note, however, that interpretation of these structures is complicated in that one must disentangle properties that may originate in the explosion from later influences related to possible post-explosion radioactive heating, ejecta interaction with the surrounding material, and effects of the reverse shock on the ejecta.

The ejecta rings are not unlike large-scale features seen in SN explosion models (e.g., Hammer *et al.* 2010) and may have origin in part to a "Ni bubble effect" having influenced the remnant's expansion dynamics shortly after the original explosion. It may be that the observed ejecta rings represent cross-sections of large cavities in the expanding ejecta created by a post-explosion input of energy from plumes of radioactive ^{56}Ni-rich ejecta. Li *et al.* (1993) have described how this input of energy might account for the high-volume filling factor of Fe in SN 1987A despite its small mass, and Basko *et al.* (1994) and Blondin *et al.* (2001) have investigated hydrodynamic simulations based on this model. In this scenario the compression of surrounding non-radioactive material by rising and expanding bubbles of radioactive ^{56}Ni-rich ejecta occupy a large fraction of the ejecta volume and give way to a "Swiss cheese" ejecta structure.

We find that the overall conical distributions of the NE and SW jet regions exhibit comparable maximum expansion velocities and are broadly anti-parallel in an orientation consistent with a bipolar jet-counterjet structure. However, the observed opening half-angles of these two flows (approximately 40 degrees) is quite broad and is not what would be anticipated in a highly-collimated (opening half-angle < 10 degrees) jet-induced explosion (e.g., Khokhlov *et al.* 1999; Wheeler *et al.* 2002; Akiyama *et al.* 2003). Future work mapping the chemical abundances of the jet knots (several thousand in number) to their kinematics might contribute to a better understanding of their nature and possible relationship to the main ejecta shell.

4. Broader Implications

Our results with Cas A may have implications for other young core-collapse remnants. Milisavljevic *et al.* (2012) summed all our main shell spectra into a single, integrated spectrum. This mimicked what the remnant would appear as as an unresolved extragalactic source. Intriguingly, close similarities were found between the integrated Cas A spectrum and several late-time optical spectra of decades-old extragalactic SNe, including SN 1979C, SN 1993J, SN 1980K, and the ultra-luminous supernova remnant in NGC 4449.

These spectra all showed pronounced blueshifted emission with conspicuous line substructure in [O I] $\lambda\lambda$6300, 6364, [O II] $\lambda\lambda$7319, 7330, [O III] $\lambda\lambda$4959, 5007, [S II] $\lambda\lambda$6716, 6731, and [Ar III] λ7136. Since the emission line substructure observed in the forbidden oxygen emission line profiles of Cas A are associated with the large-scale rings of ejecta, we suggest that similar features in the intermediate-aged SNe which have often been interpreted as ejecta ‘clumps’ or ‘blobs’ are, in fact, probable signs that similar large-scale rings of ejecta are common in SNe.

Acknowledgements

Some of this material is based upon work supported by the National Science Foundation under Grant No. AST-0908237.

References

Akiyama, S., Wheeler, J. C., Meier, D. L., & Lichtenstadt, I. 2003, *ApJ*, 584, 954
Basko, M. 1994, *ApJ*, 425, 264
Besel, M.-A. & Krause, O. 2012, *A&A*, 541, L3
Blondin, J. M., Borkowski, K. J., & Reynolds, S. P. 2001, *ApJ*, 557, 782
DeLaney, T., Rudnick, L., Stage, M. D., *et al.* 2010, *ApJ*, 725, 2038
Fesen, R. A., Hammell, M. C., Morse, J., *et al.* 2006, *ApJ*, 645, 283
Hammer, N. J., Janka, H., & Müller, E. 2010, *ApJ*, 714, 1371
Janka, H.-T. 2012, *Annual Review of Nuclear and Particle Science*, 62, 407
Khokhlov, A. M., Höflich, P. A., Oran, E. S., *et al.* 1999, *ApJl*, 524, L107
Krause, O., Birkmann, S. M., Usuda, T., *et al.* 2008, *Science*, 320, 1195
Li, H., McCray, R., & Sunyaev, R. A. 1993, *ApJ*, 419, 824
Milisavljevic, D. & Fesen, R. A. 2013, *ApJ*, submitted
Milisavljevic, D., Fesen, R. A., Chevalier, R. A., *et al.* 2012, *ApJ*, 751, 25
Reed, J. E., Hester, J. J., Fabian, A. C., & Winkler, P. F. 1995, *ApJ*, 440, 706
Rest, A., Foley, R. J., Sinnott, B., *et al.* 2011, *ApJ*, 732, 3
Thorstensen, J. R., Fesen, R. A., & van den Bergh, S. 2001, *AJ*, 122, 297
Wang, L., Wheeler, J. C., Höflich, P., *et al.* 2002, *ApJ*, 579, 671
Wheeler, J. C., Meier, D. L., & Wilson, J. R. 2002, *ApJ*, 568, 807

Discussion

UNIDENTIFIED: 1. Are the "rings" in the projection plot dominated by 1 element, or do they all have same/very similar composition? 2. Are the bigger rings moving faster? Could they have been ejected BEFORE the explosion from the progenitor?

MILISAVLJEVIC: The map shown in purely kinematics and does not show any information about chemical abundances. The material is generally all at the same radius from the center of expansion so everything being seen now should have originated at the same time of explosion.

TURATTO: Is there any alignment or connection between the so-called jets and planes in X-rays that you have shown?

MILISAVLJEVIC: Interesting question. There is no clear relationship.

MAEDA: Have you got an idea to get the 'integrated' O profile from the unshocked ejecta? You did it for the 'shocked' ejecta, but it might be affected by ISM/CSM ?

MILISAVLJEVIC: This is a good idea! One could use the "bubble-like" interior as a map to simulate a clumpy distribution of radioactive elements that could power emission from say, oxygen and calcium. One could then compare this to the late-time emission line profiles observed in supernovae 1 year after outburst.

BARTEL: What is the proper motion of the X-ray point source inferred from the difference between its position and the extrapolated position of the center of expansion. Further, you mentioned that the proper motion is almost perpendicular to the jet direction. What physical significance would that have?

MILISAVLJEVIC: The inferred proper motion is something like 350 km/s. I have not yet come up with an explanation of why this should be so. In fact, it run counter to what once would expect if the NE & SW flares were associated with a jet-induced explosion - if this were the case the motion of the neutron star should be along the jet axis.

Supernova Environmental Impacts
Proceedings IAU Symposium No. 296, 2013
A. Ray & R. A. McCray, eds.

doi:10.1017/S174392131300940X

Dust Destruction in the Cygnus Loop Supernova Remnant

Ravi Sankrit[1], William P. Blair[2], John C. Raymond[3] and Brian J. Williams[4]

[1]SOFIA Science Center, NASA Ames, M/S 232-12, Moffett Field, CA, USA
email:rsankrit@sofia.usra.edu

[2]Johns Hopkins University

[3]Smithsonian Astrophysical Observatory

[4]Goddard Space Flight Center

Abstract. The Cygnus Loop supernova remnant serves as an excellent laboratory for the study of radiative and non-radiative shocks with speeds in the 150–450 km s^{-1} range. We present results on shock-excited emission and dust destruction based on Spitzer Space Telescope observations of two well-studied regions in the remnant, (i) a non-radiative shock filament along the NE limb, and (ii) the XA region, characterized by emission from bright radiative shocks.

Keywords. dust, shock waves, supernova remnants

---

1. Introduction

The destruction of grains in supernova remnant (SNR) shock waves is a key step in the life-cycle of dust in the Galactic interstellar medium (ISM). It regulates the dust-to-gas ratio in the ISM (Draine 2009). Within the remnants, it determines the gas phase abundances of refractory elements, which are depleted in the pre-shock gas and released as the dust is destroyed. At shock speeds greater than about 80 km s^{-1} sputtering is the dominant destruction process, while destruction by grain-grain collisions become dominant for slower shocks (Jones *et al.* 1994). Most of the dust destruction happens in middle-aged SNRs in the Sedov-Taylor phase, when the shock speeds are in the range 200–500 km s^{-1}, since most of the volume swept up by a remnant happens during that stage of its evolution.

The Cygnus Loop is well suited for the study of dust destruction processes. It is nearby (540 pc, Blair *et al.* 2005), so the post-shock zone is spatially resolved by X-ray and infrared instruments. The foreground extinction is small ($E_{(B-V)} \approx 0.08$), and therefore it can be observed at ultraviolet wavelengths, which allows for the determination of the shock properties (Danforth, Blair & Raymond 2001; Raymond *et al.* 2003; Sankrit *et al.* 2007). We have used the Spitzer Space Telescope (*Spitzer*) to observe a non-radiative shock along the north-east limb of the Cygnus Loop and the XA region, a radiative shock interaction region in the south-east. In §2 we summarize the main results obtained for the non-radiative shock, which have been presented in Sankrit *et al.* (2010). Then in §3 we describe the evidence for dust destruction in a slower, radiative shock in the XA region. Some possible avenues for future work are discussed in §4.

2. Non-radiative Shock

The filament observed by *Spitzer* was chosen from among the non-radiative filaments that define the edge of the Cygnus Loop because the shock parameters had been determined from observations in the optical (Ghavamian *et al.* 2001) and the far-ultraviolet (Raymond *et al.* 2003). Images obtained with the MIPS instrument showed an intensity fall-off in the 24 μm band and an increase in the 70 μm/24 μm flux ratio with distance behind the shock. Dust models using input parameters guided by prior knowledge of the shock properties were able to match the intensity profile and the variation in flux ratio. It was found that non-thermal sputtering (i.e. due to the bulk motion of the grains relative to the gas) contributes significantly to the dust destruction. Models that included only thermal sputtering were unable to reproduce the observed flux ratios for the permissible shock parameters. The models predicted grain temperatures between 30 and 60 K.

The fiducial model for the shock (post-shock temperature = 0.20 keV, corresponding to a shock speed of about 400 km s^{-1}, and pre-shock hydrogen number density = 0.5 cm^{-3}) predicted that about 35% by mass of the grains have been destroyed in the course of about 1350 yr, the age of the shock, and that in another 1000 yr about 43% of the dust will be destroyed. In order to compare the model-predicted fluxes with observations, the dust-to-gas ratio in the pre-shock gas needs to be specified. For a dust-to-gas mass ratio of 0.75%, the model reproduces the observed 24 μm flux for a path length through the emitting region of about 0.35 pc. This is considerably lower than the path length of 0.7–1.5 pc through the shock required to match the far-ultraviolet and X-ray data (Raymond *et al.* 2003). If the path length is 0.7 pc, then the dust-to-gas ratio required falls to 0.38%. It is an open question whether such a low value is permissible, but it is consistent with our earlier findings for several remnants (Borkowski *et al.* 2006; Williams *et al.* 2006; Blair *et al.* 2007).

3. Radiative Shock: the XA region

The XA region is an indentation in the X-ray shell along the southeast edge of the Cygnus Loop, and is due to an interaction between the blast wave and a cloud (Hester & Cox 1986). Danforth *et al.* (2001) analyzed optical and ultraviolet data of the region and suggested that it was a protrusion on the surface of a larger cloud. We obtained a MIPS 24 μm image of the interaction region and high resolution IRS spectra (in "stare" mode) of a few locations in the region. Images of the region are shown in Fig. 1, with IRS Long-High aperture positions overlaid. The "cusp" is the tip of the protrusion impacted by the blast wave, and the "background" lies outside the remnant. We used the "Spectroscopic Modeling Analysis and Reduction Tool" (SMART) version 8.1.2 (Higdon *et al.* 2004; Lebouteiller *et al.* 2010) to extract background-subtracted, one-dimensional, calibrated data. The resulting spectrum is shown in Fig. 2. The spectrum shows no detectable dust continuum emission, and by convolving the spectrum with the MIPS 24 μm filter response, we find that the lines contribute over 80% of the flux detected in the image.

In order to interpret the spectrum, we compare the measured line ratios with predictions from shock model calculations, using the code originally described by Raymond (1979) with updates described in Raymond *et al.* (1997). We find at least two shock speeds are required to produce the observed spectrum. The high ionization lines, [Ne V] 24.32 μm and [O IV] 25.88 μm, require a 180 km s^{-1} shock that has not recombined (otherwise it would produce too much low ionization emission) and a slower shock, about 150 km s^{-1} that produces the [S III] 33.47 μm and [Si II] 34.80 μm lines. This is consistent with the conclusions of Sankrit *et al.* (2007) who found that a range of shock speeds were required to explain the far-ultraviolet spectrum of the region.

The observed [Si II] to [S III] flux ratio is 3.4, and the shock models indicate that this cannot be obtained for depleted gas phase abundances. For a sulphur abundance [S] = 7.51 on a logarithmic scale relative to [H] = 12.00, the observed flux ratio is obtained if

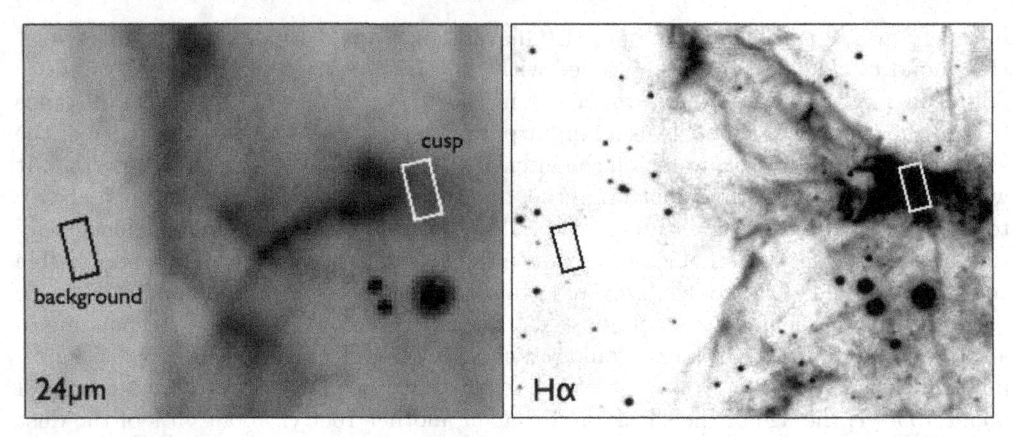

Figure 1. Images of the XA region; *left:* Spitzer MIPS 24 μm and *right:* Whipple Observatory 1.2m telescope Hα. The boxes, drawn to scale, represent the IRS Long-High aperture ($11.1'' \times 22.3''$). North is up, East to the left.

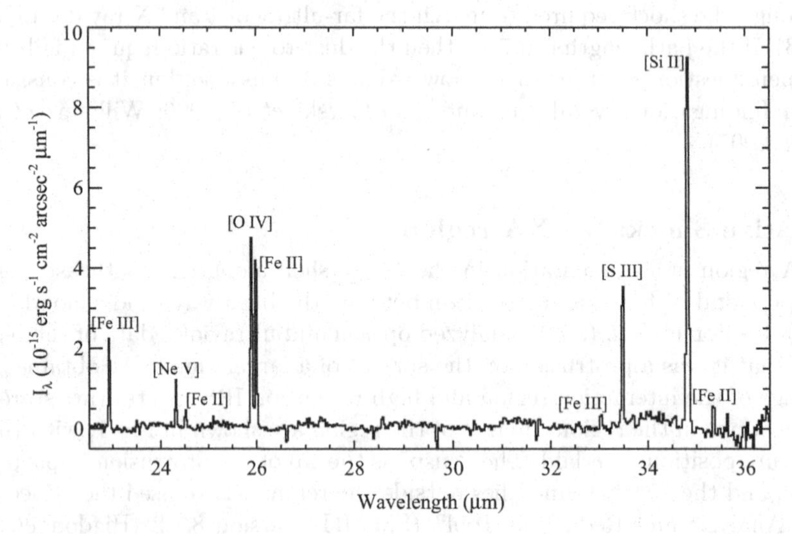

Figure 2. Background subtracted *Spitzer* IRS Long-High spectrum of the cusp (see Fig. 1).

the silicon abundance [Si] = 7.50, which is the undepleted (solar) value, and if the pre-shock magnetic field (perpendicular to the shock direction) $B_0 = 10\mu$G. The flux ratio is moderately dependent on the pre-shock magnetic field strength, with lower values of the magnetic field resulting in higher [Si II], but even for $B_0 = 1\mu$G, the silicon abundance needs to be about half the solar value. Since about 90% of the Silicon is expected to be in dust in the general ISM, our results imply that at least half of it is released into the gas phase, which in turn implies that dust has been destroyed by the shock. This conclusion is subject to the caveats that we do not have a direct measurement of the Silicon abundance in the pre-shock gas, and that we are assuming a fixed value for the Sulphur abundance. The spectrum only allows us to measure the abundance ratio in the post-shock gas.

4. Concluding Remarks

The results on the non-radiative shock presented here (and in more detail in Sankrit *et al.* 2010) suggest some avenues for exploration. To obtain a definitive answer as to the dust-to-gas ratio in the ambient ISM, it is necessary to calculate a self-consistent model of the X-ray and infrared emission. This would require following the gas phase abundances with position behind the shock. We have estimated that much deeper X-ray observations than currently available are required to reliably do so. A project currently underway is the measurement of C IV $\lambda 1550$ emission in far-ultraviolet spectra, obtained with HST/COS, to examine the sputtering of Carbon atoms behind the shock front (Raymond *et al.* in preparation).

The XA region is complex, with multiple shocks present along our line of sight. The *relative* abundances of elements may be well constrained by our data, but they do not allow us to resolve the degeneracies in other parameters such as the absolute abundances, densities and magnetic fields. In our data dust destruction is addressed only indirectly via the Silicon abundance. A promising line of study both for characterizing the shock parameters and for measuring the release of refractory elements into the gas phase is the interpretation of the [Fe II] and [Fe III] lines, which requires the calculation and use of new atomic data (Bautista, in preparation). The search for cooler dust at longer wavelengths may become viable when far-infrared instruments become available on SOFIA.

Acknowledgements: R. Sankrit thanks the organizers of the symposium for their hospitality. RS also acknowledges support from NASA contract NAS2-97001.

References

Blair, W. P., Sankrit, R., & Raymond, J. C. 2005, *AJ*, 129, 2268
Blair, W. P., *et al.* 2007 *ApJ*, 662, 998
Borkowski, K. J., *et al.* 2006 *ApJ*, 642, L141
Danforth, C. W., Blair, W. P., & Raymond, J. C. 2001 *AJ*, 122, 938
Draine, B. T. 2009, *ASP-CS*, Cosmic Dust - Near and Far, 414, 453
Ghavamian, P., Raymond, J. C., Smith, R. C., & Hartigan, P. 2001 *ApJ*, 547, 995
Hester, J. J. & Cox, D. P. 1986 *ApJ*, 300, 675
Higdon, S. J. U., *et al.* 2004 *PASP*, 116, 975
Jones, A. P., Tielens, A. G. G. M., Hollenbach, D. J., & McKee, C. F. 1994 *ApJ*, 433, 797
Lebouteiller, V., Bernard-Salas, J., Sloan, G. C., & Barry, D. J. 2010 *PASP*, 122, 231
Raymond, J. C. 1979 *ApJS*, 39, 1
Raymond, J. C., *et al.* 1997 *ApJ*, 482, 881
Raymond, J. C., Ghavamian, P., Sankrit, R., Blair, W. P., & Curiel, S. 2003 *ApJ*, 584, 770
Sankrit, R., Blair, W. P., Cheng, J. Y., Raymond, J. C., Gaetz, T. J., & Szentgyorgyi, A. 2007, *AJ*, 133, 1383
Sankrit, R., *et al.* 2010, *ApJ*, 712, 1092
Williams, B. J., *et al.* 2006 *ApJ*, 652, L33

Discussion

KOO: The low dust-to-gas ratio in preshock region that you obtained seems to be common rather than unusual. Could you comment on that?

SANKRIT: Yes, in all the cases we have looked at, including cygnus loop, Kepler and several LMC remnants, the inferred dust-to-gas ratio is low. This difference between SNR surroundings & Typical ISM regions is interesting and is an issue that needs to be explored at greater depth.

GREEN: Could you explain again the reasoning to choose the line of sight depth in your analysis.

SANKRIT: The line-of-sight depth chosen is approximately that of the filament length on the plane of the sky (taking into account the curvature). We expect that these brighter rims may actually have longer path lengths (selection effect). But that would drive the inferred dust density to lower values, and make the dust-to-gas ratio correspondingly lower. i.e. the deviation from the typical dust-to-gas ratio would increase.

FRANCE: Do you see low-ionization metals (e.g.; S:II 1526/1533, FeII 1608, etc) in your HST-COS spectra of Cygnus? If so, can you use these emissions to constrain dust destruction/gas liberation?

SANKRIT: No, We don't see low-ionization lines; the shock is non-radiative, and the swept-up post-shock gas has not had enough time to recombine. The only lines detected are the CIV doublet and HeII (1640 A)

Supernovae Environmental Impacts
Proceedings IAU Symposium No. 296, 2013
A. Ray & R. A. McCray, eds.

doi:10.1017/S1743921313009411

An Integral View of Balmer-dominated Shocks in Supernova Remnants

Sladjana Nikolić[1], Glenn van de Ven[1], Kevin Heng[2], Daniel Kupko[3], Jose Alfonso Lopez Aguerri[4], Jairo Méndez-Abreu[4], Joan Font Serra[4] and John Beckman[4]

[1]Max Planck Institute for Astronomy, Königstuhl 17, Heidelberg, Germany
email: nikolic@mpia.de

[2]University of Bern, Center for Space and Habitability, Sidlerstrasse 5, Bern, Switzerland

[3]Leibniz Institute for Astrophysics Potsdam (AIP), An der Sternwarte 16, Potsdam, Germany

[4]Instituto de Astrofísica de Canarias (IAC), Vía Láctea, La Laguna (Tenerife), Spain

Abstract. We present integral-field spectroscopic observations with the VIMOS-IFU at the VLT of fast (2000-3000 kms^{-1}) Balmer-dominated shocks surrounding the northwestern rim of the remnant of supernova 1006. The high spatial and spectral resolution of the instrument enable us to show that the physical characteristics of the shocks exhibit a strong spatial variation over few atomic scale lengths across 133 sky locations. Our results point to the presence of a population of non-thermal protons (10-100 keV) which might well be the seed particles for generating high-energy cosmic rays. We also present observations of Tycho's supernova remnant taken with the narrow-band tunable filter imager OSIRIS at the GTC and the Fabry-Perot interferometer GHaFaS at the WHT to resolve respectively the broad and narrow Hα lines across a large part of the remnant.

Keywords. Balmer emission, supernova remnants, cosmic rays, integral-field spectroscopy

---

1. Introduction

Supernova remnants are laboratories for studying optical shocks. An optical spectrum dominated by Balmer lines is seen when a fast astrophysical shock enters partly neutral interstellar gas. Balmer dominated shocks (BDSs) are characterized by velocities higher than 200 kms^{-1}, presence of two-component Hα line, absence of forbidden lines of lowly ionized metals, and general lack of non-thermal X-ray emission at the location where Hα lines are detected (Heng 2010). BDSs are observed around historical supernova remnants (SNRs) like Tycho, Kepler and SN1006.

The hydrogen lines consist of a narrow ($\sim$10 kms^{-1}) and a broad ($\sim$1000 kms^{-1}) component. The narrow component is produced by cold neutrals in the pre-shock ambient interstellar medium (ISM) that are collisionally excited by electrons and protons in the shock. While the broad component is produced by post-shock hot neutrals, created through charge exchange between incoming neutrals and hot protons in the shock. Thus, the two-component Hα lines directly yield the pre-shock and post-shock temperatures of the ISM around the remnant. The Balmer line profiles also contain signatures of shock precursors. Investigating in detail the shape of the Hα line has the potential to provide strong observational constraints on cosmic rays (CRs). In particular, the CRs will heat the cold neutrals in the ISM before they are being ionized by the shock, resulting in a narrow Hα line of which the width is broadened beyond the normal 10-20 kms^{-1} gas dispersion. The CRs can also carry away energy from the protons in the post-shock, so that the broad Hα line has a smaller width than allowed by the fast shock velocity.

Because the CR precursor is typically spatially unresolved, its additional contribution to the narrow-line Hα emission results in a decreasing broad-to-narrow intensity line ratio. The presence of CRs can also cause the shape of the broad Hα line to deviate from a Gaussian profile (Raymond *et al.* 2010).

Here we present high-spatial resolution spectro-photometric imaging of the remnants SN 1006 and Tycho, and investigate the Hα-line profiles in detail.

2. Data & Analysis

2.1. *VIMOS-IFU Observations of SN 1006*

The long-slits typically used to measure the Hα line have low spatial resolutions resulting in the contribution of multiple shock fronts to the measured Hα line. It is then unclear whether the line shape, including its width, are contaminated by this geometric effect. Optical integral-field unit (IFU) spectrographs are able to trace and distinguish multiple, projected shocks. Since BDSs have not been investigated before with IFUs, our aim in this pilot study was to demonstrate that such observations can be executed and that the scientific yields constitute a marked improvement over previous studies.

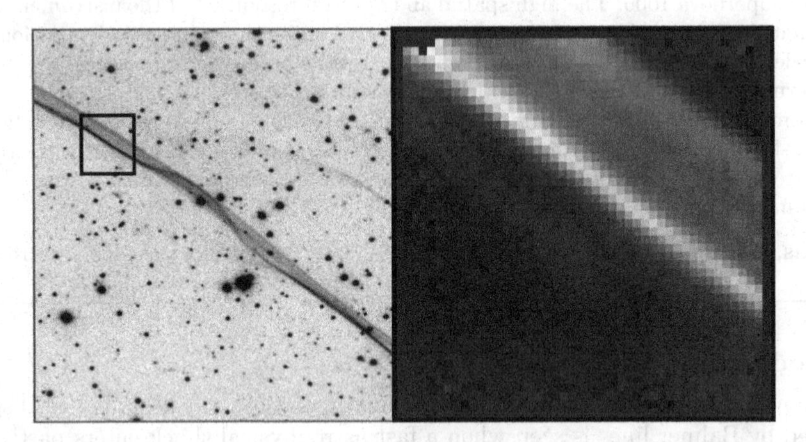

Figure 1. VIMOS-IFU spectroscopy of the shock front in the remnant of SN1006. The left panel shows a CTIO-Curtis-Schmidt narrow-band Hα image (from Winkler *et al.* 2003) of the NW rim of the remnant with the box indicating the region observed with the VIMOS-IFU. The right panel shows the reduced data cube collapsed in wavelength around the narrow Hα line, nicely recovering the shock in the image.

We used the VIMOS-IFU spectrograph on the Very Large Telescope (VLT) to investigate the fast optical shocks at the northwestern (NW) rim of the remnant of SN1006 (see Figure 1). The $27'' \times 27''$ field-of-view (FOV) of the VIMOS-IFU on a 8 m telescope enables us to collect enough photons in a reasonable time to reach the high signal-to-noise ratio (S/N) required to accurately measure the shape of the Hα line, including deviations from a Gaussian profile. The high spatial resolution of $0.''67$ ($\simeq 2\times10^{16}$ cm at the distance of $\simeq$2 kpc), in combination with the two-dimensional coverage of the VIMOS-IFU allow us to precisely trace the narrow shock front. The spectral resolution with dispersion of $\backsim$48 kms^{-1} is more than sufficient to measure the width of the broad line of 2000-3000 kms^{-1}, as well as deviations from a Gaussian profile. The width of the narrow line of $\backsim$20 kms^{-1}can not be resolved, but its intensity can be accurately measured.

As a first analysis we combined the spectra from four different regions parallel to the shock front and analyzed the resulting two-component Hα lines shown in Figure 2.

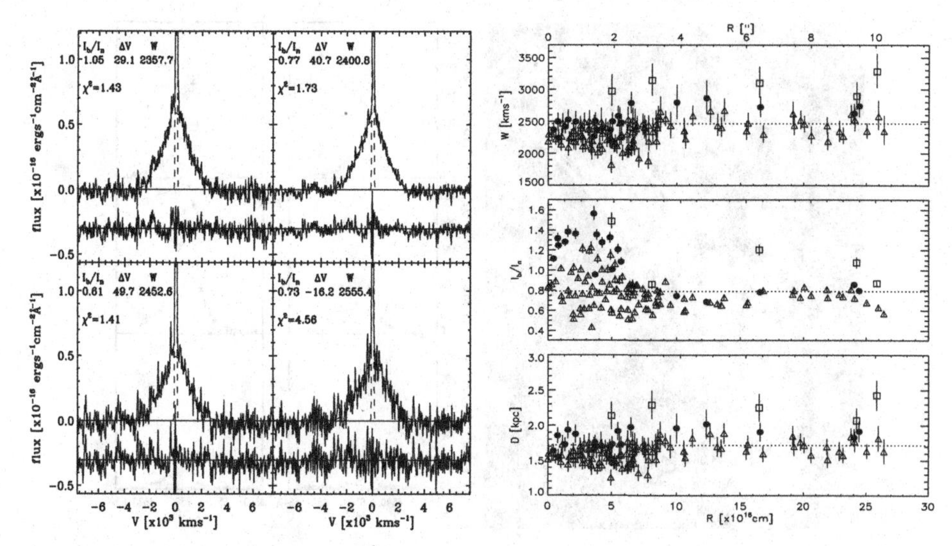

Figure 2. Two-component Hα-line profiles and spatial variations of shock parameters. The left four panels show the spectra that have been combined from four different regions parallel to the shock front. In each panel, we show the best-fit parameters: broad-to-narrow intensity ratio I, velocity offset ΔV from the narrow component (in kms^{-1}), and the FWHM of the broad component(in kms^{-1}). The three panels on the right show 133 measured (with error bars) broad line widths W, broad-to-narrow intensity ratios I_b/I_n, and heliocentric distances D from inferred shock velocities combined with measured proper motions of the shock front (Winkler *et al.* 2003). Data are ordered in increasing distance from the inner rim to the outer rim. The dashed horizontal lines indicate the measured W=2465.76 kms^{-1} and I_b/I_n=0.79 values from collapsing all spectra of the pixels on the shock front, and from there the inferred D=1.72 kpc. Data points marked with open triangles (squares) indicate that the I_b/I_n values are too low (high) to obtain a solution of the van Adelsberg *et al.* model that does not include non-thermal physics.

Performing double-Gaussian fits we extracted various parameters, including the broad-to-narrow line ratio (I_b/I_n), velocity offset (ΔV) from the narrow component and the FHWM of the broad-line component (W). Already strong variation in I_b/I_n and W, as well as hint of non-Gaussianity in the broad-line core point towards the presence of a precursor in the shock.

In order to investigate spatial variations of the physical characteristics across the shock front, we used the method of Cappellari & Copin (2003) to combine the neighboring pixels and create 133 spatial (Voronoi) bins in which the combined spectra have a minimal S/N of 40. We fitted double-Gaussian to the lines and extracted again the same parameters (per bin) as before: I_b/I_n, ΔV and W. We then converted for each bin W and I_b/I_n to a shock velocity and electron-to-proton temperature ratio (β) using the model of van Adelsberg *et al.* (2008). The variations of the intensity ratios and broad-line widths along with the distances calculated from derived shock velocities and the proper motion measurements of 280 masyr^{-1} (Winkler *et al.* 2003) are shown in the three right panels of Figure 2. Nearly 85% of the observed values (represented with empty triangles and squares) are out of the range predicted by the model, which does not include CR physics. Strong spatial variations in the intensity ratios and broad-line widths, the fact that the most of our observed values are not in the model predicted range, and the hint of the non-Gaussianity in the broad-line core all together indicate the potential presence of suprathermal protons, non-thermal particles which might be the seed particles for generating high-energy CRs (Nikolic *et al.* 2013).

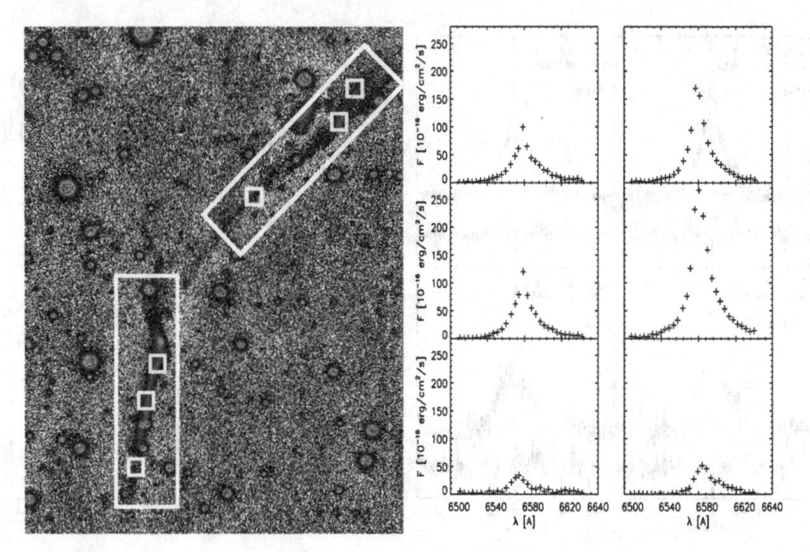

Figure 3. OSIRIS observations of Tycho's remnant. The left panel shows reduced OSIRIS narrow-band imaging data of the northeastern rim of the remnant. Small 5×5 pixel boxes (white squares) indicate six different positions along the nicely recovered shock front from which the Hα line was extracted and shown in the right panel. The first column shows the spectra from the upper white rectangular region from the top to the bottom, while the second column shows the spectra from the lower white rectangular region again in the same order.

2.2. *OSIRIS & GHaFaS Observations of Tycho*

Integral-field spectrographs are able to trace and distinguish multiple, projected shocks, but still, the field-of-view of these spectrographs typically covers only a small portion of the full remnant. Henceforth, we use OSIRIS narrow-band tunable filter imaging at GTC (Grand Telescope Canaria). The high spatial resolution of 0.″125 per pixel avoids geometric effects, while the large field-of-view of 4′×4′ allows to measure the width of the broad component of the Hα line of the shock fronts in the full northeastern part of the remnant of Tycho's supernova. Our goal is to map the changing CR acceleration efficiency along the remnant by measuring broad Hα-line widths smaller than expected from the high shock velocities, which we know independently from proper motion measurements of the shock fronts.

OSIRIS low spectral resolution prevents resolving the narrow Hα component. In order to precisely measure the intrinsic narrow Hα-line width, we use the Fabry-Perot interferometer GHaFaS at the WHT (William Herschel Telescope) which has the unique capability to scan with a spectral resolution as fine as $8\,\mathrm{km s^{-1}}$. This enables us to directly quantify the presence of the CR precursor. The high spatial resolution of 0.″2 plus large FOV 3.′9×3.′9 of GHaFaS allow us to differentiate between different individual shock fronts along Tycho's supernova remnant. Observing exactly the same parts of the Tycho's remnant with both OSIRIS and GHaFaS, we expect to place tight constrains on CR precursors in Tycho's SNR.

Acknowledgments

We would like to thank Bernd Husemann (AIP), John C. Raymond (Harvard-Smithsonian CfA), John P. Hughes (Rutgers University) & Jesús Falcón-Barroso (IAC) for their collaboration in the work of VIMOS-IFU observations of SN 1006.

References

Cappellari, M. & Copin, Y. 2003, *MNRAS*, 342, 345
Heng, K. 2010, *PASA*, 27, 23
Nikolic, S., van de Ven, G., Heng, K., Kupko, D., Husemann, B., Raymond, J. C., Hughes, J. P., & Falcon-Barroso, J. 2013, *Science Express*, February 14
Raymond, J. C., Winkler, P. F., Blair, W. P., Lee, J.-J., & Park, S. 2010, *ApJ*, 712, 901
van Adelsberg, M., Heng, K., McCray, R., & Raymond, J. C. 2008, *ApJ*, 689, 1089
Winkler, P. F., Gupta, G., & Long, K. S. 2003, *ApJ*, 585, 324

Discussion

VINK: Could the non-Gaussianity be caused by a superposition of gaussians with different width along the line of sight? It may be worthwhile to fit the Kappa-profile instead of gaussians.

NIKOLIC: Taking into account the fact that the width of the narrow H-alpha lines (even though unresolved) as well as that shifts between narrow and broad velocity centroids are much smaller than the "extra-core" size of the broad line, the non-Gaussianity can not be due to binning effect, i.e. superposition of Gaussians with different width along the line of sight. We didn't try fitting Kappa-profile because we don't see any physical meaning of it. However, we tried fitting the line with 3 Gaussians as well as fitting Gauss-Hermite polynomial, but none of these have improved the fit.

Supernova Environmental Impacts
Proceedings IAU Symposium No. 296, 2013
A. Ray & R. A. McCray, eds.

doi:10.1017/S1743921313009423

Molecular Environments of Supernova Remnants†

Yang Chen[1,2], Bing Jiang[1], Ping Zhou[1], Yang Su[3], Xin Zhou[3,2], Hui Li[4,1], and Xiao Zhang[1]

[1]Department of Astronomy, Nanjing University, Nanjing 210093, China
email: `ygchen@nju.edu.cn, bjiang@nju.edu.cn, pingzhou@nju.edu.cn, zxmysky@163.com`

[2]Key Laboratory of Modern Astronomy and Astrophysics, Nanjing University, Ministry of Education, Nanjing 210093, China

[3]Purple Mountain Observatory, 2 West Beijing Road, Nanjing 210008, China
email: `yangsu@pmo.ac.cn, xinzhou@pmo.ac.cn`

[4]Department of Astronomy, University of Michigan, 500 Church Street, Ann Arbor, MI 48109, USA
email: `hliastro@umich.edu`

Abstract. There are about 70 Galactic supernova remnants (SNRs) that are now confirmed or suggested to be in physical contact with molecular clouds (MCs) with six kinds of evidence of multiwavelength observations. Recent detailed CO-line spectroscopic mappings of a series of SNRs reveal them to be in cavities of molecular gas, implying the roles the progenitors may have played. We predict a linear correlation between the wind bubble sizes of main-sequence OB stars in a molecular environment and the stellar masses and discuss its implication for supernova progenitors. The molecular environments of SNRs can serve as a good probe for the γ-rays arising from the hadronic interaction of the accelerated protons, and this paper also discusses the γ-ray emission from MCs illuminated by diffusive protons that escape from SNR shocks.

Keywords. ISM: supernova remnants, ISM: clouds, ISM: bubbles, ISM: molecules, gamma-rays: ISM, stars: early-type

---

1. Introduction

The interplay between supernova remnants (SNRs) and their molecular environments plays an important role in many aspects of astrophysical studies. Molecular gas takes up about half mass of Galactic interstellar medium, and most core-collapse supernovae are believed to be located close to giant molecular clouds (MCs) where their progenitor stars are given birth to (e.g., Huang & Thaddeus 1986). The SNR shock waves propagating in molecular environments can compress, heat, excite, ionize, and even dissociate molecules. They are suggested to be one of the mechanisms of triggering star formation. They influence the chemical evolution of the gas and produce otherwise impossible or seldom detected molecular emission (e.g., 1720MHz OH maser, HCO^+, etc.). Shock interaction with molecular clouds can generate γ-rays as a result of neutral pion decay after p-p collision (hadronic interaction), which may serve as a probe of SNR shock acceleration of relativistic protons. Once an SNR-MC association is established, the kinematic distance of the SNR can be determined with the local standard of rest (LSR) velocity of the MC.

† Supported by the 973 Program grant 2009CB824800, the NSFC grants 11233001, 11103082, and 11203013, grants 20120091110048 and 20110091120001 from the Educational Ministry of China, and grant 2011M500963 from the China Postdoctoral Science Foundation.

This article is composed of contents on SNR-MC associations, some of our CO-line observations, bubbles in molecular environments, and probe for hadronic interaction.

2. Census of SNR-MC associations

How many among the $\gtrsim 300$ Galactic SNRs are in physical contact with MCs? There have been some endeavors in surveying and cataloging the SNR-MC associations. In an early CO-line survey toward 26 outer ($l = 70°$–$210°$) Galactic SNRs, about half were found spatially coincident with large MC complexes (Huang & Thaddeus 1986). After a survey of 1720 MHz OH maser emission towards Galactic SNRs, Frail *et al.* (1996) and Green *et al.* (1997) revealed the unique diagnostic relation between such masers and the SNR shocks in MCs and identified nearly a score of SNRs interacting with MCs. Seta *et al.* (1998) listed 26 SNRs detected in CO- and H_2-line emissions along the line of sight, but no physical evidence of shock-MC interaction was given for them.

Recently, we presented a catalog of the interacting SNRs, with a summary of six kinds of multiwavelength observational evidence for judging the contact of SNRs with MCs (Jiang *et al.* 2010). Among the six observational evidences, firstly, the 1720 MHz OH maser is the most important, which is presently widely accepted as a signpost of SNR shock-MC interaction. When the molecular gas is compressed by a C-type shock to a density of order $\sim 10^5\,\mathrm{cm}^{-3}$ and the temperature reaches 50–125 K, collisional pumping causes the population inversion of the hyperfine rotational level $^2\Pi_{3/2}(J = 5/2)$ of OH molecules, from which the 1720 MHz maser arises (Lockett *et al.* 1999). Since this OH satellite line was noted by Goss & Robinson (1968), explorations and surveys in the line have been repeatedly made toward Galactic SNRs, and the masers have been detected from 25 SNRs. Secondarily, molecular (CO, HCO^+, CS, etc.) line broadening or asymmetric profile is another important kinematic evidence of shock perturbation of the molecular gas (e.g., DeNoyer 1979). Line broadening is present in 18 SNRs listed in our table. The third kind of evidence is high high-to-low excitation line ratio in line wings (Seta *et al.* 1998). In the broad wings of the shocked ^{12}CO, where both the $J = 2$–1 and $J = 1$–0 emissions are optically thin, their ratio can be $\gtrsim 1$. The fourth is the detection of near-infrared (IR) emission, e.g., [Fe II] line or ro-vibrational lines of H_2 (such as H_2 1–0 S(1) line at 2.12 μm and H_2 0–0 S(0)–S(7) lines), due to shock excitation. The fifth is the specific IR colors suggesting molecular shocks, e.g., 3.6 μm/5.8 μm and 4.5 μm/8 μm in the Spitzer IRAC observation (see Reach *et al.* 2006 and Fig.2 therein). The sixth is the morphological agreement or correspondence of molecular features with SNR features (e.g., arc, shell, interface, etc.). Besides the OH masers, a combination of the sixth (spatial) evidence with one of the other four kinematic/physical evidences is also regarded as convincing criteria for judging the SNR-MC interaction. Thus, in our catalog, there are 34 SNRs "confirmed", 11 "probable", and 19 "possible" to be in physical contact with MCs. A most recent CO survey for $l = 60°$–$190°$ showed additional six SNRs with their radio morphology in a good spatial relation with MCs, without direct evidence for the interaction (Jeong *et al.* 2012), thus raising the number of "possible" interacting SNRs to 25.

3. Our CO observations of a series of SNRs

The OH satellite line masers have proven to be a powerful tool for the identification of interacting SNRs, nonetheless they cannot be used to investigate the detailed information of molecular environments of the SNRs. Also, a number of SNR-MC associations may be elusive in the search in the maser line because of the weak emission below the detection thresholds. For this purpose, CO, other than H_2 (without a permanent electric dipole moment), is a practical tracer of MCs and commonly used for case studies of SNR-MC interaction. Here we review our recent CO observations of a series of interacting SNRs made with the millimeter and sub-millimeter telescopes in the Purple Mountain Observatory at Delingha, Seoul Radio Astronomy Observatory, Koelner Observatory for Submillimeter Astronomy, and other observatories.

3.1. *Kes 69*

Kes 69 is morphologically characterized by the bright radio, IR, and X-ray emissions only at the southeastern boundary. The 1720 MHz OH masers are detected in both the northeast and the southeast, but with different LSR velocities (Green *et al.* 1997; Hewitt *et al.* 2008). Our ^{12}CO ($J = 1$–0) observation (Zhou *et al.* 2009) discovers a molecular arc in the LSR velocity interval 77–86 km s^{-1} (consistent with 85 km s^{-1} for the southeastern maser). The arc is in good morphological agreement with the multiwavelength partial shell of the SNR. The HCO^+ emission is detected at the position of the radio brightness peak also at the LSR velocity $V_{\rm LSR} = 85$ km s^{-1}, which is consistent with the presence of the southeastern maser, both resulting from C-shock interaction. These evidences strongly suggest that Kes 69 is physically associated with the MC at the systemic velocity 85 km s^{-1}. From this velocity, a kinematic distance 5.2 kpc is derived. We ascribe the multiwavelength emissions arising from the southeastern partial shell of the SNR to the impact of the SNR shock on a dense, clumpy patch of molecular gas, which is likely to be the cooled debris of the material swept up by the progenitor's stellar wind.

3.2. *Kes 75*

The young composite SNR Kes 75, with a pulsar wind nebula inside, displays only a half shell in the south. We find that the ^{12}CO ($J = 1$–0) line profile of the $V_{\rm LSR} = 45$–58 km s^{-1} MC is broadened in the blue wing and a molecular shell unveiled in this wing (45–51 km s^{-1}) encloses a cavity (Su *et al.* 2009). The southern part of the molecular shell is in good morphological agreement with the SNR's half shell shown in X-rays, mid-IR, and radio continuum. These spatial and kinematic evidences indicate that Kes 75 is physically associated with the MC at the systemic velocity 54 km s^{-1}. The associated large cloud has a mass $\gtrsim 10^4 M_\odot$. The distance to this SNR is thus determined to be 10.6 kpc. The presence of the dense molecular shell is interpreted to be due to the same reason as that of the molecular arc in Kes 69.

3.3. *Kes 78*

Kes 78 is a shell-type SNR, but only the eastern half is radio bright. There is a 86 km s^{-1} OH maser at the eastern boundary (Koralesky *et al.* 1998). The unidentified TeV source J1852−000 to the east of the remnant is detected by the HESS γ-ray telescope†, which implies that this SNR may have a relation with the very high energy emission. Our CO line observations toward this SNR find that it is interacting with the MC complex at the systemic velocity $V_{\rm LSR} = 81$ km s^{-1} (Zhou & Chen 2011). The ^{13}CO emission shows the presence of a dense cloud to the east, from which the maser may arise and which also seems crudely correspondent to the location of the extended TeV source. The strong ^{12}CO emission in general spatially corresponds to the eastern bright radio shell, and the spatial extent of the SNR is consistent with a ^{12}CO cavity. ^{12}CO lines are found to be broadened in some boundary regions including the eastern maser region, and ^{12}CO $J = 2$–$1/J = 1$–0 ratio is generally elevated along the SNR boundary. These are all the

† http://www.mpi-hd.mpg.de/hfm/HESS/pages/home/som/2011/02/

kinematic signatures of shock perturbation in the molecular gas. The distance to Kes 78 is estimated to be 4.8 kpc based on the systemic LSR velocity. An *XMM*-Newton X-ray spectral analysis for the northeastern boundary infers an age 6 kyr for the remnant.

3.4. *3C 396*

Analogous to Kes 75, 3C 396 is another composite SNR containing a pulsar wind nebula and is semi-circular-shaped, too. We find that the western edge, bright in radio, IR, and X-rays, is perfectly confined by a molecular wall (of mass $\sim 10^4 M_\odot$) revealed in ^{12}CO ($J = 1$–0) line at $V_{\rm LSR} \sim 84\,{\rm km\,s^{-1}}$ (Su *et al.* 2011). The CO emission gradually gets faint from west to east, which is indicative that the eastern region is of low gas density. Noticeably, a finger/pillar-like molecular structure (of mass $\sim 4 \times 10^3 M_\odot$) in the southwest intrudes into SNR edge. The pillar, which may have been shocked at the tip, should be the reason why the X-ray and radio emissions get brightened at the southwestern boundary and some IR filaments are present there. The evidences for the SNR-MC interaction also include the relatively elevated ^{12}CO $J = 2$–$1/J = 1$–0 line ratios in the southwestern pillar tip and the molecular patch at the northwestern boundary, as well as the redshifted ^{12}CO ($J = 1$–0 and $J = 2$–1) wings (86–90 km s^{-1}) of an eastern $V_{\rm LSR} \sim 81\,{\rm km\,s^{-1}}$ molecular patch. The X-ray analysis of the hot gas infers an age ~3 kyr for the remnant and the derived relative abundances of Si, S, and Ca of the ejecta are consistent with a B1–B2 progenitor star.

3.5. *3C 397*

3C 397 is a thermal composite (or mixed-morphology) SNR with an abnormal rectangular shape. Our ^{12}CO ($J = 1$–0) observation shows that the remnant is confined in a molecular gas cavity and embedded at the edge of a giant MC at the systemic velocity $V_{\rm LSR} = 32\,{\rm km\,s^{-1}}$ (Jiang *et al.* 2010). The column density of the environmental molecular gas has a gradient that increases from the southeast to the northwest and is perpendicular to the Galactic plane, in agreement with the elongation direction of the remnant. Solid evidence for the SNR-MC interaction is provided by the ^{12}CO line broadening of the $\sim 32\,{\rm km\,s^{-1}}$ component that is detected at the westmost boundary of the remnant, which is consistent with the impact of the Fe rich ejecta in the region (Jiang & Chen 2010). The systemic velocity of the molecular gas places 3C 397 at a kinematic distance of 10.3 kpc. The mean ambient molecular density is ~10–30 cm^{-3}, which can explain the high volume emission measure of the X-ray emitting gas. For the westmost line-broadened region, the density of the disturbed molecular gas is deduced to be of order $10^4\,{\rm cm^{-3}}$ based on the the pressure balance between the cloud shock and the X-ray emitting hot gas and can be ascribed to very dense clumps, implying a multi-phase gas environment there.

3.6. *Kes 79 and W 49B*

SNRs Kes 79 and W 49B have also been observed and preliminarily show interesting molecular environments.

Kes 79, harboring a central compact object, was demonstrated by the *Chandra* observation to have multi-shell structure, which implies a complicated surrounding environment (Sun *et al.* 2004). Our CO observation finds that this remnant is spatially coincident with a molecular cavity at $V_{\rm LSR} \sim 113\,{\rm km\,s^{-1}}$ (Fig. 1). The ^{12}CO $J = 2$–$1/J = 1$–0 ratio is also found to be elevated, as high as $\gtrsim 1$, in some boundary regions.

W 49B, sorted as a thermal composite, has an incomplete radio shell with centrally brightened thermal X-rays. A barrel-shaped structure with coaxial rings was revealed in the 1.64 μm [Fe II] image, and a strip of 2.12 μm shocked H_2 emission extends outside of the [Fe II] emission to the southeast; the X-ray jet-like structure is along the axis of

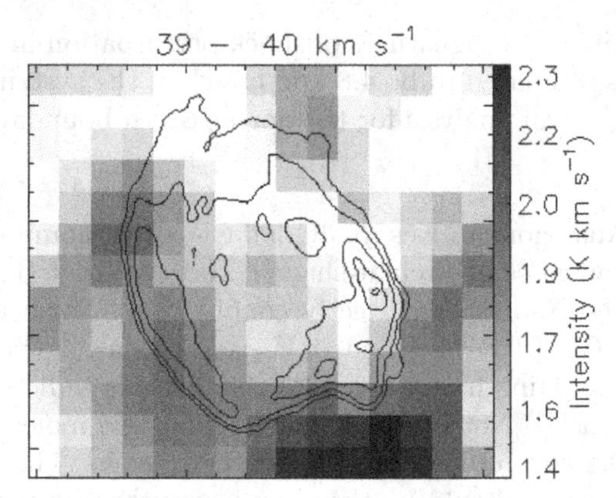

Figure 1. Left: ^{12}CO (J = 2–1) emission map at $V_{\rm LSR}$ = 112–115 km s^{-1} for Kes 79, overlaid with radio continuum contours (red) and X-ray contours (blue). The size of each grid cell is 1 arcmin. Right: ^{12}CO (J = 2–1) emission map at $V_{\rm LSR}$ = 39–40 km s^{-1} for W 49B, overlaid with radio continuum contours. The size of each grid cell is 0.5 arcmin.

the barrel (Keohane *et al.* 2007). It has been suggested that W 49B originated inside a wind-blown bubble interior to a dense MC. By CO observation, we have noticed that this SNR seems to be coincident with a molecular cavity at $V_{\rm LSR} \sim 39$ km s^{-1} (Fig. 1), but no robust kinematic evidence has been obtained yet. If this SNR-MC association is true, the kinematic distance to W 49B is 9.3 kpc.

The above observations show a common phenomenon that these interacting SNRs all evolve in molecular cavities/bubbles, which may be created by their progenitors' activities (i.e., most probably, stellar winds). The cavities/bubbles are very unlikely to be produced by the SNR shocks, because the supernova explosion energy would otherwise be much higher than the canonical budget 10^{51} erg.

4. Bubbles in molecular environment

A bubble blown by a main sequence star achieves the maximum size when the bubble is in pressure equilibrium with the ambient medium (Chevalier 1999). Chen *et al.* (2013) show a linear relation for the bubble size $R_{\rm b}$ for main-sequence OB stars: $p^{1/3}R_{\rm b} \propto M$, where p is the pressure of the surrounding interclump medium and M the stellar mass. Actually, by inserting parameters that are observationally determined and model-estimated into Chevalier's (1999) formula, they find that, for 15 exemplified main-sequence OB stars, $p^{1/3}R_{\rm b}$ does correlate linearly with M, and a good regression for the relation can be obtained as

$$p_5^{1/3}R_{\rm b} = \alpha(M/M_{\odot}) - \beta \ \text{pc}, \quad (4.1)$$

where $p_5 \equiv (p/k)/(10^5\ \text{cm}^{-3}\ \text{K})$, $\alpha = 1.21 \pm 0.05$, and $\beta = 8.98 \pm 1.76$.

In a giant MC, the mean pressure is $p_5 \sim 1$ (Krumholz *et al.* 2009), and this number is indeed needed to confine the dense clumps and supports the cloud against gravitational collapse (Blitz 1993; Chevalier 1999).

Eq. (4.1) provides a powerful way to assess the initial masses of the progenitors for SNRs associated with molecular clouds. For an initial mass below 25–30$M_{\odot}$, the stars will terminate their lives in Type II-P supernova (SN) explosion, which takes up the greatest majority of SNe II, after the red supergiant (RSG) phase, without further launching Wolf-Rayet winds. The post-main sequence stellar winds reach an extent much smaller than the main-sequence bubbles. The SN shocks will rapidly pass through the circumstellar

Table 1. Galactic SNRs with molecular shells/in molecular cavities

SNR$^{\rm ref.}$	Distance [kpc]	$R_{\rm b}$ [pc]	M [$M_\odot$]
G21.8−0.6 (Kes 69)[1]	6.2	13	~ 18
G29.7−0.3 (Kes 75)[2]	10.6	6	~ 12
G32.8−0.1 (Kes 78)[3]	4.8	17	~ 21
G33.6+0.1 (Kes 79)[4]	7	8	~ 14
G39.2−0.3 (3C 396)[5]	6.2	7	~ 13
G41.1−0.3 (3C 397)[6]	10.3	4.5–7	~ 12
G43.3−0.2 (W 49B)[7] (?)	9.3	7	~ 13
G54.4−0.3 (HC 40)[8]	3/7	18/43	$\sim$ 22/?
G263.9−3.3 (Vela)[9,19] (?)	0.29	14–19	~ 21
G347.3−0.5 (RX J1713−3946)[11,12]	1.1	9	~ 15

Notes: The "?" symbol means that the SNR-MC association needs to be confirmed.
References—1: Zhou *et al.* (2009); 2: Su *et al.* (2009); 3: Zhou & Chen (2011); 4: Sun *et al.* (2004) 5: Su *et al.* (2011); 6: Jiang & Chen(2010); 7: this paper 8: Junkes *et al.* (1992); 9: Moriguchi *et al.* (2001); 10: Dodson *et al.* (2003); 11: Fukui *et al.* (2003); 12: Moriguchi *et al.* (2005)

material and impact the massive shells of the bubbles, with drastic deceleration (Chen *et al.* 2003). Thus, such SNRs reflect mostly the bubble sizes, and the sizes can be used to infer the progenitors' masses.

Table 1 lists the interacting SNRs that are known or suggested to have molecular shells or be in molecular cavities, which include the Vela SNR and RX J1713–3946 in addition to those described in §3. The Vela SNR coincides with a molecular void at a velocity range of $V_{\rm LSR} = -5$ to $85\,{\rm km\,s^{-1}}$ (Moriguchi *et al.* 2001). It is suggested that the molecular clumps are pre-existent, rather than having been swept up by the SNR shock, and that the SNR may have been expanding in a low density medium. RX J1713.7−3946 appears to be confined in a molecular gas cavity at $V_{\rm LSR} \sim -11$ to $-3\,{\rm km\,s^{-1}}$ (Fukui *et al.* 2003; Moriguchi *et al.* 2005). It is suggested that this SNR is still in the free expansion phase and the non-decelerated blast wave is colliding with the dense molecular gas after it traveled in a low-density cavity that perhaps was produced by the stellar wind or pre-existing supernovae (Fukui *et al.* 2003).

The estimates of progenitors' masses for the SNRs are given in the last column, on the assumption that the interclump pressure to be a constant: $p_5 \approx 1$. Some of the estimates can be compared with other available independent assessments, which show considerably good consistency. For 3C 396, Su *et al.* (2011), based on *Chandra* observation, derived a progenitor mass of 13–15$M_\odot$ from the metal abundances of the X-ray-emitting gas that may be dominated by the SN ejecta. For 3C 397, a recent *XMM*-Newton X-ray study has analyzed the metal abundances of the SN ejecta and thus assessed the progenitor's mass to be 11–15$M_\odot$ (Safi-Harb *et al.* in preparation). For the Vela SNR, an SN II-P progenitor mass was suggested to be 15–20$M_\odot$ based on the moderate size of the wind-blown bubble (Gvaramadze 1999).

5. Probe for hadronic interaction

The origin of cosmic rays (CRs) has drawn more and more attention motivated by the recent decades' X- and γ-ray observations. It is generally believed that the CRs below the "knee" (3×10^{15} eV) are of Galactic origin. The candidate sites where the Galactic CRs are accelerated include SNRs, binaries, superbubbles, Galactic center, etc., and SNRs are usually regarded as the most important. Relativistic electrons accelerated up to an energy 10^{13}–10^{14} eV by the SNR shocks are evidenced by the X-ray synchrotron emission arising from them. Yet, conclusive evidence for the acceleration of relativistic protons remains poor. An indirect way is to detect the γ-ray emission generated by the hadronic interaction of the accelerated protons, namely, neutral pion decay after the collision of the protons with the baryons of the environmental dense gas. Thus, the MCs with which SNRs are associated or interacting are a perfect probe for the hadronic process.

In the second *Fermi*-LAT GeV source catalog, there are 89 sources whose 95% confidence error ranges overlap with SNRs, and among which 45% are of chance coincidences but six correspond to firmly identified γ-ray emitting SNRs (Nolan *et al.* 2012). Among the TeV sources, about ten correspond to SNRs (Holder 2012). Also, a correlation has been proposed between a class of GeV-TeV γ-ray sources that are coincident with interacting SNRs and the 1720 MHz OH masers (Hewitt *et al.* 2009). However, it is not easy to differentiate whether the γ-rays arising from the SNRs are hadronic or leptonic emission. Recent years, a progress has been made by analysing the "illuminating" effect of the nearby MCs by the SNR shock-accelerated protons (e.g., Aharonian & Atoyan 1996; Gabici *et al.* 2009; Li & Chen 2010, 2012; Ohira *et al.* 2011).

Among these work, Li & Chen (2010, 2012) established an "accumulative diffusion" model for the escaping protons and naturally interpreted the GeV-TeV spectra of nine interacting SNRs. W28 is one of the prototype interacting SNRs, in contact with a large MC in the north and harboring a large amount of 1720 MHz masers (Frail *et al.* 1994). Four TeV sources in the area are detected by the H.E.S.S. γ-ray telescope, positionally coincident with the northern large MC and three small MCs in the south (Aharonian *et al.* 2008). The three southern clouds seem to be outside the reach of the W28 blast wave, and *Fermi*-LAT detected no significant GeV emission for two of them (Abdo *et al.* 2010). They show that the various relative GeV-TeV brightness of the four γ-ray sources results from the hadronic process of the accelerated protons that escape from the shock front and bombard the nearby MCs at different radii. The "illuminating" protons are considered to be an accumulation of the diffusive protons escaping from the shock front throughout the history of the SNR expansion. For the various distances of the MCs from the SNR center, the resulting proton spectra can have prominently different shapes. Adopting different centric distances, the γ-ray spectral fit for the four sources well explains their GeV-TeV spectral properties. It is also implied that the spectral index 2.7 at the high energy side is caused by diffusion other than directly by acceleration, and the spectral break (from 2.1 to 2.7) at around 1 GeV can naturally appear due to the accumulative diffusion effect.

This model, with improvement by incorporating the finite volume of molecular clouds, is further applied to nine γ-ray emitting interacting SNRs (W 28, W 41, W 44, W 49B, W 51C, Cygnus Loop, IC 443, CTB 37A and G349.7+0.2). Like that aforementioned, the $\sim$GeV spectral breaks are commonly present in these SNRs; and most of them have a spectral ($E^2 dF/dE$) "platform" extending from the break to lower energies. This refined model perfectly explains the $\sim$ GeV spectral breaks and the "platforms", together with the available TeV data. It is also found that the index of diffusion coefficient δ derived from the spectral fit is in the range 0.5–0.7, analogous to the Galactic average, and the diffusion coefficient for the CRs around the SNRs ($\chi \sim 10^{-2}$) is essentially two orders of magnitude lower than the Galactic average, which is indicative of the suppression of CR diffusion near SNRs.

γ-rays from molecular clouds illuminated by diffusive protons that escape from interacting SNRs strongly support the scenario that hadrons accelerated by SNR shocks contribute to the Galactic CRs. A discovery of the spectral rising below $\sim$ 200 MeV for

IC 443 and W44 as an unique pion-decay signature is newly released after this Symposium (Ackermann *et al.* 2013).

So far, nearly a dozen of SNRs are proposed to emit hadronic γ-rays, including the historical SNRs Cas A and Tycho. Against the previous suggestion for tenuous medium surrounding Tycho responsible for the hadronic emission, it is argued that the γ-rays are produced from dense ambient matter, most probably MCs, too (see Zhang *et al.* 2013).

References

Abdo, A. A., Ackermann, M., Ajello, M., *et al.* 2010, *ApJ*, 718, 348

Ackermann, M., Ajello, M., Allafort, A., *et al.* 2013, *Science*, 339, 807

Aharonian, F., Akhperjanian, A. G., Bazer-Bachi, A. R., *et al.* 2008, *A&A*, 481, 401

Aharonian, F. A. & Atoyan, A. M. 1996, *A&A*, 309, 917

Blitz, L. 1993, in: E. H. Levy & J. I. Lunine (eds.), *Protostars and Planets III* (Tucson: Univ. of Arizona), 125

Chen, Y., Zhang, F., Williams, R. M., & Wang, Q. D. 2003, *ApJ*, 595, 227

Chen, Y., Zhou, P., & Chu, Y.-H. 2013, *ApJ*(Letters), in press, arXiv:1304.5126

Chevalier, R. A. 1999, *ApJ*, 511, 798

DeNoyer, L. K. 1979, *ApJ*(Letters), 232, L165

Dodson, R., Legge, D., Reynolds, J. E., & McCulloch, P. M. 2003, *ApJ*, 596, 1137

Frail, D. A., Goss, W. M., Reynoso, E. M., *et al.* 1996, *AJ*, 111, 1651

Frail, D. A., Goss, W. M., & Slysh, V. I. 1994, *ApJ*(Letters), 424, L111

Fukui, Y., Moriguchi, Y., Tamura, K., *et al.* 2003, *PASJ*, 55, L61

Gabici, S., Aharonian, F. A., & Casanova, S. 2009, *MNRAS*, 396, 1629

Goss, W. M. & Robinson, B. J. 1968, *ApL*, 2, 81

Green, A. J., Frail, D. A., Goss, W. M., & Otrupcek, R. 1997, *AJ*, 114, 2058

Gvaramadze, V. 1999, *A&A*, 352, 712

Hewitt, J. W., Yusef-Zadeh, F., & Wardle, M. 2008, *ApJ*, 683, 189

Hewitt, J. W., Yusef-Zadeh, F., & Wardle, M. 2009, *ApJ*(Letters), 706, L270

Holder, J. 2012, *Astroparticle Physics*, 39, 61

Huang, Y.-L. & Thaddeus, P. 1986, *ApJ*, 309, 804

Jeong, I.-G., Byun, D.-Y., Koo, B.-C., *et al.* 2012, *ApSS*, 342, 389

Jiang, B., Chen, Y., Wang, J., *et al.* 2010, *ApJ*, 712, 1147

Jiang, B. & Chen, Y. 2010, *Sci. China G: Phys. Astron.*, 53, 267

Junkes, N., Fuerst, E., & Reich, W. 1992, *A&AS*, 96, 1

Keohane, J. W., Reach, W. T., Rho, J., & Jarrett, T. H. 2007, *ApJ*, 654, 938

Koralesky, B., Frail, D. A., Goss, W. M., Claussen, M. J., & Green, A. J. 1998, *AJ*, 116, 1323

Krumholz, M. R., McKee, C. F., & Tumlinson, J. 2009, *ApJ*, 699, 850

Li, H. & Chen, Y. 2010, *MNRAS*(Letters), 409, L35

Li, H. & Chen, Y. 2012, *MNRAS*, 421, 935

Lockett, P., Gauthier, E., & Elitzur, M. 1999, *ApJ*, 511, 235

Moriguchi, Y., Yamaguchi, N., Onishi, T., Mizuno, A., & Fukui, Y. 2001, *PASJ*, 53, 1025

Moriguchi, Y., Tamura, K., Tawara, Y., Sasago, H., Yamaoka, K., Onishi, T., & Fukui, Y. 2005, *ApJ*, 631, 947

Nolan, P. L., Abdo, A. A., Ackermann, M., *et al.* 2012, *ApJS*, 199, 31

Ohira, Y., Murase, K., & Yamazaki, R. 2011, *MNRAS*, 410, 1577

Reach, W. T., Rho, J., Tappe, A., *et al.* 2006, *AJ*, 131, 1479

Seta, M., Hasegawa, T., Dame, T. M., *et al.* 1998, *ApJ*, 505, 286

Su, Y., Chen, Y., & Yang, J., *et al.* 2009, *ApJ*, 694, 376

Su, Y., Chen, Y., & Yang, J., *et al.* 2011, *ApJ*, 727, 43

Sun, M., Seward, F. D., Smith, R. K., & Slane, P. O. 2004, *ApJ*, 605, 742

Zhang, X., Chen, Y., Li, H., & Zhou, X. 2013, *MNRAS*(Letters), 429, L25

Zhou, P. & Chen, Y. 2011, *ApJ*, 743, 4

Zhou, X., Chen, Y., Su, Y., & Yang, J. 2009, *ApJ*, 691, 516

Supernova Environmental Impacts
Proceedings IAU Symposium No. 296, 2013
A. Ray & R. A. McCray, eds.

doi:10.1017/S1743921313009435

The molecular emission from old supernova remnants

A. Gusdorf[1], R. Güsten[2], S. Anderl[3], T. Hezareh[2], and H. Wiesemeyer[2]

[1]LERMA, UMR 8112 du CNRS, Observatoire de Paris, École Normale Supérieure, 24 rue Lhomond, F75231 Paris Cedex 05, France, email: antoine.gusdorf@lra.ens.fr

[2]Max Planck Institut für Radioastronomie, Auf dem Hügel 69, 53121 Bonn, Germany

[3]Argelander Institut für Astronomie, Universität Bonn, Auf dem Hügel 71, 53121 Bonn, Germany

Abstract. Supernovae constitute a critical source of energy input to the interstellar medium (ISM). In this short review, we focus on their latest phase of evolution, the supernova remnants (SNRs). We present observations of three old SNRs that have reached the phase where they interact with the ambient interstellar medium: W28, IC443, and 3C391. We show that such objects make up clean laboratories to constrain the physical and chemical processes at work in molecular shock environments. Our studies subsequently allow us to quantify the impact of SNRs on their environment in terms of mass, momentum, and energy dissipation. In turn, their contribution to the energy balance of galaxies can be assessed. Their potential to trigger a further generation of star formation can also be investigated. Finally, our studies provide strong support for the interpretation of γ-ray emission in SNRs, a crucial step to answer questions related to cosmic rays population and acceleration.

Keywords. ISM: supernova remnants – Shock waves – Submillimeter: ISM – ISM: individual objects: W28, IC443, & 3C391 – Stars: formation – cosmic rays.

---

1. The astrophysical importance of supernova remnants

The life of massive stars ends with a supernova explosion, characterized by an important redistribution of energy towards the interstellar medium (ISM). After a free expansion phase (ending when the swept-up mass reaches that of the envelope), and an adiabatic phase (where the energy dissipation is due to expansion), the supernova-driven shocks start radiating energy (Woltjer 1972), initially at observable optical and ultraviolet wavelengths from what has become supernova remnants (SNRs; e.g.Weiler & Sramek 1988). When SNRs encounter molecular clouds, they drive slower shock waves that compress, accelerate and heat the molecular material and result in strong infrared and sub-millimeter line emission (e.g. Neufeld *et al.* 2007, Frail & Mitchell 1998).

These relatively slow molecular shocks are found to be very similar to those observed in the jets and outflows associated with star formation (e.g. Gusdorf *et al.* 2011). However, contrary to their star-formation counterparts, SNR shocks are not expected to be contaminated by the possible UV radiation from the proto-star. Additionally, spectral lines observed in SNR shock regions do not either show any envelope or infall component (see for instance the upper right panel in Fig. 1) that would make their interpretation complicated. SNRs hence serve as clean laboratories to study the physical and chemical mechanisms that operate in shock environments.

In a further step, studying SNRs interacting with the ISM is a powerful tool to quantify their contribution to the energy balance of galaxies, through the observation of the numerous CO transitions recently allowed by the *Herschel* telescope, and presented in

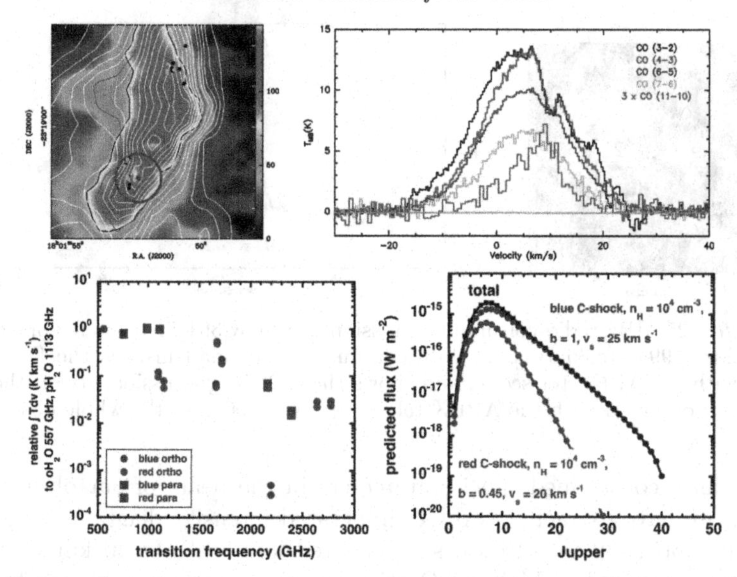

Figure 1. CO observations of W28, as presented in Gusdorf *et al.* (2012). *Top, left:* the field covered by our maps of the CO (6–5) and (3–2) transitions (colours, white contours). The small dots are the OH masers (Claussen *et al.* 1997 and Hoffman *et al.* 2005). Our shock analysis was made on the position at the center of the circle, that indicates the beam of our SOFIA observations. *Top, right:* the CO spectra (obtained with APEX, (3–2), (4–3), (6–5) and (7–6), and SOFIA, (11–10)), convolved to the SOFIA resolution, were extracted from this position. They were combined with H_2 *Spitzer* observations (Neufeld *et al.* 2007) to constrain our shock models. *Bottom, left:* the models can subsequently be used to make water emission predictions. *Bottom, right:* they can also be used to predict the emission of all of the CO transitions.

flux diagrams, the so-called CO ladders (e.g. Hailey-Dunsheath *et al.* 2012). In certain cases, their potential to trigger star formation can be investigated, see Xu *et al.* (2011).

Finally, the molecular emission from SNRs provide a valuable support for the study of cosmic rays population (CRs, hadronic or leptonic) and acceleration. Because of their large energy budget, it has indeed long been argued that SNRs are the primary sites for accelerating CRs (e.g., Blandford & Eichler 1987). Observing the γ-ray emission in SNRs environments is a way to constrain the CRs population, and acceleration mechanisms. Indeed, this radiation mainly results from three processes: π^0 decay from the interaction of the hadronic component of the CRs population with the ambient medium, Bremsstrahlung emission from the leptonic component on the ambient ISM, and inverse Compton scattering of the leptonic component on the ambient radiation field, (e.g. Frail 2011 for a short review). Although the situation is a bit more complex for old SNRs (e.g., Gabici *et al.* 2009, Bykov *et al.* 2000), whose shocks no longer accelerate CRs, an accurate knowledge of the ambient medium (density, mass of the shocked/non shocked gas, local magnetic field strength, local radiation field) remains crucial to assess the contribution from each of these three processes to the very high energy spectra observed in SNRs.

We demonstrate how the observation of the molecular emission from three old SNRs, W28, 3C391, and IC443, allow to address these diverse astrophysical topics.

2. Shock modelling in W28F

The upper panels of Figure 1 summarize the results presented in Gusdorf *et al.* (2012), dedicated to observations and models of shocked regions in the W28 SNR. By analysing the CO (from APEX and SOFIA telescopes) and H_2 (from the *Spitzer* telescope)

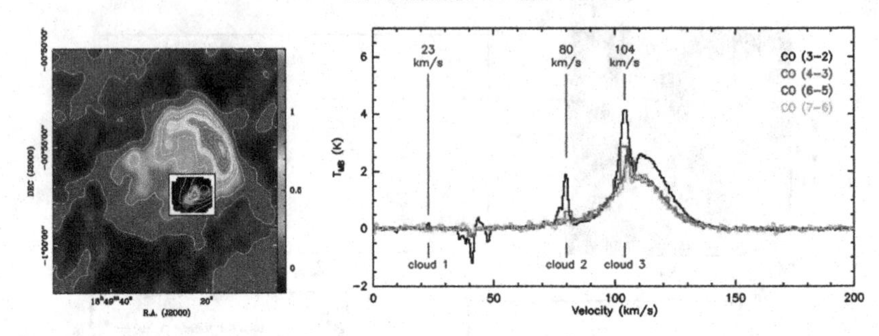

Figure 2. *Left:* 325 MHz radio-continuum emission in the 3C391 SNR, as it appeared in Moffett & Reynolds (1994) (colours and white contours). The small inset is the field of our CO observations with the APEX telescope, and shows the CO (6–5) emission. *Right:* the spectra in various CO lines observed with the APEX telescope, averaged over the whole observed field.

emission, we have constrained shock parameters in the beam of our observations. For this position, we have extracted two CO integrated intensity diagrams, one per velocity range (blue lobe, -30 to $\sim$10 km s^{-1}, and red lobe, $\sim$10 to 40 km s^{-1}). We have compared the corresponding high-J CO emission, that most unambiguously traces the shocked material, to a grid of unidimensional shock models. We hence have fitted each velocity component by a single, C-type shock wave arising from a 25″ diameter emission region, with the respective parameters: pre-shock density $n_{\rm H}$ = 10^4 cm^{-3}, magnetic field strength perpendicular to the shock front B = 100 and 45 μG, and shock velocities $v_{\rm s}$ = 20 and 25 km s^{-1}. We also checked that our models provide reasonable fits for H_2 excitation diagrams (*Spitzer* observations by Neufeld *et al.* 2007). Our final results are compatible with independent studies of the region in terms of age (Giuliani *et al.* 2010), magnetic field measurements (Claussen *et al.* 1997, Hoffman *et al.* 2005), and densities required for the excitation of observed OH masers (Lockett *et al.* 1999).

In a forthcoming publication (Gusdorf *et al.*, in prep), we will go one step further in the interpretation of the results. With the shock model parameters, and assumptions on the size of the emitting region, we can thus infer the shocked mass in the beam of our observations (from 6.2 to 18.5 $M_\odot$ depending on the adopted age, 10^4 or 3×10^4 years, respectively). Combining our shock model results with an LVG code to calculate the emission from water (Gusdorf *et al.* 2011), we are also able to predict integrated intensity diagrams for each velocity component, which can be compared for instance to observations made by the *Herschel* telescope, as can be seen on the lower left panel of the Figure 1. Additionally, we predict the excitation from all CO lines in our beam (lower right panel of the Figure 1), that can be directly compared with *Herschel* observations of the region. Such SNR shocks CO ladders can also directly be compared an help constraining their contribution to galactic ones (e.g. Hailey-Dunsheath *et al.* 2012).

3. The cases of 3C391 and IC443

Forthcoming publications will also be focused on a similar analysis of SNRs 3C391 and IC443 (Gusdorf *et al.*, in prep), based on CO and H_2 observations.

In 3C391, we find that our APEX CO observations directly allow to identify the line-of-sight clouds that lie in the vicinity of the remnant, and constitute potential targets for the cosmic rays interactions. Figure 2 shows both the field of our observations (left panel), and the averaged CO spectra we extracted (right panel). Based on our CO and ^{13}CO observations, we were able to infer the mass of the clouds that might be impacted

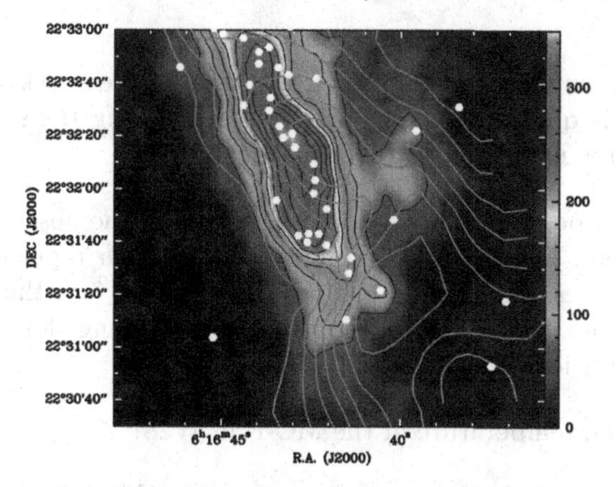

Figure 3. The G clump of the IC443 SNR, mapped with APEX in the CO (6–5) transition (in colours and black contours, Gusdorf *et al.*, in prep). The white markers indicate the position of YSO candidates in the region, selected in the 2MASS point source catalog, based on colour selection criterion by Xu *et al.* (2011). Also shown in the greyscale contour is the CO (1–0) ambient emission at 5 km s^{-1}, observed by the IRAM 30m telescope (Hezareh *et al.*, in prep).

by the cosmic rays accelerated in the SNR, the so-called clouds 2 and 3, respectively 161.3 and 43.7 $M_\odot$ - see for instance Xu *et al.* (2011) for the calculation method.

Finally, in the G clump of the IC443 SNR, our observations (Gusdorf *et al.*, in prep, and Hezareh *et al.*, in prep, and Figure 3) confirm the results obtained by Xu *et al.* (2011). As the YSO candidates in the region are older than the SNR, these authors show that in spite of the correlation between their distribution and the shock structure, star formation has probably not been triggered by the supernova-driven shock wave, but rather by the stellar winds of the massive progenitor of the remnant. Paradoxically, our observations confirm the tight correlation between the star formation and the shocked CO (6–5) gas in the region. On the other hand, our CO (1–0) observations show that the star formation has developed on the edge of a nearby ambient cloud (also see Lee *et al.* 2012), subject to the stellar winds of the SNR progenitor prior to the passage of the shock wave.

References

Blandford, R. & Eichler, D. 1987, *PR*, 154, 1
Bykov, A. M., Chevalier, R. A., Ellison, D. C., & Uvarov, Y. A. 2000, *ApJ*, 538, 203
Claussen, M. J., Frail, D. A., Goss, W. M., & Gaume, R. A. 1997, *ApJ*, 489, 143
Frail, D. A. & Mitchell, G. F. 1998, *ApJ*, 508, 690
Frail, D. A. 2011, *MemSAIt*, 82, 703
Gabici, S., Aharonian, F. A., & Casanova, S. 2009, *MNRAS*, 396, 1629
Giuliani, A., Tavani, M., Bulgarelli, A., *et al.* 2010, *A&A*, 516, L11
Gusdorf, A., Giannini, T., Flower, D. R., *et al.* 2011, *A&A*, 532, A53
Gusdorf, A., Anderl, S., Güsten, R., *et al.* 2012, *A&A*, 542, L19
Hailey-Dunsheath, S., Sturm, E., Fischer, J., *et al.* 2012, *ApJ*, 755, 57
Hoffman, I. M., Goss, W. M., Brogan, C. L., & Claussen, M. J. 2005, *ApJ*, 620, 257
Lee, J.-J., Koo, b.-C., Snell, R. L., *et al.* 2012, *ApJ*, 749, 34
Lockett, P., Gauthier, E., & Elitzur, M. 1999, *ApJ*, 511, 235
Moffett, D. A. & Reynolds, S. P. 1994, *ApJ*, 425, 668
Neufeld, D. A., Hollenbach, D. J., Kaufman, M. J., *et al.* 2007, *ApJ*, 664, 890
Weiler, K. W. & Sramek, R. A. 1988, *ARAA*, 26, 295
Woltjer, L. 1972, *ARAA*, 10, 129
Xu, J.-L., Wang, J.-J., & Miller, M. 2011, *ApJ*, 727, 81

Discussion

UNIDENTIFIED: I am curious as to how you determined the shock velocity as an input for the model. This is quite interesting because determining the shock velocity from a numerical simulation is not easy due to shock broadening.

GUSDORF: A range of possible shock velocities arise from the observations, that provide an upper limit to them. The rest is determined like the other input parameters, through a comparison of all lines observed to our models with a χ^2 method. Some theoretical considerations also exist that constrain the shock velocity range for a given set of (other) shock parameters ("critical velocities").

ZHOU P.: What is the temperature of the MC-F in W28?

GUSDORF: There are several temperature components along the line of sight/ shocked layer. The one that is probed by the CO observations presented here is typically $\sim$ 100 K, but higher lines can be excited. A strong point in our models is that we expect all these components to be accounted for, unlike in the case of a single-slab LVG model, for instance.

Supernova Environmental Impacts
Proceedings IAU Symposium No. 296, 2013
A. Ray & R. A. McCray, eds.

doi:10.1017/S1743921313009447

SRAO CO Observation of Supernova Remnants in $l = 70°$ to $190°$

Il-Gyo Jeong and Bon-Chul Koo

Astronomy Program, Department of Physics and Astronomy, Seoul National University,
Seoul 151-742, Republic of Korea
email: igjeong@astro.snu.ac.kr, koo@astro.snu.ac.kr

Abstract. We present the results of ^{12}CO $J = 1$–0 line observations of eleven Galactic supernova remnants (SNRs) between $l = 70°$ and $190°$ obtained using the Seoul Radio Astronomy Observatory (SRAO) 6-m radio telescope. We detected CO emission towards most of the remnants. In seven SNRs, molecular clouds show a good spatial relation with their radio morphology: G73.9+0.9, G84.2−0.8, G85.4+0.7, G85.9−0.6, G93.3+6.9 (DA530), 94.0+1.0 (3C 434.1), and G182.4+4.3. Two SNRs are particularly interesting. In G85.4+0.7, there is a filamentary molecular cloud aligned along the south-east boundary of the remnant. This cloud extends to the nearby HII region G84.9+0.5. If the molecular cloud is associated with both the HII region and the SNR, the distance to the SNR would be 5–7 kpc. In 3C 434.1, there is a large molecular cloud blocking the western half of the remnant where the radio continuum emission is faint. The cloud shows a very good spatial correlation with radio continuum features, which strongly suggests the physical association of the cloud with the SNR. This gives a distance of 3 kpc to the SNR. We performed ^{12}CO $J = 2$–1 line observations of this cloud using Kölner Observatorium für Sub-Millimeter Astronomie (KOSMA) 3-m telescope and found a region where the ^{12}CO $J =$ 2–1/1–0 line ratio is high. We present a hydrodynamic model showing that 3C434.1 could have resulted from a SN explosion occurred just outside the boundary of a thin, molecular cloud.

Keywords. supernova remnants, ISM: molecules, ISM: clouds

---

1. Introduction

SN explosions strongly affect the environment, while, at the same time, the evolution of a supernova remnant (SNR) is strongly affected by the environment itself. Among the environmental impacts of SN explosions, the interaction with molecular clouds (MCs) is of particular interest where we can study the microphysics of molecular shocks and the hydrodynamics of SNR blast waves. There have been many studies of individual SNRs, and about 70 out of the 274 Galactic SNRs (Green 2009) are known to be interacting with molecular clouds (Jiang *et al.* 2010). Systematic studies, however, are limited. As far as we are aware of, the only systematic study of the molecular environment of SNRs is by Huang and Thaddeus (1986), who surveyed ^{12}CO $J = 1$–0 emission lines toward Galactic SNRs from $l = 70°$ to $210°$. Their results were useful for studying the distributions of large molecular cloud complexes near SNRs, but have a limitation because of the low spatial resolution ($\sim 8'.7$) of the telescope.

We have carried out a systematic CO observation of eleven SNRs in $l = 70°$ to $190°$. In this paper, we present a summary of observations and main results, with some details on the SNR 3C 434.1. For the details of the survey results, see Jeong *et al.* (2012).

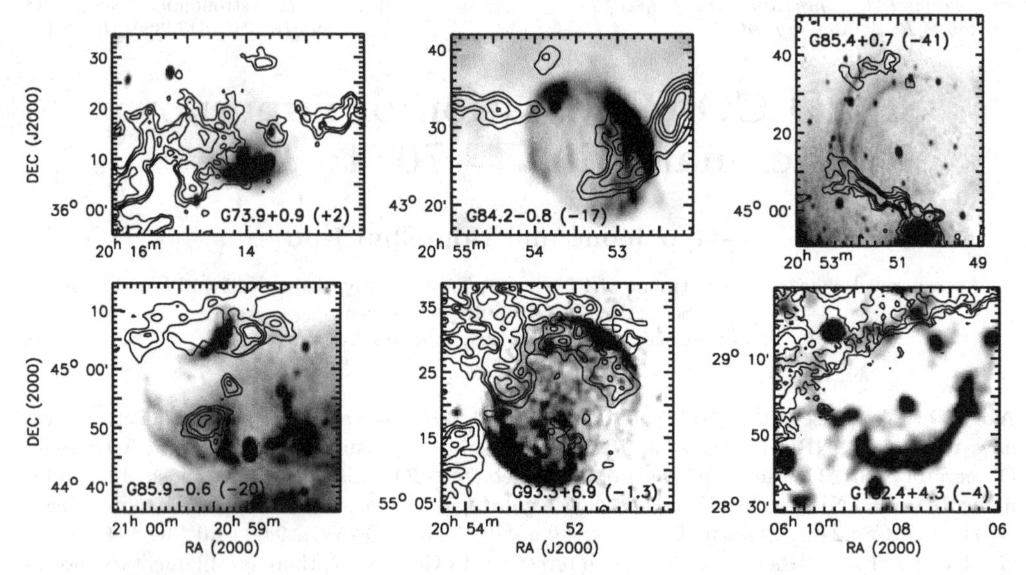

Figure 1. ^{12}CO $J = 1$–0 average intensity maps of the selected targets which show spatially-correlated features with CO molecular clouds (Jeong *et al.* 2012). The numbers in the parentheses are the central velocities of the CO maps.

2. Observation

Our target SNRs are selected among the 35 SNRs in $l = 70°$ to $190°$ (Green 2004): G73.9+0.9, G76.9+1.0, G84.2−0.8, G85.4+0.7, G85.9−0.6, G93.3+6.9 (DA530), 94.0+1.0 (3C 434.1), 166.2+2.5 (OA184), 179.0+2.6, 180.0−1.7 (S147), and G182.4+4.3. They have angular sizes between 10′ and 180′ and located outside the area of the Outer Galaxy Survey or the Galactic Ring Survey (Jackson *et al.* 2006) of the Five College Radio Astronomy Observatory (FCRAO).

The ^{12}CO $J = 1$–0 observations using SRAO 6-m telescope Seoul Radio Astronomy Observatory (SRAO) were carried out from October 2003 to May 2005. The half-power beam size of the telescope is 120″ and a main beam efficiency is 70% at 115 GHz (Koo *et al.* 2003). The telescope has a 100 GHz SIS mixer receiver with a single-side band filter and a 1024-channel auto-correlator with 50 MHz bandwidth. The typical system temperature ranged from 500 to 800 K, and the typical *rms* noise level was 0.3 K on T_{mb} scale at 1 km s^{-1} velocity resolution. To check the system performance and the pointing accuracy, we observed the bright standard source near the target every one or two hours. We mapped areas fully covering the radio morphology of individual SNRs with either half-beam (60″) or full-beam (120″) samplings.

3. Results of the Survey

We detected CO emission toward most SNRs. In seven SNRs, the CO emission showed spatially-correlated features with radio continuum emission: G73.9+0.9, G84.2−0.8, G85.4 +0.7, G85.9−0.6, G93.3+6.9 (DA530), 94.0+1.0 (3C 434.1), and G182.4+4.3. Fig. 1 shows the CO intensity maps of the 6 SNRs and a brief summary is given below for each of them. The result on 3C434.1 is presented separately in the following section.

G73.9+0.9 This is a diffuse SNR located in the complex Cygnus region. It has a partial shell-feature in south and a pulsar-wind nebula candidate inside (Kothes *et al.* 2006). We have detected a large MC at +2 km s^{-1} that appears to be blocking the eastern boundary of the SNR. But there is no obvious morphological relation between the CO and radio continuum brightnesses.

G84.2-0.8 This is a shell-type SNR of an elliptical shape. The boundary of the SNR, in particular the western boundary, shows features of enhanced radio continuum brightness. There is a MC at -17 km s^{-1} that matches very well with the bright portion of the western SNR shell. This cloud was first noted by Feldt & Green (1993), but our high-resolution map reveals its morphological correlation with the SNR clearly. If this cloud is associated with the SNR, which is likely, the distance to the SNR will be 4.9 kpc. There is also a filamentary cloud protruding the northwestern SNR shell, near another radio-continuum enhanced region, but its association is not clear.

G85.4+0.7 This is a shell-type SNR with two distinct, partial shells. According to Kothes *et al.* (2001), the inner shell is non-thermal and the outer thermal. They noted that the SNR is inside an HI bubble at -12 km s^{-1} and proposed a distance of 3.8 kpc to them. In our observation, however, there is a filamentary MC at -41 km s^{-1} which appears to have a similar curvature with the radio continuum shells and connects to their southern ends. This spatial correlation suggests that their association is likely. We mapped a larger area surrounding the SNR in ^{12}CO $J = 1$–0 line and found that this cloud extends to the compact HII region G84.9+0.5 in the south. If the MC is associated with both the HII region and the SNR, the kinematic distance to the SNR becomes 7.2 kpc according to the rotation curve of Brand and Blitz (1993) (cf. Foster *et al.* 2007 proposed a distance of 4.9 kpc to the HII region).

G85.9−0.6 This is a shell-type SNR with a kinked arc in the north. We note that there is a MC at -20 km s^{-1} superposed on this arc structure, but they do now show apparent spatial correlation. If they are associated, the distance to the SNR would be 5 kpc.

G93.3+6.9 (DA530) This is a bilateral type SNR located at high galactic latitude (Gaensler 1998). Landecker *et al.* (1999) found an HI bubble at -12 km s^{-1} which spatially coincides with the SNR, and proposed a distance of 3.5 kpc. We have detected a diffuse CO emission at the velocity of -1.3 km s^{-1} in the northern area of the remnant, but the morphology of the CO emission is not clearly correlated with the radio continuum.

G182.4+4.3 This is well-defined shell-type SNR with a bright, half shell structure in the southwest (Kothes *et al.* 1998). There is a diffuse CO emission at ~ -4 km s^{-1} that match well with the radio-faint, northwestern boundary of the SNR. There is, however, no evidence of the interaction.

4. Molecular-Blocked SNR G94.0+1.0 (3C434.1)

3C434.1 is a shell-type SNR with an asymmetric radio morphology; a bright semi-circular shell with complex filaments in the east with no distinct feature in western area (Willis 1973; Landecker *et al.* 1985). We have found that the MC at -13 km s^{-1} overlaps with the western part of the remnant. The spatial correlation between the two is clearly seen in Fig. 2 which is a channel map at -13 km s^{-1}. Note the thin filamentary structure tracing the western boundary of the SNR including a "bar-like" structure within the remnant. The above morphological correlation between the CO and radio continuum strongly suggests that the MC is interacting with the SNR. We performed ^{12}CO $J =$ 2–1 follow-up line observations of the cloud using KOSMA 3-m telescope, and found that ^{12}CO $J = 2$–1/1–0 line ratio is high in the southern molecular cloud, which supports the above conclusion. The systemic velocity of the MC yields a kinematic distance of 3.0 kpc to the SNR, which is less than the previous estimates, e.g., 4.5 kpc of Foster (2005).

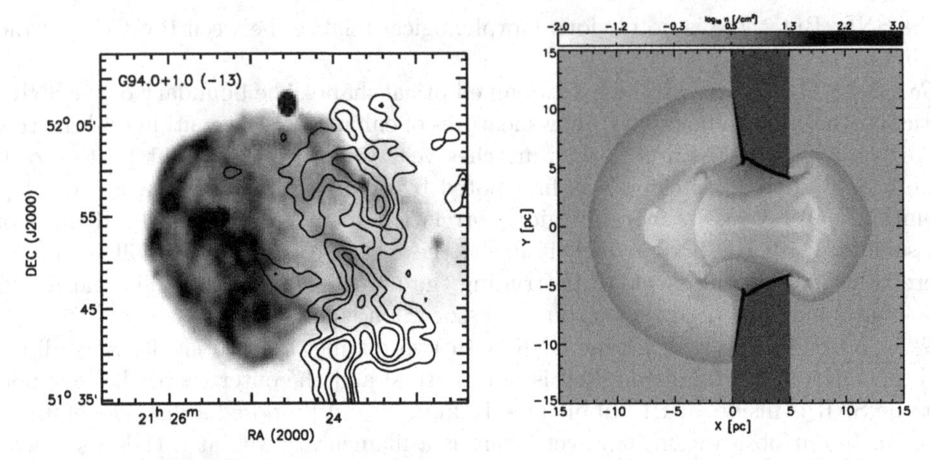

Figure 2. (left) SRAO ^{12}CO J = 1–0 channel map of SNR 3C 434.1 at −13 km s^{-1}. (right) Density structure of hydrodynamic model at t = 7,950 yrs (see text for details).

We have developed a hydrodynamic model to understand the radio morphology of the SNR (Jeong *et al.* in preparation). We assume that the SN exploded at 1 pc from the boundary of a sheet-like cloud with a thickness of 5 pc. The H-nuclei number density of the cloud is assumed to be 20 cm^{-3}, whereas the density of the diffuse medium is set to be 1 cm^{-3}. We assume that the cloud had a cylindrical hole at the center to match the radio continuum structure. The radius of the hole is assumed to be 3 pc. For the calculation, we adopt 3-dimensional hydrodynamic code developed by Harten, Lax and van Leer (HLL) with modified cooling effect (Harten *et al.* 1983). The simulation box consists of 512 × 256 × 256 grids with a spatial resolution of 1/16 pc. The result of the numerical simulations is shown in Fig. 2, which shows the density structure at 7,950 yrs. The simulation result recovers well the basic radio morphology of the SNR, except the long straight filaments that extends far out from the western part of the remnant. More details on the study of this SNR will be published elsewhere.

References

Brand, J. & Blitz, L. 1993, *A&A* 275, 67
Feldt, C. & Green, D. A. 1993, *A&A*, 274, 421
Foster, T. 2005, *A&A*, 441, 1043
Foster, T. J., Kothes, R., Kerton, C. R., & Arvidsson, K. 2007, *ApJ*, 667, 248
Gaensler, B. M. 1998, *ApJ*, 493, 781
Green, D. A. 2004, *Bulletin of the Astronomical Society of India*, 32, 335
Green, D. A. 2009, *Bulletin of the Astronomical Society of India*, 37, 45
Harten, A., Lax, P. D., van Leer, B. 1983, *SIAMR*, 25, 35
Huang, Y.-L. & Thaddeus, P. 1986, *ApJ*, 309, 804
Jackson, J. M., Rathborne, J. M., Shah, R. Y., *et al.* 2006, *ApJS*, 163, 145
Jeong, I.-G., Byun, D.-Y., Koo, B.-C., *et al.* 2012, *Ap&SS*, 342, 389
Jiang, B., Chen, Y., Wang, J., *et al.* 2010, *ApJ*, 712, 1147
Koo, B.-C., Park, Y.-S., Hong, S. S., *et al.* 2003, *J. Korean Astron. Soc.*, 36, 43
Kothes, R., Furst, E., & Reich, W. 1998, *A&A*, 331, 661
Kothes, R., Landecker, T. L., Foster, T., & Leahy, D. A. 2001, *A&A*, 376, 641
Kothes, R., Fedotov, K., Foster, T. J., & Uyanıker, B. 2006, *A&A*, 457, 1081
Landecker, T. L., Higgs, L. A., & Roger, R. S. 1985, *AJ*, 90, 1082
Landecker, T. L., Routledge, D., Reynolds, S. P., *et al.* 1999, *ApJ*, 527, 866
Willis, A. G. 1973 *A&A*, 26, 237

Discussion

JIANG B.: Among those SNRs which you showed spatial coincidence with CO emission may not be really associated with the molecular clouds. For example, DA 530, the asymetric radio shell and the distance (~ 2kpc), makes it not consistent with the molecular cloud you showed. So are you going to look for other evidence for the association?

JEONG: Yes, we will use more data to do it.

Supernova environmental impacts
Proceedings IAU Symposium No. 296, 2013
A. Ray & R. A. McCray, eds.

doi:10.1017/S1743921313009459

The Galactic distribution of SNRs

D. A. Green

Cavendish Laboratory, 19 J. J. Thomson Ave., Cambridge, CB3 0HE, U.K.
email: dag@mrao.cam.ac.uk

Abstract. It is not straightforward to determine the distribution of supernova remnants (SNRs) in the Galaxy. The two main difficulties are that there are observational selection effects that mean that catalogues of SNRs are incomplete, and distances are not available for most remnants. Here I discuss the selection effects that apply to the latest catalogue of Galactic SNRs. I then compare the observed distribution of 'bright' SNRs in Galactic longitude with that expected from models in order to constrain the Galactic distribution of SNRs.

---

1. Introduction

The distribution of supernova remnants (SNRs) within the Galaxy is of interest for a variety of reasons, not least because they are important sources of energy and high energy particles in the Galaxy. I discuss here the observational selection effects that make current catalogues of SNRs incomplete, and the difficulties in obtaining distances for most remnants. Both of these issues make it difficult to derive the Galactic distribution of SNRs directly. I present constraints on the distribution of SNRs with Galactocentric radius, by comparison of the distribution of bright remnants with Galactic longitude with those expected from simple models. These results are similar to those presented in Green (2012), but here I concentrate more on a discussion of the selection effects that apply to current SNR catalogues. In addition, the analysis presented here excludes the region near $l = 0°$, where the observational selection effects are extreme.

2. Background

I have produced several catalogues of Galactic SNRs. The earliest version, from 1984, contained 145 remnants Green (1984). The number of known remnants has almost doubled in the following 25 years, with the most recent version Green (2009a) containing 274 SNRs. Note, however, that there are many other possible and probable remnants that have also been proposed, which are briefly described in the documentation for the web version of the catalogue†. These objects are not included in the main catalogue of 274 remnants, as further observations are required to confirm their nature, or their parameters, e.g. their full extent. The largest increases in the number of identified remnants are due to large area Galactic radio surveys, e.g. the Effelsberg 2.7-GHz survey and the MOST survey, see Section 2.1.

There are two problems that make it difficult to derive the Galactic distribution of SNRs directly: (i) there are significant observational selection effects that means that the catalogue of SNRs in incomplete, and (ii) distances are not available for all SNRs. These two issues are discussed further in the next two subsections.

† See: http://www.mrao.cam.ac.uk/surveys/snrs/

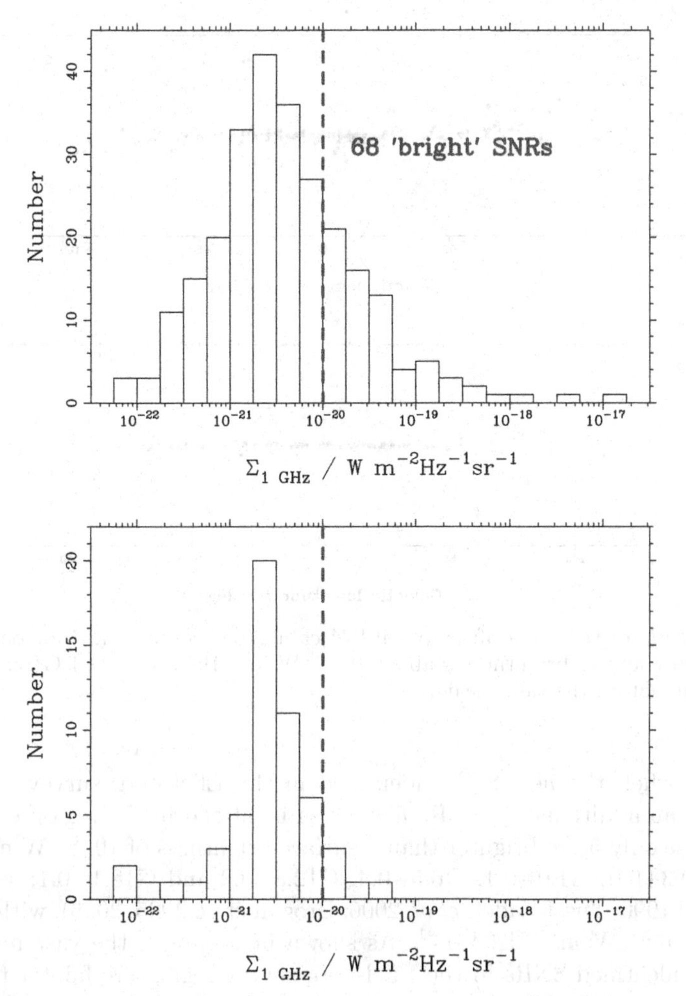

Figure 1. Histograms of the 1-GHz surface brightness of: (top) all catalogued SNRs, and (bottom) those in the area covered by the Effelsberg 2.7-GHz survey added to the catalogue since 1991.

2.1. *Selection effects*

Although some SNRs have first been identified at other than radio wavelengths, in practice the vast majority have been identified from radio observations (which, unlike the optical or X-rays, are not affected by absorption). Furthermore, it is large-area radio surveys that define the completeness of current SNR catalogues, not other (better) observations, which cover specific targets, or are of only limited areas of the Galactic plane.

For much of the Galactic plane – $358° < l < 240°$, $|b| < 5°$ – the deepest, large-scale survey is that made at 2.7-GHz with the Effelsberg 100-m telescope (Reich *et al.* 1990; Fürst *et al.* 1990). The rest of the Galactic plane has been covered by a survey at 843 MHz made with MOST. Both these surveys identified many new SNRs, see Reich *et al.* (1988) and Whiteoak & Green (1996) respectively. New remnants identified from these surveys were added to the 1991 and 1996 versions of my SNR catalogue. For a SNR to be identified it needs to be bright enough to be distinguished from the Galactic background. The approximate surface brightness limit for the Effelsberg 2.7-GHz survey is thought to be about 10^{-20} W m^{-2} Hz^{-1} sr^{-1} at 1 GHz.

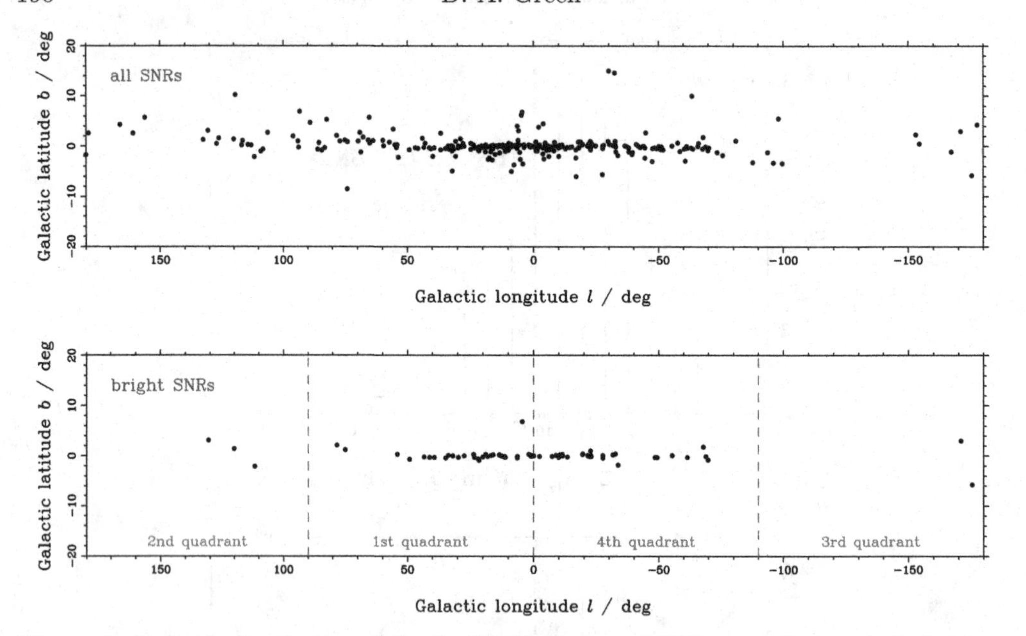

Figure 2. Galactic distribution of: (top) all 274 catalogued SNRs, and (bottom) the brighter 68 remnants, with surface brightnesses above 10^{-20} W m^{-2} Hz^{-1} sr^{-1} at 1 GHz. (Note that the l- and b-axes are not on the same scale.)

Since 1991, when the new SNRs identified in the Effelsberg survey were added to the catalogue, an additional 61 SNRs have been identified in the region covered by this survey. Of these only 5 are brighter than a surface brightness of 10^{-20} W m^{-2} Hz^{-1} sr^{-1} at 1 GHz (G0.3+0.0, G1.0−0.1, G6.5−0.4, G12.8+0.0 and G18.1−0.1; see Gray 1994; Kassim & Frail 1996; Yusef-Zadeh *et al.* 2000; Brogan *et al.* 2005, 2006), with the brightest being $\sim 3 \times 10^{-20}$ W m^{-2} Hz^{-1} sr^{-1}. As shown in Figure 1, the vast majority of the more recently identified SNRs in the Effelsberg survey region are fainter than 10^{-20} W m^{-2} Hz^{-1} sr^{-1} at 1 GHz. The numbers of catalogued remnants with a surface brightness above 10^{-20} W m^{-2} Hz^{-1} sr^{-1} at 1 GHz in the 1st and 4th Galactic quadrants are 35 and 29 respectively, which are consistent within Poisson statistics. Thus I take a surface brightness of 10^{-20} W m^{-2} Hz^{-1} sr^{-1} at 1 GHz to be the approximate effective Σ-limit of the current Galactic SNR catalogue. Figure 2 shows the observed distribution in Galactic coordinates of both (a) all catalogued SNRs, and (b) the 68† SNRs brighter than the nominal surface brightness completeness limit of 10^{-20} W m^{-2} Hz^{-1} sr^{-1} at 1 GHz. This clearly shows that taking the surface brightness selection into account – i.e. considering the brighter remnants only – the distribution of SNRs is more closely correlated towards both $b = 0°$ and the Galactic Centre than might be thought if all SNRs were considered. This is not surprising, as the lower radio emission from the Galaxy in the 2nd and 3rd quadrants, and away from $b = 0°$, means it is easier to identify faint SNRs in these regions. It is most difficult to identify SNRs close to this nominal surface brightness limit in regions of the Galactic plane with bright and complex background radio emission, i.e. close to the Galactic Centre.

† Note that in Green (2012), there was an error in the surface brightness of one SNR, so that 69 remnants were above this nominal surface brightness limit to provide a sample of 'bright' SNRs. In fact there are 68 above this limit in the 2009 SNR catalogue. This difference does not change the conclusions in Green (2012) significantly.

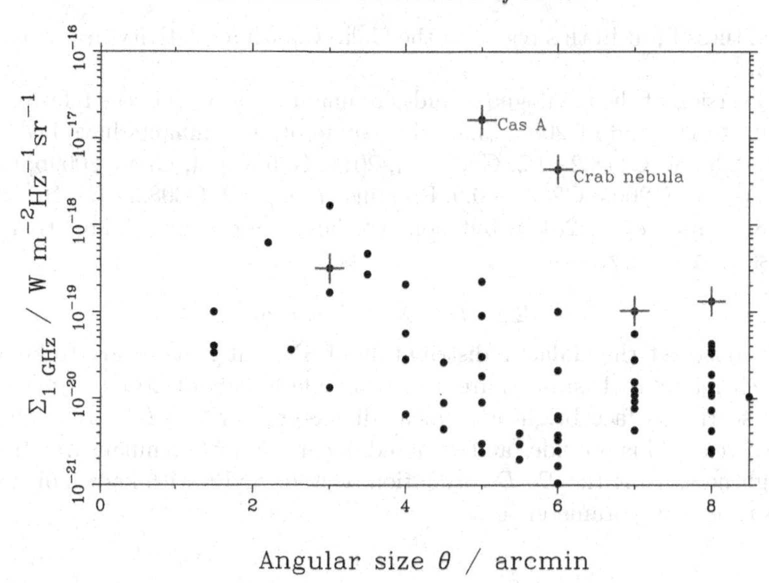

Figure 3. Surface brightness versus (mean) angular diameter for the smaller catalogued SNRs. The remnants of the known 'historical' supernovae of AD 1054 (the Crab nebulae) 1181, 1572, 1604 plus Cas A are indicated by crosses. (The remnant of the supernova of AD 1006 is not included, as it has a diameter of ≈ 30 arcmin.)

Additionally, there is a selection effect that means that some small angular size SNRs are overlooked. It is generally necessary to resolve a SNR in order to recognise it structure, but not all of the Galactic plane has been observed with sufficiently high resolution to resolve the structure of all sources. The Effelsberg 2.7-GHz survey has a resolution of 4.3 arcmin, making it difficult to recognised the structure of a remnant unless it is ~ 10 arcmin or larger in extent. This means that there is a deficit of small angular size SNRs, which is illustrated by Figure 3. This shows the surface-brightness versus angular diameter for the smaller SNRs in the current Galactic SNR catalogue. The remnants of 'historical' supernovae chronicled in the last thousand years or so are indicated. All these remnants are relatively close-by, as otherwise their parent supernova would not have been seen visibly, and therefore they sample only a small part of the Galactic disk. If these known young remnants were further away, they would have the same surface brightness, but would be smaller in angular size. Although there are some such remnants currently known – e.g. the very young SNR G1.9+0.3 (see Green *et al.* 2008; Reynolds *et al.* 2008; Carlton *et al.* 2011) – there are fewer than expected (see further discussion in Green 2005). Hence there is a selection effect against the identification of young but distant SNRs in the Galaxy. Note that most of these missing young remnants will be on the far side of the Galaxy, and therefore appear nearer $b = 0°$ and to $l = 0°$. This is the region of the Galactic plane where the background is brightest, and where there is also more likely to be confusion with other Galactic sources along the line of sight.

Of the 5 sources brighter than 10^{-20} W m^{-2} Hz^{-1} sr^{-1} – i.e. above the nominal surface brightness limit of the current SNR catalogue – which have been identified since 1991 in the Effelsberg survey area, all are close (within 20°) to the Galactic Centre. Moreover, 3 of them are small, $\lesssim 8$ arcmin in diameter. Thus, it is likely that the sample of 68 'bright' SNRs may be somewhat incomplete near the Galactic Centre, due to (i) missing young but distant remnants, and (ii) the difficulty of identifying remnants near the surface brightness limit in this region of the Galaxy, with a relativity bright and complex background.

The 2009 version of the catalogue includes remnants identified in the refereed literature published up to the end of 2008. Since then some other remnants have been identified (e.g. G25.1−2.3 and G178.2−4.2, Gao *et al.* 2011; G35.6−0.4, Green 2009b; G64.5+0.9, Hurley-Walker *et al.* 2009; G296.7−0.9, Robbins *et al.* 2012; G308.3−1.4, Hui *et al.* 2012; G310.5+0.8, Stupar *et al.* 2011), but none of these are clearly brighter than 10^{-20} W m^{-2} Hz^{-1} sr^{-1} at 1 GHz.

2.2. *The '$\Sigma - D$' relation*

To directly construct the Galactic distribution of SNRs it is necessary to know the distance to each remnant. Distances are only available for about 20% of currently known SNRs, and so the surface brightness–linear diameter – or '$\Sigma - D$' – relation has often been used instead. This provides an estimated linear size for a remnant from its *observed* surface brightness, using the $\Sigma - D$ correlation seen for SNRs with known distances. This correlation is usually parameterised as

$$\Sigma = C D^{-n}$$

as physically small SNRs tend to have larger surface brightnesses than larger ones. As is discussed in Green (2005), much of this correlation is arguably due to a D^{-2} bias due to the fact that $\Sigma \propto L/D^2$, where L is the luminosity of the remnant. In practice, however, there are several issues with the '$\Sigma - D$' relation. First, SNRs show a wide range of physical diameters for a given surface brightness, approximately an order of magnitude in range. This means that a distance derived for an individual remnant is quite inaccurate. Second, due to the observational selection effect discussed above, the range of properties of SNRs may be larger than is evident from currently identified remnants, as small angular size, faint remnants are particularly difficult to identify. Third, as has been discussed in Green (2005), some '$\Sigma - D$' studies have used inappropriate least-square straight line regressions. As there is a larger scatter in the $\Sigma - D$ plane, regressions minimising deviations in Σ give quite a different correlation than one minimising deviations in D (e.g. see Isobe *et al.* 1990). Since the $\Sigma - D$ relation is used to predict a value for D from the Σ value for an individual remnant, then minimising deviations in D should be used. Case & Bhattacharya (1998) minimised the deviations in Σ, and obtained a $\Sigma - D$ relation with $n = 2.64 \pm 0.26$ (for 37 'shell' remnants, including Cas A), whereas a significantly steeper relation with $n = 3.53 \pm 0.33$ is obtained if deviations in D are used. This means that fainter remnants – which are the majority, see Figure 1 – have their diameters, and hence distances, *overestimated* if a $\Sigma - D$ relation minimising deviations in Σ is used.

3. The Galactic distribution of SNRs

The direct approach to deriving the distribution of Galactic SNRs is to use the $\Sigma - D$ relation to derive distances to individual remnants, and then construct the 3-D distribution of remnants. However, because of the large range of diameters shown for remnants with similar surface brightnesses, the $\Sigma - D$ relation does not provide reliable distances to individual remnants. Moreover there are the observational selection effects discussed in Section 2.1, which mean that it is not possible to use treat catalogued remnants with equal weight. Instead, the approach I use is to consider only brighter remnants above the nominal surface-brightness limit, and compare their distribution in Galactic longitude with that expected from various models. This approach does not need distance estimates for individual SNRs. Because of the possible remaining selection effects close

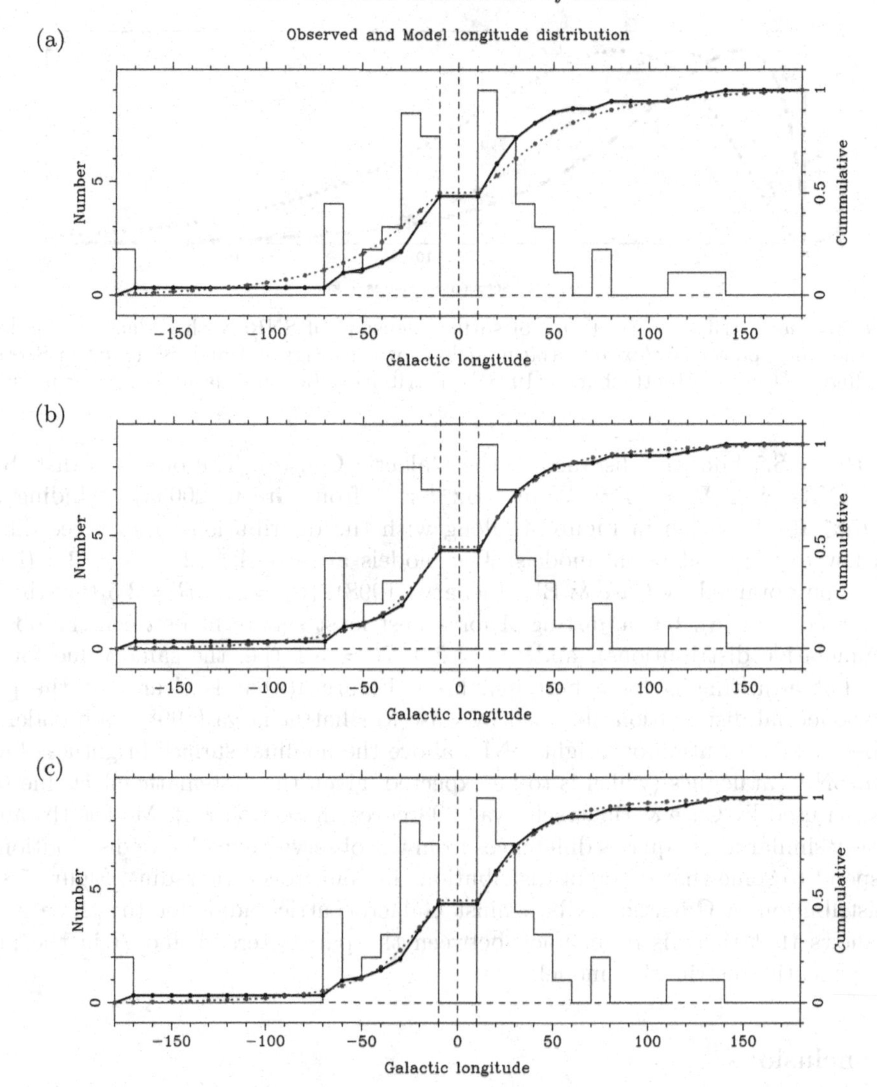

Figure 4. The l-distribution of the 56 'bright' Galactic SNRs – excluding those with $|l| \leqslant 10°$ – shown as (i) histogram (left scale), and (ii) cumulative fraction, solid line (right scale). In addition the cumulative fraction for a model distribution is also plotted, dotted line (right scale). The three models presented are for the surface density of SNRs varying with Galactocentric radius, R, as (a) $\propto (R/R_\odot)^{2.0} \exp[-3.5(R - R_\odot)/R_\odot]$ (as derived by Case & Bhattacharya 1998), (b) $\propto (R/R_\odot)^{0.7} \exp[-3.5(R - R_\odot)/R_\odot]$, and (c) $\propto (R/R_\odot)^{2.0} \exp[-5.1(R - R_\odot)/R_\odot]$.

to the Galactic centre, the region $|l| \leqslant 10°$ is excluded from the analysis presented here, leaving 56 brighter remnants.

One model from the distribution of SNRs (and other star formation tracers, e.g. pulsars and star formation regions, see Bronfman *et al.* 2000; Lorimer *et al.* 2006) is a two parameter power-law/exponential radial distribution for the density of SNRs with Galactocentric radius, R, of the form

$$\propto \left(\frac{R}{R_\odot}\right)^A \exp\left[-B\frac{(R - R_\odot)}{R_\odot}\right]$$

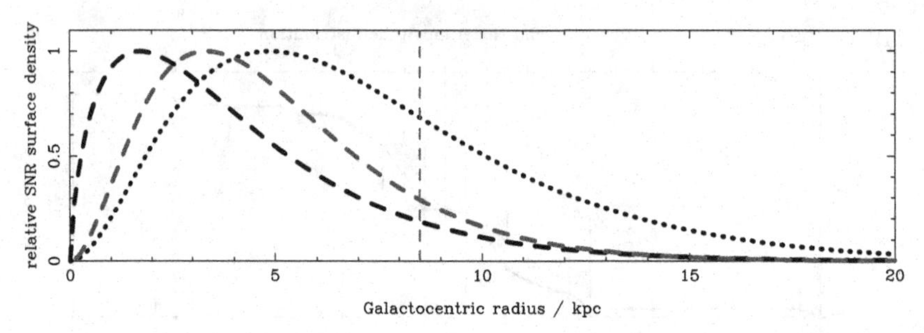

Figure 5. The distribution (in terms of surface density) of SNRs with Galactocentric radius, R, for the three power-law/exponential models shown in Figure 4 and discussed in Section 3: dotted line for Case & Bhattacharya (1998)'s distribution (a), and dashed lines for models (b) and (c).

(with $R_\odot = 8.5$ kpc, the distance to the Galactic Centre). The observed distribution in l of SNRs with $\Sigma > 10^{-20}$ W m^{-2} Hz^{-1} sr^{-1} from Green (2009a) excluding those with $|l| \leqslant 10°$ is shown in Figure 4, along with the distributions from three different power-law/exponential radial models. The models are (a) $A = 2.0$, $B = 3.5$ (i.e. the distribution obtained by Case & Bhattacharya 1998), (b) $= 0.7$, $B = 3.5$ (i.e. the same value for B as in (a), but adjusting A for a best least square fit between the observed and cumulative distributions), and (c) $= 2.0$, $B = 5.1$ (i.e. the same value for A as in (a), but adjusting B for a best fit). From Figure 4(a) it is clear that the power-law/exponential distribution obtained by Case & Bhattacharya (1998), is broader than the observed distribution of 'bright' SNRs above the nominal surface brightness limit of current SNR catalogues (which is to be expected, given the systematic effect due to the regression used by Case & Bhattacharya 1998 noted in Section 2.2). Models (b) and (c) have very similar least squares differences from the observed cumulative distribution, but correspond to somewhat different distributions in Galactocentric radius. Figure 5 shows the distribution of Galactic SNRs against Galactocentric radius for the three models. This shows that there is degeneracy between the parameters A and B in the power-law/exponential distribution model.

4. Conclusions

The lack of distances to most known Galactic SNRs, plus observational selection effects, means that it is difficult to derive the distribution of SNRs in our Galaxy directly. However, by considering 'bright' SNRs – i.e. those not strongly affected by selection effects – constraints on the Galactic distribution of SNRs can be obtained, by comparison of their l-distribution with that expected from models. This shows that the Galactic distribution of SNRs obtained by Case & Bhattacharya (1998) is too broad.

Acknowledgements

I thank Irina Stefan for useful discussions.

References

Brogan, C. L., Gaensler, B. M., Gelfand, J. D., Lazendic, J. S., Lazio, T. J. W., Kassim, N. E., & McClure-Griffiths, N. M., 2005, *ApJ*, 629, L105

Brogan, C. L., Gelfand, J. D., Gaensler, B. M., Kassim, N. E., & Lazio, T. J. W., 2006, *ApJ*, 639, L25

Bronfman, L., Casassus, S., May, J., & Nyman, L.-Å., 2000, *A&A*, 358, 521

Case, G. L. & Bhattacharya, D., 1998, *ApJ*, 504, 761
Carlton, A. K., Borkowski, K. J., Reynolds, S. P., Hwang, U., Petre, R., Green, D. A., Krishnamurthy, K., & Willett, R., 2011, *ApJ*, 737, L22
Fürst, E., Reich, W., Reich, P., & Reif, K., 1990, *A&AS*, 85, 691
Gao, X. Y., Sun, X. H., Han, J. L., Reich, W., Reich, P., & Wielebinski, R., 2011, *A&A*, 532, A144
Gray, A. D., 1994, *MNRAS*, 270, 835
Green, D. A., 1984, *MNRAS*, 209, 449
Green, D. A., 2005, *MmSAI*, 76, 534
Green, D. A., 2009a, *BASI*, 37, 45
Green, D. A., 2009b, *MNRAS*, 399, 177
Green, D. A., 2012, in: F. Aharonian, W. Hofmann & F. Rieger (eds), *High Energy Gamma-Ray Astronomy*, AIP Conference Proceedings, Volume 1505 (American Institute of Physics, Melville, New York), p. 5
Green, D. A., Reynolds, S. P., Borkowski, K. J., Hwang, U., Harrus, I., & Petre, R., 2008, *MNRAS*, 387, L54
Hui, C. Y., Seo, K. A., Huang, R. H. H., Trepl, L., Woo, Y. J., Lu, T.-N., Kong, A. K. H., & Walter, F. M., 2012, *ApJ*, 750, 7
Hurley-Walker, N., *et al.*, 2009, *MNRAS*, 398, 249
Isobe, T., Feigelson, E. D., Akritas, M. G., & Babu, G. J., 1990, *ApJ*, 364, 104
Kassim, N. E. & Frail, D. A., 1996, *MNRAS*, 283, L51
Lorimer, D. R., *et al.*, 2006, *MNRAS*, 372, 777
Reich, W., Fürst, E., Reich, P., Junkes, N., 1988, in: R. S. Roger & T. L. Landecker (eds), *Supernova Remnants and the Interstellar Medium*, Proc. IAU Colloquium No. 101 (Cambridge University Press), p. 293
Reich, W., Fürst, E., Reich, P., & Reif, K., 1990, *A&AS*, 85, 633
Reynolds, S. P., Borkowski, K. J., Green, D. A., Hwang, U., Harrus, I., & Petre, R., 2008, *ApJ*, 680, L41
Robbins, W. J., Gaensler, B. M., Murphy, T., Reeves, S., & Green, A. J., 2012, *MNRAS*, 419, 2623
Stupar, M., Parker, Q. A., & Filipović, M. D., 2011, *Ap&SS*, 332, 241
Whiteoak, J. B. Z. & Green, A. J., 1996, *A&AS*, 118, 329
Yusef-Zadeh, F., Shure, M., Wardle, M., & Kassim, N., 2000, *ApJ*, 540, 842

Discussion

SANKRIT: Are there likely to be any High-Latitude Remnants we haven't yet detected? If so, how best to search for them?

GREEN: Yes, there are some traces already, and they had to be fat, as it is possible to find fat remnants where the background is low. Future surveys will no doubt find more.

UNIDENTIFIED: Of course, what we really would like to know is the derived SN rate, care to tell us?

GREEN: As mentioned by Enrico Cappellaro, you can infer a note of a few per century from the known 'Historical' remnants. But it is difficult to be precise.

BRANDT: Can you bin the (bright) SNRs to the resolution of the Galactic arm?

GREEN: No. There are insufficient statistics only 68 'bright' remnants above the nominal surface brightness completeness limit.

FOLATELLI: Is there any other observational parameter that could be expected to help reduce the scatter in the Σ-D relation?

GREEN: There have been some attempts to identify a subset of SNRs,e.g. those interacting with a molecule cloud, which might have less scatter.

WANG: Did you compare your distribution of SNRs with that predicted by tracers such as stellar mass and SFR? I'd expect that the predicted distribution would be rather different especially in the Galactic center region.

GREEN: Yes, some tries have been made.

Supernova Environmental Impacts
Proceedings IAU Symposium No. 296, 2013
A. Ray & R. A. McCray, eds.

doi:10.1017/S1743921313009460

Observations and discoveries of supernova remnants with GMRT

Subhashis Roy[1] **and Sabyasachi Pal**[2,3]

[1]NCRA-TIFR, Pune University Campus,
Pune-411007, India
email: roy@ncra.tifr.res.in

[2]Indian Centre for Space Physics,
Kolkata-700084, India
[3]Ionospheric and Earthquake Research Centre,
Kolkata-7211154, India
email: sabya@csp.res.in

Abstract. We have measured HI absorption distance to the youngest Galactic supernova remnant G1.9+0.3. Absorption by known anomalous velocity features near the Galactic centre (GC) puts a lower limit on its distance from Sun as 10 kpc, 2 kpc further away from the GC. We have found a small diameter (1.6′) shell like structure G354.4+0.0, that shows polarised emission in the NVSS. Based on its morphology, angular size, HI distance and its spectrum between 1.4 GHz and 330 MHz, it is perhaps the second youngest SNR in the Galaxy that is expanding in a dense environment of an HII region surrounding it. Our pilot observation of the inner Galactic 4th quadrant within 337 ° $< l <$ 354° with a fixed Galactic latitude of 0.37° has confirmed G345.1−0.2 as an SNR.

Keywords. supernova remnants, radio continuum: ISM

---

1. Introduction

The number of supernova remnants (SNRs) in our Galaxy is catalogued to be 273 (Green 2009). However, the expected number of SNRs based on the OB star counts, pulsar birth rates and Iron abundance in our Galaxy, and supernova rates in Local Group galaxies is 1000 (Tammann *et al.* 1994). This deficit is likely due to the result of instrumental insensitivity towards old, faint, large remnants and the effect of poor angular resolution working against detection of small diameter SNRs in low resolution surveys(Green 1991). A lack of their detection would otherwise indicate our inability to characterise the stellar evolution in our own Galaxy or in the prediction of the timescale over which an SNR remains detectable in the ISM. The missing remnants are likely concentrated towards the inner Galaxy, where the diffuse emission causes the most confusion.

Given the expected supernova rate in the Galaxy of 2.8 per century (Li *et al.* 2011) we expect to see about 20 supernovae remnants of age less than 700 years, but only 4 are known (Tycho, Kepler, Cas A and G1.9+0.3). Among these four SNRs, G1.9+0.3 is the smallest in angular size ∼1.2′ (Green & Gull 1984). This indicated it to be comparatively young, and remarkably enough, its size measured from Chandra X-ray image was found to be larger than measured from VLA radio data taken two decades back and from its expansion over the last two decades it has been shown to be the youngest known Galactic supernova remnant of age about 150 years (Reynolds *et al.* 2008). Based on very high absorbing column density found from their X-ray observations, Reynolds *et al.* (2008) suggested it to be located at the distance of the Galactic centre (GC) of 8 kpc. This distance is, however, a lower limit, as its location beyond the GC would not change the absorption column density much. Given that its physical properties like linear size and expansion velocity depends on conversion of angular size to linear size depending on its distance, a more direct determination of its distance is important. We carried out HI absorption measurements towards it in 2004. In the next section, we present the results and determine its distance and discuss its consequent ramifications on its properties.

Most of the missing remnants in the Galaxy are likely concentrated towards the inner Galaxy. Therefore, we had used GMRT at 330 MHz to confirm candidate SNRs from Molonglo observatory synthesis telescope (MOST) survey of the GC region (Roy & Bhatnagar 2006). The main objective of these observations was to confirm the nature of certain candidate SNRs. In one of these fields, we identified a small shell like structure G354.4+0.0 that was seen just outside the primary beam of that field. In the next section, we describe its confirmation as a newly discovered young SNR in the Galaxy.

A histogram plot of the number of SNRs in the Galaxy (Green 2009) as a function of Galactic longitude peaks within $|l| \sim 0° - 20°$. The number of SNRs catalogued within $0° < l < 20°$ is 65, and between $340° < l < 360°$ is 42, indicating a clear asymmetry in SNR distribution between the 1st and 4th quadrant of the inner Galaxy. This is mostly due to the new discovery of SNRs by Brogan *et al.* (2006) within $4.5° < l < 22°$ and $|b| < 1.5°$ region of the Galaxy. Asymmetry in supernova rate or in the timescale for which SNRs remain detectable in the ISM are unlikely to be different from the 1st to the 4th quadrant of the Galaxy. Therefore, observations of the inner 4th quadrant of the Galaxy with the sensitivity and resolution attained by Brogan *et al.* (2006), will discover missing remnants (~ 23) in the Galactic 4th quadrant. Earlier observations of this region has been carried out by the Molonglo Observatory Synthesis Telescope (MOST) at 843 MHz and claimed to have reached a sensitivity of 2 mJy.beam^{-1} (Whiteoak & Green 1996). Assuming a spectral index of -0.5 for the Galactic SNRs, their sensitivity was actually 1.5 times more than Brogan *et al.* (2006) (both had similar resolution). This suggests that VLA observations of SNRs in the inner 1st quadrant of the Galaxy to have better dynamic range than the MOST survey. Since GMRT has high sensitivity at 330 MHz, and is sensitive to structures with angular scales of $\sim 30'$, we have carried out a pilot survey of this region with $337° < l < 354°$ with a fixed Galactic latitude of $0.37°$, with a plan to extend this survey in future. In the next section, we describe the outcomes.

2. Results and Discussions

2.1. *HI absorption towards G1.9+0.3*

Fig. 1 shows the HI absorption spectrum towards the source. Since the direction of the object is close to that of the GC, Galactic rotation cannot be used to constrain its distance. However, absorption by the anomalous velocity features, whose velocities and distance from GC are known could be used to constrain its distance. In the spectrum, absorption near 0 km.sec^{-1} is believed to be from local gas seen nearly perpendicular to its Galactic rotation direction. Prominent absorption is seen near velocity of -50 km.sec^{-1}. Weak absorption is also seen near velocity of 150 km.sec^{-1}. The 3-kpc arm located about 5 kpc away from Sun is known to have a velocity of about -50 km.sec^{-1} at this Galactic longitude. The absorption near 150 km.sec^{-1} is believed to be caused by the 'Feature-I' located 2 kpc further from the GC (Cohen 1975). Therefore, lower limit on the distance to G1.9+0.3 is 10 kpc from Sun. Consequently, its diameter is modified to >3.5 pc, and mean expansion velocity to 11,000 km.sec^{-1}. Assuming this SNR to have been created by a Type Ia explosion, Ksenofontov *et al.* (2010) calculated its nonthermal properties and found its TeV gamma ray energy to depend on its (distance)$^{-11}$. If located

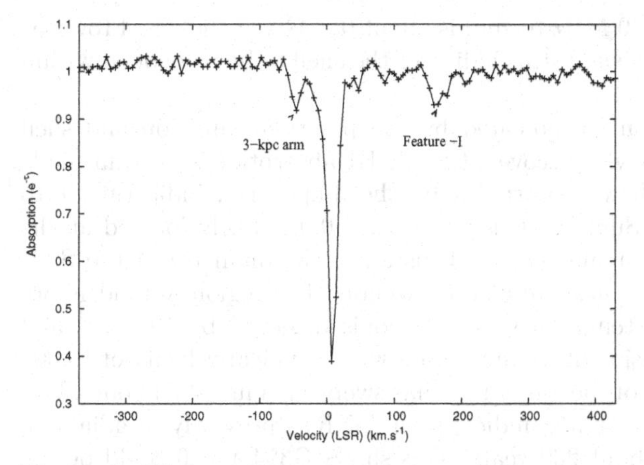

Figure 1. HI absorption spectrum towards the youngest SNR G1.9+0.3.

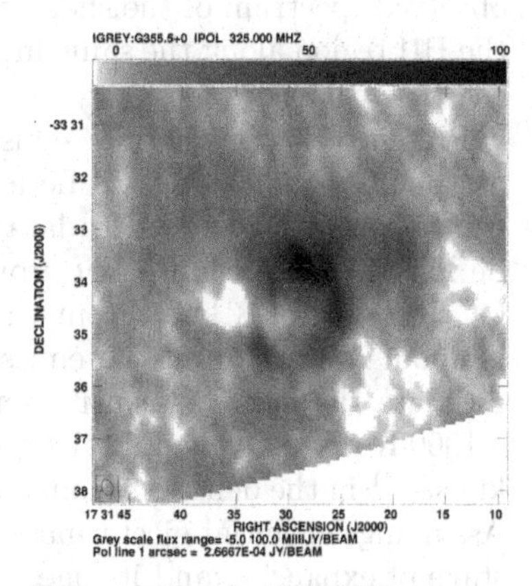

Figure 2. A 325 MHz image of G354.4+0.0. Polarisation vectors from NVSS map are overlaid. Resolution $\sim 15''$, rms noise 3 mJy.beam^{-1}.

within 5.6 kpc, they predicted its TeV gamma rays to be detectable by future instruments like Cerenkov telescope array. However, our results rule out any such possible detection in future.

2.2. *Discovery of an young SNR five degrees away from the GC*

The continuum map of the source G354.4+0.0 made from P band data having a resolution of $\sim 15''$ is shown in Fig. 2. This map is sensitive to large scale structures up to $30'$. The object shows morphology of a partial shell, and the size of it as measured by taking several cross-cuts across it and then finding their average is $1.6'$. We have searched the NRAO VLA sky survey (Condon *et al.* 1998) to detect any polarized emission from the partial shell structure and thereby confirm non-thermal emission from it. Polarization vectors from NVSS are overlaid on Fig. 2. To make the polarization total intensity and angle images from the NVSS Stokes Q and U maps (rms noise $\sim$0.3 mJy.beam^{-1}), all pixels below a signal to noise ratio of 4.5 were blanked and correction for noise bias in polarized total intensity was made. We do detect significant polarized emission from near the two brightest parts of the shell like structure with peak polarized flux densities of $\sim$2.2 mJy.beam^{-1} (Fig. 2).

We have also observed this object at 1.42 GHz with GMRT. Based on the flux density of the shell between 330 MHz and 1.4 GHz, we find its spectral index to be 0.0 ± 0.1. This flat spectral index is unexpected from a shell type SNR. To search for any free-free absorbing screen towards the object, we made a low resolution map from the L band GMRT data that shows diffuse extended emission of size $\sim 5'$ on and around the location of the shell. This diffuse emission has been catalogued by the PMN survey (Wright *et al.* 1996) at 4.8 GHz (flux density 3.7 Jy). A comparison of the flux density of this extended emission between the above two frequencies also shows a flat spectrum for it, indicating thermal emission from the extended emission. This is also confirmed from its detection in the 60 μm IRAS map. The diffuse emission itself contributes 2.7 Jy at 1.4 GHz. However, from the same region, the measured flux density due to the diffuse emission only is 0.9 Jy at 330 MHz. This shows the thermal extended emission to undergo self absorption at 330 MHz with optical depth (τ) $\sim$ 4. Assuming an intrinsic spectral index of -0.6 ($S(\nu) \propto \nu^{\alpha}$) for the shell, the required τ for the free-free absorbing medium to flatten the observed spectrum of the shell at 330 MHz to zero is about 0.9. Given the $\tau \sim 4$ towards the HII region along the same line of sight, this indicates the shell to be embedded within the HII region.

Presence of non-thermal emission as indicated by the polarised emission and shell morphology indicates it to be a newly discovered SNR. HI absorption spectrum made from the GMRT 1.4 GHz data shows absorption by the 3-kpc arm, indicating it to be located more than 5 kpc from Sun. We believe G354.4+0.0 is likely located at the GC distance of 8 kpc. From the angular size and distance, the diameter of the SNR is about 3.7 pc. From the emission measure of the extended HII region around it, we estimate the electron density and temperature of the ionised gas to be 70 cm^{-3} and $\sim$1000 K respectively. Initial expansion of the supernova with a typical velocity of 10,000 $\mathrm{km.sec}^{-1}$ in the dense environment of the ionised gas has swept up a mass of about $4M_{\odot}$. Assuming an initial ejecta mass of $\sim 1\ M_{\odot}$ indicates the SNR is presently in adiabatic stage of expansion and its age is about 260 years. This shows G354.4+0.0 could be the second youngest SNR in the Galaxy. More details shall be found in Roy & Pal (2013).

2.3. *A pilot survey of the inner 4th quadrant of the Galaxy at 330 MHz with GMRT*

Within the region of the survey mentioned in the introduction, we achieved a resolution of about $19'' \times 13''$. The rms noise attained was 6-10 $\mathrm{mJy.beam}^{-1}$, a factor of 2-3 higher than expected. We detect the following 9 known SNRs: G337.8−0.1, G338.3−0.0, G340.6+0.3, G341.8−0.2, G344.6−0.1, G346.6−0.2, G348.5+0.0, G348.7+0.3, G351.2+0.1. Whiteoak & Green (1996) suggested 2 objects of high surface brightness, G337.2+0.1, G339.6−0.6 as candidate SNRs. We could not detect them, and they are unlikely to be SNRs. One of the candidate SNRs G345.1−0.2 described by Whiteoak & Green (1996) is seen in our 330 MHz maps. Between 330 and 843 MHz, its spectral index is found to be steep, −0.6. Therefore, a shell type morphology and steep spectral index confirms G345.1−0.2 to be an SNR.

Acknowledgement

We thank the staff of GMRT that made these observations possible. GMRT is run by National Center for Radio Astrophysics of the Tata Institute of Fundamental Research. Work of SP is supported by a grant of MOES.

References

Brogan, C. L., Gelfand, J. D., Gaensler, B. M., Kassim, N. E., & Lazio, T. J. W. 2006, *ApJL*, 639, L25
Cohen, R. J. 1975, *MNRAS*, 171, 659
Condon, J. J., Cotton, W. D., Greisen, E. W., *et al.* 1998, *AJ*, 115, 1693
Green, D. A. 1991, *PASP*, 103, 209
—. 2009, *Bulletin of the Astronomical Society of India*, 37, 45
Green, D. A. & Gull, S. F. 1984, *Nature*, 312, 527
Ksenofontov, L. T., Völk, H. J., & Berezhko, E. G. 2010, *ApJ*, 714, 1187
Li, W., Chornock, R., Leaman, J., *et al.* 2011, *MNRAS*, 412, 1473
Reynolds, S. P., Borkowski, K. J., Green, D. A., *et al.* 2008, *ApJL*, 680, L41
Roy, S. & Bhatnagar, S. 2006, *Journal of Physics Conference Series*, 54, 152
Roy, S. & Pal, S. 2013, *to be submitted to ApJL*
Tammann, G. A., Loeffler, W., & Schroeder, A. 1994, *ApJS*, 92, 487
Whiteoak, J. B. Z. & Green, A. J. 1996, *AApS*, 118, 329
Wright, A. E., Griffith, M. R., Hunt, A. J., *et al.* 1996, *ApJS*, 103, 145

Discussion

SANKRIT: How many HII regions are found in the survey region?

ROY S.: I am yet to count the number of HII regions in the survey. However the number of HII regions that morphologically appear similar to shell type SNRs are $\sim$ 10, quite close to the number of known SNRs in the region.

Supernova Environmental Impacts
Proceedings IAU Symposium No. 296, 2013
A. Ray & R. A. McCray, eds.

doi:10.1017/S1743921313009472

Radio polarization observations of large supernova remnants at λ6 cm

J. L. Han[1], X. Y. Gao[1], X. H. Sun[1,2], W. Reich[2], L. Xiao[1], P. Reich[2], J. W. Xu[1], W. B. Shi[1], E. Fürst[2], and R. Wielebinski[2]

[1]National Astronomical Observatories, Chinese Academy of Sciences, Jia-20 Datun Road, Chaoyang District, Beijing 100012, China. hjl@nao.cas.cn
[2]Max-Planck-Institut für Radioastronomie, Auf dem Hügel 69, 53121 Bonn, Germany

Abstract. We have observed 79 supernova remnants (SNRs) with the Urumqi 25 m telescope at λ6 cm during the Sino-German λ6 cm polarization survey of the Galactic plane. We measured flux densities of SNRs at λ6 cm, some of which are the first ever measured or the measurements at the highest frequency, so that we can determine or improve spectra of SNRs. Our observations have ruled out spectral breaks or spectral flattening that were suggested for a few SNRs, and confirmed the spectral break of S147. By combining our λ6 cm maps with λ11 cm and λ21 cm maps from the Effelsberg 100 m telescope, we calculated the spectral index maps of several large SNRs. For many remnants we obtained for the first time polarization images, which show the intrinsic magnetic field structures at λ6 cm. We disapproved three objects as being SNRs, OA184, G192.8−1.1 and G16.8−1.1, which show a thermal spectrum and no polarization. We have discovered two large supernova remnants, G178.2−4.2 and G25.1−2.3., in the survey maps.

Keywords. supernova remnants, polarization, radio continuum: ISM

---

1. Introduction

Radio observations of supernova remnants (SNRs) probe two aspects of their physics. One aspect is particle acceleration and synchrotron radiation. Radio images show the surface brightness distribution of remnants, which in general is composed of diffuse emission and of filamentary or shell-like emission. The particles were accelerated in the shock-front of a SNR and radiate in the filamentary area. After some time, the aged particles diffuse away from the shock area and radiate over a much wider area, producing diffuse emission. The filamentary emission produced by shock-accelerated particles shows a power-law spectrum with an index of typically $\alpha = -0.4$ to -0.5 ($S \sim \nu^{\alpha}$). This is for most of observed SNRs in adiabatic expansion phase. Particles with high energies lose their energy faster than those with low energies, so that the aged particles produce extended emission with a steeper spectrum. Observations with adequate resolution should be able to distinguish filamentary emission from diffuse emission. To reveal the process of particle acceleration and radiation, multiband observations are needed to make images of the spectral index distribution. The second aspect of SNR physics that can be probed with radio observations is magnetic fields. Supernova explosions not only accelerate particles, but also compress the surrounding medium by their shocks. The magnetic fields penetrating the interstellar medium are therefore compressed too, and then act as the agent for accelerated particles to produce synchrotron radiation. Radio polarization observations can probe the magnetic field structure of SNRs. It has been found that young remnants have a radial field structure, while old remnants have a tangential field structure (e.g. Fürst & Reich 2004). Note, however, that the foreground Faraday rotation must be discounted to get the intrinsic polarization angles of radio emission, so that the intrinsic magnetic field structure of SNRs can be revealed.

To address these two aspects of physics radio observations of SNRs need 1) adequate angular resolution so that remnants can be resolved; 2) multiband intensity measurements so that an image of the spectral index distribution can be calculated; 3) multiband polarization measurements, so that Faraday rotation in the foreground can be corrected, and the intrinsic magnetic field orientation can be figured out. In addition, high enough observing frequencies, e.g. up to λ6 cm, should be selected so that radio images of a SNR do not get confused by fluctuating Galactic radio diffuse emission. Observations at shorter wavelengths suffer from less depolarization, and Faraday rotation changes position angles by only a small amount.

There are many apparently large SNRs within 2-3 kpc distance in the Galaxy, which have been objects for X-ray, Gamma-ray and optical observations. However, their radio images at high radio frequencies are extremely difficult to obtain. The huge apparent size of these SNRs, a few degrees in general, is too large for synthesis telescopes due to their limited field of view and their insensitivity to extended emission, for a single large dish telescope, such as the Effelsberg telescope, because of its small beam at high frequencies. We scanned many large SNRs at λ6 cm to obtain polarization maps with the single band polarization system of the Urumqi 25 m radio telescope of Xinjiang Observatory, which we used for the Sino-German λ6 cm polarization survey of the Galactic plane. An excellent receiver was constructed at the MPIfR and installed at the telescope in August 2004, which has a very good stability for long scan-observations.

In this invited talk, we introduce the main results of radio polarization observations of SNRs at λ6 cm by using the small Urumqi 25 m telescope. Combining our data with λ21 cm and λ11 cm observations made with the Effelsberg 100 m telescope, we obtained many new results on SNR spectra and polarization.

2. Observational results

Since August 2004, we have used the Urumqi 25 m radio telescope for the Sino-German 6cm polarization survey of the Galactic plane in the region of $10° \leqslant l \leqslant 230°$ and $|b| \leqslant 5°$ (Sun *et al.* 2007, Gao *et al.* 2010, Xiao *et al.* 2011, Sun *et al.* 2011a). We divided the survey region into many patches, and scanned them in both Galactic longitude l and latitude b. Regular observations of 3C286 or 3C295 are made for calibration purposes. During the data processing, we have to carefully fit the baselines, remove obvious interference, and suppress scanning effects. The final radio maps at λ6 cm were made in total power I and linear polarization, Stokes parameters Q and U. The latter two maps were then combined to calculate polarized intensity PI and polarization angle PA maps. The survey maps show more frequent fluctuations towards smaller Galactic longitudes in the inner region (Sun *et al.* 2011a, Xiao *et al.* 2011) and the Cygnus region, which leads to high confusion for SNRs there. Data for SNRs located in the survey region are extracted from the survey, for some of them additional observations were added for higher sensitivity. SNRs outside the survey region were observed separately.

2.1. *The first polarization image of the λ6 cm system: the Cygnus loop*

The first polarization observations of the 6 cm system were made towards the Cygnus loop (Fig. 1). This large SNR has a size of $4° \times 3°$. The polarization vector distribution shows two polarization shells with different properties (Sun *et al.* 2006), which supports the idea that the Cygnus Loop consists of two SNRs, as suggested by Uyanıker *et al.* (2002). Analysing the λ6 cm map together with λ11 cm and λ21 cm data from observations with the Effelsberg 100 m telescope, we got a spectral index map (Fig. 1). Steep spectra are seen in the central part of the SNR.

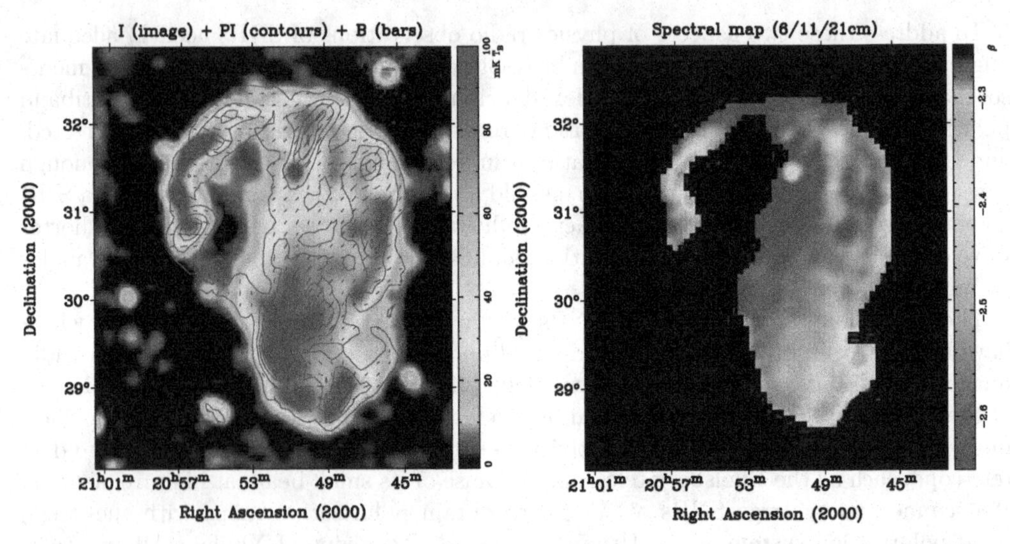

Figure 1. The first polarization map *(left)* was made for the Cygnus loop using the λ6 cm system. Combining our map with λ11 cm and λ21 cm maps from Effelsberg observations, we calculated a spectral index map *(right)*.

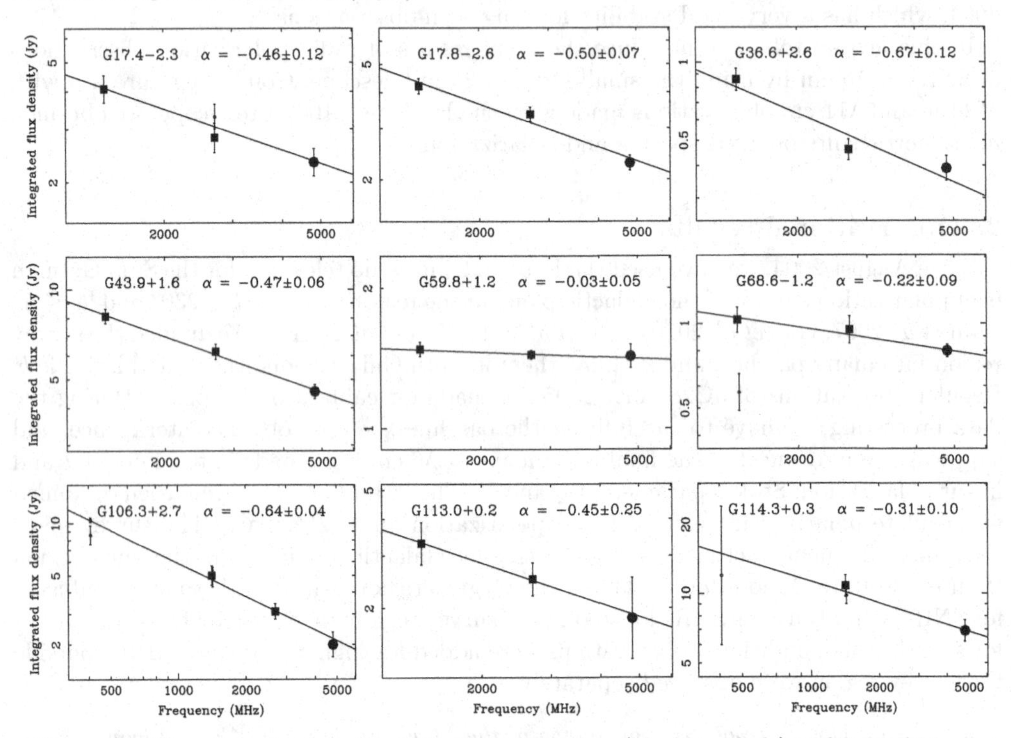

Figure 2. Examples of newly determined flux densities (black symbols) at λ6 cm, λ11 cm, and λ21 cm to obtain spectra of SNRs (Sun *et al.* 2011b, Gao *et al.* 2011a).

2.2. *New flux density measurements for integrated spectra*

From our survey maps or separate observations at λ6 cm we have measured the integrated flux densities of SNRs (Sun *et al.* 2011b, Gao *et al.* 2011a). For some SNRs, we also get new measurements of integrated flux densities from the Effelsberg λ11 cm and λ21 cm maps. Using these measurements, together with measurements at other wavelengths from the literature, we determined or improved integrated spectra of many SNRs (Fig. 2). For the SNRs G15.1−1.6, G16.2−2.7, G16.4−0.5, G17.4−2.3, G17.8−2.6, G20.4 +0.1, G36.6+2.6, G43.9 +1.6, G53.6−2.2, G55.7 +3.4, G59.8+1.2, G65.1+0.6, G68.6−1.2, G69.0 +2.7 (CTB 80), G93.7−0.2, G113.0+0.2, and G114.3+0.3, the spectra have been significantly improved (Sun *et al.* 2011b, Gao *et al.* 2011a).

2.3. *Spectral break or flattening*

A spectral break was suggested for several SNRs based on available integrated flux densities in literature (see plots in Sun *et al.* 2011b). SNR G74.9+1.2 seems to be a solid case which has a break at about 10 GHz. SNR G31.9+0.0 has a flat spectral below a few hundred MHz, probably due to the low frequency absorption. Flux density data of SNR G21.5−0.9 and G69.0+2.7 (Gao *et al.* 2011a) have very large uncertainties for claiming a spectral break. SNR G27.8+0.6 may have a break, but to verify this requires more and better high frequency data.

We disprove three claims of a possible spectral break or flattening by using our new measurements of integrated flux densities at $\lambda 6$ cm. The first one is the Cygnus loop (Sun *et al.* 2006). The small flux density at 5 GHz from previous observations (Kundu & Becker 1972) suggests a spectral break above 2.7 GHz (Uyanıker *et al.* (2004)). Our new integrated flux density at $\lambda 6$ cm rules out such a spectral steepening (see Fig. 3). The second case was G126.2+1.6. Tian & Leahy (2006) suggested a spectral break at about 1.5 GHz based on flux densities from the literature with low accuracy. Our new measurement at $\lambda 6$ cm is consistent with a single power-law for the radio spectrum (Sun *et al.* 2007). The third case is HB3. Urošević *et al.* (2007) claimed a spectral flattening above 2 GHz as an indication for radio thermal bremsstrahlung emission from a thin shell enclosing HB3. We obtained flux densities of HB3 at $\lambda 6$ cm, $\lambda 11$ cm and $\lambda 21$ cm for the region that is not confused by the nearby strong HII region W3, and found that these measurements are consistent with a single power law (Shi *et al.* 2008).

We confirmed the case for the spectral break of S147, which is a large, faint, shell-type SNR. Previous observations of S147 at 1648 MHz and 2700 MHz (Kundu *et al.* 1980)

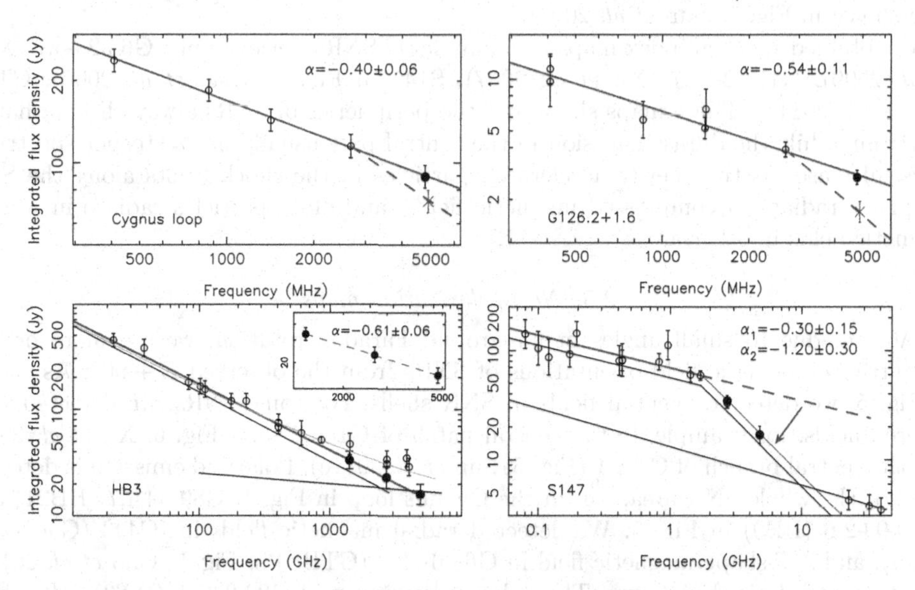

Figure 3. New integrated flux densities at $\lambda 6$ cm and $\lambda 11$ cm and $\lambda 21$ cm (black dots) rule out a spectral break previously proposed for the Cygnus loop and G126.2+1.6 and spectral flattening for HB3. We confirm the spectral break for S147.

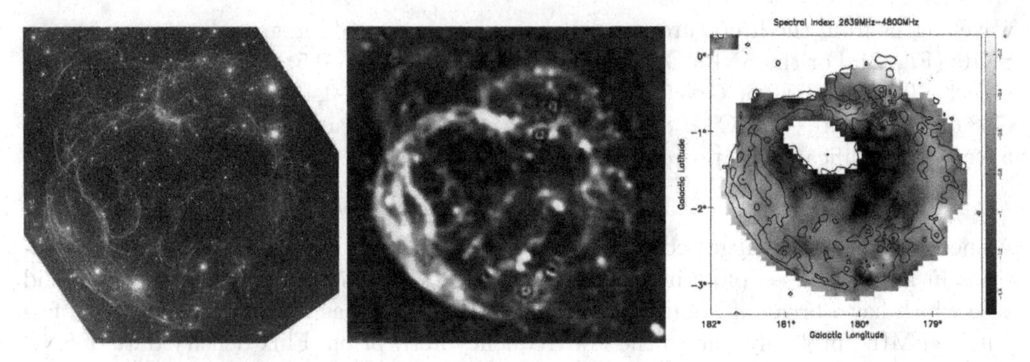

Figure 4. The filaments of SNR S147 as observed in the optical (*left*) and radio (λ11-cm total power map in the *middle*). The spectral index map (*right*) shows flat spectra in the filamentary area, and steep spectra in the diffuse area, especially in the central part.

and the southern part of the SNR at 4995 MHz (Sofue *et al.* 1980) suggested a spectral break near 1.5 GHz, with a flat spectrum at lower frequencies and a steep spectrum at higher frequencies. The break needs to be confirmed because the previous observations at 4995 MHz covered only the southern part of the remnant. Our new measurements cover the entire SNR with the Urumqi 25 m radio telescope at λ6 cm and new more sensitive Effelsberg 100 m radio telescope at λ11 cm (Xiao *et al.* 2008), and confirm a spectral break at ∼1.5 GHz. The spectral break is caused by the combination of a diffuse emission component with a steep spectrum and filamentary emission with a flat spectrum, which could be traced up to 40 GHz including WMAP data (see Fig. 3).

2.4. *Spectral index maps of large SNRs*

Our observations at λ6 cm are the first to cover many large SNRs. By combining the maps with these at λ21 cm and λ11 cm observed with the Effelsberg 100 m radio telescope or other telescopes, we can get their spectral index maps. The first object is the Cygnus loop shown in Fig. 1 (Sun *et al.* 2006).

We obtained spectral index maps for many large SNRs, for example, G65.2+5.7 (Xiao *et al.* 2009), G156.2+5.7 (Xu *et al.* 2007), S147 in Fig. 4 (Xiao *et al.* 2008), CTA1 (Sun *et al.* 2011c). These maps show that the peripheries of SNRs always have a flatter spectrum, while the diffuse emission in the central part usually has a steeper spectrum. Physical reason is that newly accelerated particles in the shock fronts along the SNR periphery radiate in compressed magnetic fields, and older particles radiate in weaker magnetic fields in other areas.

2.5. *New polarization maps*

At λ6 cm, due to small angles of foreground Faraday rotation, we see more or less the intrinsic magnetic field orientations of SNRs from the observed $\vec{\mathbf{E}} + 90°$. As shown in Fig. 5, we detect tangential fields in SNR shells. For some SNRs, we detected very ordered fields, for example, in the central patch of G156.2+5.7 (Fig. 6, Xu *et al.* 2007) and the central branch of CTA 1 (Fig. 5, Sun *et al.* 2011c). Polarized emission is detected even for the whole SNR area, e.g. in the Cygnus loop in Fig. 1, G89.0+4.7 (HB21) and G160.9+2.6 (HB9) in Fig. 5. We detected radial magnetic fields in IC443 (Gao *et al.* 2011a), and a T-shape magnetic field in G68.0+2.7 (CTB80 in Fig. 5, Gao *et al.* 2011a) and its very polarized east arm. The polarized emission of G82.2+5.3 (W63) is found to be anti-correlated with the radio total power and Hα emission (Gao *et al.* 2011a), which indicates a mixture of local thermal and nonthermal emission in the complex region.

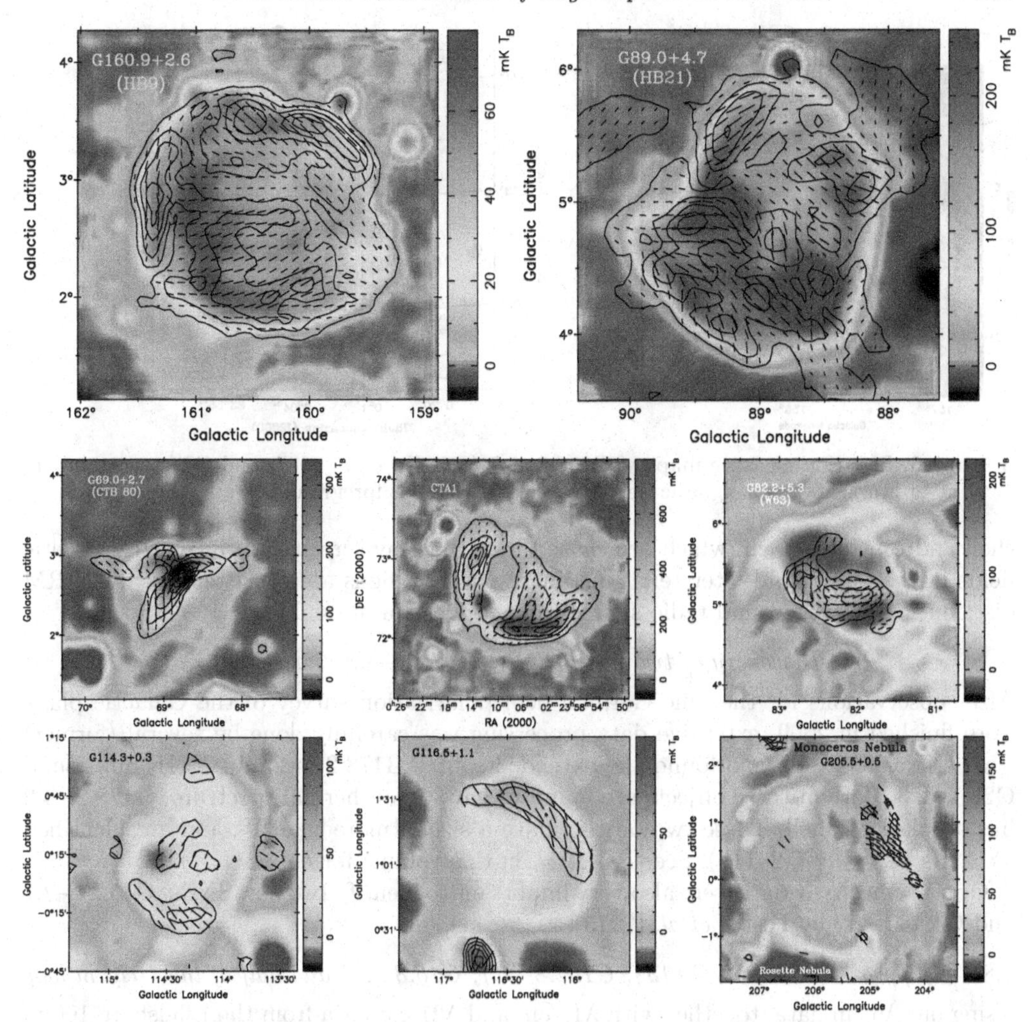

Figure 5. New polarization maps of some SNRs. Color image for total intensity, contours for the polarized intensity and vectors for **B** orientation.

For 23 of 79 observed SNRs, we get the first complete polarization images at λ6 cm. A few SNRs were never observed before in polarization (SNR G205.5+0.5 = Monoceros Nebula; G206.9+2.3; G85.9−0.6, G69.7+1.0, G16.2−2.7) until our λ6 cm measurements. For G16.2−2.7, G69.7+1.0 and G85.9−0.6, the polarized emission is detected for the first time, adding evidence that they are in fact SNRs (Sun *et al.* 2011b).

2.6. *Rotation measures in SNRs*

Comparing the polarization angle maps at λ6 cm from our Urumqi observations with those of λ21 cm and λ11 cm observations with the Effelsberg 100 m telescope, we calculated RM maps of SNRs.

Two very interesting examples are shown in Fig. 6. The RM map of G156.2+5.7 (Xu *et al.* 2007) shows decreasing RMs downwards along the left shell of the SNR and increasing RMs along the right shell. This indicates a twisted field structure probably in the SNR shell or a foreground RM gradient. The RM map of CTA 1 (Sun *et al.* 2011c) show negative RMs in the southern shell, different from the positive RMs in the northern

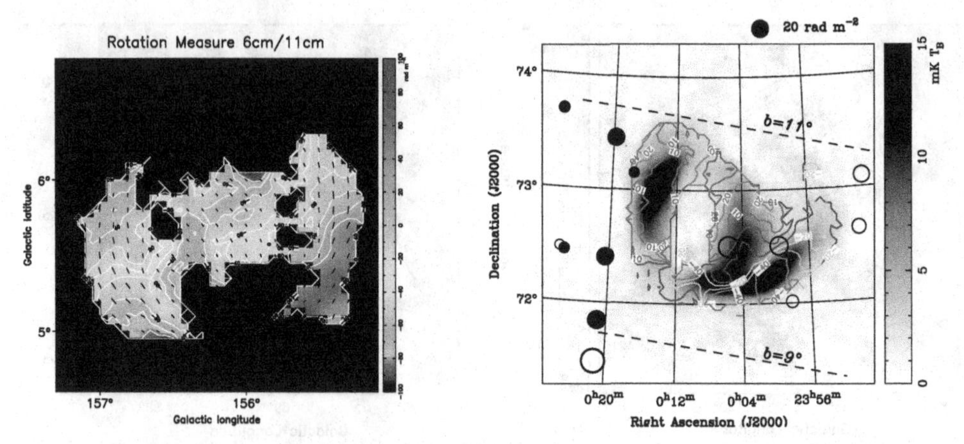

Figure 6. Rotation measure maps of G156.2+5.7 (*left*, Xu *et al.* 2007) and CTA1 (*right*, Sun *et al.* 2011c), showing magnetic field structure in SNR or foreground interstellar medium.

shell and central branch, which are caused by a Faraday screen with reversed magnetic fields in the foreground interstellar medium. The RM signs are consistent with the RM distribution of background radio sources in a wider area around CTA 1.

2.7. *Discovery of two large SNRs: G178.2−4.2 and G25.1−2.3*

After observations for the Sino-German 6cm polarization survey of the Galactic plane were finished in 2009, extensive data processing was carefully done by several (former) PhD students. X. Y. Gao found the extended source G178.2−4.2 and X. H. Sun found G25.1−2.3. Both of these objects are found to have a nonthermal spectrum. G178.2−4.2 has a polarized shell. These two objects have a size of more than 1°, and are identified as SNRs (Gao *et al.* 2011b). See Gao *et al.* in this volume for details.

Our λ6 cm survey data were also very important to identify two new SNRs, G152.4−2.1 and G190.9−2.2 by Foster *et al.* (2013).

2.8. *Disapproved "SNRs": OA184, G192.8−1.1, G16.8−1.1 and half of the Origem loop*

Using our λ6 cm data, together with λ11 cm and λ21 cm data from the Effelsberg 100 m telescope, we disapproved 3.5 "known SNRs": OA184 (Foster *et al.* 2006), G192.8−1.1 (Gao *et al.* 2011a), G16.8−1.1 (Sun *et al.* 2011a), and the lower half of the Origem loop (Gao & Han 2013).

The first disapproved SNR was OA184 (Foster *et al.* 2006). T. Foster noticed its flat spectra at low frequencies, in contrast to the nonthermal spectra of G166.0+4.3 and HB9 located in the same area of the Galactic plane. The λ6 cm observation were added in this investigation and give the final clue. It turns out that OA184 has a thermal spectra. More importantly, it appears as a depolarized extended source in the λ6 cm map, rather than showing ordered polarization as expected from a shell-type SNR.

The second disapproved SNR was G192.8−1.1 (Gao *et al.* 2011a). The bright knots on the plateau have been found to be either known background sources or known HII region, plus a newly identified HII region. The plateau itself is found to have a thermal spectrum without any associated polarized emission.

The third one is G16.8−1.1 (Sun *et al.* 2011b). This object in the very inner Galaxy appears as a depolarization source in our 6cm map embedded in a large polarization patch of the Galactic diffuse emission. It coincides well with the known HII region, SH 2-50.

In addition, the Origem loop is found to to be composed of a SNR arc in the north and HII regions in the South (Gao & Han 2013).

3. Summary

We have demonstrated that a small telescope is very useful to observe large objects. By observing large SNRs, of a few degree in size, we obtained many unique polarization images at λ6 cm. Many of these images are the first, or are the one at highest frequency so far, to reveal the intrinsic magnetic fields of SNRs. Multi-band observations are very important to calculate spectra of SNRs, or to get the spectral index images. The observations we present are very useful for studying the physical properties of SNRs on particle acceleration and magnetic fields. Using our data, we disapproved three and half "known SNRs" and dismissed the suggested spectral break of a few SNRs.

Readers can get more information from the web-page: http://zmtt.bao.ac.cn/6cm/.

Acknowledgements

We thank the SOC of IAUS 296 for their invitation to give this summary talk, and we acknowledge financial support from the National Natural Science Foundation of China (10473015, 10773016), specifically for the Sino-German λ6 cm polarization survey of the Galactic plane. The Sino-German cooperation was supported via the partner group of the MPIfR at the NAOC as part of the exchange program between the MPG and the CAS for many bilateral visits.

References

Foster, T., Kothes, R., Sun, X. H., Reich, W., & Han, J. L., 2006, *A&A* 454, 517
Foster, T., Cooper, B, Reich, W., Kothes, R., & West, J. 2013, *A&A* 549, A107
Fürst, E. & Reich, W. 2004, In: *The magnetized Interstellar Medium,* B. Uyanıker, W. Reich, and R. Wielebinski, ed., Copernicus GmbH, Katlenburg-Lindau, p. 141
Gao, X. Y. & Han, J. L. 2013, *A&A* 551, A16
Gao, X. Y., Han, J. L., Reich, W., Reich, P., Sun, X. H., & Xiao, L. 2011a, *A&A* 529, A159
Gao, X. Y., Reich, W., Han, J. L., Sun, X. H., Wielebinski, R., Shi, W. B., Xiao, L., Reich, P., Fürst, E., Chen, M. Z., & Ma, J. 2010, *A&A* 515, A64
Gao, X. Y., Sun, X. H., Han, J. L., Reich, W., Reich, P., & Wielebinski, R. 2011b, *A&A* 532, A144
Kundu, M. R. & Becker, R. H. 1972, *AJ* 77, 459
Kundu, M. R., Angerhofer, P. E., Fürst, E., & Hirth, W. 1980, *A&A* 398, 993
Sofue, Y., Fürst, E., & Hirth, W. 1980, *PASJ* 32, 1
Shi, W. B., Han, J. L., Gao, X. Y., Sun, X. H., Xiao, L., Reich, P., & Reich, W. L. 2008, *A&A* 487, 601
Sun, X. H., Han, J. L., Reich, W., Reich, P., Shi, W. B., Wielebinski, R., & Fürst, E. 2007, *A&A* 463, 993
Sun, X. H., Reich, W., Han, J. L., Reich, P., & Wielebinski, R., 2006, *A&A* 447, 937
Sun, X. H., Reich, W., Han, J. L., Reich, P., Wielebinski, R., Wang, C., & Müller, P. 2011a, *A&A* 527, A74
Sun, X. H., Reich, P., Reich, W., Xiao, L., Gao, X. Y., & Han, J. L. 2011b, *A&A* 536, A83
Sun, X. H., Reich, W., Wang, C., Han, J. L., & Reich, P. 2011c, *A&A* 535, A64
Tian, W. W. & Leahy, D. 2006, *A&A* 447, 205
Urošević, D., Pannuti, T. G., & Leahy, D. 2007, *ApJ* 655, L41
Uyanıker, B., Reich, W., Yar, A., Kothes, R., & Fürst, E. 2002, *A&A* 389, L61
Uyanıker, B., Reich, W., Yar, A., & Fürst, E. 2004, *A&A* 426, 909
Xiao, L., Fürst, E., Reich, W., & Han, J. L. 2008, *A&A* 482, 783
Xiao, L., Reich, W., Fürst, E., & Han, J. L. 2009, *A&A* 503, 827
Xiao, L., Han, J. L., Reich, W., Sun, X. H., Wielebinski, R., Reich, P., Shi, H., & Lochner, O. 2011, *A&A* 529, A15
Xu, J. W., Han, J. L., Sun, X. H., Reich, W., Xiao, L., Reich, P., & Wielebinski, R. 2007, *A&A* 470, 969

Supernova Environmental Impact
Proceedings IAU Symposium No. 296, 2013
A. Ray & R. A. McCray, eds.

doi:10.1017/S1743921313009484

Discovery of supernova remnants in the Sino-German λ6 cm polarization survey of the Galactic plane

X. Y. Gao[1], X. H. Sun[1], J. L. Han[1], W. Reich[2], P. Reich[2] and R. Wielebinski[2]

[1]National Astronomical Observatories, Chinese Academy of Sciences ,Jia-20 Datun Road, Chaoyang District, Beijing 100012, China. bearwards@gmail.com
[2]Max-Planck-Institut für Radioastronomie, Auf dem Hügel 69, 53121 Bonn, Germany

Abstract. The Sino-German λ6 cm polarization survey has mapped in total intensity and polarization intensity over an area of approximately 2 200 square degrees in the Galactic disk. This survey provides an opportunity to search for Galactic supernova remnants (SNRs) that were previously unknown. We discovered the new SNRs G178.2−4.2 and G25.1−2.3 which have non-thermal spectra, using the λ6 cm data together with the observations with the Effelsberg telescope at λ11 cm and λ21 cm. Both G178.2−4.2 and G25.1−2.3 are faint and have an apparent diameter greater than 1°. G178.2−4.2 shows a polarized shell. HI data suggest that G25.1−2.3 might have a distance of about 3 kpc. The λ6 cm survey data were also very important to identify two other new SNRs, G152.4−2.1 and G190.9−2.2.

Keywords. Radio continuum: ISM – ISM: supernova remnants – Polarization

1. Introduction

Supernova explosions have a substantial impact on the interstellar environment. Supernova remnants (SNRs) are post-explosion relics, and are formed when shocks from the explosion sweep up and interact with the surrounding medium. Large-scale radio surveys are ideal hunting grounds for new SNRs. In the most frequently used Galactic SNR catalogue compiled by Dave Green (Green 2009), mainly based on radio continuum observations, there are 274 SNRs. Ferrand & Safi-Harb (2012) recently made a new Galactic SNR catalogue by including new detections from high energy observations. The number of known Galactic SNRs is now 312. This quantity is still far less than the theoretical predictions (e.g. Tammann *et al.* 1994), because of two major limitations: the sensitivity and the angular resolution of the observations.

The Sino-German λ6 cm polarization survey of the Galactic plane (Sun *et al.* 2007, Gao *et al.* 2010, Sun *et al.* 2011, Xiao *et al.* 2011) was conducted between the years 2004 and 2009, observing the Galactic disk in the range of $10° \leqslant l \leqslant 230°$ and $|b| \leqslant 5°$. The angular resolution is 9.5′, and the average sensitivity (1σ noise) of the survey is about 0.8 mK T_b in total intensity I and 0.5 mK T_b in linear polarization U and Q. Although the angular resolution is coarser in comparison with synthesis telescopes and large single dishes, the system is more suitable for observing SNRs with large extent, and the high sensitivity of the Sino-German survey enables us to discover SNRs as faint as G156.2+5.7, the SNR with the lowest surface brightness until recently. One of the major goals of the λ6 cm survey is to study and identify Galactic SNRs (see Han *et al.* 2013, this volume). In this talk, we present the discovery of two new SNRs G178.2−4.2 and G25.1−2.3 in our λ6 cm survey (Gao *et al.* 2011).

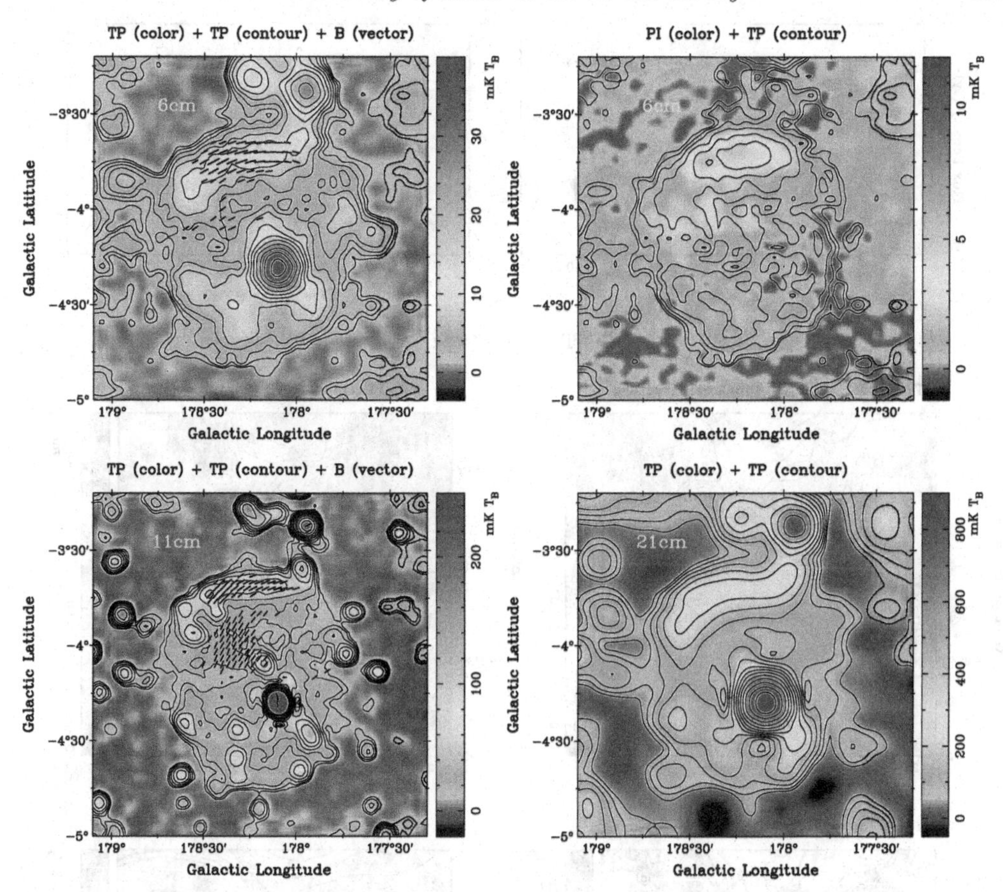

Figure 1. Total intensity and polarization intensity images of the new SNR G178.2−4.2 measured at λ6 cm, λ11 cm and λ21 cm. The top right panel shows the total intensity contours after subtracting point sources.

2. Identification of two new SNRs

Considering the limitation in angular resolution, we search for shell-type objects as SNR candidate. Shell-type SNRs often appear more extended than the crab-like ones, and are easier to identify due to three characteristics: 1) shell or partial shell structures, 2) associated polarized emission within the shell, and 3) the non-thermal spectrum with a spectral index around $\beta \sim -2.5$ ($S_\nu \sim \nu^\beta$, $\alpha = \beta + 2$), as expected for adiabatic expansion with a compression factor of 4. These are the three criteria for our SNR identifications. Note that polarization may not be detected due to Faraday depolarization. We successfully identified two new shell-type SNRs G178.2−4.2 and G25.1−2.3 in the λ6 cm survey.

G178.2−4.2 is located in the anti-center region of the Galaxy (Fig. 1). It has a circular shape with an apparent diameter of around 1°. A prominent shell is seen in its northern part. The un-related, unresolved double-sided radio source 3C139.2 is near the center of G178.2−4.2. Polarized emission is seen in the northern shell at both λ6 cm and λ11 cm. B-field vectors ($\vec{E}$+90°) are found to be tangential within the shell at λ6 cm. We observed G178.2−4.2 at λ11 cm with the Effelsberg 100-m telescope in March, 2009, and we extracted the λ21 cm data from the Effelsberg λ21 cm survey of the Galactic plane (Reich *et al.* 1997) and the Effelsberg λ21 cm medium latitude survey (Reich *et al.* 2004).

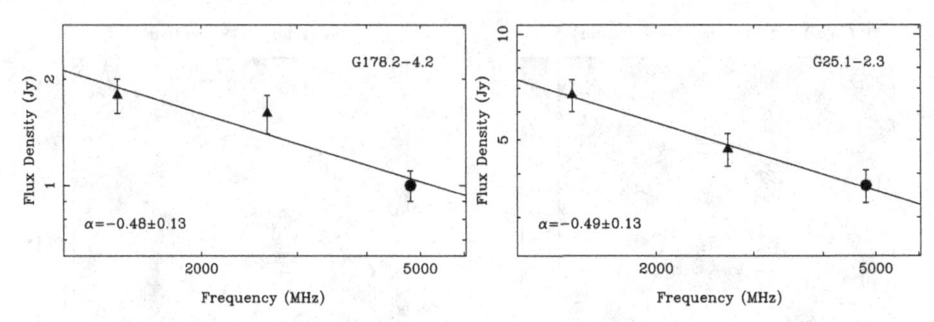

Figure 2. Integrated radio spectrum of G178.2−4.2 and G25.1−2.3.

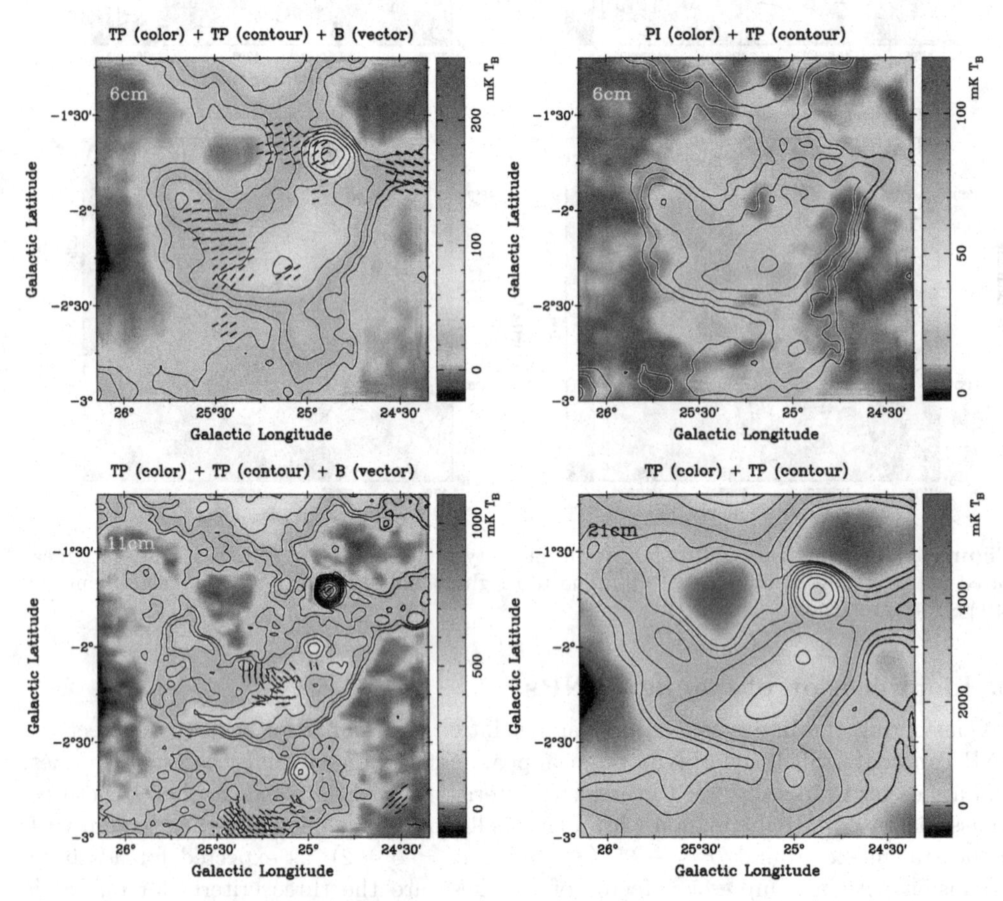

Figure 3. The same as in Fig. 1, but for G25.1−2.3.

The flux density is integrated over the same area of G178.2−4.2 at λ6 cm, λ11 cm and λ21 cm, after removing the contribution from extra-Galactic sources and background emission. We measured $S_{6cm} = 1.0 \pm 0.1$ Jy, $S_{11cm} = 1.6 \pm 0.2$ Jy and $S_{21cm} = 1.8 \pm 0.2$ Jy, yielding an integrated spectral index of $\alpha = -0.48 \pm 0.13$ (Fig. 2). This value indicates the non-thermal nature of G178.2−4.2. In summary, the shell structure, the polarized emission and the non-thermal nature strongly indicate that G178.2−4.2 is a SNR. From the integrated flux density, we calculated the surface brightness of the new SNR G178.2−4.2 to be $\Sigma_{1\ \mathrm{GHz}} = 7.2 \times 10^{-23}\,\mathrm{Wm^{-2}Hz^{-1}sr^{-1}}$. This small value places it among the faintest SNRs known in the Galaxy.

G25.1−2.3 is found in the inner part of the Galaxy. It is elusive until we filter out the confusion from the strong diffuse Galactic emission. We examined G25.1−2.3 using the data from the Urumqi λ6 cm survey, the Effelsberg λ11 cm survey (Reich *et al.* 1990a) and the Effelsberg λ21 cm survey (Reich *et al.* 1990b), and found that G25.1−2.3 has only one shell curving to the south. Polarization patches are detected within the shell at λ6 cm and λ11 cm, but they seem to be un-correlated with G25.1−2.3 (see Fig. 3). We determined that the integrated flux density of G25.1−2.3 is S_{6cm} = 3.7±0.4 Jy, S_{11cm} = 4.7±0.5 Jy, and S_{21cm} = 6.7±0.7 Jy, respectively. The spectrum that we fitted to these data has a spectral index of $\alpha = -0.49 \pm 0.13$ (Fig. 2). SNR G25.1−2.3 has a surface brightness of $\Sigma_{1\ \mathrm{GHz}} = 5.0 \times 10^{-22}\mathrm{Wm^{-2}Hz^{-1}sr^{-1}}$, which makes it one of the fainter SNRs in Green's sample (Green 2009, see his Fig. 1).

From a possibly associated cavity found in the neutral atomic gas, we estimate a distance of 3.1 kpc to the new SNR G25.1−2.3. If this is true, the distance of G25.1−2.3 can explain the absence of polarized emission coming from this object, since polarized emission originated beyond 3 kpc might not be detected at λ6 cm in this direction of the Galactic plane (Sun *et al.* 2011).

3. Other SNRs discovered with the λ6 cm data

Based on high angular resolution synthesis observations, Foster *et al.* (2013) recently identified two other new SNRs, G152.4−2.1 and G190.9−2.2, which are even fainter than G156.2+5.7. The λ6 cm total intensity and polarization data from the Sino-German λ6 cm survey were incorporated in their study and provide strong support and evidence for the identifications.

Acknowledgements

We are grateful for financial support from the National Natural Science Foundation of China (10473015, 10773016) for the Sino-German λ6 cm polarization survey of the Galactic plane. The Sino-German cooperation was supported via partner group of the MPIfR at the NAOC in the frame of the exchange program between the MPG and the CAS for many biliteral visits.

References

Ferrand, G. & Safi-Harb, S. 2012, *Adv. Space Res.*, 49, 1313
Foster, T., Cooper, B, Reich, W., Kothes, R., & West, J. 2013, *A&A* 549, A107
Gao, X. Y., Reich, W., Han, J. L., Sun, X. H., Wielebinski, R., Shi, W. B., Xiao, L., Reich, P., Fürst, E., Chen, M. Z., & Ma, J. 2010, *A&A* 515, A64
Gao, X. Y., Sun, X. H., Han, J. L., Reich, W., Reich, P., & Wielebinski, R. 2011, *A&A* 532, A144
Green, D. A. 2009, *Bull. Astron. Soc. India*, 37, 45
Reich, W., Reich, P., & Fürst, E. 1990b, *A&AS*, 83, 539
Reich, W., Fürst, E., Reich, P., & Reif, K. 1990a, *A&AS*, 85, 633
Reich, P., Reich, W., & Fürst, E. 1997, *A&AS*, 126, 413
Reich, W., Fürst, E., Reich, P., *et al.* 2004, *The Magnetized Interstellar Medium*, 45
Sun, X. H., Han, J. L., Reich, W., Reich, P., Shi, W. B., Wielebinski, R., & Fürst, E. 2007, *A&A* 463, 993
Sun, X. H., Reich, W., Han, J. L., Reich, P., Wielebinski, R., Wang, C., & Müller, P. 2011, *A&A* 527, A74
Tammann, G. A., Loeffler, W., & Schroeder, A. 1994, *ApJS*, 92, 487
Xiao, L., Han, J. L., Reich, W., Sun, X. H., Wielebinski, R., Reich, P., Shi, H., & Lochner, O. 2011, *A&A* 529, A15

Supernova Environmental Impacts
Proceedings IAU Symposium No. 296, 2013
A. Ray & R. A. McCray, eds.

doi:10.1017/S1743921313009496

Infrared [Fe II] and Dust Emissions from Supernova Remnants

Bon-Chul Koo

Department of Physics and Astronomy, Seoul National University,
Seoul 151-747, KOREA
email: koo@astro.snu.ac.kr

Abstract. Supernova remnants (SNRs) are strong thermal emitters of infrared radiation. The most prominent lines in the near-infrared spectra of SNRs are [Fe II] lines. The [Fe II] lines are from shocked dense atomic gases, so they trace SNRs in dense environments. After briefly reviewing the physics of the [Fe II] emission in SNR shocks, I describe the observational results which show that there are two groups of SNRs bright in [Fe II] emission: middle-aged SNRs interacting with molecular clouds and young core-collapse SNRs in dense circumstellar medium. The SNRs belonging to the former group are also bright in near-infrared H_2 emission, indicating that both atomic and molecular shocks are pervasive in these SNRs. The SNRs belonging to the latter group have relatively small radii in general, implying that most of them are likely the remnants of SN IIL/b or SN IIn that had strong mass loss before the explosion. I also comment on the "[Fe II]-H_2 reversal" in SNRs and on using the [Fe II]-line luminosity as an indicator of the supernova (SN) rate in galaxies. In the mid- and far-infrared regimes, thermal dust emission is dominant. The dust in SNRs can be heated either by collisions with gas species in a hot plasma or by radiation from a shock front. I discuss the characteristics of the infrared morphology of the SNRs interacting with molecular clouds and their dust heating processes. Finally, I give a brief summary of the detection of SN dust and crystalline silicate dust in SNRs.

Keywords. shock waves, ISM: supernova remnant, infrared: ISM

---

1. Introduction

Infrared (IR) covers 3 decade logarithmic scales in wavelength, from 1 to 1000 μm. This is the waveband in which we observe emission from dust, forbidden fine-structure lines from various metallic atoms and ions, molecular lines, and H-recombination lines. These diverse and unique emission features, together with their relatively small extinctions, make the IR band particularly useful for studying various physical and astrophysical processes related to shocks and supernova remnants (SNRs).

During the past 10 years, significant progress has been made in the IR study of SNRs as a result of space missions equipped with mid- and far-IR instruments and the development of wide-field IR cameras and broadband spectrometers. In this paper, I shall talk about two particular spectral features often found in SNRs: (1) [Fe II] emission lines in the near-IR (NIR) band, which is the most prominent NIR spectral feature in SNRs, and (2) dust continuum emission in mid- and far-IR spectra. For the [Fe II] lines, I briefly review the basic physics, summarize observational results, and then discuss the characteristics of [Fe II]-bright SNRs along with some related issues. For the dust emission, as there are other papers on this topic in this volume, I simply present some recent topics that are relevant to supernovae (SNe) and SNR environments.

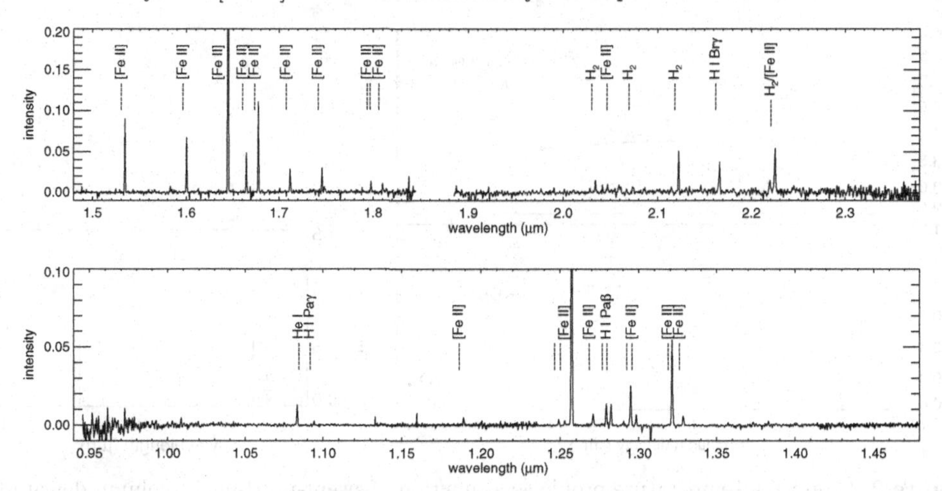

Figure 1. TripleSpec spectrum of the SNR G11.2-0.3, showing numerous [Fe II] lines and other emission lines (Courtesy of D.-S. Moon). The extinction to the source is large ($A_V = 16$ mag), so that the observed line intensities can be significantly different from the intrinsic ones.

2. NIR [Fe II] Emission from SNRs

2.1. *NIR [Fe II] Emission Lines and J Shocks*

In the NIR spectra of SNRs, [Fe II] emission lines are usually the most prominent unless there is heavy-element-enriched SN ejecta (Fig. 1). This contrasts with photoionized HII regions where H recombination lines are much stronger; i.e., [Fe II] 1.257 μm/Paβ ($\sim 3\times$ [Fe II] 1.644 μm/Paα)=2–8 in SNRs whereas it is 0.013 in Orion (Oliva *et al.* 1989; Mouri *et al.* 2000). Such a large difference arises because Fe atoms in photoionized gas are in higher ionization stages and also probably because the Fe abundance in shocked gas is enhanced by dust destruction. Therefore, [Fe II] emission can be used as a tracer of fast radiative atomic shocks, although strong [Fe II] lines may be observable in sources ionized by X-rays, e.g., in active galactic nuclei (Mouri *et al.* 2000).

The Fe^+ ion has four ground terms, each of which has 3–5 closely-spaced levels to form a 16 level system (Pradhan & Nahar 2011). The energy gap between the ground level and the excited levels is less than 1.3×10^4 K, and thus, these levels are easily excited in the postshock cooling region. The emission lines resulting from the transitions among these levels appear in the visible to far-infrared bands (Fig. 2 right). In the NIR JHK bands, 10–20 [Fe II] lines are visible; these include the two strongest lines at 1.257 and 1.644 μm. The ratios of these lines provide a very good density diagnostic and an accurate measure of extinction to the emitting region.

[Fe II] emission lines in SNRs are mostly emitted from cooling gas behind radiative atomic shocks. Figure 2 (top left) shows the temperature profile in the postshock cooling layer of a radiative shock. At $N_{\rm H} \sim 5\times10^{17}$ cm^{-2} the cooling becomes important and the temperature abruptly drops to $\sim 8,000$ K. Then the temperature remains constant over an extended region, where the heating is maintained by UV radiation generated from the hot gas immediately behind the shock front. The corresponding profiles of H nuclei and electron densities are shown in the bottom left frame of Figure 2, together with the Fe^+ fraction profile. Note that, since the ionization potential of the iron atom is 7.90 eV, far-UV photons from the hot shocked gas can penetrate far downstream to maintain the ionization state of Fe^+ where H atoms are primarily neutral. Most of the [Fe II] emission, however, originates from the temperature plateau region where the ionization fraction is not too low, as shown in Figure 2. Numerical shock models covering some parameter

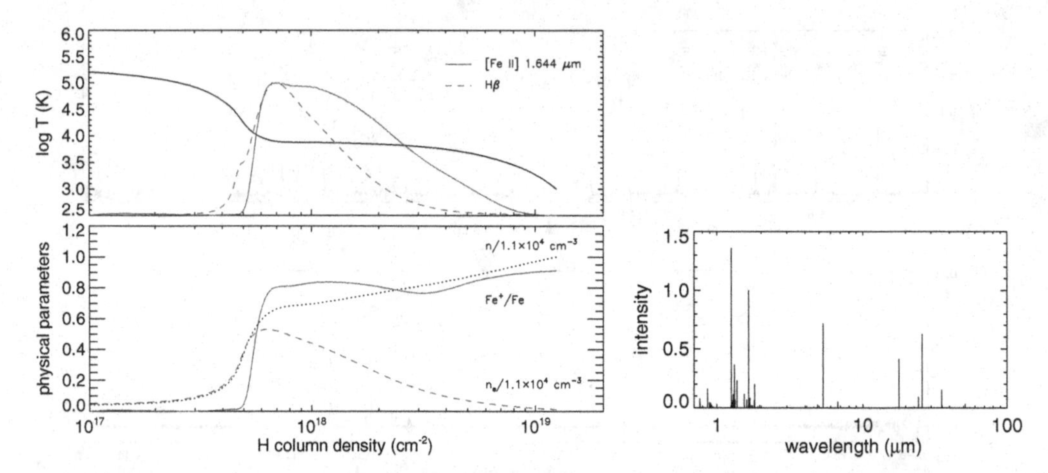

Figure 2. (Top left) Temperature profile as a function of swept-up H-nuclei column density for a 150 km s^{-1} shock propagating into an ambient medium of $n_0 = 100$ cm^{-3} and $B_0 = 10$ μG. [Fe II] 1.644 μm and Hβ line emissivities are overplotted in an arbitrary linear scale. (Bottom left) Profiles of H nuclei density (n), electron density (n_e) and fraction of Fe in Fe$^+$ (Fe$^+$/Fe) for the same shock. (Right) Synthesized IR spectrum of [Fe II] lines from the shock, normalized to the [Fe II] 1.644 μm line intensity. The calculation is done by using the Raymond code.

spaces are available in Hollenbach *et al.* (1989), Mouri *et al.* (2000), and Allen *et al.* (2008). A grid of shock models with updated atomic parameters is in preparation by the author of this paper.

2.2. *NIR [Fe II] Observations of SNRs*

The first detection of the [Fe II] 1.644 μm line in an SNR was reported by Seward *et al.* (1983) on MSH 15−52. After that, about a dozen Galactic and LMC SNRs have been observed in NIR [Fe II] lines. This number will increase with the completion of the UWIFE (UKIRT Wide-field Infrared Survey for Fe$^+$) project, which is an unbiased survey of the [Fe II] 1.644 μm line of the inner Galactic plane ($\ell = 7°$ to $65°$; $|b| \leqslant 1.3°$) using the UKIRT 4-m telescope. The UWIFE project is a "cousin" of the UWISH2 project, which covers the same area in the H$_2$ $v = 1 \rightarrow 0$ S(1) line at 2.122 μm (Froebrich *et al.* 2011). Lee, Y.-H. *et al.* (this volume) introduce the two projects and present preliminary results on SNRs. In short, there are 77 SNRs in this area, and about 20%–30% of them are detected in [Fe II] and/or H$_2$ lines, more than half of which are new detections.

The SNRs bright in [Fe II] emission lines may be divided into two groups: (1) middle-aged SNRs interacting with dense molecular (or atomic) clouds, and (2) young SNRs interacting with the dense circumstellar medium (CSM).

Middle-aged SNRs bright in [Fe II] emission. Prototypical SNRs belonging to this category are W44, 3C391, and IC 443. All of them are interacting with molecular clouds (MCs). An indication of the MC interaction in the NIR band is the presence of H$_2$ ro-vibrational emission lines that arise from slow, non-dissociative C shocks propagating into dense molecular gas of low-fractional ionization. Therefore, these middle-aged SNRs that are bright in [Fe II] emission are also bright in H$_2$ emission (Fig. 3).

Figure 3 shows [Fe II] 1.644 μm and H$_2$ 2.122 μm line images of W44; we see that the overall morphologies of the SNR in the two lines are similar, although the former appears rather diffuse and confined to the SNR boundary, whereas the latter is considerably filamentary and fills the entire SNR (see also Reach *et al.* 2005). A detailed inspection shows that, in some areas, there is a good spatial correlation between the two, whereas in other areas, the correlation is less significant. The detection of both [Fe II] and H$_2$

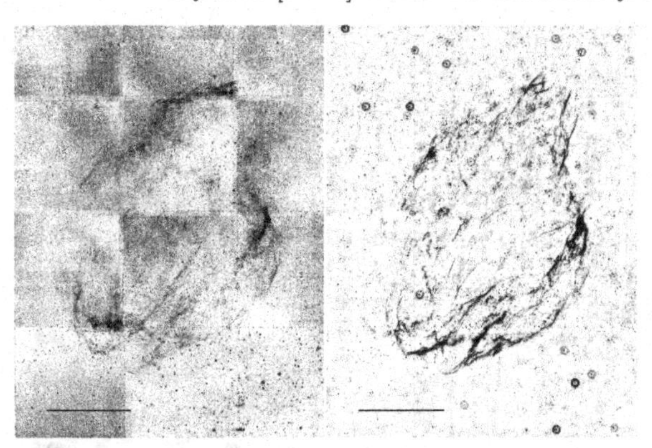

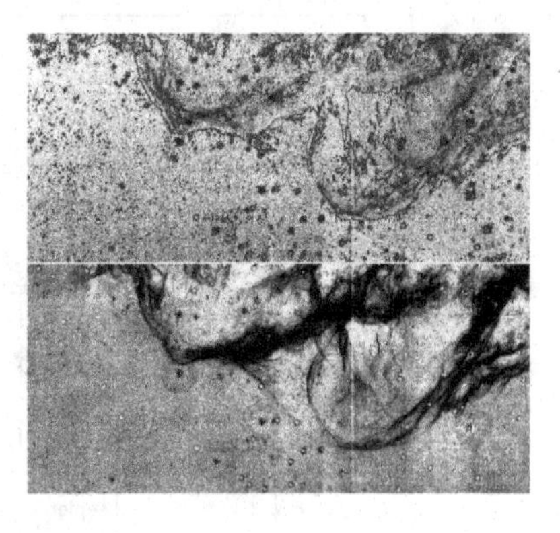

Figure 3. UWIFE [Fe II] 1.644 μm (left) and UWISH2 H_2 2.122 μm (middle) images of W44. The scale bar corresponds to 10′. The images on the right show zoomed-in views of the southern area in the [Fe II] (top) and H_2 (bottom) emissions, respectively. The contours of the H_2 emission are overlaid on the top right [Fe II] image.

emission lines is consistent with the general consensus that a molecular cloud is clumpy, being composed of dense clumps embedded in a rather diffuse interclump gas. An SNR produced inside a MC expands into the interclump medium while engulfing dense clumps. In late stages, the SNR shock in the interclump medium becomes radiative, so that the SNR becomes surrounded by a fast-expanding ($\sim$ 100 km s^{-1}) atomic shell, which is observable in HI 21-cm line (e.g., Koo & Heiles 1995; Chevalier 1999). The [Fe II] emission in such SNRs can originate in two regions; either from radiative atomic shocks in the interclump medium or from reflected shocks in the SNR shell generated by the interaction with molecular clumps. A detailed, comparative study of [Fe II] and H_2 emissions should reveal more detailed information about the structure of molecular clouds.

Young SNRs bright in [Fe II] emission. There are also young SNRs bright in [Fe II] emission. Prototypical ones are Cas A, G11.2−0.3, RCW 103, and W49B, all of which are core-collapse SNRs (CCSNRs) with a central stellar remnant, except W49B, where no central source has been detected. One way to infer the environment of these [Fe II]-bright young SNRs is to inspect diagrams such as Figure 4, where we compare the radii and ages of young SNRs. (For a discussion of SN types of young SNRs, see Chevalier 2005.) As in the figure, there is a trend of increasing SNR size with age, but with a large scatter, indicating diverse SN environments and also possibly diverse SN explosion energies. Among Type Ia SNRs, SN 1006 is located about 500 pc above the Galactic plane where the ambient density is ~ 0.05 cm^{-3}, whereas the Kepler SNR is interacting with a relatively dense ($\gtrsim 1$ cm^{-3}) medium. Young CCSNRs are interacting either with CSM or a wind bubble created in the main-sequence lifetime of their progenitors. Cas A, which is Type IIb SN, for example, is interacting with a dense red supergiant (RSG) wind. The solid line is a model for Cas A from Chevalier & Oishi (2003), but assuming $\dot{M}_w = 3 \times 10^{-5} M_\odot \mathrm{yr}^{-1}$, $v_w = 15$ km s^{-1}, $M_{\rm ej} = 5 M_\odot$, and $E_{\rm SN} = 10^{51}$ ergs where $\dot{M}_w$ is the wind mass-loss rate, v_w is the wind speed, $M_{\rm ej}$ is the ejecta mass, and $E_{\rm SN}$ is the explosion energy. Cas A follows this "Cas A-like" line as long as it continues to interact with the dense CS wind. There are CCSNRs which fall well below the Cas A-like line, i.e., G11.2−0.3, RCW 103, and G292.0+1.8. Note that the first two SNRs are bright in [Fe II] lines. (G292.0+1.8 has not been observed in [Fe II] emission.) Their relatively small sizes could be due to either a small explosion energy ($< 10^{50}$ ergs) or/and dense CSM. The strong [Fe II] lines in these SNRs suggest that it is more likely because of dense CSM and that the CSM is much denser than that of Cas A. Hence, they are likely the

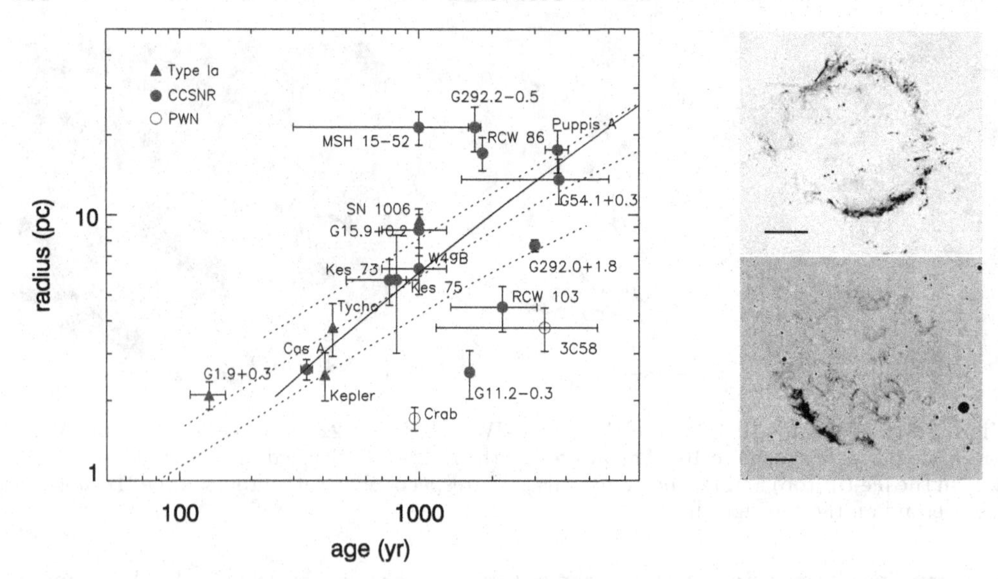

Figure 4. (Left) Radius versus age of young SNRs. The dotted lines are models for SNRs in uniform ambient media of $n_0 = 1$, 0.1, and 0.01 cm^{-3}, respectively (Truelove & McKee 1999; $n = 7$ ejecta model with $M_{\rm ej} = 5\ M_\odot$ and $E_{\rm SN} = 10^{51}$ ergs). The solid line is for an SNR in the RSG wind case (Chevalier& Oishi 2003; see text for the parameters of the model.) Note that the ones marked by empty circles are pulsar wind nebulae, so they do not represent true sizes of SNRs. (Right) [Fe II] images of Cas A and G11.2-0.3 from top to bottom. The scale bars correspond to 1 pc at the distances of the SNRs.

remnants of massive SN IIL/b or SN IIn. On the other hand, there are remnants much larger than Cas A: MSH 15−52, G292.2−0.5, and RCW 86. These remnants might have exploded inside a large bubble, and thus, they are candidates for SN Ib/c, although a large SN explosion energy ($> 10^{52}$ ergs) could be another possibility. Figure 4 suggests that all four [Fe II]-bright young CCSNRs mentioned at the beginning of this paragraph are SN IIL/b candidates interacting with dense CSM. (For W49B, however, the bipolar Type Ib/c SN origin has been suggested. See Lopez *et al.* 2013 and references therein.)

The [Fe II] emission in these young SNRs originate from both shocked CSM and shocked SN ejecta. In Cas A, it is well known that there are two types of knots detected in the visible waveband: quasi-stationary flocculi (QSFs), which are dense CS knots moving at a few hundred kilometers per second, and fast moving knots (FMKs), which are metal-rich SN ejecta knots moving at several thousand kilometers per second. Lee, Y.-H. *et al.* (this volume) show that there are also fast-moving [Fe II] knots that lack other metallic lines, which could be pure Fe ejecta synthesized in the innermost SN region. In G11.2−0.3, which is known as a cousin of Cas A because of its similar morphology (Fig. 4), the knots in the central area have radial velocities of $\sim 1,000$ km s^{-1}, which suggests that they might be SN ejecta (Moon *et al.* 2009). Again, their spectra do not show metallic lines other than Fe. The bright filament in the southeast of G11.2−0.3, on the other hand, appears to be composed of mostly dense CSM. For the [Fe II] emission features in RCW 103 and W49B, detailed spectroscopic studies are yet to be performed (cf. Oliva *et al.* 1999; Keohane *et al.* 2007).

2.3. *Some Issues*

[Fe II] - H_2 reversal. Since the early days of NIR observations of SNRs, it has been known that there are SNRs with H_2 filaments lying beyond [Fe II] filaments, i.e., further out from the SNR center, which is not easily explained by shock models (Graham *et al.*

1991; Oliva, Moorwood, & Danziger 1990; Burton & Spyromilio 1993). We now have more sources showing similar patterns, e.g., G11.2-0.3, W49B, and 3C396 (Koo *et al.* 2007; Keohane *et al.* 2007; Lee *et al.* 2009). W44 in Figure 3 is another example. Hence, we need an explanation for this "[Fe II]-H_2 reversal"; i.e., we need to know what the exciting mechanisms of the H_2 emission is and how they excite the H_2 gas beyond the SNR. Some proposed mechanisms are fluorescent UV excitation, X-ray heating, magnetic precursors, and reflected shocks, but high-resolution NIR spectroscopic studies are needed to address the issue.

[Fe II] luminosity as an SN rate indicator . Can the [Fe II] 1.257 or 1.644 μm luminosities be used as an indicator of galactic SN rates? Several groups have addressed this for starburst galaxies and have derived a conversion factor of SN rate = 0.03–0.1 $L_{[FeII]}/10^6 L_\odot$ yr^{-1} where $L_{[FeII]}$ is the [Fe II] 1.644 μm luminosity (Morel *et al.* 2002; Alonso-Herrero *et al.*2003; Rosenberg *et al.* 2012). In the case of nearby starburst galaxies M82 and NGC 253, however, 70%–80% of the [Fe II] emission is known to be diffuse emission of unknown origin, although it is speculated to be related to the SN activity therein (Greenhouse *et al.* 1997; Alonso-Herrero *et al.* 2003). In the Galaxy, the [Fe II]-bright SNRs represent 20%-30% of the *known* $\sim$ 300 SNRs, which occupy a small fraction of the entire set of SNRs present in the Galaxy. It is clear that we need to have better understanding of the population of [Fe II]-bright SNRs and also the origin of the diffuse [Fe II] emission to obtain a more reliable relation between $L_{[FeII]}$ and the SN rate in galaxies.

3. Dust IR Emission from SNRs

3.1. *Dust Heating in SNRs Interacting with MCs*

Dust in SNRs can be heated either collisionally or radiatively. In SNRs with fast, non-radiative shocks, dust grains are heated by collisions with gas particles, mainly electrons, in a hot plasma behind the shocks (e.g., Dwek *et al.* 2008). Many SNRs have mid- and far-IR morphology almost identical to that of X-ray, which suggests that the IR emission in these SNRs is primarily from collisionally-heated dust grains.

SNRs interacting with MCs generally have IR morphology different from the X-ray morphology; they appear shell-like in IR whereas they are centrally brightened in X-rays (Fig. 5). The dust in these SNRs cannot be collisionally-heated by X-ray emitting gas. Instead the likely source of the heating is radiation from the shock front. In radiative

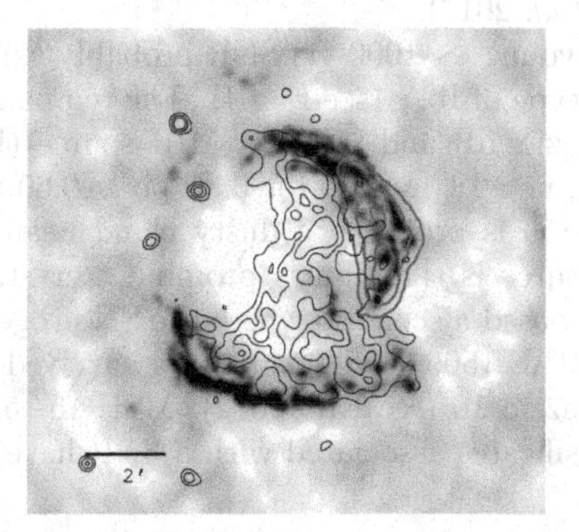

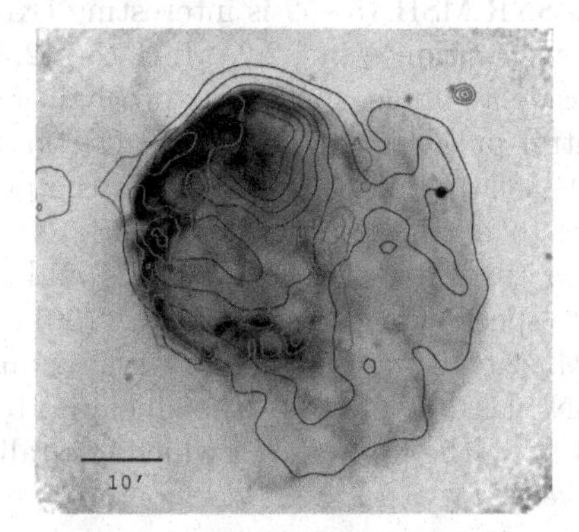

Figure 5. (Left) ATCA 20-cm image of Kes 17 with overlaid XMM 0.2–12 keV X-ray (thin) and AKARI 65 μm (thick) contours. (See also Lee *et al.* 2011.) (Right) VLA 20-cm image of IC 443 with overlaid ROSAT 0.2–2.4 keV X-ray (thin) and AKARI 90 μm (thick) contours.

shocks, the UV radiation from the cooling postshock gas could be much stronger than the general interstellar radiation field (e.g., McKee *et al.* 1987). This UV radiation, dominated by trapped Lyα photons, heat the dust in the cooling layer, and, subsequently, the infrared radiation from these hot dust grains heats the dust at larger column densities (e.g., Hollenbach *et al.* 1979). The far-IR bright regions in MC-interacting SNRs (Fig. 5) are probably where the radiation field is strong and the ambient density is high. Andersen *et al.* (2011) carried out a systematic study of the dust emission from MC-interacting SNRs found in the GLIMPSE survey, and derived dust temperatures of 29–66 K from the Spitzer MIPS spectral energy distribution (60–90 μm). There could be a dust component at a lower temperature, however, because the SNRs interacting with MCs are bright in the far-IR waveband beyond the MIPS coverage (e.g., see Lee *et al.* 2011). A systematic study of MC-interacting SNRs, including the AKARI and Herschel far-IR data, will be useful to understand the heating mechanisms and also the processing of dust grains in these SNRs.

3.2. *Star Dust in SNRs*

SN dust in young CCSNRs. The dense, metal-rich, cooling SN ejecta can effectively provide an environment for dust to condense. In the high-redshift galaxies, where low-mass stars do not have enough time to evolve to AGB stars, SNe could be the main contributors of dust, depending on the dust yield (Dwek & Cherchneff 2011). Theoretical studies have shown that as much as 1 $M_\odot$ of different dust species can form in SN IIP with massive H envelopes, whereas in SN IIL/b or SN Ia with little or no H envelopes only a limited amount of dust could form (Nozawa *et al.* 2010; Nozawa *et al.* 2011).

In observational studies, however, only a very small amount of dust in SNe has been detected, i.e., $\lesssim 10^{-3} M_\odot$. (see Gall *et al.* 2011 and references therein). It is only toward the LMC SN 1987A and some young Galactic SNRs where a significant amount of SN dust has been detected: In 1987A, Matsuura *et al.* (2011) reported detection of 0.4–0.7 $M_\odot$ of dust, whereas, in the Galaxy, 0.1–0.2 $M_\odot$ of dust has been detected in Cas A and the Crab nebula which are SNIIb and SN IIP(?), respectively. In another SN IIP candidate, G54.1+0.3, a dust ring of 0.58–0.86 $M_\odot$ has been detected around its pulsar wind nebula, but the nature of the ring is not yet conclusively identified (Koo 2012).

Crystalline silicate dust in MSH 15−52. Essentially all dust grains in the ISM are amorphous. Crystalline silicate dust grains have been found mainly in evolved stars and young stellar objects, indicating that they form in situ in circumstellar disks and/or outflows of these objects (Henning 2010). In this regard, the detection of crystalline silicates in the SNR MSH 15−52 is interesting (Koo *et al.* 2011).

As we mentioned in § 2.2, MSH 15−52, is a young ($\sim$ 1000 yr) SNR probably expanding inside a bubble, suggesting progenitor SN type of Ib/c (see Fig. 4). The remnant has a central pulsar, and there is an O star (Muzzio 10) and a bright MIR source (IRAS 15099−5856) lying very close from the pulsar, i.e., 18″ & 31″ (or 0.35 pc & 0.60 pc at 4 kpc) to north, respectively. IRAS 15099−5856 is probably a dusty cloud heated by Muzzio 10. What is special to this mid-IR source is that it has prominent crystalline silicate spectral features. Koo *et al.* (2011) proposed a scenario where the SN progenitor and Muzzio 10 were in a binary system and IRAS 15099−5856 is a CSM survived from the SN blast wave due to the shielding by Muzzio 10 (see Koo 2012). MSH 15−52 appears to be the first case in which crystalline silicates associated with a SNR have been observed.

Acknowledgments I wish to thank Lee, Y.-H., Jeong, I.-G, and Moon, D.-S. for their help with figures. My research is supported by Basic Science Research Program through the National Research Foundation of Korea (NRF) funded by the Ministry of Education, Science and Technology (NRF-2011-0007223).

References

Allen, M. G., Groves, B. A., Dept, M. A., Sutherland, R. S., & Kewley, L. J. 2008, *ApJS*, 178, 20
Alonso-Herrero, A., Rieke, G. H., Rieke, M. J., & Kelly, D. M. 2003, *ApJ*, 125, 1210
Andersen, M., Rho, J., Reach, W. T., Hewitt, J. W., & Bernard, J. P. 2011, *ApJ*, 742, 7
Burton, M. & Spyromilio, J. 1993, *Proceedings of the Astronomical Society of Australia*, 10, 327
Chevalier, R. A. 1999, *ApJ*, 511, 798
Chevalier, R. A. 2005, *ApJ*, 619, 839
Chevalier, R. A. & Oishi, J. 2003, *ApJL*, 593, L23
Dwek, E., Arendt, R. G., Bouchet, P., *et al.* 2008, *ApJ*, 676, 1029
Dwek, E. & Cherchneff, I. 2011, *ApJ*, 727, 63
Froebrich, D., Davis, C. J., Ioannidis, G., *et al.* 2011, *MNRAS*, 413, 480
Gall, C., Hjorth, J., & Andersen, A. C. 2011, *A&ARv*, 19, 43
Graham, J. R., Wright, G. S., Hester, J. J., & Longmore, A. J. 1991, *AJ*, 101, 175
Greenhouse, M. A., Satyapal, S., Woodward, C. E., *et al.* 1997, *ApJ*, 476, 105
Henning, T. 2010, *ARAA*, 48, 21
Hollenbach, D. J., Chernoff, D. F., & McKee, C. F. 1989, *Infrared Spectroscopy in Astronomy*, 290, 245
Hollenbach, D. & McKee, C. F. 1979, *ApJS*, 41, 555
Keohane, J. W., Reach, W. T., Rho, J., & Jarrett, T. H. 2007, *ApJ*, 654, 938
Koo, B.-C. 2012, *Publication of Korean Astronomical Society*, 27, 225
Koo, B.-C. & Heiles, C. 1995, *ApJ*, 442, 679
Koo, B.-C., McKee, C. F., Suh, K.-W., *et al.* 2011, *ApJ*, 732, 6
Koo, B.-C., Moon, D.-S., Lee, H.-G., Lee, J.-J., & Matthews, K. 2007, *ApJ*, 657, 308
Lee, H.-G., Moon, D.-S., Koo, B.-C., *et al.* 2011, *ApJ*, 740, 31
Lee, H.-G., Moon, D.-S., Koo, B.-C., Lee, J.-J., & Matthews, K. 2009, *ApJ*, 691, 1042
Lopez, L. A., Ramirez-Ruiz, E., Castro, D., & Pearson, S. 2013, *ApJ*, 764, 50
Matsuura, M., Dwek, E., Meixner, M., *et al.* 2011, *Science*, 333, 1258
McKee, C. F., Hollenbach, D. J., Seab, G. C., & Tielens, A. G. G. M. 1987, *ApJ*, 318, 674
Moon, D.-S., Koo, B.-C., Lee, H.-G., *et al.* 2009, *ApJ*, 703, L81
Morel, T., Doyon, R., & St-Louis, N. 2002, *MNRAS*, 329, 398
Mouri, H., Kawara, K., & Taniguchi, Y. 2000, *ApJ*, 528, 186
Nozawa, T., Kozasa, T., Tominaga, N., *et al.* 2010, *ApJ*, 713, 356
Nozawa, T., Maeda, K., Kozasa, T., *et al.* 2011, *ApJ*, 736, 45
Oliva, E., Moorwood, A. F. M., & Danziger, I. J. 1989, *A&A*, 214, 307
Oliva, E., Moorwood, A. F. M., & Danziger, I. J. 1990, *A&A*, 240, 453
Oliva, E., Moorwood, A. F. M., Drapatz, S., Lutz, D., & Sturm, E. 1999, *A&A*, 343, 943
Pradhan, A. K. & Nahar, S. N. 2011, *Atomic Astrophysics and Spectroscopy* Cambridge University Press: Cambridge and New York
Reach, W. T., Rho, J., & Jarrett, T. H. 2005, *ApJ*, 618, 297
Rosenberg, M. J. F., van der Werf, P. P., & Israel, F. P. 2012, *A&A*, 540, A116
Seward, F. D., Harnden, F. R., Jr., Murdin, P., & Clark, D. H. 1983, *ApJ*, 267, 698
Truelove, J. K. & McKee, C. F. 1999, *ApJS*, 120, 299

Discussion

PODSIADLOWSKI: What is the estimated mass of the O-star companion in MSH15-52?

KOO: It's spectral type is O4.5 III and the mass is 45 MSun

Supernova Environmental Impacts
Proceedings IAU Symposium No. 296, 2013
A. Ray & R. A. McCray, eds.

doi:10.1017/S1743921313009502

The First Systematic Multi-wavelength Survey of Extragalactic Supernova Remnants

I. Leonidaki[1], P. Boumis[1] and A. Zezas[2,3,4]

[1]Institute of Astronomy, Astrophysics, Space Applications & Remote Sensing, National Observatory of Athens, I. Metaxa & Vas. Pavlou St., Palaia Penteli GR-15236 Athens, Greece
email: ileonid@noa.gr, ptb@noa.gr

[2]University of Crete, Physics Department & Institute of Theoretical & Computational Physics, 71003 Heraklion, Crete, Greece

[3]Foundation for Research and Technology-Hellas, 71110 Heraklion, Crete, Greece

[4]Harvard-Smithsonian Center for Astrophysics, 60 Garden St., Cambridge, MA 02138, USA
email: azezas@cfa.harvard.edu, azezas@physics.uoc.gr

Abstract. We present the largest sample of multi-wavelength Supernova Remnants (SNRs) in six nearby galaxies, based on *Chandra* archival data and deep optical narrow-band Hα and [S II] images as well as spectroscopic observations. We have identified 37 X-ray selected thermal SNRs, 30 of which are new identifications and $\sim$ 400 optical SNRs, for 67 of which we spectroscopically verified their shock-excited nature. We discuss the properties of the X-ray/optically detected SNRs in different types of galaxies and hence different environments, in order to address their dependence on their Interstellar Medium (ISM). We also discuss the SNR populations in the context of the star formation rate of their host galaxies. We cross-correlate parameters of the optically detected SNRs with parameters of coincident X- ray emitting SNRs in order to understand their evolution and investigate possible selection effects.

Keywords. supernova remnants, ISM, star formation, galaxies

---

1. Introduction

Supernova Remnants (SNRs) provide a significant fraction of the mechanical energy that heats, shapes, and chemically enriches the Interstellar Medium (ISM). Therefore, SNRs can yield significant information on the global properties of a galaxy's ISM (Blair & Long 2004). Furthermore, being the endpoints of core-collapse massive stars ($M > 8M_{\odot}$) they can be used as proxies for measurements of massive star formation rate (SFR) and studies of stellar evolution (Condon & Yin 1990).

About 274 SNRs are known to exist in our Galaxy (Green 2009) and a large number of them has been studied in detail (e.g. radio: Green 2009; optical: Boumis *et al.* 2009, Fesen & Milisavljevic 2010; X-rays: Reynolds *et al.* 2009, Slane *et al.* 2002; infrared: Reach *et al.* 2006). However, these studies are impeded by Galactic absorption and distance uncertainties, hampering the investigation of SNRs in a wide variety of environments. On the other hand, extragalactic studies of SNRs offer several advantages: we can achieve larger samples in determined distances with much fewer observations, they cover a broader range of metallicities and ISM parameters than our Galaxy while internal Galactic absorption effects are minimized, especially on face-on galaxies.

Detecting large samples of SNRs in a multi-wavelength context can provide several key aspects of the physical processes taking place during their evolution. For example, the blast waves of newly formed SNRs can heat the material behind the shock front to temperatures up to 10^8 K producing thermal X-rays. Synchrotron emission in radio wavelengths is radiated from the vicinity of the shock as well as from the cooling filaments and is easily detectable throughout the life of the remnant (e.g. Charles & Seward 1995). Optical filaments are a sign of older SNRs since they form in the cooling regions behind the shock (e.g. Stupar & Parker 2009). Therefore, multi-wavelength studies can surmount possible selection effects inherent in 'monochromatic' samples of SNRs and provide a more complete picture of their nature and evolution as well as their interplay with the ISM and their correlation with star forming activity.

2. Sample of Galaxies

We have embarked in an extensive multi-wavelength investigation of the SNR populations in six nearby galaxies (NGC 2403, NGC 3077, NGC 4214, NGC 4395, NGC 4449 and NGC 5204), involving X-ray and optical data. These galaxies were selected from the Third Catalog of Bright Galaxies (RC3; de Vaucouleurs *et al.* 1995) to fulfill the following criteria: **(a)** to be late-type (T > 4, Hubble Type); **(b)** close (< 5 Mpc) in order to minimize source confusion; **(c)** at low inclination ($\leqslant 60°$) in order to minimize internal extinction and projection effects and **(d)** above the Galactic plane ($|b| > 20°$). From this pool of objects drawn from the aforementioned selection criteria, we selected galaxies with *Chandra* archival data with exposure times long enough to achieve a uniform detection limit of 10^{36} erg s^{-1}.

3. X-ray SNR investigation

The first step of this attempt (Leonidaki *et al.* 2010) was to detect in a self-consistent way X-ray thermal SNRs, that is sources with soft, thermal spectrum (kT $\leqslant$ 2 keV). For that reason we analyzed *Chandra* archival data for the six galaxies in our sample. We detected 244 discrete X-ray sources down to a limiting flux of 10^{15} erg s^{-1}cm^{-2}. The X-ray colors of all the detected sources were calculated and illustrated on plots one for each galaxy (see Leonidaki *et al.* 2010). On the same plots we added grids for power-law and thermal plasma models for different values of temperature (kT), absorbing H I column density (N_H), and photon index Γ. Sources that appear to have temperatures below 2 keV and mainly lie on the thermal grid, as well as those which are consistent with the thermal grid within their error bars, are *candidate SNRs.* In order to verify the thermal or non-thermal emission of these sources, we performed a spectroscopic investigation for all candidate X-ray SNRs with adequate number of counts. Sources with spectra fitted only with thermal plasma model(s) are considered *thermal X-ray SNRs.* In total, 37 X-ray thermal SNRs were identified by this study, 30 of which are new discoveries. Since core-collapse SNRs are the end points of the evolution of the most massive stars, they are good indicators of the current Star Formation Rate (SFR). In order to investigate the X-ray properties of SNRs in different star-forming environments, we correlated the average X-ray luminosity of the SNRs in each galaxy against the total FIR luminosity of their host galaxy. As expected, we do not see any trend between those two properties since the latter is a global indicator. What is intriguing though is that SNRs in irregular galaxies tend to be more luminous than those in spirals. This indicates a difference of the SNR population characteristics between the two samples. We attribute that either to the typically lower metallicities of irregular galaxies than those in spirals or to the higher local densities of the ISM which is often the case in irregular galaxies.

Furthermore, someone would expect a linear relation between the number of X-ray-selected SNRs and SFR. To verify this connection, we plot the number of SNRs against the integrated FIR luminosity of each galaxy. A linear correlation coefficient of 0.72 shows that this is a significant correlation.

In order to examine our results in the context of SNR populations detected in other galaxies, we compare the luminosity distributions of X-ray SNRs in different types of galaxies with the number of X-ray-detected SNRs in the studied sample. Therefore we test if the numbers of SNRs in the irregular galaxies in our sample are consistent with those expected, by simply rescaling the SNR X-ray luminosity Functions (XLFs) of the Magellanic Clouds. We find that the numbers of SNRs in irregular galaxies are more consistent with an MC-like SNR X-ray luminosity function, while those of spiral galaxies are more consistent with the SNR-XLF of the spiral M33.

4. Optical SNR investigation

The second attempt of this multiwavelength, extragalactic SNR survey was based on a detailed optical spectro-photometric study (Leonidaki *et al.* 2013). We obtained optical images through [S II] 6716 & 6731 Å and Hα filters for the galaxies in our sample, using the 1.3m Skinakas (Crete, Greece) telescope. We performed photometry on all detected sources and the SNR classification was based on the well-established criterion of [S II]/H$\alpha >$ 0.4, pioneered by Mathewson & Clarke (1973). This study revealed ~400 photometric SNRs down to a limiting Hα flux of 10^{-15} erg sec^{-1} cm^{-2}. We extracted long-slit and multi-slit spectra for a large number (134) of photometric SNRs using the 1.3m Skinakas (Crete, Greece) and 4m Mayall (Arizona, USA) telescopes. Using the aforementioned emission line flux criterion of [S II]/H$\alpha >$ 0.4 for SNRs, 67 sources spectroscopically verified their shock-excited nature.

We calculated various emission line ratios (e.g. log(Hα/([N II] 6548, 6584 Å), log(Hα/([S II] 6716, 6731 Å) and [S II] (6716)/[S II] (6731)Å) of the spectroscopically detected SNRs in order to place them in the diagnostic plots of Sabbadin *et al.* (1977) and Garcia *et al.* (1991) (see Leonidaki *et al.* 2013). These diagrams can help us distinguish the excitation mechanism of the emission lines (photoionization for HII regions and Planetary Nebulae or collisional excitation for SNRs). The most interesting result at these plots is where the Hα/[N II] ratio is present: there is a trend for irregular galaxies to have lower [N II]/Hα ratios than those in spirals. Since [N II]/Hα is a very sensitive metallicity indicator (more than [S II]/Hα), mainly due to its secondary nucleosynthesis, this indicates a difference of the SNR populations between different types of galaxies due to metallicity.

In order to investigate the optical properties of SNRs in different star-forming environments and derive safe conclusions on their connection with SFR, we plot the number of photometric SNRs against the integrated Hα luminosity of each galaxy. We find a linear correlation coefficient of 0.87 which shows that this is a significant correlation.

Mining for optical SNRs within the six nearby galaxies of our sample revealed 18 sources that are associated with X-ray selected SNRs from Leonidaki *et al.* (2010). In order to examine whether the optical properties of SNRs are good predictors of X-ray SNRs, we compared their Hα luminosities as well as their [S II]/Hα ratios with their X-ray luminosities derived in Leonidaki *et al.* (2010). However, we do not find any statistically significant correlation between these properties. This indicates the existence of different materials in a wide range of temperatures: the X-ray emission originates from hot material behind the shock front and long cooling timescales while the Hα emission comes from cooling regions of dense recombining gas around the edges of the remnant and short cooling timescales.

In the context of multi-wavelength SNR correlation, we compared the number of optical, X-ray and radio-selected SNRs in the form of Venn diagrams for all galaxies in our sample. This comparison not only provides information on the different evolutionary stages of these sources but also depicts various selection effects. For example, the poor match rates between optical and X-ray/radio SNRs could be due to the lack of deep radio surveys for half of our galaxies. The trend of detecting more easily older optical SNRs gives rise to the large difference in the match rate between optical and X-ray SNRs. Furthermore, optical searches are more likely to detect SNRs located in low-diffused regions, while radio/X-ray searches are more likely to detect SNRs in regions of high optical confusion. All these facts highlight the importance of multi-wavelength surveys for the study of extragalactic SNR populations.

5. Conclusions

This comprehensive multi-wavelength investigation of SNRs in six nearby galaxies revealed 37 X-ray thermal SNRs, 30 of which are new identifications. In the optical band ~400 photometric SNRs were detected (~350 of which are new identifications), 67 of which spectroscopically verified their nature. Some of the most interesting results derived from this study are: (a) We find that X-ray selected SNRs in irregular galaxies tend to be more luminous than those in spirals, i (b) There is a trend of SNRs in irregular galaxies to present lower [N II]/Hα ratios than those in spiral galaxies, (c) We do not see a correlation between L_X - $L_{H\alpha}$ or L_X - [S II]/Hα, possibly indicating the presence of matter in a wide range of temperatures, (d) There is a linear relation between the number of luminous X-ray/optical SNRs and SFR and (e) There is indication for different luminosity distributions in the SNR populations between spiral and irregular galaxies.

References

Blair, W. P. & Long, K. S. 2004, *ApJS*, 155, 101

Boumis, P., Xilouris, E. M., Alikakos, J., Christopoulou, P. E., Mavromatakis, F., Katsiyannis, A. C., & Goudis, C. D. 2009, *A&A*, 499, 789

Charles, P. A. & Seward, F. D. 1995, *Exploring the X-ray Universe (Cambridge Univ. Press)*

Condon, J. J. & Yin, Q. F. 1990, *ApJ*, 357, 97

de Vaucoulers, G., de Vaucoulers, A., Corwin, H. G., Buta, R. J., Paturel, G., & Fouque, P. 1995, *Third Reeference Catalog of Bright Galaxies, Springer, New York*

Fesen, R. A. & Milisavljevic, D. 2010, *AJ*, 140, 1163

Garcia-Lario, P., Manchado, A., Riera, A., Mampaso, A., & Pottasch, S. R. 1991, *A&A*, 249, 223

Green, D. A. 2009, *Bulletin of the Astronomical Society of India*, 37, 45. *(See: arxiv:0905.3699)*

Leonidaki, I., Zezas, A., & Boumis, P. 2010, *ApJ*, 725, 842

Leonidaki, I., Boumis, P., & Zezas, A. 2013, *MNRAS*, 429, 189

Mathewson, D. S. & Clarke, J. N. 1973, *ApJ*, 180, 725

Reach, W. T., Rho, J., Tappe, A., Pannuti, T. G., Brogan, C. L., Churchwell, E. B., Meade, M. R., Babler, B., Indebetouw, R., & Whitney, B. A. 2006, *AJ*, 131, 1479

Reynolds, S. P., Borkowski, K. J., Green, D. A., Hwang, U., Harrus, I., & Petre, R. 2009, *ApJ*, 695, 149

Sabbadin, F., Minello, S., & Bianchini, A. 1977, *A&A*, 60, 147

Slane, P., Smith, R. K., Hughes, J. P., & Petre, R. 2002, *ApJ*, 564, 284

Stupar, M. & Parker, Q. A. 2009, *MNRAS*, 394, 1791

Discussion

MILISAVLJEVIC: Did any of your optical spectra show evidence of broad features indicative of young remnants?

LEONIDAKI: Oxygen rich SNRs with broad emission lines (e.g. [OIII] 5007 A) were excluded from our SNR list since they have different detection techniques.

Supernova Environmental Impacts
Proceedings IAU Symposium No. 296, 2013
A. Ray & R. A. McCray, eds.

doi:10.1017/S1743921313009514

Thermal X-ray Spectra of Supernova Remnants

Patrick Slane

Harvard-Smithsonian Center for Astrophysics
60 Garden Street
Cambridge, MA 02138 USA
email: slane@cfa.harvard.edu

Abstract. The fast shocks that characterize supernova remnants heat circumstellar and ejecta material to extremely high temperatures, resulting in significant X-ray emission. The X-ray spectrum from an SNR carries a wealth of information about the temperature and ionization state of the plasma, the density distribution of the postshock material, and the composition of the ejecta. This, in turn, places strong constraints on the properties of the progenitor star, the explosive nucleosynthesis that produced the remnant, the properties of the environment into which the SNR expands, and the effects of particle acceleration on its dynamical evolution. Here I present results from X-ray studies SNRs in various evolutionary states, and highlight key results inferred from the thermal emission.

Keywords. (ISM:) supernova remnants, X-rays: ISM

---

1. Introduction

The very fast shocks formed in young and middle-aged supernova remnants (SNRs) act to heat matter to temperatures exceeding many millions of degrees. As a result, these systems are copious emitters of thermal X-rays. This emission, characterized by bremsstrahlung continuum accompanied by line emission from recombination and de-excitation in the ionized gas, can originate from three distinct regions of the SNR: (1)behind the forward shock (FS), where interstellar or circumstellar material is swept-up and heated; (2)interior to the SNR boundary, where cold ejecta are heated by the reverse shock (RS); and (3)outside a central pulsar wind nebula (PWN), if one exists, where the slow-moving central ejecta are heated by the expanding nebula. Thermal X-ray spectra in an SNR thus provide crucial information on the surrounding environment, which may have been modified by strong stellar winds from the progenitor, as well as the ejecta that bears the imprint of the stellar and explosive nucleosynthesis in the progenitor star.

Studies of the thermal X-ray emission from SNRs reveal details on the temperature, density, composition, and ionization state of the shocked plasma. These, in turn, can provide specific information on the density structure for both the ejecta and the surrounding circumstellar material, the nature of the supernova explosion (core-collapse vs. Type Ia), the age and explosion energy of the SNR, the mass of the progenitor star, and the distribution of metals in the post-explosion supernova. Here I present a summary of results from several studies of the thermal spectra of SNRs. This review is brief; space prohibits an in-depth discussion of many aspects and results from this rich field. For an excellent review of X-ray emission from SNRs, including a detailed discussion of the thermal X-ray emission, the reader is referred to Vink *et al.* (2012).

2. X-rays from SNRs

Spectral analysis of thermal X-ray emission can provide measurements of the electron temperature and the relative abundances in the shocked plasma. These quantities are important for studies of the dynamical evolution of the SNR and the composition of the shocked gas. However, the low density environments in which SNRs evolve result in crucial effects that affect the interpretation of the spectra. Of particular importance are the relatively long timescales for electron-ion temperature equilibration and ionization equilibrium.

2.1. *Electron-Ion Temperature Equilibration*

In the simplest picture, and for an ideal gas, all particles that go through the SNR shock attain a velocity $v = 3v_s/4$ where v_s is the shock velocity. Since the associated kinetic temperature scales with mass, the result is that the electron temperature is initially much lower than that of the ions. Because the ions comprise the bulk of the mass swept up by the SNR, it is the ion temperature that characterizes the dynamical evolution. However, the temperature determined from X-ray measurements is that of the electrons. Thus, in connecting X-ray measurements to the SNR evolution, modeling of the electron-ion temperature ratio is required.

On the slowest scales, Coulomb collisions between electrons and ions will bring the populations into temperature equilibrium. More rapid plasma processes may result in faster equilibration, and often the two extreme cases of instantaneous temperature equilibration and Coulomb equilibration are considered in order to investigate the boundaries of the problem. SNR expansion velocities can be used to estimate the ion temperatures for some remnants, and the estimated temperatures are much higher than the observed electron temperatures, suggesting slow temperature equilibration. However, the effects of cosmic-ray acceleration can result in ion temperatures lower than are indicated by the shock velocity (see Section 5), making more direct measurements of both T_e and T_i crucial.

In a small number of cases, measurements of Balmer lines at the forward shock can provide the proton temperature and the degree of electron-ion equilibration, and there is a significant trend for larger T_e/T_p values with lower shock velocities, potentially consistent with electron heating by lower hybrid waves (Ghavamian *et al.* 2007). Future high-resolution X-ray measurements of thermal line broadening offer the promise of direct measurements of both T_e and T_p in SNRs.

2.2. *Ionization Effects*

Because the density of the X-ray emitting plasma in SNRs is extremely low, the ionization state of the gas takes considerable time to reach the equilibrium state that is characteristic of its temperature. The ionization parameter $\tau = n_e t$ determines how far the ionization has progressed. The plasma reaches collisional ionization equilibrium (CIE) for $\tau \gtrsim 10^{12.5}$ cm^{-3} s; for smaller values the plasma is in a nonequilibrium ionization (NEI) state. This is illustrated in Figure 1. The left panel shows a portion of the soft X-ray spectrum for a plasma at a temperature of $10^{6.5}$ K for different values of $n_e t$. Of particular note is the relative strengths of the O VII triplet and the Lyα and Lyβ lines from O VIII. It is clear from this that determination of the plasma temperature based on relative line strengths must thus account for NEI conditions in the plasma. In the right panel, we plot the same spectra folded through the *Chandra* ACIS-S detector response. While the limited spectral resolution provided by CCD detectors obviously obscures details of specific line ratios, spectral fits to the spectrum can easily identify gross differences of the ionization state.

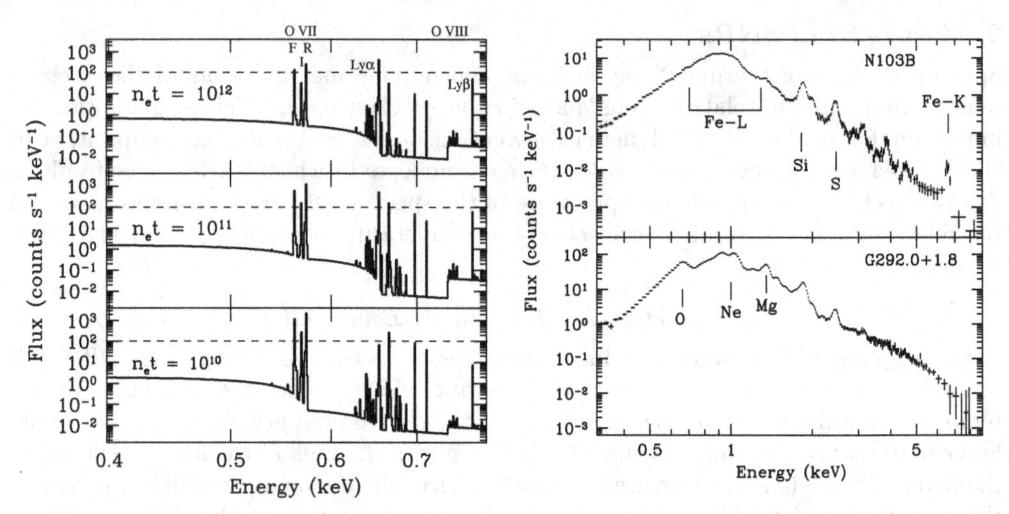

Figure 1. Left: Evolution of X-ray spectrum with ionization age, emphasizing lines from O VII and O VIII. The plasma temperature is $10^{6.5}$ K, and the units for $n_e t$ are cm^{-3} s^{-1}. The dashed line at 10^2 counts s^{-1} keV^{-1} is simply for comparison purposes. Right: Spectra at left folded through the *Chandra* ACIS-S detector response.

The progression of the ionization state in the evolving postshock gas in SNRs is readily observed. Early *Chandra* observations of 1E 0102.2−7219, for example, show that the peak of the O VII emission is found at a smaller radius than that of O VIII, consistent with the expectation that the ionization state of ejecta most recently encountered by the inward-propagating RS lags behind that of the ejecta that have been shocked earlier (Gaetz *et al.* 2000). The same effect is evident in shocked circumstellar material in G292.0+1.8, where the ionization state of the emission directly behind the FS is observed to be considerably lower than that for regions further downstream (Lee *et al.* 2010). Such measurements thus provide constraints on the thermal history of the shocked material as well as on variations in the density structure.

While conditions of underionization are found in many SNRs, recent X-ray observations have identified exactly the opposite situation in several remnants. Kawasaki *et al.* (2005) used the relative intensity of H-like and He-like lines of Ar and Ca from *ASCA* observations of W49B to determine an ionization temperature that is higher than T_e, indicating an *overionized* plasma. Ozawa *et al.* (2009) used *Suzaku* observations to identify a distinct radiative recombination continuum (RRC) feature that confirms this overionized state (Figure 2, left). Using *XMM* observations, Miceli *et al.* (2010) find that the overionization in W49B appears to be related to regions in which rapid expansion and adiabatic cooling have lowered the plasma temperature to a value below its ionization state. Similar results have been observed for IC 443 (Yamaguchi *et al.* 2009). A semi-quantitative analysis by Moriya (2012) suggests that the progenitors of such overionized SNRs may be massive RSG stars for which the wind-driven circumstellar environment is sufficiently dense to promote rapid early ionization followed by a drop in temperature as the SNR expands into the lower density regions of the wind shell. This is illustrated in Figure 2 (right) where the electron density encountered by a 10,000 km s^{-1} shock traveling through a CSM formed by a stellar wind is shown. The dashed line follows regions for which the plasma will reach approximate ionization equilibrium. The shaded region corresponds to mass-loss rates typical of a massive RSG progenitor, and shows that for early times, the plasma may be left in an overionized state. Such conditions

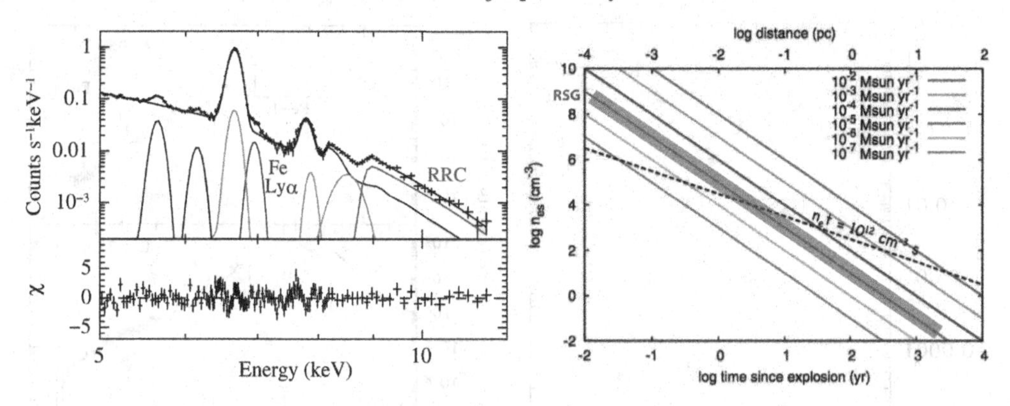

Figure 2. Left: *Suzaku* spectrum for W49B, showing RRC features and enhanced H-like emission from Fe, indicative of overionized plasma. (From Ozawa *et al.* 2009.) Right: Electron density distribution for wind model as a function of time (and distance) traveled by 10,000 km s^{-1} SNR shock. See text for description. (From Moriya 2012. Reproduced by permission of the AAS.)

appear to be connected to the mixed-morphology class of SNRs, making further study of the dynamical conditions leading to such overionized states of particular interest.

3. Studies of SNR Ejecta

The very different stellar evolution histories and explosion processes for Type Ia and core-collapse (CC) SNe result in distinct signatures in the shock-heated ejecta of SNRs. Type Ia events, corresponding to the complete disruption of a C/O white dwarf star, produce more than $0.5M_\odot$ of Fe-group elements, accompanied by a significant contribution of intermediate mass elements. CC SNe, on the other hand, are dominated by materials synthesized during the stellar evolution of the massive progenitor – particularly O – with additional products from explosive nucleosynthesis in the innermost regions surrounding the collapsed core. As illustrated in Figure 3 (left), where we plot the mass distributions for key nucleosynthesis products for characteristic Type Ia and CC events (Iwamoto *et al.* 1999), the former are dominated by Fe while the latter contain much larger amounts of O. For comparison, the total mass of these elements contained in $10M_\odot$ of swept-up material with solar abundances is also shown. Particularly at young ages when the total amount of mass swept up by the FS is not exceedingly high, the thermal X-ray spectra from such remnants provide rich information about supernova ejecta.

In Figure 3 (right), we compare the spectra from N103B (top), a Type Ia SNR (Lewis *et al.* 2003) with that from the CC SNR G292.0+1.8 (bottom). The dominant flux just below 1 keV in N103B is largely from Fe-L emission, characteristic of the large amount of Fe created in such events, while the spectrum from G292.0+1.8 shows strong emission features from O and Ne (Park *et al.* 2004). Identification and modeling of such spectral features has been used to identify the SN type that produced numerous SNRs, notably Kepler's SNR, which Reynolds *et al.* (2007) identified as a Type Ia remnant based on the dominant Si, S, and Fe emission. Subsequent detection and modeling of the Mn and Cr lines in Kepler imply a high-metallicity progenitor star (Park *et al.* 2013).

A particularly important example of SNR "typing" through X-ray spectra is that of SNR 0509−67.5. Hughes *et al.* (1995) used *ASCA* spectra to argue for a Type Ia progenitor for this remnant. By comparing different Type Ia explosion models with spectra from *XMM* and *Chandra* observations, Badenes *et al.* (2008) argued that 0509−67.5 is

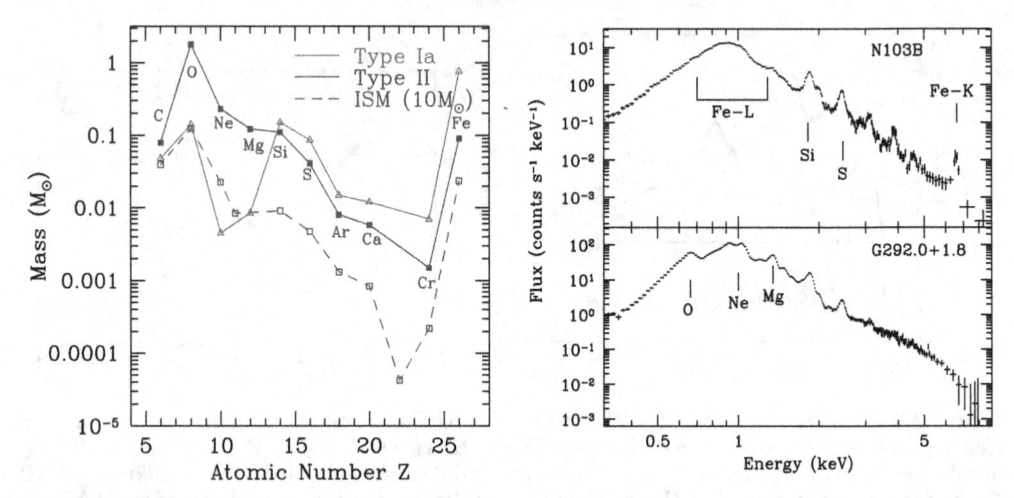

Figure 3. Left: Mass of ejecta components, by atomic number, in representative core-collapse and Type Ia SNe. For comparison, the mass contained in $10 M_{\odot}$ of solar-abundance material is also shown. Right: Thermal X-ray spectra from a Type Ia remnant (N103B) and core-collapse remnant (G292.0+1.8), illustrating the significant difference in Fe and O/Ne content.

the result of an energetic, high-luminosity SN 1991T-like event, a result subsequently confirmed with light echo spectra from the original event (Rest *et al.* 2008).

The early development of SN explosions can imprint signatures on the SNRs that they form. Studies of the spatial distribution of ejecta can thus provide evidence of asymmetries and mixing in these events. X-ray studies of Cas A, for example, show distinct evidence of Fe ejecta in the outermost regions of the remnant (Hughes *et al.* 2000a), despite the expectation that Fe is produced in the regions closest to the remnant core. This large-scale disruption of the ejecta layers has been studied in detail by Hwang and Laming (2012) who performed fits to over 6000 X-ray spectra in Cas A and find distinct examples of regions where Fe is accompanied by other products of incomplete Si burning, along with others that are nearly pure Fe, presumably produced in regions of α-rich freezeout during complete Si burning. They conclude that nearly all of the Fe in Cas A is found outside the central regions of the remnant, apparently the result of hydrodynamic instabilities in the explosion.

The large-scale distribution of SNR ejecta has been investigated for a large number of SNRs by Lopez *et al.* (2011), who find that the thermal X-ray emission from remnants of Type Ia events shows a higher degree of spherical symmetry and mirror symmetry than for CC remnants. This may indicate that CC SNe evolve in more asymmetric environments, or perhaps that the events themselves are asymmetric.

4. Studies of Shocked Circumstellar Material

Remnants of CC SNe initially evolve in the circumstellar environment left by their associated progenitors. For progenitors with significant pre-explosion wind phases, the remnants initially evolve into density profiles with $\rho \propto r^{-2}$. The composition of the circumstellar material from progenitors with strong RSG or WR wind episodes is expected to show the signatures of the CNO cycle, which leads to an environment with an enhanced N/O ratio. Studies of thermal X-ray emission from behind the FS can provide constraints on the shocked CSM, and thus on the late-phase properties of the progenitor star.

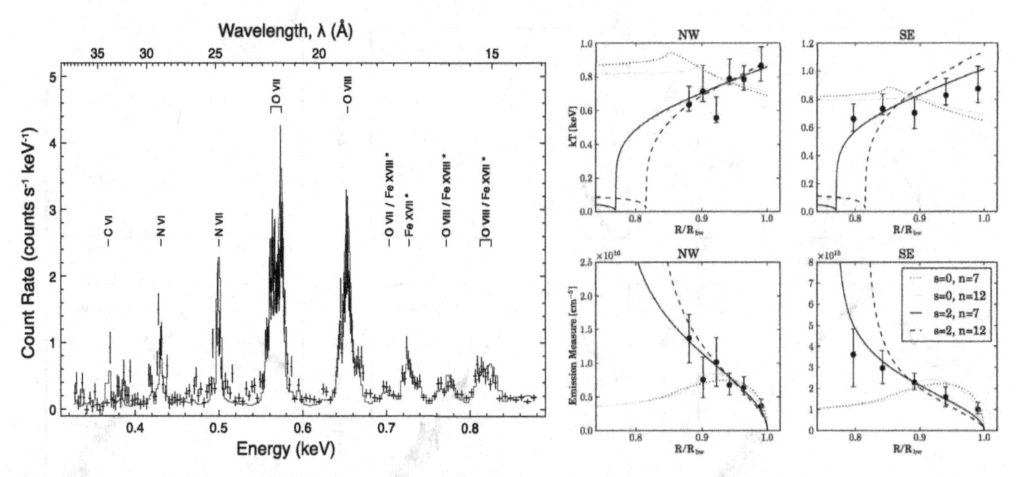

Figure 4. Left: *XMM* RGS spectrum from G296.1−0.5 showing N and O line features from shocked circumstellar wind. (From Castro *et al.* 2011.) Right: Temperature and emission measure profiles of FS regions in G292.0+1.8 compared with power law density models for uniform ($s = 0$) and wind-like ($s = 2$) surroundings, and ejecta profiles typical of Type Ia ($n = 7$ and CC ($n = 12$) events. (From Lee *et al.* 2010.) [Figures reproduced by permission of the AAS.]

Detection of enhanced N in a shocked CSM environment is complicated by both interstellar absorption and the modest spectral resolution provided by typical X-ray CCD detectors. *XMM* spectra of G296.1−0.5 from the MOS and pn detectors reveal emission from regions just behind the FS that indicate a low column density and weak evidence for an overabundance of N and an underabundance of O, as expected from CNO-cycle products found in winds of massive stars. High resolution spectroscopy using the *XMM* RGS (Figure 4, left) confirm these results, clearly establishing the remnant as the result of a CC event from a fairly massive progenitor (Castro *et al.* 2011). Using the similarity solution of Chevalier (2005), the inferred swept-up wind mass is $\sim d_2^{5/2} 19 M_\odot$, and the SNR age is $\sim 2800 E_{51}^{-1/2} d_2^{9/4}$ yr.

G292.0+1.8 is an O-rich SNR with an identified pulsar and PWN, clearly establishing it as the result of a CC event. X-ray spectral studies suggest a progenitor mass of $\sim 20 - 40 M_\odot$ (Hughes & Singh 1994; Gonzalez & Safi-Harb 2003; Park *et al.* 2004), making it likely that the remnant has evolved in the stellar wind density profile of its progenitor. *Chandra* studies of the thermal X-ray emission from the outer regions of the SNR shell (Figure 4, right) reveal a temperature and emission measure structure consistent with predictions of the similarity solutions from Chevalier (2005) for evolution in a medium with $\rho \propto r^{-2}$, with a steep ejecta profile typical of CC events (Lee *et al.* 2010). The overall kinematics are consistent with evolution in an RSG wind comprising a mass of more than 15 $M_\odot$. More recently, the same technique has been applied to the thermal X-ray emission from Cas A, revealing similar evidence of a circumstellar environment dominated by an RSG wind (Lee *et al.* – these proceedings).

5. Constraints on Particle Acceleration

Particle acceleration in SNRs has long been suggested as a primary source for production of Galactic cosmic-rays, and X-ray measurements have provided some of the most important constraints on this process. While much of this evidence is provided by synchrotron radiation, whose presence indicates the presence of multi-TeV electrons, significant information is provided by the thermal X-ray emission as well. In 1E 0102.2−7219, the

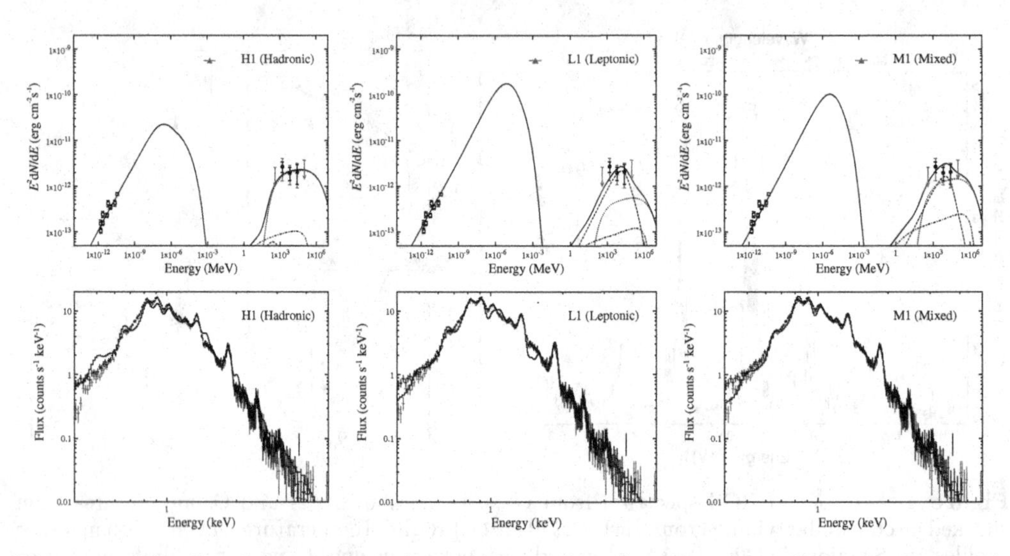

Figure 5. Broadband (upper) and thermal X-ray (lower) spectra from CTB 109 along with models in which the γ-ray emission is dominated by hadrons (left), electrons (center), and a mixture (right). The thermal X-ray spectra provides a best fit for the mixed scenario. (From Castro *et al.* 2012. Reproduced by permission of the AAS.)

temperature derived from X-ray measurements is much lower than that implied by the shock velocity determined from expansion measurements, even accounting for the slowest possible equilibration between the electrons and ions (Hughes *et al.* 2000b). This result is consistent with a picture in which a significant fraction of the shock energy has gone into the acceleration of particles instead of heating the gas. In Tycho's SNR, the separation between the FS and contact discontinuity (and also RS) is much smaller than predicted from dynamical models of the SNR evolution unless a significant amount of energy has been lost to some nonthermal process, such as particle acceleration (Warren *et al.* 2005). It is crucial to note that, since cosmic-ray protons outnumber electrons by a factor of ~ 100, the significant energy in relativistic particles inferred from these studies provides strong evidence for acceleration of cosmic-ray *ions* in SNRs.

Observations of γ-ray emission from SNRs provide additional compelling evidence for particle acceleration. Because γ-rays can be produced by both electrons (though inverse-Compton scattering and nonthermal bremsstrahlung) and protons (though the production of neutral pions, which decay to γ-rays), modeling of the broadband emission is required to determine the origin of the γ-rays, and thus the total efficiency with which the systems are able to accelerate particles. In RX J1713.7$-$3946, hydrodynamical modeling of the SNR evolution that includes particle acceleration, and follows the ionization history of the postshock gas, shows that the *absence* of observed thermal X-ray emission places strong constraints on the ambient density, effectively ruling out significant π^0-decay γ-rays (Ellison *et al.* 2010, 2012) for scenarios with expansion into a uniform medium or wind-driven cavity, although evolution in a medium with dense clumps may provide sufficient target material for a pions to provide a dominant contribution (Inoue *et al.* 2012).

In CTB 109, for which γ-ray emission is also observed, self-consistent modeling (Castro *et al.* 2012) of the evolution can produce broadband spectra that adequately reproduce the observed radio and γ-ray emission for scenarios in which hadrons dominate the γ-rays (Figure 5, left), or in which the γ-rays originate primarily from electrons (Figure 5, center). However, the thermal X-ray emission predicted by both of these models fails to reproduce the observed *XMM* spectrum. The high density required for hadrons to dominate results in an overproduction of high ion states for Mg and Si, while the low density required for electrons to dominate results in an underprediction of the same ion states. A mixed scenario in which the density is sufficiently high for electrons and protons to contribute nearly equally to the γ-ray flux yields excellent agreement with the thermal X-ray emission.

6. Conclusions

As indicated in this brief review, thermal X-ray spectra from SNRs provide information on the nature, environments, and dynamical evolution of these systems. Current X-ray observatories continue to uncover new and important properties of both bright, nearby SNRs and the fainter population within the Galaxy and beyond. Advances in plasma codes, MHD simulations, and upcoming high spectral resolution capabilities that will become available with *ASTRO-H*, hold particular interest for enhancing our abilities to probe the detailed physics of SNRs and their evolution.

This work was carried out under support from NASA Contract NAS8-03060.

References

Badenes, C. *et al.* 2008, *ApJ*, 680, 1149
Castro, D. *et al.* 2011, *ApJ*, 734, 86
Castro, D. *et al.* 2012, *ApJ*, 756, 88
Chevalier, R. A. 2005, *ApJ*, 619, 839
Ellison, D. C. *et al.* 2010, *ApJ*, 712, 287
Ellison, D. C. *et al.* 2012, *ApJ*, 744, 39
Gaensler, B. M. & Slane, P. 2006, *ARAA*, 44, 17
Gaetz, T. J. *et al.* 2000, *ApJ*, 534, L47
Ghavamian, P. *et al.* 2007, *ApJ*, 654, L69
Gonzalez, M. & Safi-Harb, S. 2003, *ApJ*, 583, L91
Hughes, J. P. & Singh, K. P. 1994, *ApJ*, 422, 126
Hughes, J. P. *et al.* 1995, *ApJ*, 444, L81
Hughes, J. P. *et al.* 2000a, *ApJ*, 528, L109
Hughes, J. P. *et al.* 2000b, *ApJ*, 543, L61
Hwang, U. & Laming, J. M. 2012, *ApJ*, 746, 130
Inoue, K. *et al.* 2012, *ApJ*, 744, 71
Iwamoto, K. *et al.* 1999, *ApJS*, 125, 439
Kawasaki, M. *et al.* 2005, *ApJ*, 631, 935
Lee, J. J. *et al.* 2010, *ApJ*, 711, 861
Lewis, K. T. *et al.* 2003, *ApJ*, 582, 770
Lopez, L. A. *et al.* 2011, *ApJ*, 732, 114
Miceli, M. *et al.* 2010, *A&A*, 514, L2
Moriya, T. J. 2012, *ApJ*, 750, L13
Ozawa, M. *et al.* 2009, *ApJ*, 706, L71
Park, S. *et al.* 2004, *ApJ*, 602, L33
Park, S. *et al.* 2013, *ApJ* (in press; arXiv:1302.5435)
Rest, A. *et al.* 2008, *ApJ*, 680, 1137
Reynolds, S. P. *et al.* 2007, *ApJ*, 668, L135
Vink, J. 2012, *A&AR*, 20, 49
Warren, J. S. *et al.* 2005, *ApJ*, 634, 376
Yamaguchi, H. *et al.* 2009, *ApJ*, 705, L6

Discussion

SURNIS: What is the ideal range of radio frequencies to observe for constraining the PWN Physics? Do the low frequency points help?

SLANE: The entire radio band is actually quite important. At the lowest frequencies, one can probe the low-energy end of the particle population, for example. In some cases, there seems to be evidence for a low-energy population of electrons that produce Gamma-rays through IC emission. Observations in the mm-wave band may be important to see where that population cuts off.

RAY: Could you elaborate on the constraining of the hadronic vs. Leptonic scenarios of gamma-ray production in CTB 109 by X-ray spectrum?

SLANE: The observed gamma-ray emission, and radio emission, can be adequately described by models in which the gamma-rays originate primarily from hadronic or primarily from leptons. These scenarios require different ambient densities, and result in quite different ionization states for the X-ray gas. In CTB 109, the hadronic case predicts more emission from highly ionized states than we actually observe. The leptonic case under predicts the higher ionization states. We find that an intermediate case, with density, age, and electron-to-proton ratio's that are able to reproduce the X-ray and radio spectra, also produces the observed X-ray ionization states.

LOPEZ: In CTB 109, is there contamination in the gamma-ray emission from the anomalous X-ray pulsar at the center of the supernova remnant?

SLANE: To date, no gamma-ray emission has been detected from magnetars. However they can verify the emission is not from the AXP by seeing how the emission varies with time and if it co-relates with the period of the AXP.

Supernova Environmental Impacts
Proceedings IAU Symposium No. 296, 2013
A. Ray & R. A. McCray, eds.

doi:10.1017/S1743921313009526

X-ray imaging and spectroscopic study of the SNR Kes 73 hosting the magnetar 1E 1841–045

H. S. Kumar[1], S. Safi-Harb[1,2], P. O. Slane[3], and E. V. Gotthelf[4]

[1]Department of Physics & Astronomy, University of Manitoba, Winnipeg, Canada
email: harsha@physics.umanitoba.ca
[2]Canada Research Chair
[3]Harvard-Smithsonian Center for Astrophysics, 60 Garden Street, Cambridge, MA 02138, USA
[4]Columbia Astrophysics Laboratory, Columbia university, New York, NY 10027, USA

Abstract. We present the first detailed *Chandra* and *XMM-Newton* study of the young Galactic supernova remnant (SNR) Kes 73 associated with the anomalous X-ray pulsar (AXP) 1E 1841–045. Images of the remnant in the radio (20 cm), infrared (24 μm), and X-rays (0.5–7 keV) reveal a spherical morphology with a bright western limb. High-resolution *Chandra* images show bright diffuse emission across the remnant, with several small-scale clumpy and knotty structures filling the SNR interior. The overall *Chandra* and *XMM-Newton* spectrum of the SNR is best described by a two-component thermal model with the hard component characterized by a low ionization timescale, suggesting that the hot plasma has not yet reached ionization equilibrium. The soft component is characterized by enhanced metal abundances from Mg, Si, and S, suggesting the presence of metal-rich supernova ejecta. We discuss the explosion properties of the supernova and infer the mass of its progenitor star. Such studies shed light on our understanding of SNRs associated with highly magnetized neutron stars.

Keywords. ISM: individual (SNR Kes 73) – pulsars: individual (AXP 1E 1841–045) – supernova remnants – X-rays: ISM

1. Introduction

Kes 73 (G27.4+0.0), hosting the anomalous X-ray pulsar (AXP) 1E 1841–045, is a shell-type supernova remnant (SNR) located at a distance of 8.5 kpc. At radio wavelengths, the remnant shows a small diameter ($\sim$5′) spherical shell characterized by a steep spectral index ($\alpha \sim 0.68$) and a flux density of 6 Jy at 1 GHz (Kriss *et al.* 1985). The SNR was clearly detected in the 24 μm infrared band with an estimated dust mass of 0.11 $M_\odot$ (Pinheiro Gonçalves *et al.* 2011). At X-ray energies, the system reveals the bright magnetar at its center. AXP 1E 1841–045 is the slowest (P=11.8 s) known magnetar with a surface dipole magnetic field B = 7.1×10^{14} G and a characteristic age of $\sim$4.7 kyr (Vasisht & Gotthelf 1997).

Previous X-ray observations of Kes 73 with *ASCA* have shown that the SNR emission is dominated by lines from Mg, Si, and S. The SNR spectrum was best described by a thermal bremsstrahlung model ($kT \sim 0.6$ keV), with an estimated age of $\leqslant$2000 years (Gotthelf & Vasisht 1997). An *XMM-Newton* study of Kes 73 using only the MOS data argued against the millisecond proto-neutron star model for magnetar formation (Vink & Kuiper 2006). Another recent study by Lopez *et al.* (2011) made use of one of the *Chandra* observations of Kes 73, where the SNR spectrum was fitted by a one-component VPSHOCK model ($kT = 0.84 \pm 0.49$ keV) and reported the detection of enhanced abundances from Mg, Si, and S. Since all these studies lacked a thorough investigation of Kes

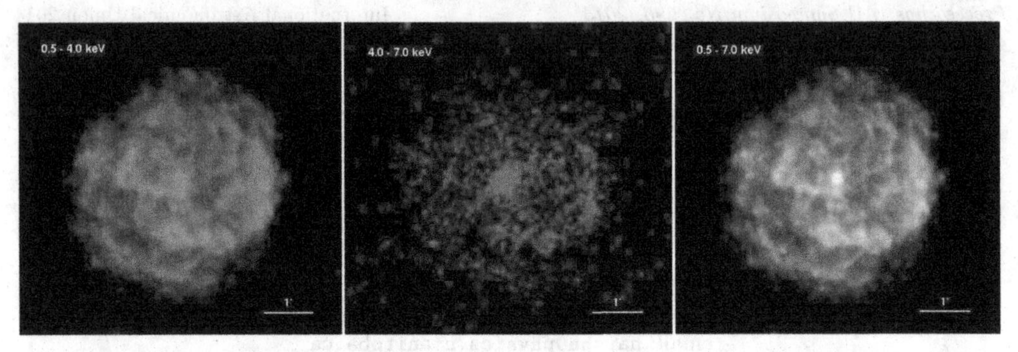

Figure 1. *Chandra* images of Kes 73 in different energy bands: 0.5–4 keV (*left*), 4–7 keV (*middle*), and 0.5–7 keV (*right*). The images from both *Chandra* data were exposure-corrected, merged, and smoothed using a Gaussian with σ=1″–2″ for a significance of detection 2 to 5.

73 using all available data, we performed a detailed X-ray imaging and spectroscopic study to address the remnant's multi-wavelength morphology, constrain the SNR spectral parameters and explosion properties, and infer the mass of its progenitor star. We summarize these results briefly below. A detailed analysis and discussion can be found elsewhere (Kumar *et al.* 2013).

2. Observations and data analysis

The SNR Kes 73 was observed with *Chandra* ACIS on 2000 July 23 (ObsID: 729) and 2006 July 30 (ObsID: 6732). The data were reduced using the standard CIAO routines, yielding a total effective exposure time of ~54 ks. *XMM-Newton* observed Kes 73 on 2002 October 5 and 7 (ObsIDs: 0013340101 and 0013340201) with the European Photon Imaging Camera (EPIC). The data from both the MOS and pn instruments were analyzed using the standard XMMSAS tasks, yielding a total exposure time of ~27 ks. X-ray spectra were extracted from various regions across Kes 73 using all datasets, and the background was selected from an adjacent source-free region to the north-east of the remnant. We also extracted the whole SNR region of radius 150″ with the emission from the pulsar (~25″ from *XMM-Newton* and ~10″ from *Chandra*) plus other point sources removed. The background was extracted from a ring surrounding the SNR, extending from 150″–180″.

3. Morphology of Kes 73

In Figure 1, we show the combined *Chandra* images of Kes 73 in different energy bands: 0.5–4 keV, 4–7 keV, and 0.5–7 keV. The remnant has a circular morphology with an apparent diameter of $\leqslant 5'$ and reveals small-scale clumpy and bright knotty structures filling the SNR interior. These structures are clearly noticeable in the 0.5–4 keV band, which has an overall morphology similar to that of the broadband (0.5–7 keV) image. The 4–7 keV image shows significant hard X-ray emission originating from the pulsar, in addition to emission from most of the SNR particularly enhanced in the west. An incomplete and outer shell-like feature runs from east to south. The interior region to the east of the AXP shows bright diffuse emission, which extends towards the southern side of the SNR to merge with the bright western limb. The northern portion of the SNR is relatively faint in X-rays, with no distinct shell-like feature.

Figure 2 shows the remnant's 20 cm radio image obtained with MAGPIS (*left*) and the 24 μm band infrared image obtained with MIPSGAL (*right*) overlaid on the broadband

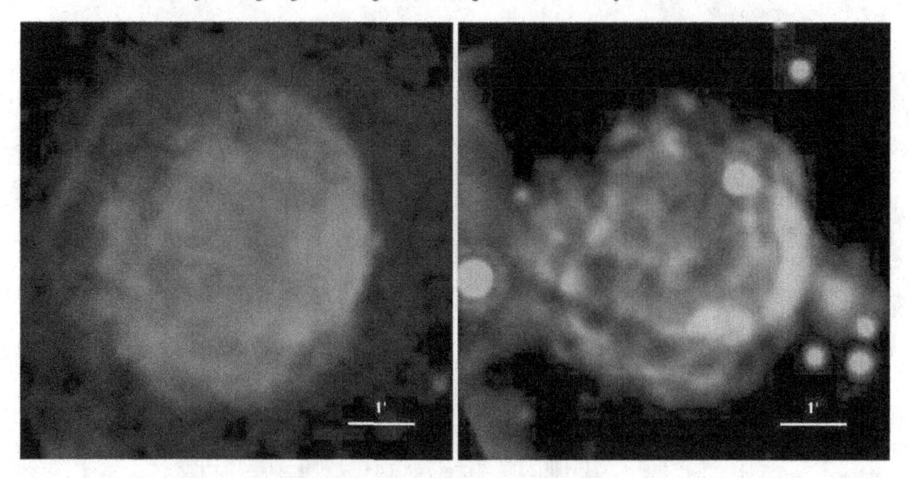

Figure 2. Multi-wavelength morphology of SNR Kes 73. *Left*: MAGPIS 20 cm radio image (red) overlaid on the 0.5–7 keV *Chandra* image (blue). *Right*: MIPSGAL 24μm band infrared image (green) overlaid on the 0.5–7 keV *Chandra* image (blue). The radio and IR images have a Gaussian smoothing of σ=3″ and the *Chandra* image has a Gaussian smoothing of σ=1″–2″.

Chandra image. The radio shell is clearly more extended on the north-eastern side of the SNR, displaying a thin filamentary structure along the edges (seen in pure red). This feature is likely the expanding outer blast wave which is not detected in X-rays. The X-ray emission seems to dominate the SNR interior filling the radio shell. The radio shell extending from the eastern limb to the southern edge of the SNR also matches with the corresponding outer shell-like structure in the X-ray image. The *Spitzer* MIPS image reveals a morphology more comparable to that of the X-ray image with arc-like features and bright infrared emission filling the SNR. The remnant's western limb is very bright in the infrared as well, precisely outlining the radio and X-ray emission. We do not see any significant infrared emission from the north-eastern boundary of the SNR.

4. Spectral properties and progenitor of Kes 73

A two-component thermal model best describes the X-ray spectra originating from the small-scale regions (VPSHOCK+VPSHOCK) and the global SNR spectrum (VSEDOV+VPSHOCK; Figure 3). We do not find any significant spatial variations of the spectral parameters across the SNR, except for a slightly elevated column density N_H (which ranges between $\sim$(2.6–3.3)$\times 10^{22}$ cm^{-2} from east to west) on the western side with an estimated difference of $\sim 7\times 10^{21}$ cm^{-2} between the eastern and western sides. However, there is no evidence of any cloud or nearby sources interacting with the SNR. The soft components show plasma temperatures $\sim$0.3–0.5 keV and ionization timescales $\gtrsim 10^{12}$ cm^{-3} s for most of the regions with enhanced metal abundances, while the hard components exhibit plasma temperatures $\sim$1.1–1.7 keV and low ionization timescales $\sim$(0.5–2.8)$\times 10^{11}$ cm^{-3} s with solar abundances (except for a slightl y enhanced Fe abundance for some regions). These results indicate that the soft-component arises from the reverse-shocked ejecta with most regions showing a plasma that has reached ionization equilibrium, while the hard-component originates from the shocked interstellar or circumstellar medium. The detection of enhanced Si and S abundances across the remnant suggests that the X-ray emission from the SNR Kes 73 is ejecta-dominated.

Using the global fit to the SNR's spectrum, we infer an SNR age ranging between $\sim$1.1 $D_{8.5}$ kyr for the free expansion phase (assuming an expansion velocity of 5000 km s^{-1})

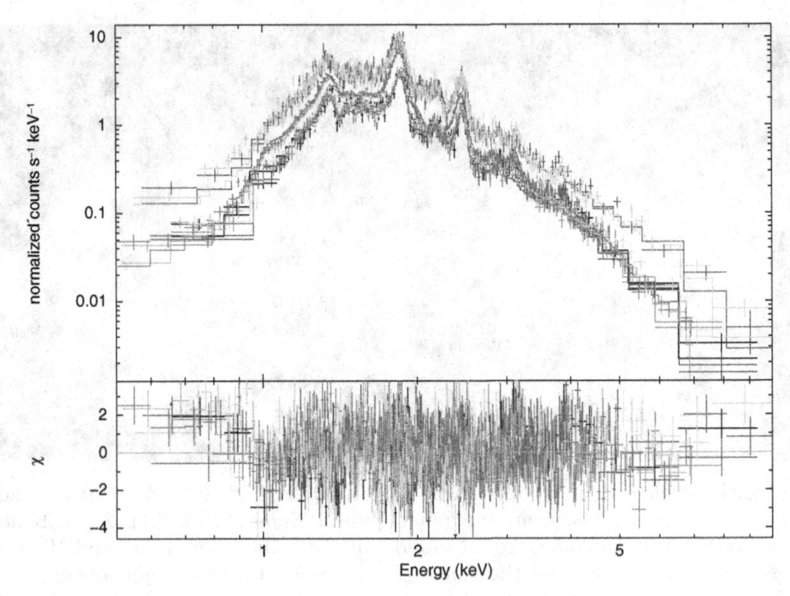

Figure 3. Best-fit VSEDOV+VPSHOCK model to the global SNR spectra using *Chandra* and *XMM-Newton* data [pn: red and cyan (top spectra), *Chandra*: pink and green (middle spectra), and MOS1+2: blue, yellow, orange, and black (bottom spectra)].

and ~2.1 $D_{8.5}$ kyr assuming a Sedov phase of evolution. Under the assumption of an explosion in a uniform ambient medium, the Sedov phase yields a shock velocity of ~1000 km s^{-1}, an average ambient density of ~0.5 $f^{-1/2}$ cm^{-3}, and an explosion energy of ~2.0×10^{50} $f^{-1/2}$ ergs (where f is the volume filling factor of the hot component). We compare the derived metal abundances from our spectral fits to the core-collapse nucleosynthesis model yields (e.g., Woosley & Weaver 1995) and infer a progenitor mass of ~(25–30) $M_{\odot}$, supporting the earlier prediction of a SN type IIL/b star for Kes 73 (Chevalier 2005). This estimated progenitor mass is comparable to that determined for a few other magnetars and high-magnetic field radio pulsars (Safi-Harb & Kumar 2012), suggesting very massive progenitors for the majority of highly magnetized neutron stars.

This research made use of NASA's ADS and HEASARC. S. Safi-Harb acknowledges support from the CRC program, NSERC, CFI, and CSA.

References

Chevalier, R. G. 2005, *ApJ*, 619, 839

Pinheiro Gonçalves, D., Noriega-Crespo, A., Paladini, R., Martin, P. G., & Carey, S. J. 2011, *AJ*,142, 47

Gotthelf, E. V. & Vasisht, G. 1997, *ApJ*, 486, L133

Kriss, G. A., Becker, R. H., Helfand, D. J., & Canizares, C. J. 1985, Apj, 288, 703

Kumar, H. S., Safi-Harb, S., Slane, P. O., & Gotthelf, E. V. 2013, *ApJ*, submitted

Lopez, L. A., Ramirez-Ruiz, E., Huppenkothen, D., Badenes, C., & Pooley, D. A. 2011, *ApJ*, 732, 114

Safi-Harb, S. & Kumar, H. S. 2012, *Proceedings of IAUS 291*, Cambridge University Press, Vol 291, 480 (arXiv:1211.5261)

Vasisht, G. & Gotthelf, E. V. 1997, *ApJ*, 486, L129

Vink, J. & Kuiper, L. 2006, *MNRAS*, 370, 1, L14

Woosley, S. E. & Weaver, T. A. 1995, *ApJSS*, 101, 181

Supernova Environmental Impacts
Proceedings IAU Symposium No. 296, 2013
A. Ray & R. A. McCray, eds.

doi:10.1017/S1743921313009538

What Shapes Supernova Remnants?

Laura A. Lopez
MIT-Kavli Institute for Astrophysics and Space Research,
77 Massachusetts Ave., 37-664H, Cambridge, MA 02139
email: `lopez@space.mit.edu`

Abstract. Evidence has mounted that Type Ia and core-collapse (CC) supernovae (SNe) can have substantial deviations from spherical symmetry; one such piece of evidence is the complex morphologies of supernova remnants (SNRs). However, the relative role of the explosion geometry and the environment in shaping SNRs remains an outstanding question. Recently, we have developed techniques to quantify the morphologies of SNRs, and we have applied these methods to the extensive X-ray and infrared archival images available of Milky Way and Magellanic Cloud SNRs. In this proceeding, we highlight some results from these studies, with particular emphasis on SNR asymmetries and whether they arise from "nature" or "nurture".

Keywords. supernova remnants — X-rays: ISM — infrared: ISM — methods: data analysis

1. Introduction

In the past decades, evidence has mounted that supernova explosions (SNe) can have significant deviations from spherical symmetry. Spectropolarimetry studies – the measure of the polarization of light as a function of wavelength as it is scattered through the debris layers of expanding SNe – demonstrate that both Type Ia and core-collapse (CC) SNe are aspherical near maximum brightness (e.g., Wang & Wheeler 2008; Kasen *et al.* 2009). Line profiles in nebular spectra (100–200 days after explosion) of SNe show similar evidence of these ejecta asymmetries (e.g., Mazzali *et al.* 2001; Maeda *et al.* 2010); confirmation of SN asymmetries is also possible at much later times via comparison of SN light echo spectra from different perspectives (e.g., Rest *et al.* 2011). Furthermore, proper motion studies demonstrate that pulsars have velocities up to $\sim$1000 km s^{-1} (Lyne & Lorimer 1994; Faucher-Giguère & Kaspi 2006), consistent with "kicks" imparted to newly-forming neutron stars in asymmetric SNe. The asymmetries may be inherent to the explosion mechanisms, and several mechanisms have been proposed that reproduce the necessary degrees of asymmetry: e.g., acoustic power generated in the proto-neutron star (Burrows *et al.* 2007), the non-spherically symmetric standing accretion shock instability (Blondin & Mezzacappa 2007), and neutrino heating (Scheck *et al.* 2004).

Supernova remnants (SNRs) retain imprints from the geometry of their progenitors' explosions as well. For example, the three-dimensional projections of Cassiopeia A reveal complex, asymmetric features which are attributed to the explosion (DeLaney *et al.* 2010). However, SNRs are also shaped by their environments; in particular, inhomogeneities in the circumstellar medium (CSM) structure affects the SNR morphology as well (e.g., Kepler's SNR: Reynolds *et al.* 2007). The relative role of the explosion and environment in shaping SNRs remains an outstanding question, and it has even been cited as the biggest challenge in modern SNR research (Canizares 2004).

Recently, we have developed techniques to quantify the morphological properties of SNRs (Lopez *et al.* 2009a), and we have applied these methods to archival *Chandra X-ray Observatory* and *Spitzer Space Telescope* images to assess the influence of "nature" versus "nurture" in SNR dynamics and evolution. In this proceeding, we highlight some results from these morphological studies to characterize SNR asymmetries.

2. Method

The method we employed to characterize the symmetry of SNRs is a power-ratio method (PRM). The PRM has been used extensively to quantify the morphologies of galaxy clusters (Buote & Tsai 1995, 1996; Jeltema *et al.* 2005). We extended the technique to compare the distribution of elements in individual SNRs (Lopez *et al.* 2009a) and to examine the symmetry of X-ray and IR emission in Type Ia and CC SNRs (Lopez *et al.* 2009b; Lopez *et al.* 2011; Peters *et al.* 2013). We refer the reader to these papers for a detailed derivation and description of the method; we provide a basic summary below.

The PRM measures asymmetries via calculation of the multipole moments of emission in a circular aperture. It is derived similarly to the expansion of a two-dimensional gravitational potential, except an image's surface brightness replaces the mass surface density. The powers P_{m} of the expansion are obtained by integrating the magnitude of each term Ψ_{m} over the aperture radius R. We divide the powers P_{m} by P_0 to normalize with respect to flux, and we set the origin position in our apertures to the centroids of our images so that the dipole power P_1 approaches zero. In this case, the higher-order terms reflect the asymmetries at successively smaller scales. The quadrupole power ratio P_2/P_0 reflects the ellipticity or elongation of a source, and the octupole power ratio P_3/P_0 quantifies the mirror asymmetry of a source.

3. Results

We applied the PRM to *Chandra* Advanced CCD Imaging Spectrometer (ACIS) and *Spitzer* Multiband Imaging Photometer (MIPS) images of Milky Way, LMC, and SMC SNRs, and we highlight the results and implications of these analyses below.

Type Ia and CC SNRs have distinct symmetries. Using the PRM on X-ray line images of seventeen SNRs, we found that the emission from shock-heated metals in Type Ia SNRs is more spherical and more mirror symmetric than in CC SNRs (Lopez *et al.* 2009b).

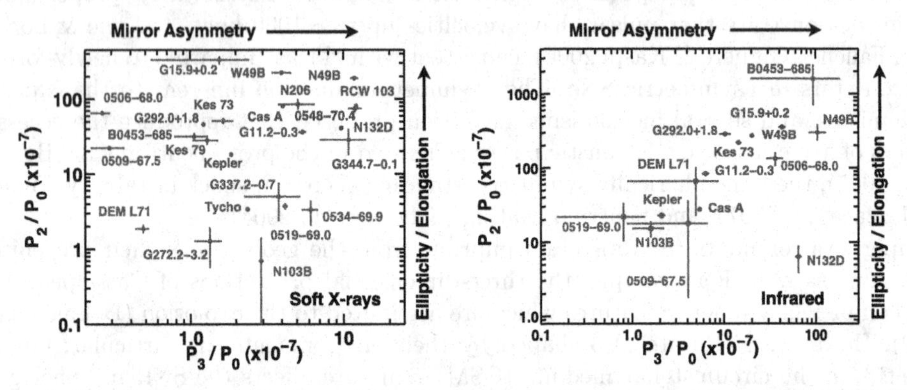

Figure 1. Results from application of the power-ratio method to 24 galactic and LMC SNRs: quadrupole power ratio P_2/P_0 (which measures ellipticity/elongation) vs. octupole power ratio P_3/P_0 (which quantifies mirror asymmetry) of the soft X-ray band (0.5–2.1 keV) images (left) and of the *Spitzer* 24 μm images (right). Type Ia SNRs are plotted in red, and the CC SNRs are in blue (classified by abundance ratios). One source, 0548−70.4, is in purple because of its anomalous abundance ratios. The Type Ia SNRs separate naturally from the CC SNRs in this diagram. Figures are adapted from Lopez *et al.* (2011) and Peters *et al.* (2013).

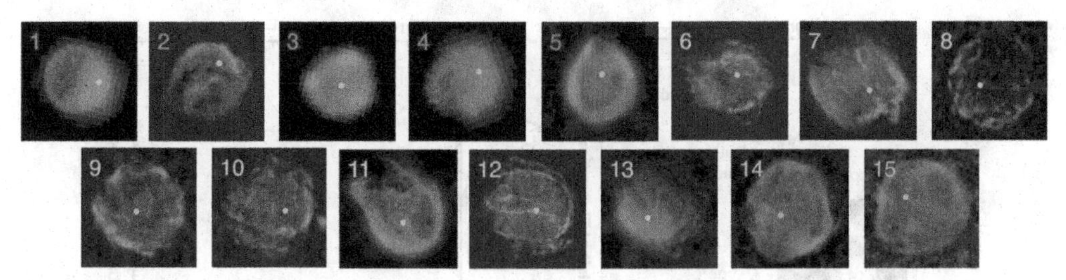

Figure 2. Images (adapted from Peters *et al.* 2013) of the *Spitzer* 24 μm emission (red) and *Chandra* soft X-ray (0.5–2.1 keV) emission of the 15 SNRs analyzed in Figure 1 (right). The thermal X-rays trace the ejecta material heated by the reverse shock, while the IR originates from circumstellar dust heated by interaction with the blast wave. Red numbers denote Type Ia SNRs; blue numbers denote CC SNRs. SNRs are as follows: [1] 0509−67.5; [2] Kepler; [3] 0519−69.0; [4] N103B; [5] DEM L71; [6] Cas A; [7] W49B; [8] G15.9+0.2; [9] G11.2−0.3; [10] Kes 73; [11] N132D; [12] G292.0+1.8; [13] 0506−68.0; [14] N49B; [15] B0453−685.

Furthermore, this result holds for the X-ray bremsstrahlung emission of Type Ia and CC SNRs as well (see Figure 1; Lopez *et al.* 2011). The ability to distinguish between the two SN classes based on bremsstrahlung emission morphology alone suggests the potential to type SNRs with weak X-ray lines or with low spectral resolution X-ray observations.

We further extended the technique to 24 μm *Spitzer* images of the same sample, and we found that the warm dust emission of Type Ia and CC SNRs also has distinct symmetries (Peters *et al.* 2013; see Fig. 1). Similar to the X-ray results, Type Ia SNRs have more circular and mirror symmetric IR emission than CC SNRs. The two wavelength regimes probe distinct emitting regions (see Figure 2): the thermal X-rays trace the ejecta material heated by the reverse shock, while the IR originates from circumstellar dust heated by interaction with the blast wave. Yet both are sensitive to the shape of the contact discontinuity, where the ejecta are impacting the shocked ISM. As the contact discontinuity is shaped by the explosion geometry as well as the structure of the surrounding medium, the distinct symmetries of Type Ia and CC SNRs in the X-ray and IR wavebands reflect both the different explosion mechanisms and the environments of the two classes.

It is noteworthy that SNRs with bright neutron stars/pulsars (e.g., G11.2−0.3, Kes 73, B0453−685) are among the most circular of the CC SNRs. Based on our preliminary investigations, CC SNRs with neutron stars appear more spherical and symmetric than CC SNRs without neutron stars. However, strict limits on the presence/absence of neutron stars in SNRs are necessary before the trend can be established statistically.

Elements within individual SNRs have distinct symmetries. Our quantitative approach also enables us to compare images of different emission features (like the morphologies of the shock-heated metals) within individual remnants. For example, Figure 3 (left) plots the symmetry diagram for seven X-ray emission line images of Cassiopeia A (O VIII, Mg XI, Si XIII, S XV, Ar XVII, Ca XIX, and Fe XXV). The O has the most symmetric distribution, while the Fe is comparatively more asymmetric and elongated than the other metals. By contrast, the intermediate-mass elements (Mg, Si, S, Ar, and Ca) appear to have similar morphologies. Thus, despite relatively efficient mixing of the ejecta (Lopez *et al.* 2011), the distinct symmetries of the metals reflect imprints of the explosion geometry.

Environment Shapes Large-Scale SNR Morphology. To examine the role of environment in shaping SNRs, we have compared systematically the morphological properties of SNRs in different ISM conditions. For example, we have measured the symmetry of SNRs thought to be interacting with molecular clouds (see Figure 3, right), based on

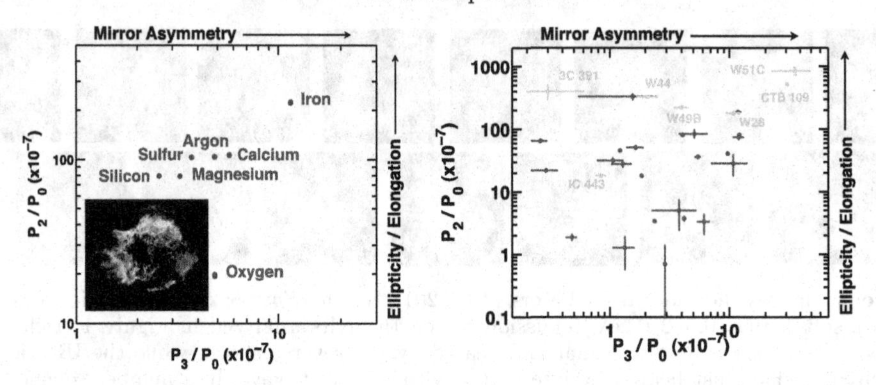

Figure 3. *Left*: Quadrupole power ratio P_2/P_0 vs. octupole power ratio P_3/P_0 for 7 X-ray line images of Cassiopeia A (O VIII, Mg XI, Si XIII, S XV, Ar XVII, Ca XIX, and Fe XXV). The inset image shows the distribution of O VIII (green), Si XIII (red), and Fe XXV (blue). Plot and images were produced using data in Lopez *et al.* 2011). *Right*: P_2/P_0 vs. P_3/P_0 for SNRs interacting with molecular clouds (in cyan); non-interacting SNRs are from Fig. 1, with Type Ia SNRs in red and CC SNRs in blue. Interacting SNRs are more elliptical than non-interacting SNRs.

coincidence with OH masers (which indicate the presence of shocked, molecular gas: Wardle & Yusef-Zadeh 2002). The interacting SNRs are the most elliptical of the CC SNRs, evidence that environment has a dramatic large-scale effect on SNR morphologies.

4. Conclusions

We have begun to address the "nature" vs. "nurture" conundrum in SNR science using a systematic approach on available archival data. Many exciting questions and prospects remain, including application of the techniques to other wavelength images, investigation of SNRs in other galaxies, characterization of SNR morphological evolution with age, and comparison to hydrodynamical model predictions.

References

Blondin, J. M., & Mezzacappa, A. 2007 *Nature*, 445, 58
Buote D. A. & Tsai, J. C. 1995, *ApJ*, 452, 522
Buote D. A. & Tsai, J. C. 1996, *ApJ*, 458, 27
Burrows, A., Livne, E., Dessart, L., Ott, C. D., & Murphy, J. 2007, *ApJ*, 655, 416
Canizares, C. R. 2004, *Frontiers of X-ray astronomy*, 107–116
DeLaney, T., Rudnick, L., Stage, M. D., *et al.* 2010, *ApJ*, 725, 2038
Faucher-Giguère, C.-A. & Kaspi, V. M. 2006, *ApJ*, 643, 332
Jeltema, T. E., Canizares, C. R., Bautz, M. W., & Buote, D. A. 2005, *ApJ*, 624, 606
Kasen, D., Röpke, F. K., & Woosley, S. E. 2009, *Nature*, 460, 869
Lopez, L. A., Ramirez-Ruiz, E., Pooley, D. A., & Jeltema, T. E. 2009a, *ApJ*, 691, 875
Lopez, L. A., Ramirez-Ruiz, E., Badenes, C., Huppenkothen, D., Jeltema, T. E., & Pooley, D. A. 2009b, *ApJ* (Letters), 706, L106
Lopez, L. A., Ramirez-Ruiz, E., Huppenkothen, D., Badenes, C., & Pooley, D. A. 2011, *ApJ*, 732, 114
Lyne, A. G. & Lorimer, D. R. 1994, *Nature*, 369, 127
Maeda, K., Taubenberger, S., Sollerman, J., *et al.* 2010, *ApJ*, 708, 1703
Mazzali, P. A., Nomoto, K., Patat, F., & Maeda, K. 2001, *ApJ*, 559, 1047
Peters, C. L., Lopez, L. A., Ramirez-Ruiz, E., Stassun, K. G., & Figueroa-Feliciano, E. 2013, submitted to *ApJ* (Letters)
Rest, A., Foley, R. J., Sinnott, B., *et al.* 2011, *ApJ*, 732, 3

Reynolds, S. P., Borkowski, K. J., Hwang, U., *et al.* 2007, *ApJ* (Letters), 668, L135
Scheck, L., Plewa, T., Janka, H.-T., Kifonidis, K., & Müller, E. 2004, *Phys. Rev. Lett.*, 92, 011103
Wang, L. & Wheeler, J. C. 2008, *ARAA*, 46, 433
Wardle, M., & Yusef-Zadeh, F. 2002 *Science*, 296, 2350

Discussion

SANKRIT: Where would the Cygnus Loop, in X-ray, Optical, & IR fall on the quadrupole / octupole Plot? Where would a filamentary optical SNR like S147 fall? It would be useful to explore what the technique yields for middle-aged remnants.

LOPEZ: I haven't analyzed the cygnus loop because my sample so far has only been on young, ejecta dominated sources. It would be interesting to compare the symmetry of old & young SNRs to see how their morphology evolves. I have not applied the technique to the optical, but the optical emission traces the forward shock, so it may have different symmetry properties than the X-ray & IR, which trace reverse shocked material.

UNIDENTIFIED: 1. How is the "edge of the SNR defined in this study? 2. Crab SNR is highly asymmetric any comment?

LOPEZ: 1. We ensured that all the X-ray extension of a source was enclosed in the aperture used in the multiple expansion. 2. Although the symmetry result holds for thermal emission, it does not appear to hold for non-thermal emission (that is, the finding that type 1a SNRs are more circular than core collapse SNRs is only true for their thermal emission). As the crab is all non-thermal, the symmetry rule doesn't apply.

WANG: You probably want to explore the dependence on SNR ages. A proxy that may be used is the physical sizes of SNRs.

LOPEZ: Yes, that is a parameter that we can explore.

BARTEL: I am wondering how much your definition of a symmetry is biased because of sensitivity effects. Suppose there is a highly circularly symmetric image of a SNR shell with source modulation around the rim. A poor telescope may just detect a bright spot here and another there. The SNR could appear highly asymmetric. However, a mac sensitive telescope would pick up the radiation from the entire shell, and the SNR would appear highly symmetric. Would this not cause a bias in your analysis?

LOPEZ: The sensitivity of the observation influences the measured symmetry, but as long as the extension is resolved, it does not vary substantially. Since the bright features dominate anyway (and dim features do not contribute much). One of the selection criteria for our sample was that it meet a cut in surface brightness.

HAN: Because of geometrical effects of our location in the Milky Way, the distances of SNRs are not well determined and hence we do not have clarity about the ISM environment of SNRs, which is probably a limitation of your analysis of dependence of SNR shape on the environment.

LOPEZ: Yes there are issues with projection effects. However, there are several signatures that reveal whether an SNR is interacting with dense molecular material (e.g. OH

Masers). Furthermore, we can study SNRs in face on galaxies (like Magellanic clouds) where we do not have line of sight confusion.

PODSIADLOWSKI: From a theoretical point of view (also Ken talked about this), you do not expect energetic supernovae when you form black holes from a non-rotating star. This should produce at best a wimpy supernova remnant. Do you have an example?

LOPEZ: Some SNRs are thought to have had explosion energies down to 10^{50} erg, but I do not know of any wimpy looking remnants.

Supernova Environmental Impacts
Proceedings IAU Symposium No. 296, 2013
A. Ray & R. A. McCray, eds.

doi:10.1017/S174392131300954X

Nonthermal X-rays from supernova remnants

Anne Decourchelle
Laboratoire AIM, CEA-IRFU/CNRS/ Univ. Paris Diderot, Service d'Astrophysique,
L'Orme des Merisiers, CEA Saclay
F-91191 Gif sur Yvette, France
email: anne.decourchelle@cea.fr

Abstract. Since the discovery of nonthermal X-rays in the shell-type supernova remnant SN1006 almost 20 years ago, the field has developed considerably, owing significant progress to our understanding of particle acceleration. Key to the characterization of the nonthermal emission is the ability of current satellites, *XMM-Newton* and *Chandra*, to perform spatially resolved spectroscopy at a relatively small spatial scale.

In this review, I intend to present the main contributions of the study of nonthermal X-rays from supernova remnants to the understanding of particle acceleration.

Keywords. Supernova remnant, Particle acceleration, X-rays, Synchrotron emission

---

1. Introduction

Following the explosion, the supernova material (ejecta) is ejected at high velocity (a few tens of thousands of km/s) in the circumstellar medium and experiences a rapid expansion. As the expansion velocity is much larger than the sound speed, the ejecta are preceded by a shock, called forward shock that propagates outwards and compresses, heats and accelerates the ambient medium. After an initial free expansion phase, the ejecta are decelerated giving rise to a reverse shock, which propagates inwards and compresses, heats and decelerates the ejecta. The result of the interaction of high-velocity ejecta with the surrounding medium is thus a two-shock structure as is illustrated in Fig. 1. The interface between the ejecta and the ambient medium is distorted due to hydrodynamic instabilities (notably, Rayleigh-Taylor). The two shocked media are expected to be powerful X-ray emitters, as they have temperatures of tens of million degrees and above. It is in this early ejecta-dominated phase (for remnants younger than typically 1000 years) that the elements synthesized in the supernovae are best observable, notably in X-rays, as the thermal emission is dominated by that of the shocked ejected matter (see Slane 2013, this proceedings).

On theoretical grounds, strong shocks in supernova remnants are expected to be responsible for the acceleration of Galactic cosmic rays up to the knee at about 3 10^{15} eV (Blandford & Eichler 1987). This is the only galactic population fitting the energetic requirement to maintain the pool of Galactic cosmic rays, assuming that about 10 % of their kinetic energy is taken by the acceleration process. The presence of GeV accelerated electrons at the shocks in supernova remnants is known since the discovery of synchrotron radio emission in Tycho in 1954 (Hanbury Brown 1954) and is now observed in almost all SNRs. The spectral index of their radio synchrotron spectrum is close to the theoretical expectation. But the search for observational proofs of electrons accelerated close to the knee has failed for a long time in the high-energy domain.

Indeed, synchrotron X-rays were observed only 40 years later by the *ASCA* Japanese satellite in the shell-like remnant of SN 1006 (Koyama *et al.* 1995), thanks to the achievement of CCD X-ray spectro-imagers. It has allowed obtaining the first evidence of electrons accelerated up to TeV energies in the remnant of SN 1006. The bright X-ray and radio limbs were indeed revealed to be featureless in X-rays, and interpreted as X-ray synchrotron emission from TeV electrons. Conversely, the emission in the faint areas (center of the remnant and faint limbs), exhibiting emission lines, are of thermal origin. This result, as well as the subsequent discovery of two other synchrotron-dominated supernova remnants, have shown the new potential of X-ray observations to observe the highest-energy electrons that shocks can accelerate in supernova remnants (SNRs).

As illustrated in Fig. 1, accelerated particles give rise to a number of emission processes through their interaction with the magnetic field (synchrotron, from radio to X-rays), with the ambient photon field (inverse Compton scattering, in the gamma-ray band) and with the ambient medium (pion decay, in the gamma-ray domain). A full multi-wavelength nonthermal approach is important to consistently derive the characteristics of the accelerated particles and of their environment. This includes using combined radio and X-ray data to constrain the synchrotron spectrum, as well as very high-energy gamma rays whose inverse-Compton emission arises from the X-ray synchrotron emitting electron population. The very high-energy gamma-ray domain (see Lemoine-Goumard 2013, this proceedings) has considerably developed in the last ten years, with the implementation of stereoscopic ground-based Cherenkov detectors. H.E.S.S. has revealed the morphology of RX J1713.7-3946 SNR in the TeV domain, providing the first image of a TeV source (Aharonian *et al.* 2004), well correlated with the X-ray synchrotron morphology (Aharonian *et al.* 2006). However, the TeV gamma-ray emission can be produced either by accelerated electrons (Inverse Compton) or by accelerated protons (through pion decay following proton/proton collisions). The nature of which is a key question for particle acceleration as their respective energetic requirement is considerably different; a consistent modeling of the overall non-thermal spectrum is crucial to disentangle between the two possible origins.

2. Nonthermal X-rays in Supernova remnants

Nonthermal X-ray emission is observed in a number of supernova remnants over a relatively large range of parameters: from very young supernova remnants (like G1.9+0.3, 110 years old) to evolved remnants of above a thousand years (like RXJ1713-3946), from type Ia to core-collapse supernova remnants, from thermally-dominated remnants (like Tycho's SNR) to synchrotron-dominated remnants (like RXJ1713-3946).

2.1. *Synchrotron X-ray emission in young ejecta-dominated supernova remnants*

Chandra and *XMM-Newton* spatially-resolved spectroscopic studies of galactic historical supernova remnants have allowed to distinguish the emission arising from the shocked ejecta from that of the shocked interstellar medium. This was not affordable with previous X-ray instruments.

The first discovery has been on the nature of the emission associated with the shocked interstellar medium. While thermal emission was expected, the spectra of the blast wave were revealed to be featureless in Tycho's SNR (Hwang *et al.* 2002) and the morphology of the emission filamentary (see Fig. 2). In the framework of a thermal emission, the observed spectra implied a strongly out of ionization equilibrium plasma (to avoid line emission) with a high electronic temperature (of about 2 keV) all along the rim. For shock velocities of about 3000-4000 km/s, this was implying an electron-to-mean temperature ratio of 0.1-0.2, which requires additional electron heating than solely through Coulomb interaction. On the other hand, a non-thermal synchrotron origin was consistent with the good agreement with the radio synchrotron morphology. The spectra, fitted by an exponential cut-off synchrotron model, gave values of the cut-off frequency of the order of 3 to 7 10^{16} Hz. As shown in Figure 2, such featureless spectra were observed in the other historical young supernova remnants like Kepler's SNR (Cassam-Chenaï *et al.* 2004a) and Cas A (Vink & Laming 2003).

On the basis of solely available spectra, it was not possible to distinguish between both possible interpretations (thermal vs nonthermal emission). The answer came from the interpretation of the morphology of the emission, notably the fact that the emission appears as narrow filaments, which was more naturally explained in the context of a nonthermal synchrotron emission. Such features of the blast wave X-ray emission were revealed by *Chandra* observations in a number of young supernova remnants: Cas A (Hughes *et al.* 2000, Gotthelf *et al.* 2001); Tycho (Hwang *et al.* 2002, Warren *et al.* 2005), Kepler (Bamba *et al.* 2005), SN 1006 (Bamba *et al.* 2003, Long *et al.* 2003). They were interpreted as synchrotron loss-limited filaments implying an amplified post-shock magnetic field (Vink & Laming 2003, Ballet 2006, Völk, Berezhko & Ksenofontov 2005, Parizot *et al.* 2006). Pohl, Yan & Lazarian (2005) put forward an alternative interpretation, namely the damping of the magnetic field behind the shock. Both interpretations require significant magnetic field amplification at the shock, which is supported by theoretical studies (Lucek & Bell 2000, Bell & Lucek 2001, Bell 2004).These models predict slightly different synchrotron morphologies and spectral shapes in X-rays (Cassam-Chenaï *et al.* 2007, Rettig & Pohl 2012).

The second discovery was to realize that the forward shock was lying much closer to the contact discontinuity than was expected from pure hydrodynamical models (Decourchelle 2005, Warren *et al.* 2005), implying that efficient particle acceleration was on going at the forward shock in young supernova remnants like Tycho and Kepler, as expected from

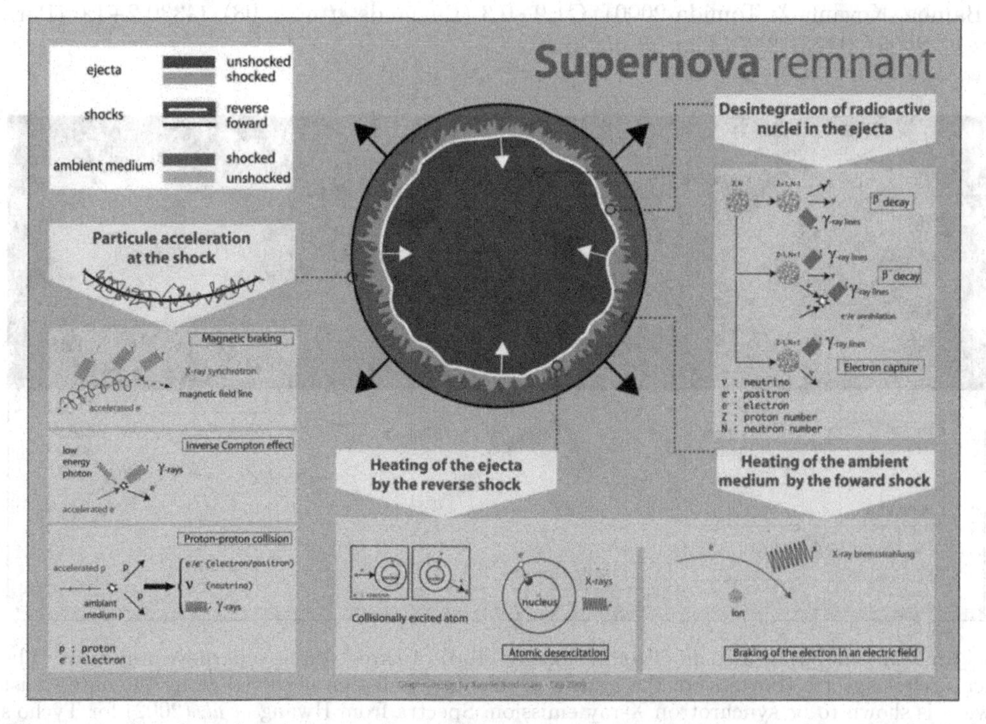

Figure 1. Sketch of the structure of a young supernova remnant, showing the main physical processes at work and the associated emission processes.

hydrodynamical models including efficient particle acceleration at the shock (Chevalier 1983, Decourchelle, Ellison & Ballet 2000).

2.2. *Synchrotron-dominated supernova remnants*

The first signature of TeV accelerated electrons in shell-type supernova remnants was obtained in SN 1006 with *ASCA* (Koyama *et al.* 1995). A second more extreme shell-type supernova remnant, RXJ 1713-3946, exhibiting only synchrotron emission was discovered in 1997 with *ASCA* (Koyama *et al.* 1997). The source was first identified as a supernova remnant in 1996 by Pfeffermann & Aschenbach (1996), based on *ROSAT* observations. The *ASCA* observations (Koyama *et al.* 1997, Slane *et al.* 1999) revealed a featureless X-ray spectrum well fitted by a power law model ($\gamma \simeq 2.4$). No thermal emission was detected anywhere in the SNR implying a very low-density medium ($\leqslant 0.28$ cm^{-3}, assuming a distance of 6 kpc). Assuming a magnetic field of 10 μG, the observed X-ray synchrotron emission arises from accelerated electrons at energies of 300 TeV. Further studies with *XMM-Newton* have enabled a complete characterization of the properties of the nonthermal emission. Combined with radio CO and HI data, absorption studies lead to reevaluate the distance of RXJ 1713-3946 to about 1 kpc (see Cassam-Chenaï *et al.* 2004b) and were confirmed by dedicated CO observations (e.g., Fukui *et al.* 2003, Moriguchi *et al.* 2005). Combined X-ray and TeV gamma-ray spectral analysis were performed on this unique source to constrain the origin (leptonic versus hadronic) of the gamma-ray emission (Acero *et al.* 2009). A non-linear correlation between the X-ray and TeV gamma-ray flux were suggested by the data, which is easier to reproduce in the framework of a leptonic model. While SN 1006 and RXJ 1713-3946 are the best examples of this category of supernova remnants, X-ray synchrotron emission has been discovered in a number of other cases: RX J0852.0-4622 (Slane *et al.* 2001), RCW 86 (Bamba, Koyama & Tomida 2000), G1.9+0.3 (Reynolds *et al.* 2008), G330.2+1.0 (Park *et al.* 2009), J1731-347 (Tian *et al.* 2010).

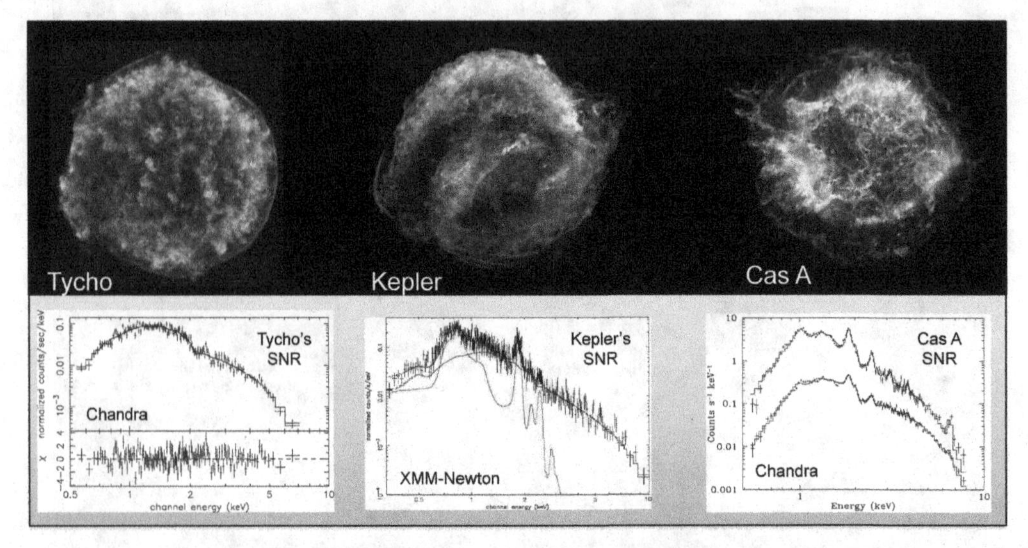

Figure 2. Composite 3-color *Chandra* images (top) of three young supernova remnants. The featureless spectra (bottom) of the narrow filamentary emission observed along the outer blast wave is shown to be synchrotron X-ray emission. Spectra from Hwang *et al.* (2002) for Tycho's SNR, Cassam-Chenaï *et al.* (2004a) for Kepler's SNR, and Vink & Laming (2003) for Cas A SNR.

3. Constraints on particle acceleration from nonthermal X-ray emission in supernova remnants

Since the discovery that cosmic rays are extra-terrestrial by Victor Hess in 1912, extensive observational and theoretical efforts have been made to understand their nature and origin. Their spectrum extends with relative regularity over 12 orders of magnitude in energy to about 10^{20} eV. Cosmic rays are mainly constituted of ions (87% protons, 12% helium, 1% heavier elements) and only 2% of them are electrons. While a number of characteristics of this population are now well known, the understanding of their origin remains a recalcitrant problem. It raises a number of distinct questions related to the origin of the energy, the acceleration mechanism at work, the maximum energy (Lagage & Cesarsky 1983) and spectral form produced by this mechanism. The study of supernova remnants is crucial to answer these questions for cosmic rays below the knee at about $3\,10^{15}$ eV. In the X-ray domain, the information comes primarily from synchrotron emission from accelerated electrons spiraling in the remnant magnetic field. While electrons are only a small fraction of cosmic rays, they can reveal a lot on the mechanism of diffusive shock acceleration, as they are accelerated almost like protons. Synchrotron X-rays arise from the highest-energy electrons (10^{13} eV and more), and thus probe the cut-off energy in the electron distribution, the downstream magnetic field and indirectly the geometry of the ambient magnetic field. It allows to investigate a number of key questions: level of magnetic field amplification at the shock, maximum energy of the accelerated particles, efficiency of particle acceleration, as well as where particle acceleration occurs preferentially (quasi-perpendicular versus quasi-parallel acceleration), and how particle acceleration depends on ambient magnetic field orientation.

3.1. *Magnetic field*

The magnetic field is a crucial parameter for understanding particle acceleration, for deriving the maximum energy of accelerated particles from synchrotron cut-off frequency and for interpreting the origin of TeV gamma rays in term of leptonic versus hadronic emission.

It has been realized in the last ten years that the magnetic field could be significantly amplified at supernova remnant shocks. This is supported by theoretical approaches which predict magnetic field amplification at shocks, due to instabilities occurring for efficient particle acceleration (Lucek & Bell 2000, Bell & Lucek 2001, Bell 2004) or alternatively due to a preexisting turbulent medium (Giacalone & Jokipii 2007). There are also a number of observational indications that the magnetic field is indeed amplified to the range $50 - 500\,\mu$G:
- the sharpness of the synchrotron filaments at the forward shock (see Sect. 2.1).
- the fast variability of the brightness of some of these filaments (e.g., Patnaude & Fesen 2009, Uchiyama *et al.* 2007, Uchiyama & Aharonian 2008), although see Bykov *et al.* (2008) for an alternative interpretation.
- the broadband modeling of the nonthermal emission (e.g., Berezhko, Ksenofontov & Völk 2012, Berezhko, Pühlhofer & Völk 2009, Völk, Ksenofontov & Berezhko 2008), see for an alternative Ellison *et al.* (2012).

The morphology of the synchrotron emission provides strong constraints on the geometry of the acceleration (Reynolds 1998). Reynolds (1996) proposed a model in which the asymmetry in SN 1006 is due solely to a large scale magnetic field aligned southeast to north-west. The *Chandra* data on the northeastern limb were interpreted in this framework where cosmic rays are accelerated identically everywhere (Long *et al.* 2003). Using *XMM-Newton* observations, the very low brightness of the emission in the interior above 2 keV (mostly from synchrotron emission) was shown to be incompatible with an equatorial belt geometry and rather indicates that the bright non-thermal limbs are polar caps (Rothenflug *et al.* 2004). Berezhko, Ksenofontov & Völk (2002, 2003) theoretically proposed such geometry of the upstream field. This implies that particles are accelerated where the magnetic field is parallel to the shock velocity in SN1006. 3D MHD simulations of the radio morphology are consistent with this interpretation (Bocchino *et al.* 2011).

3.2. *Maximum energy of accelerated particles*

From the measurements of the maximum energy of accelerated electrons and the downstream magnetic field, we can theoretically determine the maximum energy of accelerated protons. The measurement of the rolloff photon energy is related to the energy of the exponential cutoff in the electron spectrum, $E_{\rm max} = 39(h\nu_{\rm roll}/B_{10})^{1/2}$TeV (e.g., Reynolds *et al.* 2008), where $h\nu_{\rm roll}$ is in keV and B_{10} in units of 10 μG.

The highest measured rolloff frequency is observed in the youngest known galactic supernova remnant, G1.9+0.3. With a shock velocity of $\simeq$ 14000 km/s and an age of about 110 years (Carlton *et al.* 2011), the featureless synchrotron emission reveals a rolloff frequency of 2.2 keV (Reynolds *et al.* 2009). Assuming a moderate magnetic field of 10 μG, the maximum energy reaches about 70 TeV.

High latitude SNRs evolving in a uniform interstellar magnetic field, like SN 1006, offer the possibility to investigate the dependence of the process of particle acceleration with magnetic field orientation (see Sect. 3.1). *XMM-Newton* observations suggested that the maximum energy of electrons was much larger at the bright limbs than elsewhere, so that the maximum energy reached by particles (not only their number) is higher at the bright limbs than in the southeast and northwest (Rothenflug *et al.* 2004, Miceli *et al.* 2009). The observed averaged azimuthal variation of the rolloff frequency along the shock is very strong (factor 10) and cannot be explained by variations of the magnetic compression alone. A more detailed study is however required to get a proper estimate of the rolloff energy of the synchrotron spectrum at the spatial scale of the filamentary structures all along the shock (using radio data to anchor the spectrum at low energy). This is one of the objective of a large *XMM-Newton* program of observation of SN1006. First results on the shape of the cutoff in the synchrotron spectrum indicates that a loss-limited model provides the best match to the data (Miceli *et al.* 2013).

3.3. *Efficiency of particle acceleration*

Depending on the efficiency of particle acceleration at the shock, the synchrotron spectrum is expected to differ in shape. Test-particle acceleration models predict a power-law synchrotron emission, while nonlinear diffusive shock acceleration predict a curvature of the particle spectra (e.g., Ellison & Reynolds 1991). A few attempts have been made to search for curvature in the synchrotron spectra of supernova remnants. Some indications of curvature were found combining radio and X-ray data in SN 1006 (Allen *et al.* 2008) and RCW 86 (Vink *et al.* 2006), using infrared data in Cas A (Jones *et al.* 2003) and from radio data for Tycho and Kepler (Reynolds & Ellison 1992).

4. Conclusion

Nonthermal X-ray emission from supernova remnants provides crucial diagnostics to constrain the magnetic field intensity and geometry, and determine the properties of particle acceleration at their shock. Such studies require spatially resolved X-ray spectroscopy. Large observing programs, with *XMM-Newton* or *Chandra* satellites, are required to obtain sufficient statistics at the relevant spatial scale to characterize the properties of the narrow shock structure and its possible variability. There is a clear need for the extension of the operations of *XMM-Newton* and *Chandra* satellites, as well as a need for the preparation of the next X-ray observatory *Athena+*.

A consistent approach, including both thermal and nonthermal X-rays as well as broadband nonthermal observations from radio to TeV gamma rays, is required to fully constrain the parameters of the acceleration of particles at shocks in supernova remnants. These are key elements to understand the origin of Galactic cosmic rays.

References

Acero, F., Ballet, J., Decourchelle, A., Lemoine-Goumard, M., Ortega, M. *et al.* 2009, *A&A*, 505, 157

Aharonian, F., Akhperjanian, A. G., Aye, K.-M., Bazer-Bachi, A. R., Beilicke, M. *et al.* 2004, *Nature*, 432, 75

Aharonian, F. A., Akhperjanian, A. G., Bazer-Bachi, A. R., Beilicke, M., Benbow, W., Berge, D., Bernlhr, K., Boisson, C., Bolz, O., Borrel, V. *et al.* 2006, *A&A*, 449, 223

Allen, G. E., Houck, J. C., & Sturner , S. J. 2008, *ApJ*, 683, 773

Ballet, J. 2006, *AdSR*, 37, 1902

Bamba, A., Koyama, K., & Tomida, H. 2000, *PASJ*, 52, 1157

Bamba, A., Yamazaki, R., Ueno, M., & Koyama, K. 2003, *ApJ*, 589, 827

Bamba, A., Yamazaki, R., Yoshida, T., Terasawa, T., & Koyama, K. 2005, *ApJ*, 621, 793

Bell, A. R. 2004, *MNRAS*, 353, 550

Bell, A. R. & Lucek, S. G. 2001, *MNRAS*, 321, 433

Berezhko, E. G., Ksenofontov, L. T., & Völk, H. J. 2002, *A&A*, 395, 943

Berezhko, E. G., Ksenofontov, L. T., & Völk, H. J. 2003, *A&A*, 412, L11

Berezhko, E. G., Ksenofontov, L. T., & Völk, H. J. 2012, *ApJ*, 759, 12

Berezhko, E. G., Pühlhofer, G., & Völk, H. J. 2009, *A&A*, 505, 641

Blandford, R. & Eichler, D. 1987, *Phys.Rep.*, 154, 1

Bocchino, F., Orlando, S., Miceli, M., & Petruk, O. 2011, *A&A*, 531, 129

Bykov, A. M., Uvarov, Y. A., & Ellison, D. C. 2008, *ApJ*, 689, L133

Carlton, A. K., Borkowski, K., Reynolds, S. P., Hwang, U., Petre, R., Green, D. A., Krishnamurthy, K., & Willett, R. 2011, *ApJ*, 737, L22

Cassam-Chenai, G., Decourchelle, A., Ballet, J., Hwang, U., Hughes, J. P., & Petre, R. 2004, *A&A*, 414, 545

Cassam-Chenai, G., Decourchelle, A., Ballet, J., Sauvageot, J.-L., Dubner, G., & Giacani, E., 2004, *A&A*, 427, 199

Cassam-Chenaï, G., Hughes, J. P., Ballet, J., & Decourchelle, A. 2007, *ApJ*, 665, 315

Chevalier, R. A. 1983, *ApJ*, 272, 765

Decourchelle, A. 2005, in: L.O. Sjouwerman & K. K Dyer (eds.), *X-Ray and Radio Connections*(Published electronically by NRAO, http://www.aoc.nrao.edu/events/xraydio), E4.02, 10 pages

Decourchelle, A., Ellison, D. C., & Ballet, J. 2000, *ApJ*, 543, L57

Ellison, D. C. & Reynolds, S. P. 1991, *ApJ* 382, 242

Ellison, D. C., Slane, P., Patnaude, D. J., & Bykov, A. M. 2012, *ApJ*, 744, 39

Fukui, Y., Moriguchi, Y., Tamura, K., Yamamoto, H., Tawara, Y. *et al.* 2003, *PASJ*, 55, L61

Giacalone, J. & Jokipii, J. R. 2007, *ApJ*, 663, L41

Gotthelf, E. V., Koralesky, B., Rudnick, L., Jones, T. W., Hwang, U., & Petre, R. 2001, *ApJ*, 552, L39

Hanbury Brown, R. 1954, *The Observatory*, 74, 185

Hughes, J. P., Rakowski, C. E., Burrows, D. N., & Slane, P. O. 2000, *ApJ*, 528, L109

Hwang, U., Decourchelle, A., & Holt, S. S., Petre R. 2002, *ApJ*, 581, 1101

Jones, T. J., Rudnick, L., DeLaney, T., & Bowden, J. 2003, *ApJ*, 587, 227

Koyama, K., Kinugasa, K., Matsuzaki, K., Nishiuchi, M., Sugizaki, M. *et al.* 1997, *PASJ*, 49, L7

Koyama, K., Petre, R., Gotthelf, E. V., Hwang, U., Matsuura, M., Ozaki, M., & Holt, S. S. 1995, *Nature*, 378, 255
Lagage, P. O. & Cesarsky, C. J. 1983, *A&A* 125, 249
Lemoine-Goumard, M. 2013, *this proceedings*
Long, K. S., Reynolds, S. P., Raymond, J. C., Winkler, P. F., Dyer, K. K., & Petre, R. 2003, *ApJ* 586, 1162
Lucek, S. G. & Bell, A. R. 2000, *MNRAS*, 314, 65
Miceli, M.; Bocchino, F., Decourchelle, A., Vink, J., Broersen, S., & Orlando, S. 2013, *A&A* in press, astro-ph ariv:1306.6048
Miceli, M., Bocchino, F., Iakubovskyi, D., Orlando, S., Telezhinsky, I., Kirsch, M. G. F., Petruk, O., Dubner, G., & Castelletti, G. 2009, *A&A* 501, 239
Moriguchi, Y., Tamura, K., Tawara, Y., Sasago, H., Yamaoka, K. *et al.* 2005, *ApJ*, 631, 947
Parizot, E., Marcowith, A., Ballet, J., & Gallant, Y. A. 2006, *A&A*, 453, 387
Park, S., Kargaltsev, O., Pavlov, G. G., Mori, K., Slane, P. O., Hughes, J. P., Burrows, D. N., & Garmire, G. P. 2009, *ApJ*, 695, 431
Patnaude, D. J. & Fesen, R. A. 2009, *ApJ*, 697, 535
Pfeffermann, E. & Aschenbach, B. 1996, in: Zimmermann, Trümper and Yorke (eds.), *Roentgenstrahlung from the Universe*, 267
Pohl, M., Yan, H., & Lazarian, A. 2005, *ApJ*, 626, L101
Rettig, R. & Pohl, M. 2012, *A&A*, 545, 47
Reynolds, S. P. 1996, *ApJ*, 459, L13
Reynolds, S. P. 1998, *ApJ*, 493, 375
Reynolds, S. P., Borkowski, K. J., Green, D. A., Hwang, U., Harrus, I., & Petre, R. 2008, *ApJ*, 680, L41
Reynolds, S. P., Borkowski, K. J., Green, D. A., Hwang, U., Harrus, I., & Petre, R. 2009, *ApJ*, 695, L149
Reynolds, S. P. & Ellison, D. C. 1992, *ApJ* 399, L75
Rothenflug, R., Ballet, J., Dubner, G., Giacani, E., Decourchelle, A. *et al.* 2004, *A&A*, 425, 121
Slane, P. 2013, *this proceedings*
Slane, P., Gaensler, B. M., Dame, T. M., Hughes, J. P., Plucinsky, P. P., & Green, A. 1999, *ApJ*, 525, 357
Slane, P., Hughes, J. P., Edgar, R. J., Plucinsky, P. P., Miyata, E., Tsunemi, H., & Aschenbach, B. 2001, *ApJ*, 548, 814
Tian, W. W., Li, Z., Leahy, D. A., Yang, J., Yang, X. J. *et al.* 2010, *ApJ*, 712, 790
Uchiyama, Y., Aharonian, F. A., Tanaka, T., Takahashi, T., & Maeda, Y. 2007, *Nature*, 449, 576
Uchiyama, Y. & Aharonian, F. A. 2008, *ApJ*, 677, L105
Vink, J. & Laming, J. M. 2003, *ApJ*, 584, 758
Vink, J., Bleeker, J., van der Heyden, K., Bykov, A., Bamba, A. *et al.* 2006, *ApJ*, 648, L33
Völk, H. J., Berezhko, E. G., & Ksenofontov, L. T. 2005, *A&A*, 433, 229
Völk, H. J., Ksenofontov, L. T., & Berezhko, E. G., 2008, *A&A*490, 515
Warren, J. S., Hughes, J. P., Badenes, C., Ghavamian, P., McKee, C. F., Moffett, D., Plucinsky, P. P., Rakowski, C., Reynoso, E., & Slane, P. 2005, *ApJ* 634, 376

Discussion

MITRA: What is the expression for the maximum energy of cosmic rays? How do you get the magnetic field?

DECOURCHELLE: The expression of the maximum energy of electrons is given in sect. 3.2. It is derived from synchrotron physics and assumes an exponential cutoff in the electron spectrum. The maximum energy of cosmic rays depends on the limiting mechanism. If radiative losses cutoff the electron spectrum, accelerated protons can have much higher maximum energy. The magnetic field is derived from the width of the narrow X-ray synchrotron loss-limited filaments along the shock (Vink & Laming 2003).

Supernova Environmental Impacts
Proceedings IAU Symposium No. 296, 2013
A. Ray & R. A. McCray, eds.

doi:10.1017/S1743921313009551

Supernovae driven galactic outflows

Biman B. Nath

Raman Research Institute, Sadashivanagar, Bangalore 560080, India
email: bimanOrri.res.in

Abstract. Outflows from galaxies play a crucial role in the evolution of galaxies and also affect the surrounding medium. The standard scenario of explaining these outflows with the help of supernovae driven wind has recently come under criticism, and other processes such as radiation pressure and cosmic-rays have been invoked. We examine the relative importance of supernovae as the driving mechanism of galactic outflows in light of these competing processes.

Keywords. ISM: general– shock waves – galaxies: starburst – intergalactic medium – evolution

---

1. Introduction

Cosmological studies have informed us of the formation of different kinds structures in the universe, primarily through the influence of dark matter, which is the most dominant type of matter in the universe. However, our knowledge of the baryonic component of these structures remain incomplete to a large extent. A proper understanding of the evolution of galaxies requires a study of the complicated physical processes that include star formation, evolution of the ISM and interaction with the intergalactic or intracluster medium. These processes are challenging to study because of the feedback processes that connect one process with another in a complicated manner. These studies require a multi-wavelength approach to observations in order to understand the importance of variation of physical conditions in different phases of gas. On the theoretical side, feedback processes make it difficult to disentangle the effects of different processes. In numerical simulations, the constraint on dynamic range makes it difficult to study together the dynamics of gas inside galaxies as well as its effect on the gas that pervades the intergalactic medium. This constrains one to either focus on the large scale effects after assuming the effects of physical processes working at small scales.

One problem that surfaced in the early studies of galaxy formation and evolution in the cold dark matter cosmologies is that of 'overcooling', as simulations and semi-analytical calculations predicted galaxies which are denser, smaller and contained far too much baryons in the form of stars than was observed. It is now believed that galactic outflows provide an important feedback mechanism for the evolution of galaxies, as well as for the evolution of the intergalactic medium. Outflows can in principle solve the 'overcooling' problem by removing gas from the central region and regulating the process of star formation. It was suggested that low mass galaxies would be more vulnerable to loss of gas through outflows because of their shallow potential well, and therefore the evolution of low mass galaxies would be different from that of massive galaxies (Dekel & Silk 1986). This could explain the mass-metallicity relation observed in ellipticals, removing the metals preferentially from low mass galaxies and making them metal poor. Outflows were incorporated in the study of metallicity evolution of ellipticals, and the enrichment of the intracluster medium (ICM). After the discovery of metals in the intergalactic medium (IGM), it has been suggested that outflows, especially from low mass galaxies that were abundant at earlier cosmological epochs, could enrich the IGM (Nath & Trentham 1997).

More detailed studies of the enrichment of the IGM have found that the prescriptions used for outflow speed and mass loss rate are crucial in understanding the evolution of IGM metallicity (Oppenheimer & Davé 2006). However, studies aimed at cosmological implications of outflows cannot resolve small scale physical processes and have to depend on different ansatz for outflow parameters. The hot gas in the outflowing matter can also substantially distort the cosmic microwave background through Sunyaev-Zel'dovich effect (e.g, Majumdar *et al.* 2001) Recently, galactic outflows have also been invoked to solve the problem of reconciling the steep dark matter density profile predicted from simulations and observed profiles with core (e.g., Governato *et al.* 2010).

However a proper understanding of the effects of outflows requires a detailed knowledge of the mechanisms and the interplay between different physical processes responsible for the outflows. Detailed observations of local galaxies, and spectroscopic studies on high redshift galaxies, show that galactic outflows have far more complicated structure than that of simple steady winds. (Veilleux, Cecil & Bland-Hawthorn 2005, and references therein). These outflows harbor gas at different phases, from cold and molecular, to warm neutral to partially ionized clouds (that are detected through line radiation or absorption) and which are embedded in a hot medium which shines in x-rays. The standard scenario assumes that multiple supernovae in a star forming galaxy heats up the interstellar medium (ISM), which produces the hot phase of outflows, and that the thermal pressure of the hot gas drives the outflowing gas. The consideration of heating and cooling processes in the ISM heated by supernovae led Dekel & Silk (1986) to the result that outflows are likely to be more important for low mass galaxies with velocity dispersion $\leqslant 100$ km s^{-1}, i.e., dwarf galaxies. Since the temperature of the hot phase of outflowing gas is found to be ~ 1 keV (Martin 2005), this implies a near constant speed of the outflowing gas, of order the thermal speed. Chevalier & Clegg (1985) considered supersonic outflowing wind from starbursts such as M82 (ignoring the effect of gravity), and suggested a outflow speed in excess of 1000 km s^{-1}. Wang (1995) included the effect of dark matter potential and found that gravity can considerably slow down the wind speed. A number of simulations have also studied the dynamics of outflowing gas, especially in comparison with the observations of M82 (e.g., Strickland & Stevens 2000).

There are however a number of problems with this scenario of thermal pressure driven outflows. Firstly, there is a problem of the survival of the cold/warm clouds in the hot gas. Observations find these clouds at large distances (several kpc) from the central star forming regions of the galaxy. Assuming thermal speed of a 1 keV gas, the time taken by the clouds to reach such distances is ~ 10 Myr. This long travel time makes it difficult for the clouds to survive against several instabilities at work, e.g., Kelvin-Helmholtz instability or evaporation by thermal conduction, which have a time scale of order $\leqslant 1$ Myr (Marcolini *et al.* 2005). Secondly, recent observations show a correlation between the maximum speed of cold/warm clouds (at 10^4 K) embedded in the hot medium and the galaxy rotation speed (Martin 2005). Also, observations of Lyman break galaxies at high redshift show that the outflow speed is correlated with the star formation rate (Shapley *et al.* 2003), as well as with the reddening due to dust. Ferrara and Ricotti (2006) found that the observed correlation of outflow speed with star formation rate is difficult to explain in the standard SN-driven wind scenario.

Murray *et al.* (2005) pointed out the importance of radiation pressure in outflows and showed that it can be comparable to the ram pressure of the hot gas on the cold clouds. They worked out the dynamics of radiatively driven shells of gas and dust, and Martin (2005) found that this model explains the observations of cold clouds better than the thermal pressure scenario. Nath & Silk (2009) discussed a hybrid model of outflows from

Lyman break galaxies with ram and radiation pressure. In this scenario the radiation pressure from massive stars in a stellar population pushes a shell of gas and dust out to a large distance, and then the ram pressure from SNe heated gas pushes it further. The shell fragments due to the acceleration and can explain the existence of clouds at large distances (see also Murray *et al.* 2011).

2. Ram and radiation pressure

The effect of radiation pressure for a disk galaxy can be estimated in terms of an Eddington parameter, $\Gamma_0 = \kappa I/(2cG\Sigma)$, where κ is the opacity of a mixture of dust and gas (in the units of cm^2 g^{-1}), I and Σ are the surface brightness and surface mass density of the disk respectively. Recently, Sharma *et al.* (2011) considered a general case of wind driven by radiation pressure on dust grains from a disk galaxy and calculated the outflow speed as a function of the galaxy rotation speed. They argued that although the details of the wind structure depends on various parameters, the terminal speed of the outflowing gas can be estimated in terms of the parameters near the disk by using the Bernoulli equation. They considered a halo with Navarro-Frenk-White density profile, a spherical bulge and a disk with constant surface density, and assumed an instantaneous burst of star formation. They found that the minimum value of the Eddington parameter Γ_0 for outflows to escape the virial radius of a halo to be ~ 2, which corresponds to a mass to light ratio (in the solar units) of $\Upsilon_B \leqslant 0.04$ (for typical NFW parameters of a galaxy, and for $\kappa \sim 200$ cm^2 g^{-1}). Although this is much smaller than the observed values of disk galaxies ($\Upsilon_B \sim 1.5$), starburst galaxies can have a much lower value. They also found that the ratio of the outflow terminal speed to the galaxy rotation speed is ~ 2–4 for a period of ~ 10 Myr, consistent with observations. Also, simulations of IGM enrichment claim to better match with data when they use a terminal speed of order three times the galactic rotation speed (Oppenheimer & Davé 2006).

This result was confirmed with a 2-D hydrodynamic simulation, which further enabled one to study the structure of the wind in the presence of angular momentum of the gas lifted from a rotating disk (Chattopadhyay *et al.* 2012). They found that the combined effect of gravity and radiation pressure from an exponential, rotating disk can drive the outflowing gas out to a distance of ~ 5 kpc in ~ 40 Myr for typical galactic parameters. The outflow speed was found to increase rapidly for $\Gamma_0 \geqslant 1.5$. The wind was seen to develop a conical structure because of the angular momentum of the gas, and the rotation speed of the wind was less than the galaxy rotation speed. The structure of the rotating wind was similar to that in M82, and another such example has been recently observed by Vogt *et al.* (2013) in the case of HCG16.

In order to compare the relative roles of ram and radiation pressure in driving the clouds embedded in outflows, Sharma & Nath (2012) considered outflows from disk galaxies with continuous star formation, and included the effects of gravity due to an exponential disk, a spherical bulge and dark matter halo. They presented the results in the parameter space of the star formation rate (SFR) and the rotation speed of galaxies, and in terms of the ratio of outflow speed in the units of the galaxy rotation speed, so that the results could be compared with data from outflows. According to their results, the dynamics of clouds in outflows from galaxies with rotation speed $v_c \leqslant 200$ km s^{-1} and SFR $\leqslant 100$ M$_\odot$ yr^{-1} is dominated by ram pressure, and radiation pressure becomes more important in more massive galaxies and with with larger SFR. Further, they found that the ratio of mass loss rate in the wind and star formation rate scales as $v_c^{-1}\Sigma_g^{-1}$, where Σ_g is the gas column density of the disk. Recent simulations confirm such a division in the regimes of thermal and radiation pressure dominated outflows in terms of galactic mass, and the dependence of mass loss rate with the galactic rotation speed.

A crucial parameter in this calculation is the assumption of the speed of the hot gas in the outflows. As mentioned earlier, this has been assumed to be either of order thermal speed appropriate for a gas with temperature $\sim$ 0.5–1 keV (for which the isothermal speed is $\sim$ 300–400 km s^{-1}), or a larger speed corresponding to a supersonic case. It therefore becomes important to study the transonic solutions of the hot gaseous wind, especially in the presence of both thermal and radiation pressure. There is also the possibility of radiation pressure from an accreting, luminous black hole in the central regions of massive galaxies. Sharma & Nath (2012) considered steady spherical wind driven by energy, mass and momentum injection processes that are confined to a central region of size $\sim$ 200 pc. One can write a general equation for the transonic wind in terms of the Mach number, which reduces to the Chevalier & Clegg (1985) solution in the case of zero gravity.

Sharma & Nath (2012) showed that the dynamics of winds is determined by the interplay of three velocity scales: (a) $v_* \sim (\dot{E}/2\dot{M})^{1/2}$ is related to the starburst activity, with $\dot{E}$ and $\dot{M}$ as the energy and mass injection rate in the central region; (b) $v_{bh} \sim (GM_{bh}/2R)^{1/2}$ reflects the gravitational effect of a central black hole on gas in the central region, and (c) $v_s = (GM_h/2\mathcal{C}r_s)^{1/2}$, which is related to the galactic rotation speed for an NFW halo for halo mass M_h, where r_s is the scale length and $\mathcal{C}$ is a function of the concentration parameter. The values of v_* range from $\sim$ 200–500, depending on the efficiency of supernovae to deposit energy in the ISM, with a larger value being appropriate for starburst galaxy, and lower values for Milky Way type galaxies with quiescent SFR. This follows naturally from the definition of v_* that $v_* \sim 560\sqrt{\alpha}$ km s^{-1}, where α denotes the efficiency of SN energy thermalisation, and the two limiting values of v_* mentioned above follow from the assumption of $\alpha \sim 0.1$–1.

The generalized formalism then allows one to determine the terminal velocity of outflows to be $(4v_*^2 + 6(\Gamma - 1)v_{bh}^2 - 4v_s^2)^{1/2}$, where Γ is the ratio of force due to radiation pressure to gravity of the central black hole. The hot wind speed in the low to intermediate mass galaxies was determined to be in the range$\sim$ 400–1000 km s^{-1}. On the contrary, massive galaxies with central black holes radiating at Eddington limit have outflows with speed in excess of 1000 km s^{-1}. Observations of terminal speed of outflows can therefore be a useful diagnostic of whether or not it is driven by supernovae or active galactic nuclei (AGN). This limiting speed of $\sim$ 1000 km s^{-1} is supported by recent observations (Tremonti *et al.* 2007; Sturm *et al.* 2011).

Sharma & Nath (2012) further showed that winds from quiescent star forming galaxies cannot escape for intermediate size galaxies ($10^{11.5} M_\odot \leqslant M_h \leqslant 10^{12.5} M_\odot$), and that these galaxies should harbor a large amount of gas in their halos. This is consistent with the observations of Tumlison *et al.* (2011) in which halos of galaxies in this mass range were reported to contain a large amount of hot gas.

They derived a ratio $[2v_*^2 - (1-\Gamma)v_{bh}^2]/v_c^2$ which is proportional to the amount of gas lost through the winds. The behavior of this ratio with galactic mass therefore determines the ability of galaxies to expel or retain their gas reservoir, and this function can be evaluated by assuming an appropriate relation between the central black hole mass and the galactic mass, and an appropriate opacity of dust grains in the infrared (K) band. It is interesting that the value of this ratio has a minimum at the galactic mass scale of $\sim 10^{12-12.5} M_\odot$, which signifies the change in regime of supernovae to AGN domination in outflow properties. Used with a simple chemical evolution model of galaxies, they showed that this result is consistent with the observed trend of stellar mass to galactic mass ratio, and that the above mentioned range of hot gas wind (400–1000 km s^{-1}) can explain the observed scatter in the observed ratio of stellar to halo masses.

3. Cosmic ray driven wind

Besides the effect of thermal and radiation pressure, another important physical processes that has been invoked to be responsible for driving outflows is that involving cosmic rays. It is believed that when cosmic rays stream with respect to the ambient medium, they excite Alfvén waves on which they scatter, and consequently limit the effective bulk speed of cosmic rays. The damping of Alfvén waves excited by streaming cosmic rays can lead to the heating of the ISM and subsequently to outflows. Recently, a detailed numerical simulation that includes the physics of cosmic ray streaming, it was found that this process can drive galactic winds from galaxies with halo masses $\leqslant 10^{11}$ $M_\odot$ (Uhlig *et al.* 2012). This process also suppresses the star formation rate, particularly in dwarf galaxies (10^9 $M_\odot$), in which $\sim 60\%$ of the baryonic matter is expelled, with a mass loss rate reaching a factor of 5 more than the star formation rate.

These results are also supported by analytical estimates done with the help of the Bernoulli equation, as in the case of radiation pressure driven winds. Furthermore, the simulations showed that in higher mass galaxies, the winds develop a bi-conical structure. It is possible that the early galactic halos, which were mostly low mass galaxies, were affected by cosmic ray pressure driven winds, since radiation pressure would have been of less importance because of lack of metals required to produce dust grains. Also, there are a few examples of galactic outflows which occupy a region of the parameter space of SFR and galactic rotation speed, as in Sharma & Nath (2012), that are neither explained by thermal or radiation pressure adequately, with very low SFR ($\leqslant 0.1$ $M_\odot$ yr^{-1}) and low galactic rotation speed ($\leqslant 60$ km s^{-1}). Interestingly, the examples simulated by Uhlig *et al.* (2012) fall in this range of parameters, and it would be curious to study such cases with more detailed simulations in the future.

4. Threshold condition for supernovae driven outflows

Another important problem to study is the conditions required to launch galactic outflows from disk galaxies. Although it is believed that in general galaxies with high SFR excite such outflows, Heckman (2002) argued that the determining condition is related to the surface density of SFR, and not the average SFR in the galaxy. He found a threshold SFR of ~ 0.1 $M_\odot$ yr^{-1} kpc^{-2} for the existence of galactic winds. Another set of observations, aimed at studying the extraplanar gas in edge-on disk galaxies, found a minimum value of energy injection that were required for the existence of synchrotron emitting gas. The threshold surface density of energy injection is $\sim 10^{-4}$ erg s^{-1} cm^{-2} (Dahlem *et al.* 1995). Tüllmann *et al.* (2006) further considered the conditions for the existence of X-ray, radio and far-infrared emission from extended halo gas in disk galaxies, and found the minimum energy injection rate to be $\sim 10^{-3}$ erg s^{-1} cm^{-2}.

Recently, it has been argued that these threshold conditions can be understood from the requirement of multiple SNe driven superbubbles to compensate against radiative loss and be able to break out of the disk with sufficient Mach number in order for the hot interior gas to reach a considerable height above the disk (Roy *et al.* 2013). The basic argument can be illustrated with the help of a simple estimate. Consider a SN remnant, which enters the radiative phase when the shock speed becomes ~ 70 km s^{-1}, corresponding to a post-shock temperature of $\sim 10^5$ K, where the cooling function peaks. We denote this epoch by $t_{rad} \sim 3 \times 10^5$ yr $E_{51}^{1/3}$ $n^{1/3}$, where $E = 10^{51} E_{51}$ erg is the typical energy of a SN, and n is the particle number density of the ISM in the units of cm^{-3}. The corresponding radius is $R_{rad} \sim 50\,\mathrm{pc}\,E_{51}^{1/3}\,n^{1/3}$. Writing the volume rate of SN as ν_{SN}, the condition that multiple SN can inject enough energy for the superbubble not to stall after reaching the radiative phase can be written as, $(4\pi/3)R_{rad}^3 t_{rad}\nu_{SN} \geqslant 1$, which translates to $\nu_{SN} > 6\times 10^{-12}(n/E_{51})^{4/3}$ SNe yr^{-1} pc^{-3}. The surface density of SNe can be found by multiplying this by the scale height (which can be shown to be comparable to the size of the bubble in the disk plane during the break out phase), and this leads to 6×10^{-4} SNe yr^{-1} kpc^{-2}, for a scale height of 100 pc. Using Salpeter IMF, and considering stellar masses in the range of 1–100 $\mathrm{M}_\odot$, the corresponding surface density of SFR becomes $\sim 0.1\ \mathrm{M}_\odot\ \mathrm{yr}^{-1}\ \mathrm{kpc}^{-2}$, consistent with observations.

Roy *et al.* (2013) has worked out the dynamics of superbubbles triggered by multiple SNe in a disk galaxy, first analytically and then with hydrodynamic simulations, and they found that this simple estimate is supported by detailed calculations. Furthermore, they studied the clumping and fragmentation of the superbubbles during its evolution, and found that thermal instability plays an important role in clumping of the shell material, and in subsequent seeding of Rayleigh-Taylor instability. Without the aid of thermal instability, Roy *et al.* (2013) showed that the clumping due to Rayleigh-Taylor instability usually takes place at a time scale that is longer than the main sequence life time of O stars, the main source of ionizing radiation. This problem is therefore important not only for the hot interior gas to be released into the halo region, but also for the escape of ionizing radiation from massive stars. The latter issue is relevant for the problem of reionization of the IGM.

5. Summary

Although thermal pressure of supernovae heated gas remains the dominant mechanism for driving outflows from galaxies, other physical processes such as radiation and cosmic ray pressure can also play important roles. Recent analytical and simulations show that radiation pressure becomes important for massive galaxies either with high SFR or with an AGN component. The signature of AGN driven winds can be diagnosed from the outflow speed which in this case can exceed 1000 km s^{-1}. Cosmic ray pressure can be important for low mass galaxies. Intermediate sized galaxies with quiescent SFR are predicted to contain a large amount of gas in their halo. Also, consideration of radiative cooling is important to understand the threshold conditions for excitation of galactic winds from disk galaxies.

References

Bisnovatyi-Kogan, G. S. & Silich, S. A. 1995, *Rev Mod Phys*, 67, 661

Chattopadhyay, I., Sharma, M., Nath, B. B., & Ryu, D. 2012, *MNRAS*, 423, 2153

Chevalier, R. A. & Clegg, A. W. 2009, *Nature*, 317, 44

Dekel, A. & Silk, J. 1986, *ApJ*, 303, 39

Ferrara, A. & Ricotti, M. 2006, *MNRAS*, 373, 571

Governato, F. *et al.* 1986, *Nature*, 463, 203

Heckman, T. M. 2002, in: L. S. Mulchaey & J. Stocke (eds.), *ASP Conf. Ser. 254, Extragalactic Gas at Low Redshift* (San Fransisco, CA: ASP), p. 292

Majumdar, S., Nath, B. B., & Chiba, M. 2001, *MNRAS*, 324, 537

Marcolini, A. Strickland, D. K., D'Ercole, A., Heckman, T. M., & Hoopes, C. G. 1986, *MNRAS*, 362, 626

Martin, C. L. 2005, *ApJ*, 621, 227

Murray, N., Quataert, E., & Thompson, T. A. 2005, *ApJ*, 618, 569

Murray, N., Ménard, B., & Thompson, T. A. 2011, *ApJ*, 735, 66
Nath, B. B. & Trentham, N. 1997, *MNRAS*, 291, 505
Nath, B. B. & Silk, J. 2009, *MNRAS*, 396, L90
Oppenheimer, B. D. & Davé, R. 2006, *MNRAS*, 373, 1265
Roy, A., Nath, B. B., Sharma, P., & Shchekinov, Y. 2013, *arxiv:1303.2664*
Shapley, A. E., Steidel, C. C., Pettini, M., & Adelberger, K. L. 2003, *ApJ*, 588, 65
Sharma, M., Nath, B. B., & Shchekinov, Y. 2011, *ApJL*, 763, 27
Sharma, M. & Nath, B. B. 2012, *ApJ*, 750, 55
Sharma, M. & Nath, B. B. 2013, *ApJ*, 763, 17
Strickland, D. K. & Stevens, I. R. 2000, *ApJ*, 314, 511
Sturm, E. *et al.* 2011, *ApJL*, 733, 16
Tremonti, C. A., Moustakas, J., & Diamnond-Stanic, A. M. 2007, *ApJL*, 663, 77
Tumlinson *et al.* 2011, *Science*, 334, 948
Uhlig, M., Pfrommer, C., Sharma, M., Nath, B. B., Enßlin, T. A., & Springel, V. 2012, *MNRAS*, 423, 2374
Veilleux, S., & Gerald, C. Bland-Hawthorn, J. 2005, *ARAA*, 43, 769
Wang, B. *et al.* 1995, *ApJ*, 444, 590

Discussion

WANG: One possibility is that condensations can be important when seeds of cool gas clouds are available which can be driven out by radiation pressure, for example.

NATH: Yes, this possibility needs to be explored.

HENSLER: Two comments: 1. Galactic winds models usually do not take the external gas into account, e.g. Hot holo-gas, surrounding gas disks in star burst galaxies, both hampering the outflow. 2. HVC should be self-gravitating according to their mass. This suppresses Rayleigh Taylor instability. In addition when passing through hot halo-gas, HVCs experience thermal conduction. Vieser & Hensler (2007, A & A, 475, 251) have shown that this thermal conduction suppresses Kelvin Helmholtz instability so that such HVCs can survive for 100 Myrs and longer.

NATH: Thank you for the comments. Perhaps the issue of cloud survival in hot wind needs more work.

NTORMOUSI: Regarding the growth of RT and KH instabilities on the surfaces of cold clumps embedded in a hot medium. What is the relationship between the size of the clouds and the mean free path of the particles in the hot gas? (a large mfp would mean no hydro instabilities).

NATH: The mean free path is smaller than typical cloud sizes.

SINGH: If supernova outflows can create a global thermal equilibrium in scales as large as intracluster medium of galaxy clusters, can the mass outflow alter the NFW profile of these relaxed halos?

NATH: These outfows are thought to affect the dark matter profile at very small scales (< 1 kpc) but not at larger scales.

Supernova Environmental Impacts
Proceedings IAU Symposium No. 296, 2013
A. Ray & R. A. McCray, eds.

doi:10.1017/S1743921313009563

Reprise of the Supershells

Sayan Chakraborti[1] and Alak Ray[2]

[1]Harvard University, Cambridge MA 02138, USA
email: `schakraborti@fas.harvard.edu`

[2]Tata Institute of Fundamental Research, Mumbai 400005, India
email: `akr@tifr.res.in`

Abstract. Neutral hydrogen cavities have been detected in the Milky Way for decades and more recently in other nearby star-forming galaxies. It has been suggested that at least a fraction of them may be expanding supershells driven by the combined mechanical feedback from multiple supernovae occurring in an OB association. Yet most extragalactic cavities had neither a demonstrated expansion velocity nor an identified OB association inside them. In this presentation, we will outline how new multiwavelength observations are providing us with systems to test the theory of supershells driven by the mechanical feedback from multiple supernovae. We shall also discuss the consequences of these recent results in the context of supernova feedback, propagating star formation and particle acceleration.

Keywords. supernovae: general — Galaxies: individual (M101) — ISM: Bubbles — Radio lines: ISM

---

1. Intorduction

Most massive stars in spiral galaxies are born in OB associations and other young stellar clusters. They contain a few to thousands of massive stars of spectral class O and B. O stars are short-lived, and die as supernovae within a few megayears. These associations last for few tens of megayears because all their massive stars exhaust their nuclear fuel within this duration. OB associations ionize the interstellar medium (ISM) in galaxies and produce localized HII regions as well as diffuse ionized gas. According to Oey *et al.* (2002) supernovae, exploding from the massive stars that form in these associations, pressurize and churn the ISM. Expanding neutral hydrogen structures, called supershells were identified in the Milky Way by Heiles (1979). McCray and Kafatosw (1987) have shown that supershells form and evolve due to the continuous mechanical energy injection by stellar winds and SNe in massive parent OB associations. The outer shocks of the superbubbles sweep up the ambient ISM into a thin cool shell. These supershells may also play a role in the process of star formation and favor propagating star formation, as pointed out by McCray and Kafatosw (1987), Palous *et al.* (1994). Hopkins *et al.* (2008), point out that the neutral and molecular gas replenishment in the walls of the supershells may provide the trigger for collapse and further star formation. In NGC 300, Blair and Long (1997) reported several cavities, much larger than galactic supernova remnants (SNRs). Payne *et al.* (2004) suggest that these may represent the above mentioned supershells.

Cavities have been found in neutral hydrogen (HI) surface density maps of several nearby galaxies. For a classical review and inventory of HI holes, see Tenorio-Tagle and Bodenheimer (1988). For a recent review and survey of HI holes see Bagetakos *et al.* (2011). Also seeKamphuis *et al.* (1991), Chakraborti and Ray (2011) for supershells in

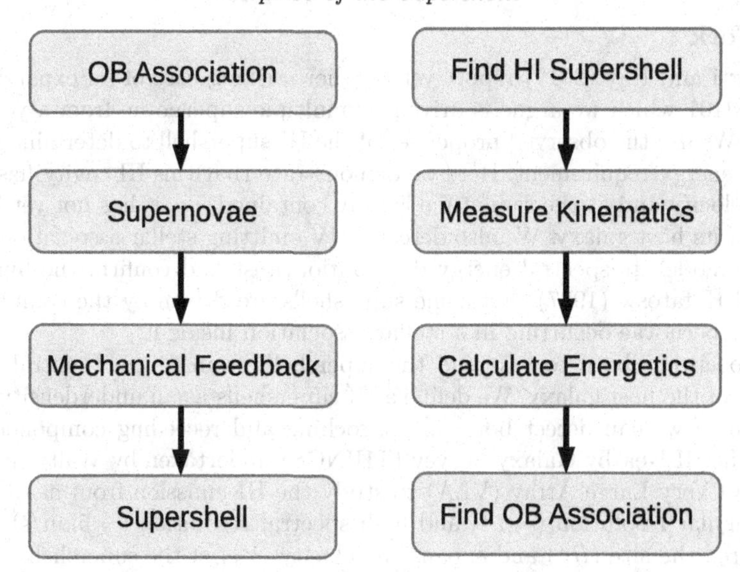

Figure 1. The flowchart on the left describes how a supershell is formed. It starts with an initial OB association which finally powers the supershells. The flowchart on the right describes how one can test this paradigm. It starts by finding a superhsell and looks for its energy source, namely an OB association. Note that the steps in the two charts essentially go in reverse.

M101. However, the presence of a cavity alone is insufficient evidence for the existence of a supershell. It may simply be a low density region between higher density ones, created by turbulent distribution of gas. This pitfall is illustrated by the case of Holmberg II. In that galaxy Puche *et al.* (1992) used underdensities in HI column density map to create a catalog of HI cavities and postulated that they have been evacuated by multiple SNe. However Rhode *et al.* (1999) rule out a multiple supernova origin for most of these holes since they were unable to find any young cluster or even trace of recent star formation activity which would have led to the SNe responsible for these HI-holes. Stewart *et al.* (2000) have since then re-examined the Rhode *et al.* (1999) results and concluded that the energy available from massive stars inside HI shells indicates that energy deposited into the ISM from supernovae and stellar winds is sufficient to account for the HI morphology. Silich *et al.* (2006) have found that the observed properties of the most prominent kpc-scale structure in IC 1613 and the level of the detected star formation activity are inconsistent with the hypothesis that they are formed by energy injection from multiple supernovae.

The first unambiguous case of an extragalactic expanding HI shell associated with a hole in the HI surface density was reported by Kamphuis *et al.* (1991) in observations of M101 using the Westerbork Synthesis Radio telescope. Weisz *et al.* (2009) have investigated the creation of a supergiant HI shell in a dwarf irregular galaxy IC 2574 and its role in triggering star formation around it. Star formation studies such as those reviewed by Kennicutt (1998) in most nearby galaxies indicate that sites of recent star formation are correlated with sites of higher HI surface densities, leading to several recipes for star formation. The feedback processes described above should however clear out the gas and lead to young stellar associations in regions of low gas density surrounded by higher density swept up shells. Therefore we began searching for cavities in the HI surface density, occupied by young clusters.

2. Our Work

Chakraborti and Ray (2011) report yet another striking case of an expanding HI supershell in M101 which we argue is driven by multiple supernovae from a young stellar association. We use the observed properties of the HI supershell to determine its dynamical age and energy requirement. Here we demonstrate that this HI cavity has measured expansion velocity and at the same time is fully contained, i.e. it has not yet broken out of the disk of its host galaxy. We also detect a UV emitting stellar association inside the HI hole. We model its spectral energy distribution, test and confirm the hypothesis of McCray and Kafatosw (1987) that some supershells are driven by the combined action of multiple supernovae occurring in a stellar association inside it.

Our approach has been to first find the supershells from 21 cm (neutral Hydrogen) observations of the host galaxy. We define a HI supershells as an underdensity of neutral hydrogen where we can detect both an approching and receeding component. We use data from The HI Nearby Galaxy Survey (THINGS) undertaken by Walter *et al.* (2008) at the NRAO Very Large Array (VLA) to study the HI emission from nearby galaxies. The high angular resolution ($\sim 7''$) and high spectral resolution ($\sim$ 5km/s) data allow us to measured the size (R_S) and expansion velocity (V_S) of the supershell.

We then inverted the McCray and Kafatosw (1987) model, as R_S and the V_S are directly observable from the HI data. In this work, we therefore re-frame the equations as

$$t_7 = (R_S/97\ \mathtt{pc})(V_S/5.7\ \mathtt{km\ s}^{-1})^{-1} \tag{2.1}$$

and

$$(N_* E_{51}/n_0) = (R_S/97\ \mathtt{pc})^2(V_S/5.7\ \mathtt{km\ s}^{-1})^3 \tag{2.2}$$

to express the variables, which characterize the supershell, purely in terms of the observable quantities. These equations will allow us to estimate the age ($t = t_7 \times 10^7$ years) and mass (where N_* is the number of massive stars) of an association which will suffice to reproduce the observed size and expansion velocity of any given supershell. We used data from Far and Near Ultraviolet observations are available from GALEX Nearby Galaxies Survey, conducted by Martin *et al.* (2005) with the Galaxy Evolution Explorer (GALEX). The ultraviolet emission provided us with a simple tool for identifying the young stellar association, which was supplied the energy budget for the expanding supershell.

3. The way ahead

It is important to understand supernova feedback in galaxies at from the level of individual supernovae to the whole galaxy. Supershells a a very important link in this chain and Chakraborti and Ray (2011) have shown that the dynamics of a supernova driven supershell can explain the kinematics of an observed supershell. Chakraborti (2011) shows that self-regulated star formation, driven by the competition between gravitational instabilities and mechanical feedback from supernovae, can explain the nearly constant neutral hydrogen surface density across galaxies. Warren *et al.* (2011) studied the formation of kiloparsec-scale HI holes in dwarf galaxies and concluded that large HI holes are likely formed from multiple generations of star formation. This complicates the story of a single coeval star cluster powering a single supershell. SN 2008jb, discovered by Prieto *et al.* (2012), was the first supernova inside a supershell. This is our opportunity to see supernova feedback live in action. Suad *et al.* (2012) reported the discovery of GS100-02-41, yet another large HI shell in the outer Milky Way and discussed the evidence for trigerred star formation. This goes on to show that new supershells can still be found in our own galaxy and as a result studied in great detail. Cold filamentary structures formed in supershells, simulated by Ntormousi *et al.* (2011), could be responsible for trigerred star formation. The study of Continuum Halos in Nearby Galaxies: An EVLA Survey (CHANG-ES) led by Irwin *et al.* (2012) may open new windows into magnetized outflows from large supershells bursting out of the cold gas disk of starforming galaxies. Roy *et al.* (2013) outline the conditions necessary for such break outs.

References

M. S. Oey, B. Groves, L. Staveley-Smith, & R. C. Smith, *AJ* 123, 255–268 (2002), arXiv:astro-ph/0112057

C. Heiles, *ApJ* 229, 533–537 (1979)

R. McCray, & M. Kafatos, *ApJ* 317, 190–196 (1987)

J. Palous, G. Tenorio-Tagle, & J. Franco, *MNRAS* 270, 75 (1994)

A. M. Hopkins, N. M. McClure-Griffiths, & B. M. Gaensler, *ApJ* 682, L13–L16 (2008), 0806.0662

W. P. Blair, & K. S. Long, *ApJS* 108, 261 (1997)

J. L. Payne, M. D. Filipović, T. G. Pannuti, P. A. Jones, N. Duric, G. L. White, & S. Carpano, *A&A* 425, 443–456 (2004)

G. Tenorio-Tagle, & P. Bodenheimer, *ARA&A* 26, 145–197 (1988)

I. Bagetakos, E. Brinks, F. Walter, W. J. G. de Blok, A. Usero, A. K. Leroy, J. W. Rich, & R. C. Kennicutt, Jr., *AJ* 141, 23 (2011), 1008.1845

J. Kamphuis, R. Sancisi, & T. van der Hulst, *A&A* 244, L29–L32 (1991)

S. Chakraborti, & A. Ray, *ApJ* 728, 24 (2011), 1011.5232

D. Puche, D. Westpfahl, E. Brinks, & J.-R. Roy, *AJ* 103, 1841–1858 (1992)

K. L. Rhode, J. J. Salzer, D. J. Westpfahl, & L. A. Radice, *AJ* 118, 323–336 (1999), arXiv:astro-ph/9904065

S. G. Stewart, M. N. Fanelli, G. G. Byrd, J. K. Hill, D. J. Westpfahl, K.-P. Cheng, R. W. O'Connell, M. S. Roberts, S. G. Neff, A. M. Smith, & T. P. Stecher, *ApJ* 529, 201–218 (2000)

S. Silich, T. Lozinskaya, A. Moiseev, N. Podorvanuk, M. Rosado, J. Borissova, & M. Valdez-Gutierrez, *A&A* 448, 123–131 (2006), arXiv:astro-ph/0510812

D. R. Weisz, E. D. Skillman, J. M. Cannon, F. Walter, E. Brinks, J. Ott, & A. E. Dolphin, *ApJ* 691, L59–L62 (2009), 0812.2036

R. C. Kennicutt, Jr., *ARA&A* 36, 189–232 (1998), arXiv:astro-ph/9807187

F. Walter, E. Brinks, W. J. G. de Blok, F. Bigiel, R. C. Kennicutt, Jr., M. D. Thornley, & A. Leroy, *AJ* 136, 2563–2647 (2008), 0810.2125

D. C. Martin, J. Fanson, D. Schiminovich, P. Morrissey, P. G. Friedman, T. A. Barlow, T. Conrow, R. Grange, P. N. Jelinsky, B. Milliard, O. H. W. Siegmund, L. Bianchi, Y.-I. Byun, J. Donas, K. Forster, T. M. Heckman, Y.-W. Lee, B. F. Madore, R. F. Malina, S. G. Neff, R. M. Rich, T. Small, F. Surber, A. S. Szalay, B. Welsh, & T. K. Wyder, *ApJ* 619, L1–L6 (2005), arXiv:astro-ph/0411302

S. Chakraborti, *ApJ* 732, 105 (2011), 1103.2763

S. R. Warren, D. R. Weisz, E. D. Skillman, J. M. Cannon, J. J. Dalcanton, A. E. Dolphin, R. C. Kennicutt, Jr., B. Koribalski, J. Ott, A. M. Stilp, S. D. Van Dyk, F. Walter, & A. A. West, *ApJ* 738, 10 (2011), 1105.4117

J. L. Prieto, J. C. Lee, A. J. Drake, R. McNaught, G. Garradd, J. F. Beacom, E. Beshore, M. Catelan, S. G. Djorgovski, G. Pojmanski, K. Z. Stanek, & D. M. Szczygieł, *ApJ* 745, 70 (2012), 1107.5043

L. A. Suad, S. Cichowolski, E. M. Arnal, & J. C. Testori, *A&A* 538, A60 (2012), 1112.0516

E. Ntormousi, A. Burkert, K. Fierlinger, & F. Heitsch, *ApJ* 731, 13 (2011), 1011.5751

J. Irwin, R. Beck, R. A. Benjamin, R.-J. Dettmar, J. English, G. Heald, R. N. Henriksen, M. Johnson, M. Krause, J.-T. Li, A. Miskolczi, S. C. Mora, E. J. Murphy, T. Oosterloo, T. A. Porter, R. J. Rand, D. J. Saikia, P. Schmidt, A. W. Strong, R. Walterbos, Q. D. Wang, & T. Wiegert, *AJ* 144, 43 (2012), 1205.5694

A. Roy, B. B. Nath, P. Sharma, & Y. Shchekinov, *ArXiv e-prints* (2013), 1303.2664

Discussion

KOO: The (P, V) diagram that you showed clearly shows that there is a bubble at the position of the supershell. But it is not obvious to me if it shows that these are expanding portions of the shell because these are comparable HI emission at the same velocity outside the supershell.

CHAKRABORTI: Yes, so the point is that this supershell is not so big that it has yet punctured out of the disk of the galaxy. There is an example of a supershell in M101, by Kamphuis *et al.* (1991 A & A), in which these components have gone out and detached themselves from the rest of the galaxy. So, that would be an example of a supershell which is so big that it has punctured out of the disk of the galaxy and has fragmented the top and bottom parts bue to Rayleigh Taylor instabilities, possibly, which was discussed by Dr. Biman Nath (see paper by B. Nath at same symposium), but this shell is not yet so big.

FRANCE: Comment: UV spectroscopic observations of hot star wind lines (e.g. CIII λ 1177,NV 1240, SiIV 1400, CIV 1550 $\AA$) will give you a good measure of both age and metallicity when compared with star bursts 99 models.

CHAKRABORTI: There is not much extinction which would stop the UV. So sure, that would be very nice.

WANG: M101 has been well observed in X-ray. This should give you a direct check of your scenario.

CHAKRABORTI: Yes, that is an excellent suggestion. This galaxy has mega-seconds of Chandra observations. We looked at this region of the data. I cant tell you for sure whether there is extended emission there? There seem to be a few point sources. It would be very interesting to be able to detect the extended emission from the coronal gas inside. The point sources would most probably be HMXBs. But this is work which is ongoing and very important to do.

Supernova Environmental Impacts
Proceedings IAU Symposium No. 296, 2013
A. Ray & R. A. McCray, eds.

doi:10.1017/S1743921313009575

The Supernova – ISM/Star-formation interplay

Gerhard Hensler

Department of Astrophysics, University of Vienna, Tuerkenschanzstr. 17, 1180 Vienna, Austria
email: gerhard.hensler@univie.ac.at

Abstract. Supernovae are the most energetic stellar events and influence the interstellar medium by their gasdynamics and energetics. By this, both also affect the star formation positively and negatively. In this paper, we review the complexity of investigations aiming at understanding the interchange between supernova explosions with the star-forming molecular clouds. Commencing from analytical studies the paper advances to numerical models of supernova feedback from superbubble scales to galaxy structure. We also discuss parametrizations of star-formation and supernova-energy transfer efficiencies. Since evolutionary models from the interstellar medium to galaxies are numerous and are applying multiple recipes of these parameters, only a representative selection of studies can be discussed here.

Keywords. ISM: supernova remnants, ISM: kinematics and dynamics, ISM: bubbles, ISM: structure, stars: formation, galaxies: evolution, galaxies: ISM

1. Introduction

Stars form from the cool gas of the interstellar medium (ISM) and couple to their environment during their lives already by stellar mass and energy release, the latter comprising radiation and stellar wind energy. Nevertheless, the most vehement effect to the ISM and whole galaxies is contributed at their deaths when massive stars explode as supernovae type II (SNeII) and intermediate-mass stars expel planetary nebulae or die as SNeIa from binary systems. Only a minor fraction of the initial star mass is retained as remnants, almost 10–15% for the massive and 20-30% for intermediate-mass stars (Weidemann & Koester 1983) and zero in the SNIa case. The rest refuels the ISM. This cosmic matter cycle acts on intra-galactic scales and contributes not only energy but also nucleosynthesis products to the ISM (e.g. Hensler & Recchi 2010). The processes which determine this galactic ecosystem seem to be fine-tuned in a manner that e.g. galactic gas disks are mostly in energy balance in which the vertically integrated star-formation rate (SFR) $\Sigma_{\rm SFR}$ (in units of $M_\odot yr^{-1} pc^{-2}$) correlates with the gas surface density $\Sigma_{\rm g}$ over orders of magnitude, known as Kennicutt-Schmidt law (Kennicutt 1998). More precisely, this relation holds for the molecular gas $\Sigma_{\rm mol}$ (Schruba *et al.* 2011) what can be understood because the molecular gas fraction is determined by the ISM pressure and, by this, also the star-formation efficiency (SFE) $\epsilon_{\rm SF}$ (Leroy *et al.* 2008).

If the SFR is simply determined by the molecular gas reservoir and by the free-fall time $\tau_{\rm ff}$ of molecular clouds (Elmegreen 2002), it would exceed the observed one in the Milky Way by up to two orders of magnitude (e.g. Hensler 2011), energetic processes have to intervene and to stretch the star-formation (SF) timescale with respect to the dynamical timescale implying the SFE such that $\delta\rho_{\rm SF}/\tau_{\rm SF} = \epsilon_{\rm SF} \cdot [\rho_{\rm g}/\tau_{\rm ff}]$. This equilibrium on disk scales requires that heating processes balance the inherent cooling of the ISM. Besides multiple heating processes from dissipation of dynamics, as e.g. differential disk rotation, gas infall, tidal interactions, etc., the above-mentioned local and immediate feedback by formed stars themselves is the most favourable mechanism of SF regulation. Köppen, Theis, & Hensler (1995) demonstrated already that the SFR achieves a dependence on ρ_g^2, if the stellar heating is compensated by collisional-excited cooling emission (e.g. Böhringer & Hensler, 1989).

2. Supernova feedback

2.1. *Supernovae and the Matter Cycle with Star Formation*

The most efficient stellar energy power is exerted by SNe, of which those SNeII accumulate to superbubbles because of their local concentration still in the star-forming sites and their short lifetimes, while SN type Ia occur as isolated effects on long timescales (Matteucci & Recchi 2001) and are more distributed over the ISM on larger scales.

After the confirmation of the existence of a hot ISM phase as predicted by Spitzer (1956) and an observational baseline of SN remnants (SNR) over decades (Woltjer 1972), the importance to understand SNR (Chevalier (1974)) and their relevance for the energy, dynamics, and mass budget of the ISM phases (McKee & Ostriker 1977) and for the non-dynamical matter cycle as interplay of gas phases (see e.g. Habe, Ikeuchi, & Takaka (1981), Ikeuchi & Tomita (1983)) moved into the focus of ISM and galactic research in the 70's. With a toy model consisting of 6 ISM components and at least 10 interchange processes Ikeuchi, Habe, & Tanaka (1984) included SF from giant molecular clouds after their formation from cool clouds, which are swept-up and condensed in SN shells. By this, SF and SN explosions together with different gas phases form a consistent network of interaction processes. As a reasonable effect of this local consideration the SF oscillates with timescales determined by the gas density and the interaction strengths.

Taking single SNR models into account, e.g. Chevalier (1974), Cioffi & Shull (1991) modelled the evolution and volume filling of randomly distributed and temporally exploding SNe with SNR cooling and expansion and with mutual transitions between the warm and cool gas phases. Since it is obvious that this hot SN gas dominates energeticly and kineticly the ISM, the detailed understanding of its interaction with the cool gas, its cumulative effect as superbubbles, and the dynamical structuring of galactic gas disks (comprehensively reviewed by Spitzer, 1990) is of vital importance for the evolution of galaxies. Because SN interactions happen on largely different length and time scales such studies have to cover a large variety of aspects (Chevalier 1977) reaching from the stirr-up of the ISM by turbulence (this sect.), by this, regulating the galactic SF, to the triggering of SF (sect. 2.2), and at least to the gas and element loss from galaxies (sect. 3).

First dynamical approaches to the structure evolution of the ISM and gas disks, aiming at understanding the disk-halo connection were performed by Rosen & Bregman (1995). As heating sources they took the energy of massive stellar winds only into account, but overestimated their impact on the ISM because highly resolved numerical studies reveal surprisingly low energy transfer efficiencies (Hensler 2007). Nonetheless, their models show the compression of gas filaments and the expulsion of gas vertically from the disk even under self-gravity.

The influence of SNe on the SF can be imagined by two measures: the expansion of hot gas, its deposit of turbulent energy (Mac Low & Klessen 2004), and its evaporation of embedded cool gas should, at first, lead to the suppression of SF, while vice versa the sweep-up and condensation of surrounding gas in the shells of SNRs and superbubbles could trigger SF. That hot SN gas regulates SF thermally has been demonstrated by Köppen, Theis & Hensler (1998) who analysed the equations of a multi-component system when thermal conduction accounts for gas-phase transitions (see fig. 1).

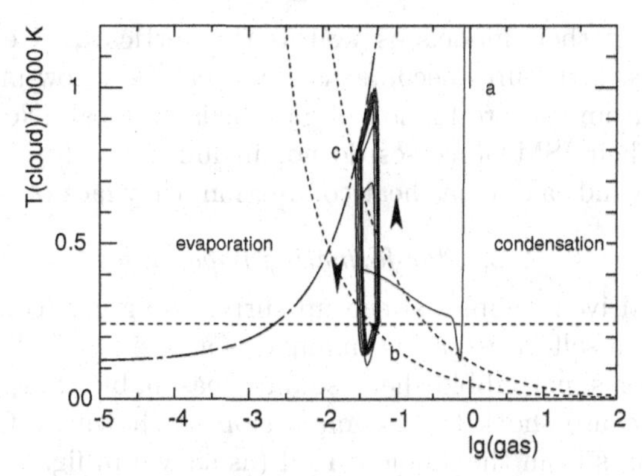

Figure 1. Evolution of the full system of cold+hot gas from the initial state ('a') until the completion of the first few oscillations (solid line). The dashed lines descending to the right are the loci where the system switches from evaporation to condensation (lower curve, at 'b') and vice versa (at 'c'). The horizontal dashed line is the locus of the evaporation funnel and the dot-dashed curve depicts the condensation funnel. *(for details see Köppen, Theis & Hensler (1998)).*

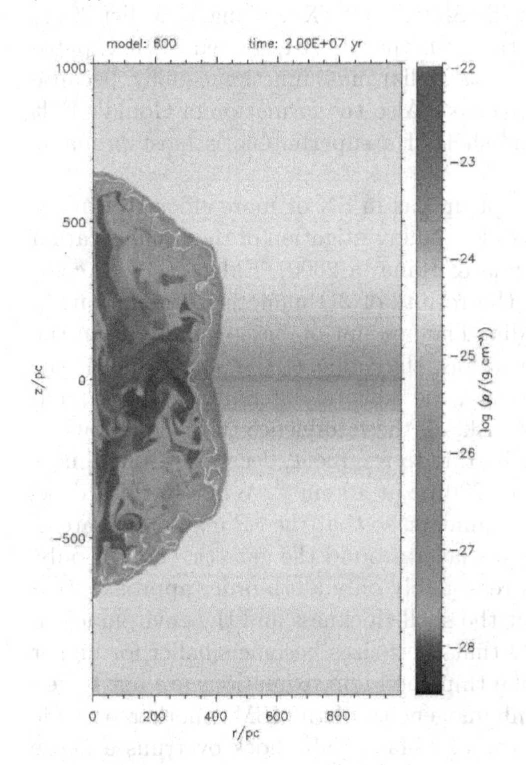

Figure 2. Density distribution of a superbubble after almost 20 Myrs. The superbubble results from 100 supernova typeII explosions of stars in the mass range between 10 and 100 $M_\odot$. The temporal sequence of explosions happens according to the lifetimes of massive stars with a Salpeter IMF. The star cluster is located at the origin of the coordinates. The galactic disk is vertically composed of the three-phase interstellar medium, cool, warm, and hot phase, respectively, with (central density ρ_0 [$in\,g\,cm^{-3}$]; temperature T [*in K*]; scaleheight H [*in pc*]) of (2×10^{-24}; 150; 100), (5×10^{-25}; 9000; 1000) (1.7×10^{-27}; 2×10^{6}; 4000). The density varies from almost $10^{-23}\,g\,cm^{-3}$, in the densest part of the shell to $2\times10^{-28}\,g\,cm^{-3}$ in the darkest bubble interiors. *(from Gudell 2002)*

In more comprehensive numerical investigations of the ISM evolution Slyz *et al.* (2005) studied the influence of SF and SN feedback on the SFR with self-gravity of gas and stars. As the main issues one can summarize that feedback enhances the ISM porosity, increases the gas velocity dispersion and the contrasts of T and ρ, so that smaller and more pronounced structures form, and, most importantly, that the SFR is by a factor of two higher than without feedback.

At the same time, de Avillez & Breitschwerdt (2004) simulated the structure evolution of the solar vicinity in a $1\times1\times10\,kpc^3$ box and identified the Local Bubble and its neighbouring Loop I in their models as well as the vertical matter cycle. In addition, filamentary neutral gas structures become visible where SF of low-mass star clusters (see next sect.) resulting from the production of gas shells by local SNe express the positive SN feedback. Since their ISM processes do not include SF self-regulation processes by stellar radiation and winds as well as heat conduction, they lack of negative SF feedback.

2.2. *Star-formation triggering*

SN and stellar wind-driven bubbles sweep up surrounding gas, condense it, and could, by this, trigger SF in a self-propagating manner as a positive feedback. The perception of SF trigger in SN or superbubble shells sounds reasonable from the point of view of numerical models because shock front compression as shown by Chevalier (1974) and sufficient swept-up mass from the ambient ISM (as shown in fig. 2 preferably in the gas disk itself) lead to cooling and gravitational instabilities. This mechanism, however, is not generally confirmed by observations. Shell-like distributions of young stars, are e.g. found in G54.4-0.3, called *sharky* (Junkes, Fürst, & Reich 1992), in the Orion-Monoceros region (Wilson *et al.* 2005), more promising in the Orion-Eridanus shell (Lee & Chen 2009), and in several superbubbles in the Large Magellanic Cloud, as e.g. Henize 206 (Gorjian *et al.* 2004). The SF associated with the SNR IC443 (Xu, Wang, & Miller 2011) raises e.g. the already above-mentioned question, whether SNR-triggered SF is capable to lead to massive star clusters which fill the whole stellar mass function equally, because here only about $10^4 M_\odot$ of molecular gas is involved. Also the formation of Gould's Belt as site of low-mass stellar associations in the shell of a superbubble is most probable (Moreno, Alfaro, & Franco 1999).

Such SF trigger by the condensation of swept-up gas in SN or more efficiently in superbubbles (see fig. 2) can be explored in detail by the investigation of the fragmentation timescale (see e.g. Ehlerova *et al.* 1997, Fukuda & Hanawa 2000). Ehlerova *et al.* compared a self-similar analytical solution with the results of 3D numerical simulations of superbubble expansions in homogeneous media. The amount of energy supply from the final number of young stars in an OB association, the value of the sound speed, the stratification and density of the ambient medium, the galactic differential rotation, and the vertical gravitational force in the galactic disk, all these influence the fragmentation. The typical superbubble radius, at which shells start to fragment, decreases from almost 700 pc at an ambient gas density of 1 cm^{-3} to 200 pc at 10 cm^{-3}. While in thick disks like they exist in DGs nearly the whole shell fragments, so that the SF may propagate in all directions, in thin disks it is restricted to gas layer around the galactic equator only. Since the applied thin shell approximation is reasonably only a 0th-order approach, in a recent paper Wünsch *et al.* (2010) clarify that the shell thickness and the environmental pressure influences the fragments in the sense that their sizes become smaller for higher pressure. Nevertheless, the deviations from the thin-shell approximations are not large.

Since these studies do not allow for the inhomogeneity of the ISM, another possible feedback effect by SN can be caused, when the ultra-fast SNR shock overruns a dense interstellar cloud, so that the clouds are quenched (Orlando *et al.* 2005). Stars should be formed instantaneously by such cloud crushing, and the cloud mass determines the star cluster mass.

2.3. *Supernova energy impact*

Although numerical simulations have been performed to understand the heating (or energy transfer) efficiency ϵ_{SN} of SNe (Thornton *et al.* 1998), superbubbles (e.g. Strickland *et al.* 2004), and starbursts (Melioli & de Gouveia Dal Pino 2004), they are yet too simplistic and mostly spatially poorly resolved to account for quantitative results. Thornton *et al.* derived an efficiency $\epsilon_{\rm SN}$ of 0.1 from 1D SN simulations as already applied by chemo-dynamical galaxy models (Samland, Hensler & Theis 1997), while unity is also used in some galaxy models (see sect. 3), but seems far too large.

3. Supernovae and Galaxy Evolution

3.1. *Supernova parametrization in numerical models*

Because of their high power in various forms, massive stars are usually taken into account as the only heating sources for the ISM in galaxy evolution models and here mostly only the energy deposit of SNII explosions alone. Recently, Stinson *et al.* (2013) envoked indeed the necessity of short-term SF feedback which is naturely implied by massive stellar radiation and winds but was already included in the chemo-dynamical prescription (see e.g. Samland, Hensler & Theis 1997). It must be explored, however, how effectively this energy is transferred into the ISM as turbulent and finally as thermal energy. Although it is generally agreed that the explosive energy E_{SN} of an individual SN lies around 10^{51} ergs with significant uncertainties of probably one order of magnitude, the energy deposit is still more than unclear, but is one of the most important ingredients for galaxy formation and evolution (e.g. Efstathiou 2000, Silk 2003). Massive stars do not disperse from their SF site and thus explode within the stellar associations, by this, contributing significantly to the ISM structure formation e.g. by cavities and holes in the HI gas and chimneys of hot gas. On large scales SNeII trigger the matter cycle via galactic outflows from a gaseous disk. By this, also the chemical evolution is affected thru the loss of metal-enriched gas from a galaxy (for observations see e.g. Martin, Kobulnicki & Heckman (2002), for models e.g. Recchi & Hensler (2006b)).

Although numerical experiments of superbubbles and galactic winds are performed, yet they only demonstrate the destructive effect on the surrounding ISM but lack of self-consistency and a complex treatment. Simulations of the chemical evolution of starburst DGs by Recchi *et al.* (2002, 2006a), that are denoted to reproduce the peculiar abundance patterns in these galaxies by different SF episodes, found, that $\epsilon_{\rm SN}$ can vary widely: it starts with 10% drops and increases successively but not above 18% (Hensler 2011). For the subsequent SNIa explosions, always single events, $\epsilon_{\rm SN}$ is much smaller, i.e. below 1%. Moreover, if a closely following SF episode (might be another burst) drives its SNII explosions into an already existing chimney of a predecessor superbubble, the hot gas can more easily escape without any hindrance and thus affects the ISM energy budget much less. Recchi & Hensler (2006a) found that depending on the galactic HI density the chimneys do not close before a few hundred Myrs.

Since $\epsilon_{\rm SF}$ must inherently depend on the local conditions so that it is high in bursting SF modes, but of percentage level in the self-regulated SF mode, numerical simulations often try to derive the "realistic" SFE by comparing models of largely different $\epsilon_{\rm SF}$ with observations, as e.g. to reproduce gas structures in galaxy disks and galactic winds (e.g. undertaken by Tasker & Bryan (2006, 2008) with $\epsilon_{\rm SF} = 0.05$ and 0.5). In addition, specific SN energy deposit $\epsilon_{\rm SN} \times E_{SN}/M_*$ by them is fixed to 10^{51} erg per 55 $M_\odot$ of formed stars (1.8×10^{49} erg $M_\odot^{-1}$ by Dalla Vecchia & Schaye (2008)), but their results cannot ye tbe treated quantitatively, since they also mismatch with the Kennicutt relation. Stinson *et al.* (2006) performed a comprehensive study about the influence of $\epsilon_{\rm SN}$ over a large scale of values and also of the dependence on the mass resolution of their SPH scheme GASOLINE of galaxy evolution. Although their results demonstrate the already expected trends, more realistic treatments must adapt $\epsilon_{\rm SN}$ self-consistently to the local state of

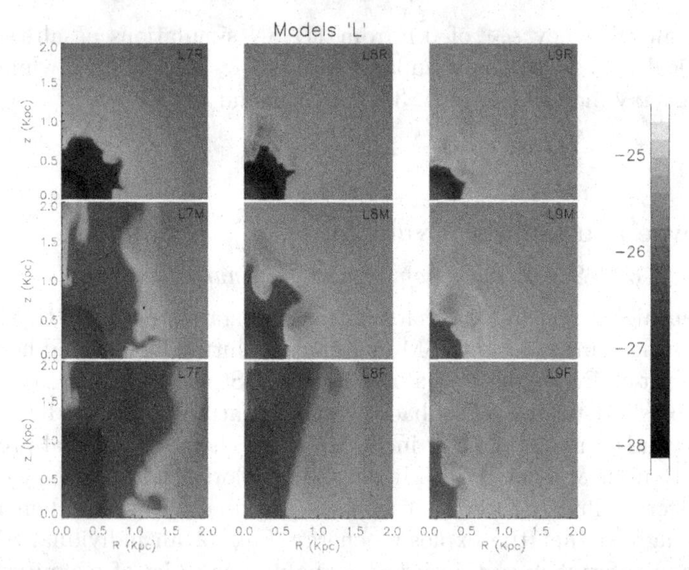

Figure 3. Gas density distribution for the 9 models of the of 60%-initial gas fraction (L) after 200 Myr of evolution. The first column represents models with $10^7\,M_\odot$ of initial baryonic mass; the middle column shows the gas distribution of $10^8\,M_\odot$ models, and the right column the $10^9\,M_\odot$ models. The top row models are characterized by a roundish initial gas distribution (R; with b/a = 5), the middle row by b/a = 1 (M), and finally the bottom row represents flat disks with b/a = 0.2 (F). At the top-right corner of each panel the model designation is also indicated. The right-hand strip shows the (logarithmic) density scale (in g cm^{-3}). *(from Recchi & Hensler 2013)*

the ISM what requires much larger numerical capacity for more intensive high-resolution numerical simulations.

Theoretical studies by Elmegreen & Efremov (1997) achieved a dependence of ϵ_{SF} on the external pressure, while Köppen, Theis, & Hensler (1995) explored a temperature dependence of the SFR both effects affecting the SFR. Furthermore, most galaxy evolutionary models at present lack of the appropriate representation of the different ISM phases allowing for their dynamics and their direct interactions by heat conduction, dynamical drag, and dynamical instabilities thru forming interfaces, not to mention resolving the turbulence cascade.

3.2. *Superbubbles and galactic winds*

A superbubble expanding from a stellar association embedded in a HI disk has, at first, to act against the surrounding medium, by this, is cooling due to its pressure work and radiation, but compresses the swept-up shell material and implies turbulent energy to the ISM. How much the superbubble expansion is efficiently hampered depends on the surrounding gas density and pressure, the HI disk shape (Recchi & Hensler 2013), and the energy loss by radiative cooling. From fig. 3 it is discernible that only flat gas disks of preferably low-mass galaxies allow a hot wind to escape from the galaxy. Consequences for the chemical evolution are in the focus of those models (Recchi & Hensler 2013). And finally, observed superbubbles also reveal a mismatch of their spatial extent and the energy content required to drive the expansion with the observed X-ray luminosity (Hensler *et al.* 1998). This fact can be only explained by a significant energy loss.

4. Conclusions

The dominating influence of SN explosions on structure, dynamics, and energy budget of the ISM are obvious and agreed. Signs and strengths of these feedback effects are, however, widely uncertain. Whether the feedback is positive (trigger) or negative (suppression) can be understood analyticly from first principles, but because of the nonlinearity and the complexity level of the acting physical plasma processes clear results cannot be quantified reliably. In addition, the temporal behaviour varies by orders of magnitude because of the changing conditions. In summary, the energy transfer efficiency of SN energy to the ISM energy modes is much below unity and must not be overestimated, but also depends on the temporal and local conditions.

Acknowledgements

The author is grateful for numerous discussions with Simone Recchi and with the participants of the Aspen Center for Physics summer program 2012 on "Star-formation regulation" which was supported in part by the NSF under Grant No. PHYS-1066293.

References

Böhringer, H. & Hensler, G. 1989, *A&A*, 215, 147
Chevalier, R. A. 1974, *ApJ*, 188, 501
Chevalier, R. A. 1977, *ARA&A*, 15, 175
Cioffi, D. F. & Shull, J. M. 1991, *ApJ*, 367, 96
Dalla Vecchia, C. & Schaye, J. 2008, *MNRAS*, 387, 1431
de Avillez, M. A. & Breitschwerdt, D. 2004, *A&A*, 425, 899
Efstathiou, G. 2000, *MNRAS*, 317, 697
Ehlerova, S., Palous, J., Theis, C., & Hensler, G. 1997, *A&A*, 328, 111
Elmegreen, B. G. 2002, *ApJ*, 577, 206
Elmegreen, B. G. & Efremov, Y. N. 1997, *ApJ*, 480, 235
Fukuda, N. & Hanawa, T. 2000, *ApJ*, 533, 911
Gorjian, V., Werner, M. W., Mould, J. R., *et al.* 2004, *ApJS*, 154, 275
Gudell, A. 2002, diploma thesis, University of Kiel
Habe, A., Ikeuchi, S., & Tanaka, Y. D. 1981, *PASJ*, 33, 23
Hensler, G. 2007, in: E. Ensellem *et al.* (eds.), *Chemodynamics: from first stars to local galaxies* Proc. CRAL-Conference Series I, EAS Publ. Ser. No. 7, p. 113
Hensler, G. 2009, in: J. Andersen, J. Bland-Hawthorn, & B. Nordstroem (eds.), *The Galaxy Disk in Cosmological Context*, Proc. IAU Symp. No. 254, p. 269
Hensler, G. 2011, in: J. Alves, B. Elmegreen, & V. Trimble (eds.), *Computational Star Formation*, Proc. IAU Symp. No. 270, p. 309
Hensler, G. & Recchi, S. 2010, in: K. Cunha, M. Spite & B. Barbuy (eds.), *Chemical Abundances in the Universe: Connecting First Stars to Planets*, Proc. IAU Symp. No. 265, p. 325
Hensler, G., Dickow, R., Junkes, N., & Gallagher, J. S. 1998, *ApJ*, 502, L17
Ikeuchi, S. & Tomita, H. 1983, *PASJ*, 35, 56
Ikeuchi, S., Habe, A., & Tanaka, Y. D. 1984, *MNRAS*, 207, 909
Junkes, N., Fürst, E., & Reich, W. 1992, *A&A*, 261, 289
Kennicutt, R. J. 1998, *ApJ*, 498, 541
Köppen, J., Theis, C., & Hensler, G. 1995, *A&A*, 296, 99
Köppen, J., Theis, C., & Hensler, G. 1998, *A&A*, 328, 121
Krumholz, M. R. & McKee, C. F. 2005, *ApJ*, 630, 250
Lee, H.-T. & Chen, W. P. 2009, *ApJ*, 694, 1423
Leroy, A. K., Walter, F., Brinks, E., *et al.* 2008, *AJ*, 136, 2782
Mac Low, M.-M. & Klessen, R. S. 2004, Rev. Mod. Phys., 76, 125
Martin, C. L., Kobulnicki, H. A., & Heckman, T. M. 2002, *ApJ*, 574, 663
Matteucci, F. & Recchi, S. 2001, *ApJ*, 558, 351

McKee, C. F. & Ostriker, J. P. 1977, *ApJ*, 218, 148
Melioli, C. & de Gouveia Dal Pino, E. M. 2004, *A&A*, 424, 817
Moreno, E., Alfaro, E. J., & Franco, J. 1999, *ApJ*, 522, 276
Oey, M. S., Watson, A. M., Kern, K., & Walth, G. L. 2005, *AJ*, 129, 393
Orlando, S., Peres, G., Reale, F., *et al.* 2005, *A&A*, 444, 505
Recchi, S. & Hensler, G. 2006a, *A&A*, 445, L39
Recchi, S. & Hensler, G. 2006b, *Rev. Mod. Astronomy*, 18, 164
Recchi, S. & Hensler, G. 2007, *A&A*, 476, 841
Recchi, S. & Hensler, G. 2013, *A&A*, 551, A41
Recchi, S., Hensler, G., Angeretti, L., & Matteucci, F. 2006, *A&A*, 445, 875
Recchi, S., Matteucci, F., D'Ercole, A., & Tosi, M. 2002, *A&A*, 384, 799
Rosen, A. & Bregman, J. N. 1995, *ApJ*, 440, 634
Samland, M., Hensler, G., & Theis, C. 1997, *ApJ*, 476, 544
Schruba, A., Leroy, A. K., Walter, F., *et al.* 2011, *AJ*, 142, 37
Slyz, A. D., Devriendt, J. E. G., Bryan, G., & Silk, J. 2005, *MNRAS*, 356, 737
Silk, J. 2003, *MNRAS*, 343, 249
Spitzer, L. 1956, *ApJ*, 124, 20
Spitzer, L. 1990, *ARA&A*, 28, 71
Stinson, G. S., Seth, A., Katz, N., *et al.* 2006, *MNRAS*, 373, 1074
Stinson, G. S., Brook, C., Maccio, A. V., *et al.* 2013, *MNRAS*, 428, 129
Strickland, D. K., Heckman, T. M., Colbert, E. J. M., *et al.* 2004, *ApJ*, 606, 829
Tasker, E. J. & Bryan, G. L. 2006, *ApJ*, 641, 878
Tasker, E. J. & Bryan, G. L. 2008, *ApJ*, 673, 810
Thornton, K., Gaudlitz, M., Janka, H.-Th., & Steinmetz, M. 1998, *ApJ*, 500, 95
Weidemann, V. & Koester, D. 1983, *A&A*, 121, 77
Wilson, B. A., Dame, T. M., Masheder, M. R. W., & Thaddeus, P. 2005, *A&A*, 430, 523
Woltjer, L. 1972, *ARA&A*, 10, 129
Wünsch, R., Dale, J. E., Palous, J., & Whitworth, A. P. 2010, *MNRAS*, 407, 1963
Xu, J.-L., Wang, J.-J., & Miller, M. 2011, *ApJ*, 727, 81

Discussion

CESARSKY: Cosmic Ray and magnetic fields potentially play a big role in some of the process you were discussing, fragmentation, bubble formation and especially galactic winds. Were they considered in some of the models you discussed?

HENSLER: Not yet. The actually treated complexity of the ISM processes of multi phases, SF, and star-gas interactions is already challenging the high-performance computing capacity.

ZHOU, P.: Is there any clear observational evidence to prove that star formation could be triggered by SNR? How to distinguish the SNR impacts from progenitor wind impact?

HENSLER: This observational evidence seems to exist, but obviously not in general. This means that local conditions determine the possibility of SF triggering. From models of wind-blown and radiation-driven HII regions the stellar wind impact seems not to sweep-up and compress the surrounding gas sufficiently; see e.g. Freyer, Hensler, & Yorke, 2003, *ApJ*, 594, 888 and 2006, *ApJ*, 638, 262.

Supernova Environmental Impacts
Proceedings IAU Symposium No. 296, 2013
A. Ray & R. A. McCray, eds.

doi:10.1017/S1743921313009587

Supernovae and the Galactic Ecosystem

Q. Daniel Wang
Astronomy Department,
619-E, LGRT, University of Massachusetts, 710 N. Pleasant St., Amherst, MA 01003, USA
email: wqd@astro.umass.edu

Abstract. Supernovae are the dominant source of stellar feedback, which plays an important role in regulating galaxy formation and evolution. While this feedback process is still quite uncertain, it is probably not due to individual supernova remnants as commonly observed. Most supernovae likely take place in low-density, hot gaseous environments, such as superbubbles and galactic bulges, and typically produce no long-lasting bright remnants. I review recent observational and theoretical work on the impact of such supernovae on galaxy ecosystems, particularly on hot gas in superbubbles and galactic spheroids.

Keywords. ISM: general, galaxies: ISM, X-rays: galaxies, (ISM:) supernova remnants

1. Introduction

Supernovae (SNe) are a major source of the mechanical energy input in the interstellar medium (ISM). On average, an SN releases about 10^{51} ergs energy, driving a blastwave into the ambient medium. How far this blastwave goes and how fast the energy is dissipated depend sensitively on the density and temperature of the medium. Ironically, commonly-known and well-studied supernova remnants (SNRs), though looking spectacular, are atypical products of SNe, and typically occur in relatively dense media. Most SNe are expected to explode in low-density hot environments, e.g., inside superbubbles (SBs) and/or simply in the inter-cloud hot ISM (for core-collapsed SNe) and in the Galactic halo and bulge (for Type Ia SNe). The resultant remnants are typically too faint to be well observed individually. But collectively such "missing" SNRs are probably more important than those in dense environments, in terms of both heating and shaping the global ISM. In the following, I will first briefly review the behavior of SN blastwaves in hot gas and will then focus on discussing how discrete SN events affect hot gas properties in the two types of environments, SBs and galactic spheroids.

2. Supernova blastwaves in hot gas

As shown by Tang & Wang (2005), the evolution of an SNR in a low-density hot medium has several distinct properties: 1) The blastwave always moves at a speed greater than, or comparable to, the sound wave and can thus reach a much larger radius than that predicted by the Sedov solution; 2) The swept-up thermal energy is important, affecting the evolution of both the blastwave and the interior structure; 3) Because the Mach number of the blastwave is typically small, its heating is subtle and over a large volume; 4) The blastwave can hardly dissipate until it meets cool gas. These properties of SNRs in a low-density hot medium make them an important ingredient in regulating large-scale environments in galaxies.

To study the collective effect of SNe on the ISM, we need to conduct simulations, which needs to cover a large dynamic range from the evolution of individual SNRs to a possible galaxy-wide outflow, for example. It would be computationally very expensive, if even possible, to simulate the evolution of each SNR on sub-parsec scales in such a simulation. The SNR evolution in general cannot be described by the self-similar Sedov Taylor solution, which neglects the SN ejecta and assumes a cool ambient medium (hence with no energy content). In fact, the evolution depends on both the density and temperature of the ambient medium (Tang & Wang 2005). Tang & Wang (2009) have further shown that one can adaptively embed individual structured SNR seeds into the 3D simulation grid on such scales that their thermal and chemical evolution are adequately represented by detailed 1D simulations. The SNR seed embedding, worked with an adaptive mesh refinement scheme, can effectively extend such a 3D simulation to include the subgrid evolution of SNRs. In fact, using a scaling law, which is applicable to SNRs evolving in hot gas, individual seeds can be adaptively generated from a library of templates. Each consists of the radial profiles of density, temperature and velocity when the SNR has a certain shock front radius or age. These templates can be obtained from a single 1-D simulation of an SNR evolving in a uniform ambient medium of certain density and temperature. With this scheme, 3D hydrodynamic simulations of the supernova-dominated ISM can be readily conducted.

3. Superbubbles

Core-collapsed SNe represent the end of massive stars, the bulk (if not all) of which form in OB associations. The energy release from such an association, highly correlated in space and time, has a great impact on the surrounding medium. Initially, the energy release from the association is primarily in the form of intense ionizing radiation from very massive stars, which tends to homogenize the surrounding medium and to make it puff up. After several million years, fast stellar winds start to play a major role in heating and shaping the medium, creating a low-density hot bubble (e.g., Weaver *et al.* (1977)), before the explosion of the first SN in the association. Later, after about 5×10^6 yr since the star formation (if more or less coeval), core-collapsed SNe become the dominant source of the mechanical energy input into the already hot surroundings (e.g., Monaco 2004). This combination of the concerted feedbacks, lasting for $\sim 5 \times 10^7$ yr — the lifetime of an 8 $M_\odot$ star — leads to the formation of a so-called superbubble of low-density hot gas enclosed by a supershell of swept-up cool gas (e.g., Mac Low & McCray 1988). The expansion of such a superbubble is expected to be substantially faster than typical OB association internal velocities of a few kilometers per second. Therefore, a majority of core-collapsed SNe ($\sim 90\%$) should occur inside their parent superbubbles (e.g., Higdon *et al.* 1998; Parizot *et al.* 2004).

Fig. 1 shows examples of SBs in various evolutionary stages. The 30 Dor nebula is probably the youngest and most energetic one, while 30 Dor C is the oldest and least powerful. The diffuse X-ray-emitting plasma, heated by fast stellar winds and possible SNe of massive stars, fills various cavities traced by Hα-emitting shells. Such SBs will eventually blow out from galactic disks and vent the hot plasma into galactic halos, even into the intergalactic space.

While the evolution model for SBs, analogous to that for stellar wind bubbles, has been developed for many years (e.g., Weaver *et al.* 1977; Mac Low & McCray 1988). Comparisons of this model with observations have revealed two major discrepancies. The observed X-ray luminosities of some SBs seem to be substantially brighter (by a factor of up to ~ 10) than predicted by the model (Chu & Mac Low 1990; Wang & Helfand 1991; Jaskot *et al.* 2011 and references therein). Various processes have been proposed, including SN metal enrichment, additional mass loading due to the destruction of embedded, dense cloudlets of the ISM that have been overrun by expanding supershells.

But it appears that the most important process is the blastwave heating of supershells by sporadic SNe, especially off-center ones (Jaskot *et al.* (2011) and references therein).

The blastwave heating of supershells may also be responsible for another apparent discrepancy between the model prediction and the observations: The energetics of SBs, observationally accounting for both thermal and kinetic energies, is generally much less than expected from the injection from massive stars enclosed (e.g., Yamaguchi *et al.* (2010)). Or the growth rate of an SB is typically much less than what would be expected from the model (e.g., Oey (2009)). The energy injection from stars is assumed as being continuous, which may be a reasonable approximation for an evolved SB of a large dimension, which the model was intended to applyd to. But for a relatively young and small SB, as typically observed, an SN-induced shockwave or sound wave can propagate through the hot interior, eventually reach the surrounding cool shells, and steepen again into a blastwave, somewhat like an earth quake-induced Tsunami (Tang & Wang (2005)). This blastwave, or its corresponding reverse-shock, can sometimes be strong enough to produce enhanced X-ray emission, as observed. Furthermore, much of the SN energy may be lost radiatively, probably outside the X-ray band. Indeed, as shown by Cho *et al.* (2008), based on 1-D simulations of such young SBs, discrete SNe can lead to a loss of 80-90% of the SN mechanical energy via radiative cooling in supershells, and only about 10% remains as the thermal energy in the enclosed hot gas, while the kinetic energy is $\lesssim$ 10%. Of course, such a substantial energy loss can significantly change the growth of SBs, and potentially their chance to blow out from galactic disks. Therefore, it should be interesting to extend such simulations to higher dimensions and to later evolutionary stages of SBs. Interfaces between the hot and cold gases can be enlarged, for example, due to various 3-D instabilities, which can potentially strongly affect the energetics of SBs (e.g., Krause *et al.* (2013)).

In addition to X-ray emission, SBs may loss energy via other channels. Substantial amounts of dust have been observed in various Galactic wind-blown bubbles (e.g., Everett & Churchwell 2010). Dust can be a significant sink of heat in hot gas. But the survival time of dust in hot gas is short, due to sputtering. So dust needs to be replenished, possibly by clouds overrun by expanding shells and by dense stellar winds from massive stars such as Carbon-rich Wolf-Rayets and Luminous Blue variables (e.g., Rajagopal *et al.* 2007, Dong *et al.* (2012)).

In addition to the radiation loss, SN energy may also escape from SBs via cosmic rays (CRs). SBs have been proposed to be the acceleration site of Galactic CRs (e.g., Higdon & Ligenfelter 2005, Fisk & Gloeckler 2012; Parizot *et al.* 2004; Bykov 2001; Bykov &

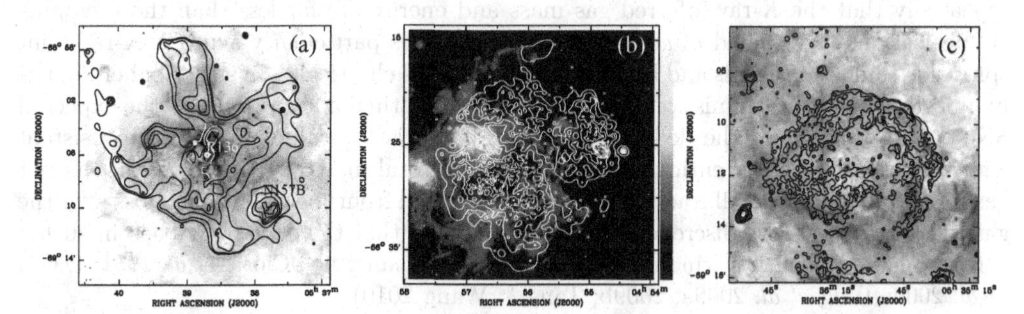

Figure 1. Examples of superbubbles in the LMC: (a) 30 Dor, (b) N11, and (c) 30 Dor C. Overlaid on the Hα images are X-ray intensity contours constructed in the *Chandra* ACIS-I 0.5-2 keV band for (a) and (b) and in the *XMM-Newton* MOS 0.4-1 keV band for (c). Discrete sources, as marked by*crosses*, have been excised in (b).

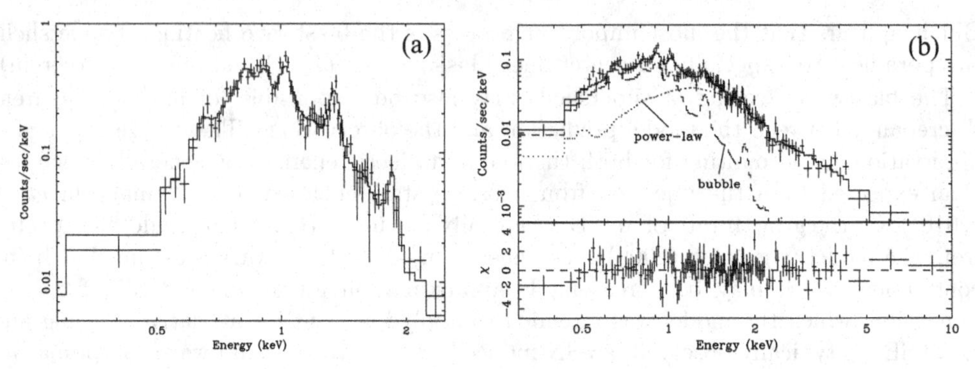

Figure 2. (a) Comparison of XMM-Newton MOS spectra of (a) 30 Dor, excluding R136 and SNR N157B (Fig. 1a) and (b) 30 Dor C.

Fleishman 1992). The energy efficiency of the CR-acceleration in SBs could reach up to about 30% at their early evolutionary stages (ages of a few 10^6 years; e.g., Butt & Bykov (2009)) Nonthermal X-ray emission has indeed been observed in a number of SBs, including 30 Dor C (Smith & Wang 2004, Yamaguchi *et al.* (2010) in the LMC, IC131 in M33 (Tüllmann *et al.* 2009). Fig. 2 compares X-ray spectra of 30 Dor and 30 Dor C. While the former is dominated by thermal emission, the latter exhibits a substantially hard (apparently nonthermal) component (Smith & Wang 2004).

All these processes still need to be carefully studied to determine their actual roles in regulating the energetics of SBs, which could have strong implications for our understanding the impact of massive stars on galaxy ecosystems.

4. Hot gas in galactic spheroids

Spheroids, including bulges of spiral galaxies and elliptical galaxies, are primarily comprised of old stars, which account for more than half of the stellar mass in the local Universe (Fukugita, Hogan, & Peebles 1998). These stars collectively generate a substantial feedback in form of gradual stellar mass-loss and energetic Type Ia SNe. Understanding how this relatively gentle but long-lasting feedback affects galaxy evolution is one of the fundamental questions in astrophysics (e.g., Wang 2010). In order to address this issue, one needs an affective tracer of the ejected mass and energy in spheroids. While they typically contain little cool gas, one would expect that the mass should be mostly in hot gas, heated by the SNe. Indeed, diffuse optically-thin thermal X-ray emission has been observed in elliptical galaxies and in galactic bulges. However it has been shown repeatedly that the X-ray-inferred gas mass and energy are far less than those empirical predictions (e.g. David *et al.* 2006), which becomes particularly acute in X-ray-faint spheroids (bulges of spirals and low/intermediate-mass ellipticals). In such a spheroid, the luminosity of the X-ray emission accounts for no more than a few percent of the expected SNe energy input, and the deduced iron metallicity is typically sub-solar, inconsistent with the expected Ia SN enrichment. The inferred total mass of the X-ray-emitting hot gas represents only a small fraction of what is expected from the stellar mass loss over the galaxy's lifetime. These discrepancies clearly indicate that the stellar feedback in such a spheroid has gone with a wind or outflow (e.g., Bregman 1980; Ciotti *et al.* 1991; David *et al.* 2006; Tang *et al.* 2009a, 2009b; Tang & Wang 2010).

The study of hot gas in galactic spheroids has quite a long history. Early works were based primarily on 1-D galactic wind modeling. Except for massive ellipticals, very hot (>1 keV) and fast ($v > 800$ km/sec) spheroidal wind was predicted (e.g., Mathews & Baker 1971; Ciotti *et al.* 1991). For such a hot wind, the inferred X-ray emission should be much less than 10^{36} ergs s^{-1} (e.g., Tang *et al.* 2009b), which is two orders of magnitude lower than the recently observed value of the galactic spheroid ($\sim 3 \times 10^{38}$ ergs s^{-1} for the M31 bulge; Li & Wang 2007; Bogdan & Gilfanov 2008).

One important fact that the 1-D supersonic wind modeling does not account for is the discreteness of Ia SN heating. X-ray emission, proportional to the emission measure, is sensitive to the detailed structure of the hot gas. We have therefore conducted 3-D simulations of SN-driven galactic winds (Tang *et al.* 2009b), in which the energy is assumed to come mainly from sporadic SNe and the mass mainly from stellar winds. The results are illustrated in Fig. 3. Our simulations confirm that Ia SNe produce a spheroidal wind, and more importantly, reveal its substructures. The bulk of the X-ray emission originates from the relatively low-temperature and low-abundance gas shells associated with SN blastwaves; the SN ejecta are not well mixed with the ambient medium. These results are qualitatively consistent with the apparent lack of evidence for iron enrichment in X-ray-faint galaxies. Compared to the 1-D wind model, the non-uniformity of simulated gas density, temperature, and metallicity in the 3-D simulations increases the diffuse X-ray luminosity by a factor of a few, narrowing the discrepancy between the theory-predicted and the observed X-ray emission. However, the resultant diffuse luminosities in the 3-D simulations is still more than one order of magnitude less than those observed in the Galactic and M31 bulges, indicating that gas in these spheroids is in a subsonic outflow state, probably due to additional mass loading to the hot gas and/or due to energy input rate that is substantially lower than the current estimate.

A subsonic outflow from a spheroid is intimately related to its formation and evolution history (Ciotti *et al.* (1991), Tang *et al.* (2009a)). We have developed a 1-D model of the spheroid outflow and galaxy accretion interplay. This model is based on 1-D hydrodynamic simulations and on an approximated dark matter accretion history of galaxies (Tang *et al.* 2009a). In our model, the outer boundary condition is the (supersonic) Hubble flow, which avoids the boundary problem. The spheroid wind/outflow occurs well inside the spatial range of the simulations. The feedback is assumed to consist of two primary phases: 1) an initial burst during the spheroid formation and 2) a subsequent long-lasting mass and energy injection from stellar winds and Ia SNe of evolved low-mass stars. These two phases of the feedback re-enforce each-other's impact on the gas dynamics. An outward blastwave is initiated by the burst and is maintained and enhanced by the long-lasting stellar feedback. For an M31-like bulge, for example, this blastwave can heat the surrounding medium not only in the galactic halo, but also in regions beyond the virial radius. The long-lasting feedback forms a galactic spheroid wind initially, which is reverse-shocked at a large radius, and may later evolve into a stable subsonic outflow as the energy injection gradually decreases with time. In a subsonic outflow, the properties of hot gas depend sensitively on the environment and formation history, which can explain the large dispersion of L_X/L_K in early-type galaxies with similar L_K as well as the missing stellar feedback problem (Wang 2010 and references therein). Therefore, the understanding of the interplay between the hot gas outflows and the large-scale galaxy environment is essential to the correct interpretation of observed diffuse X-ray emission in and around galactic spheroids.

Tang & Wang (2010) have further conducted 3-D simulations based on the 1-D subsonic solution to explore combined effects. In addition to the expected enhanced X-ray emission, they find that SN reverse shock-heated iron ejecta is typically found to have a very high temperature and low density, hence producing little X-ray emission. Such hot ejecta, driven by its large buoyancy, can quickly reach a substantially higher outward velocity than the ambient medium, which is dominated by mass-loss from evolved stars. The

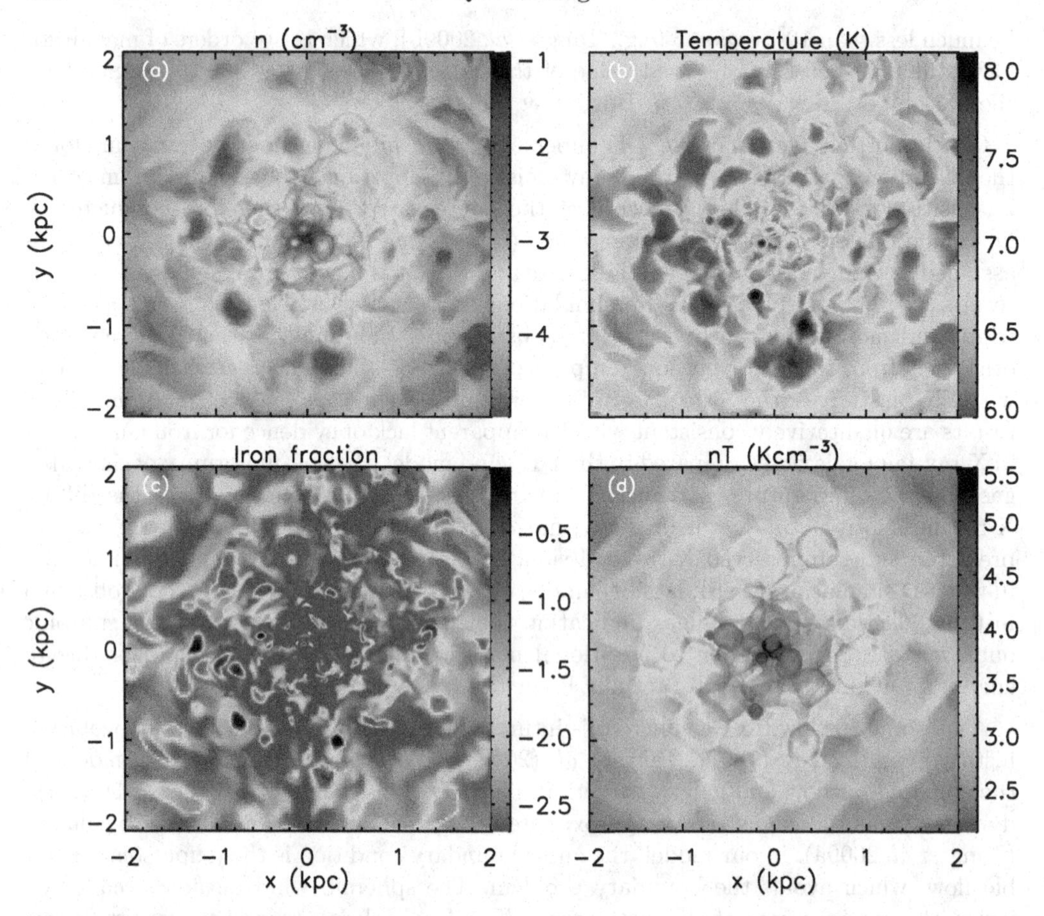

Figure 3. Density, temperature, iron mass fraction, and pressure of gas in a cut through a simulated spheroidal wind. The image values are scaled logarithmically. The upper-right octant has a higher resolution in the simulation, allowing for testing various resolution effects.

ejecta is gradually and dynamically mixed with the medium at large galactic radii. The ejecta is also slowly diluted and cooled by in situ mass injection from evolved stars. These processes together naturally result in the observed positive gradient in the average radial iron abundance distribution of the hot gas, even if mass weighted. This trend is in addition to the X-ray measurement bias that tends to underestimate the iron abundance for the hot gas with a temperature distribution.

One limitation of our previous simulations is that the mass loss from evolved stars takes only the form of continuous input, following the average stellar distribution of a galactic spheroid. In reality, however, the continuous wind, mainly from constant mass loss of numerous giant stars, contributes only about half of the stellar mass loss in the spheroid (e.g. Buckley & Schneider 1995). The other half of mass loss occurs in the planetary nebula (PN) stage of intermediate and low mass stars. This mass loss is an approximately impulsive mass-loss event at the end of the life of such a star and lasts only $\sim 10^3$ years. The ejecta forms an expanding ($\sim$ 10 km/s) and cold ($\lesssim 10^5$ K) gas shell. Naturally, one wants to know whether or not the PN ejecta can be heated by shocks and/or by thermal mixing with the hot ambient gas, or how the discreteness of the mass input can affect the diffuse X-ray emission of a spheroid.

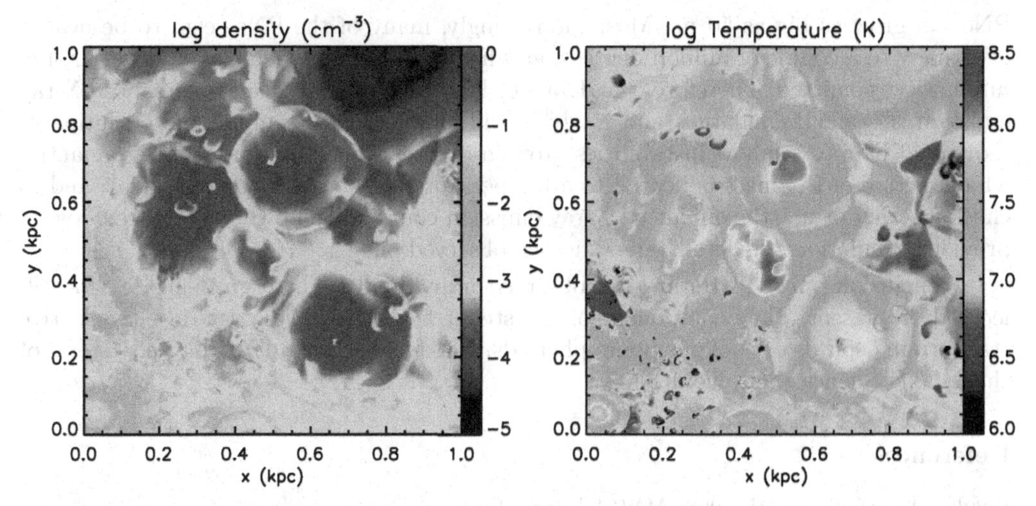

Figure 4. A snapshot of our pilot simulation, including discrete PNe. Presented here are density (left) and temperature (right), viewed in a cut through the central 1kpc × 1kpc region of a spheroid. The simulation, which includes all three types of feedback (SNe + stellar winds + PNe), has evolved 1.3 Myr after discrete PN seeds start to be planted.

Bregman & Parroit (2009) have conducted 2-D hydrodynamic simulations of an individual PN undergoing ram-pressure stripping of ambient hot gas in a typical spheroid environment. They show that fluid instabilities result in mixing and heating of about half of the PN ejecta to $10^5 - 10^6$ K, a temperature range that is still at least a factor of several lower than that of the hot gas, while the other half remains cold ($< 10^5$ K) and flows out of the simulation box (25 pc long in the opposite direction of the motion). These results provide useful insights into the PNe feedback processes. But clearly the simulations have several major limitations, in terms of the explored volume, strength/mode of fluid instabilities, and interplay among various objects/processes in a spheroid.

In a real galactic spheroid, the environment can be very violent because of the presence of SN shockwaves and the collisions among randomly moving PNe, as well as the global spheroid-wide outflow, and hence the large pressure and density gradients. We have recently performed a pilot study, which for the first time simulates a galactic spheroid with discrete PN feedback. While the basic setting is the same as that described in Tang & Wang (2010), we make the PN mass contribution discrete, after the SN-driven galactic outflow becomes stable (Tang & Wang 2010). The positions and velocities of the imbedded PN seeds, with an occurrence rate of 4.5×10^4 Myr^{-1}, are determined in a separate N-body simulation using the software package Zeno (developed by Joshua Barnes). Each embedded PN seed has a mass of 0.3 $M_\odot$, an expansion velocity of 20 km s^{-1}, and a uniform temperature of 10^5 K (artificially set to achieve a sufficiently high initial resolution). The size of the seed is determined by a rough local pressure balance at the embedded position. The refinement level reaches down to ~ 1 pc. The embedded PNe then continue to expand and interact with their environments.

Fig. 4 illustrates the result of the simulation, showing a diversity of PN morphologies on scales of a few pc to a few tens of pc, while SN shock waves are typically on larger scales. Close examination of individual PNe suggests that some of them are morphologically similar to that in the 2-D simulations (Bregman & Parroit 2009), but others appear to be very different, demonstrating complicated interaction of PNe with the environments. In the inner most region where the gas density and stellar number density are high, PNe only grow to several pc in size, while in outer regions or inside low-density SNRs

PNe can grow as big as 50 pc. Most interestingly, many of the PNe seem to be heated eventually to a couple of million degrees, at which the bulk of the observed diffuse X-ray luminosity is emitted. Therefore, the debris of PNe can naturally lead to enhanced X-ray emission. Detailed simulation and analysis will enable us to examine the evolution of individual PNe to see how instabilities grow in different local environments, to quantify what fraction of PN mass is heated to a temperature high enough to emit X-ray, and to check whether or not the enhanced X-ray emission could make up the difference between previous simulation predicted value and the observed one.

In conclusion, most SNRs are not observed individually. These "missing" SNRs collectively represent a key component of the stellar feedback in galaxies, in either active star-forming spirals or cool gas-poor spheroids. The feedback regulates the ecosystems of the galaxies, hence their evolution.

References

Bogdan A. & Gilfanov M. 2008, *MNRAS*, 388, 56
Bregman, J. N. 1980, *ApJ*, 237, 280
Bregman, J. N & Parriott, J. R. 2009, *ApJ*, 699, 923
Buckley, D. & Schneider, S. E. 1995, *ApJ*, 446, 279
Butt, Yousaf M., & Bykov, Andrei M. 2008, *ApJ* (Letters) 677, 21
Bykov, A. M. & Fleishman, D. G. 1992, *MNRAS*, 255, 269
Cho, H. & Kang, H., 2008, *New Astron.*, 13, 163
Chu, Y.-H. & Mac Low, M. 1990, *ApJ*, 365, 510
Ciotti, L., Pellegrini, S., Renzini, A., & D'Erocole, A. 1991, *ApJ*, 376, 380
David L. P., Jones C., Forman W., Vargas I. M., & Nulsen P., 2006, *ApJ*, 653, 207
Dong, H., Wang, Q. D., & Morris, M. R. 2012, *MNRAS*, 425, 884
Everett, J. E. & Churchwell, E. 2010, *ApJ*, 713, 592
Fisk, L. A. & Gloeckler, G. 2012, *ApJ*, 744, 127
Fukugita M., *et al.* 1998, *ApJ*, 503, 518
Higdon, J. C. & Ligenfelter, R. E. 2005, *ApJ*, 628, 738
Jaskot, A. E. *et al.* 2011, *ApJ*, 729, 28
Krause, M. *et al.* 2013, *A&A*, 550, 49
Li. Z. Y., & Wang, Q. D. 2007, *ApJ*, 668, L39
Mac Low, M.-M. & McCray, R. 1988, *ApJ*, 324, 776
Mathews, W. G. & Baker, J. C. 1971, *ApJ*, 170, 241
Monaco, P. 2004, *MNRAS*, 354, 151
Oey, M. S. 2009, in *AIP Conf. Ser. 1156, The Local Bubble and Beyond II*, ed.: R. K. Smith, S. L. Snowden, & K. D. Kuntz (Melville, NY: AIP), 295
Parizot, E., *et al.* 2004, *A&A*, 424, 747
Rajagopal, J. *et al.* 2007, *ApJ*, 671, 2017
Smith, D. A. & Wang, Q. D. 2004, *ApJ*, 611, 881
Tüllmann, R., *et al.* 2009, *ApJ*, 707, 1361
Tang, S. K. & Wang, Q. D. 2005, *ApJ*, 628, 205
Tang, S. K. & Wang, Q. D. 2009, 397, 2106
Tang, S. K., Wang, Q. D., Lu, Y., & Mo, H. J. 2009a, *MNRAS*, 392, 77
Tang, S. K., Wang, Q. D., Joung, M. K. R., & Mac Low, M. M. 2009b, 398, 1468
Tang, S. K. & Wang, Q. D. 2010, *MNRAS*, 408, 1011
Wang, Q. & Helfand, D. J. 1991, *ApJ*, 373, 497
Wang, Q. D. 2010, *PNAS*, 107, 7168
Weaver, *et al.* 1977, *ApJ*, 218, 377
Yamaguchi, H., Sawada, M., & Bamba, A. 2010, *ApJ*, 715, 412

Discussion

PODSIADLOWSKI: What was the assumed SN Ia note in your MB1 simulation? I guess for a X-ray efficiency of $\sim$1 % you need about 1 SN every 10^3 yr?

WANG: The standard SN Ia rate as a fraction of stellar mass is used.

Supernova Environmental Impacts
Proceedings IAU Symposium No. 296, 2013
A. Ray & R. A. McCray, eds.

doi:10.1017/S1743921313009599

Formation of cold filaments from colliding superbubbles

Evangelia Ntormousi[1,2,3], **Katharina Fierlinger**[1,4], **Andreas Burkert**[1,2,4] **and Fabian Heitsch**[5]

[1]University Observatory Munich,
Scheinerstr. 1, D-81679 München, Germany
email: eva.ntormousi@cea.fr

[2]Max-Planck Institut für Extraterrestrische Physik
Giessenbachstrasse 1 85748 Garching

[3]Service d'Astrophysique, CEA/DSM/IRFU
Orme des Merisiers, Bat 709 Gif-sur-Yvette, 91191 France

[4]Excellence Cluster Universe
Boltzmannstr. 2 D-85748 Garching

[5] Department of Physics and Astronomy
University of North Carolina Chapel Hill
Chapel Hill, NC 27599-3255

Abstract. We present results from numerical simulations of expanding and colliding supershells. These large-scale spherical shocks, created by the combined feedback from several OB stars, are unstable to a number of hydrodynamical instabilities, so they quickly fragment into cold and highly structured clumps. A collision between two large shells can organize these small clumps into very filamentary structures, of tens of parsecs length and less than a parsec thick. In simulations where the flow of stellar material is followed with a tracer quantity, cold structures practically do not contain any enriched material from the OB associations at the time of their creation. In this context then, the clumps are created almost exclusively out of diffuse ISM material, containing almost no wind or supernova matter. Although the mechanism presented here is possibly not the only route for filament creation, this predicted property may help identify regions of sequential star formation.

Keywords. ISM, hydrodynamics, stars:formation

---

1. Introduction

Massive stars deposit large amounts of energy and considerable amounts of mass to their surroundings during their lifetimes in the form of winds. Their deaths as supernova explosions are also a powerful source of energy and new metals for the Interstellar Medium (ISM).

It is well known by now that the collective feedback from several OB stars creates compressions in the ISM of scales that range from tens to hundreds of parsecs. These huge spherical shocks around OB associations are typically observed to fragment into molecular clouds and lead to new events of star formation; as a result the conditions for large-scale shock fragmentation have been the focus of extensive theoretical work. Although a full summary is not possible here, we refer the reader to the analysis by Vishniac (1983), who provided dispersion relations for the growth of dynamical instabilities on spherical shocks, Whitworth *et al.* (1994), who showed that the fragments become self-gravitating when the thin shock layer is still confined by ram pressure and Mac Low & Norman

(1993), who performed numerical studies of the Vishniac instability, as most relevant to the contents of this paper.

The purpose of this work is to explore the nonlinear behavior of the gas in large spherical thin interstellar shocks as the result of relevant hydrodynamical instabilities, namely as the Vishniac instability, ,the Kelvin-Helmholtz and the Thermal instability, by means of numerical simulations. We show that this interaction can create cold, dense structures in a variety of physical states and. with supersonic internal velocities. The collision between two spherical shells creates very elongated cold filaments due to the large-scale shear.

The feedback from young OB associations is modeled as a time-dependent mass and energy source taken from stellar population synthesis models. In addition, the flow of the (expectedly metal-enriched) material from the hot to the cold phase is followed with a passive hydrodynamical quantity.

2. Numerical Simulations

High-resolution numerical simulations (maximum resolution of 0.1 pc) have been performed with the hydrodynamical Adaptive Mesh refinement (AMR) code RAMSES Teyssier(2002), in two and three dimensions. The time-dependent energy and mass input from OB associations, as calculated by Voss *et al.* (2009) is implemented in the code as a source term in the energy and in the mass equations following a reference table. In the same way, the heating and cooling rates typical of a solar metallicity gas are tabulated for various densities and temperatures and added or subtracted accordingly from the energy budget of each cell at every time step. In some of the simulations an additional continuity equation is solved for a tracer quantity, allowing us to follow the flow of wind and supernova material into the rest of the fluid.

2.1. *Filament formation*

All simulations start with diffuse, warm (n_H = 1 cm^{-3}, T=8000 K) gas, which corresponds to the warm thermally stable phase of the ISM as defined by the used cooling and heating rates. Two identical feedback regions representing OB associations of 20 or 50 stars each are placed on either side of the computational box. Thin dense shocks form rapidly around these regions. The simulations are allowed to continue well after the two shocks collide at the middle of the computational box and are stopped only when boundary effects start to become important. The typical morphology of a simulation is shown in Figure 1.

As the shocks decelerate through their interaction with the surrounding gas, the Vishniac instability magnifies any perturbations on the shock surface, resulting in characteristic triangular ripples. The shear inside the shock triggers the Kelvin-Helmholtz instability, while the condensations at the tips of the ripples are thermally unstable. The net result is the formation of small, cold and dense clumps. These clumps are at lower pressure than their environment, typical of the Thermal Instability, so many of them are observed to be condensing.

When the shells collide, the small clumps are brought together and at the same time sheared into a filamentary morphology. This is illustrated in the left panel of Figure 1. It is worth pointing out that it takes less than 1 Myr for the clumps to form, a timescale much shorter that those typically estimated for the gravitational fragmentation of the shell alone.

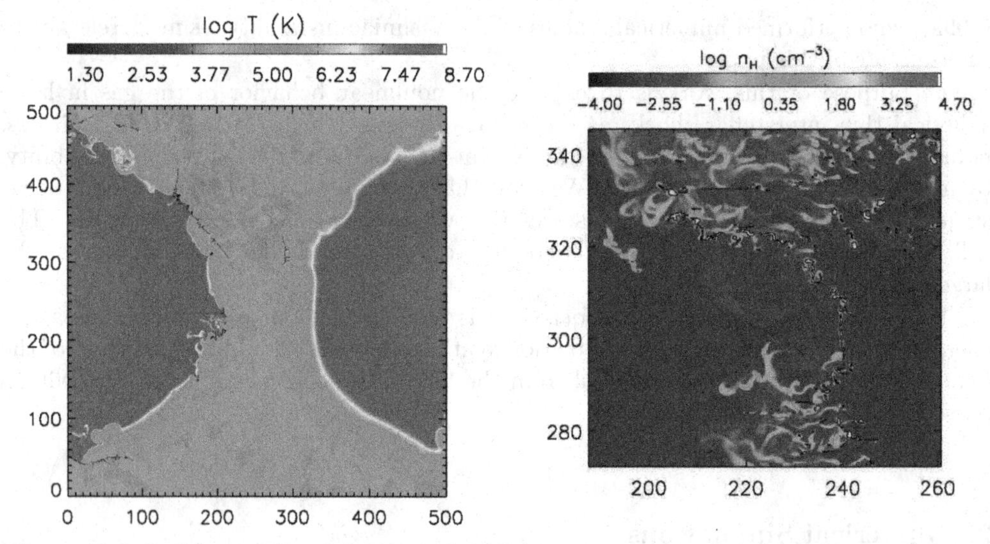

Figure 1. Left: Contour plot of the logarithm of the gas temperature at an early stage (about 3 Myrs after star formation started in the OB associations) of a 2D supershell collision simulation in a turbulent diffuse environment. Right: A zoomed-in region of a 2D supershell simulation at the shock collision interface, about 7 Myrs since star formation started in the associations. Shown here are contours of the logarithm of the hydrogen number density. The contour for $n_H = 50$ cm^{-3} is shown in black. The small-scale clumps are grouped in a filamentary configuration due to the large-scale shear. The axes are in parsecs. These Figures can also be found in Ntormousi *et al.* (2011).

2.2. *Flow of material from the hot to the cold gas*

The flow of material from the hot and metal-enriched gas to the cold clumps formed around the associations is traced in the simulations by adding an arbitrary amount of a tracer quantity in the feedback regions at each timestep. The behavior of the flow is illustrated by the plots in Figure 2. It is evident that the clumps are formed out of the warm diffuse phase of the ISM compressed by the shocks. Surprisingly, the hot gas is not mixed into them even at late times, when the shells collide and the clumps are found in the very turbulent interaction region. Since accretion due to gravity has not been modeled, we cannot draw strong conclusions for the future metallicity evolution of the clumps, but what is evident from these simulations is that at the time of their formation the clumps are composed purely out of the pre-existing diffuse gas.

3. Conclusions

A series of numerical simulations exploring the nonlinear evolution of the gas inside and around these large-scale spherical shocks show that cold filamentary structures of a few parsecs size form naturally and rapidly around OB associations. Collisions between such shocks create much longer filaments, of the order of tens of parsecs.

In those simulations where we can trace the flow of hot gas we observe that the cold clumps are formed at the outer part of the shock and thus contain no wind or supernova material from the stars that triggered their existence. The turbulence created at the collision interface does not help mixing into the coldest gas, although it does help mixing between the hot and the warm gas. In order to estimate the final metallicities of the clumps further investigation is needed, including modeling of thermal conductivity to

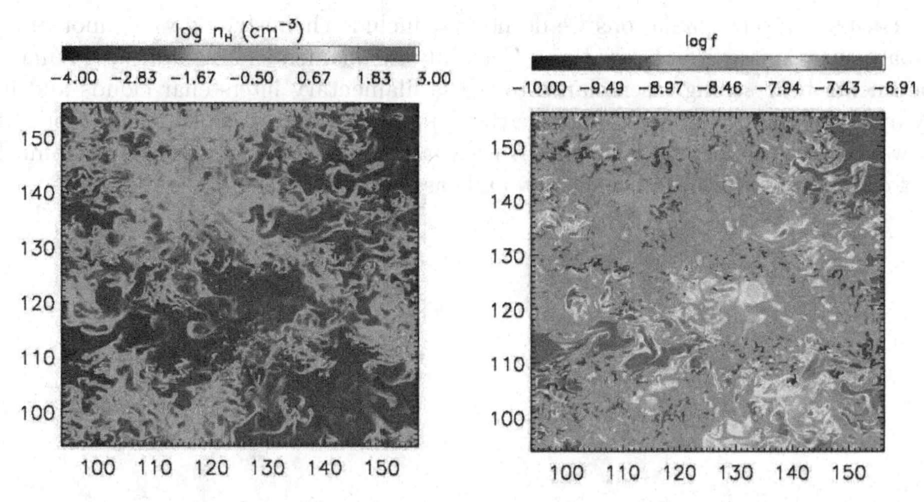

Figure 2. Contour plots of the logarithm of hydrogen number density (left) and the logarithm of relative tracer content (right) at the shock collision interface in a 2D simulation, about 3 Myrs since star formation started. The axes are in parsecs, like above, although this simulation was done in a smaller domain. It is clear that dense regions contain practically no new material. This Figure can also be found in Ntormousi & Burkert (2011).

study clump evaporation, possible inhomogeneities in the metal input and gravitational accretion from the environment.

References

Mac Low, M.-M. & Norman, M. L. 1993, *ApJ*, 407, 207

Ntormousi, E. & Burkert, A. 2011, arXiv:1111.1859

Ntormousi, E., Burkert, A., Fierlinger, K., & Heitsch, F. 2011, *ApJ*, 731, 13

Teyssier, R. 2002, *A&A*, 385, 337

Vishniac, E. T. 1983, *ApJ*, 274, 152

Voss, R., Diehl, R., Hartmann, D. H., *et al.* 2009, *A&A*, 504, 531

Whitworth, A. P., Bhattal, A. S., Chapman, S. J., Disney, M. J., & Turner, J. A. 1994, *A&A*, 290, 421

Discussion

WANG: Do you include thermal conduction in your simulations?

NTORMOUSI: Unfortunately the presence of very hot gas in these simulations would make the timesteps extremely small if thermal conduction was included. However, we are currently resimulating smaller regions of the larger volume at higher resolution, including thermal conductivity. Qualitatively, we expect the effect of thermal conduction to be cloud evaporation. This would essentially dilute the metal content of the warm gas rather than lead to any enrichment of the dense gas. We will report detailed results after the new simulations are completed.

CESARSKY: What do these clouds become? Are the filaments you see in your simulations related to the filaments observed by Philippe Andre and others with Herschel in star forming and non-star forming interstellar clouds?

NTORMOUSI: In our simulations we do not yet include chemistry, so we cannot directly compare our results with observations. Certainly the filaments formed in these numerical experiments bear strong resemblance to dense filamentary interstellar clouds and have very dynamical nature. I could imagine the Pipe nebula, for example, to have formed in this way, since it is located near an OB association, the same for the Lupus-Ophiuchus cloud complexes, which are between two OB associations.

Supernova Environmental Impacts
Proceedings IAU Symposium No. 296, 2013
A. Ray & R. A. McCray, eds.

doi:10.1017/S1743921313009605

Gamma-ray observations of supernova remnants

Marianne Lemoine-Goumard

Centre d'Études Nucléaires de Bordeaux Gradignan
Université Bordeaux 1, CNRS/IN2P3
33175 Gradignan, France
E-mail: lemoine@cenbg.in2p3.fr
Funded by contract ERC-StG-259391 from the European Community

Abstract. In the past few years, gamma-ray astronomy has entered a golden age. At TeV energies, only a handful of sources were known a decade ago, but the current generation of ground-based imaging atmospheric Cherenkov telescopes has increased this number to more than one hundred. At GeV energies, the *Fermi* Gamma-ray Space Telescope has increased the number of known sources by nearly an order of magnitude in its first 2 years of operation. The recent detection and unprecedented morphological studies of gamma-ray emission from shell-type supernova remnants is of great interest, as these analyses are directly linked to the long standing issue of the origin of the cosmic-rays. However, these detections still do not constitute a conclusive proof that supernova remnants accelerate the bulk of Galactic cosmic-rays, mainly due to the difficulty of disentangling the hadronic and leptonic contributions to the observed gamma-ray emission. In the following, I will review the most relevant results of gamma ray astronomy concerning supernova remnants (shell-type and middle-age interacting with molecular clouds).

Keywords. cosmic-rays; supernova remnants; gamma-rays

1. The cosmic-ray mystery

1.1. *The link between cosmic-rays and supernova remnants*

The association between supernova remnants (SNRs) and Galactic cosmic rays (CRs) is very popular since 1934, when Baade and Zwicky argued that this class of astrophysical objects can account for the required CR energetics (W. Baade and F. Zwicky (1934)). Indeed, in order to maintain the cosmic-ray energy density in the Galaxy, about 3 supernovae per century should transform 10 percent of their kinetic energy in cosmic-ray energy. This argument has also been supported by E. Fermi's proposal of a very general mechanism for particle acceleration, which is very efficient if applied at SNR shocks (A. R. Bell (1978)). The extremely interesting point of the diffusive shock acceleration (DSA) mechanism is that it naturally yields power-law spectra for the energy distribution of accelerated particles. However, until recently there were absolutely no observational evidence concerning the acceleration of protons and nuclei in SNRs. Indeed, through their interaction with the interstellar magnetic fields, the charged particles arriving on Earth have lost all directional information and cannot be used to pinpoint the sources. That is why, almost 100 years after their discovery by V. Hess, the origins of the cosmic-rays and their cosmic accelerators remain unknown.

Astronomy with gamma-rays provides a means to study these sources of high energy particles. Indeed, cosmic rays (ionized nuclei of all species, but mostly protons, plus a small fraction of electrons) can interact with ambient matter and photons producing gamma-rays via two different channels. One mechanism invokes the interaction of accelerated protons at supernova remnants shocks with interstellar material generating neutral pions which in turn decay into gamma rays. We call this mechanism the hadronic scenario. A second competing channel exists in the inverse Compton scattering of the photon fields in the surroundings of the SNR by the same relativistic electrons that generate the synchrotron X-ray emission. This is the leptonic scenario. Being of leptonic or hadronic origin, these gamma-rays are not affected while they travel to Earth and can therefore be used to pinpoint the cosmic accelerators in our Galaxy.

1.2. *Gamma-ray experiments*

Two major breakthroughs occurred recently in gamma-ray astronomy. Firstly, after more than 20 years of development, the first source of very high energy gamma-rays, the Crab Nebula, was discovered in 1989 by the Whipple telescope. Since this date the technical progresses in this field have led to important scientific results, especially by the Cherenkov telescopes H.E.S.S., VERITAS and MAGIC. These ground-based experiments for gamma-ray astronomy rely on the development of cascades (air-showers) initiated by astrophysical gamma-rays. Such cascades only persist to ground-level above 1 TeV and only produce significant Cherenkov light above a few GeV, setting a fundamental threshold to the range of this technique. Today, more than 160 gamma-ray sources have been detected with high significance, 23 being associated to supernova remnants or molecular clouds.

Second, in space, the Large Area Telescope (LAT) onboard the *Fermi* satellite has considerably improved our knowledge of the 0.1–100 GeV gamma-ray sky with 1873 objects detected in only two years of observation (P. L. Nolan *et al.* (2012)). It has moved the field from the detection of a small number of sources to the detailed study of several classes of Galactic and extragalactic objects. A complete study of association of the 1873 sources detected show that $\sim 4\%$ of them are associated to supernova remnants (P. L. Nolan *et al.* (2012)).

Thanks to the observations of radio and X-ray synchrotron from a large number of supernova remnants, there is no doubt today that supernova remnants can accelerate efficiently particles up to 10^{14} eV. The question is whether these particles are protons or electrons and if they can be accelerated up to the knee of the cosmic-ray spectrum (10^{15} eV). A direct signature of accelerated protons is expected through pion decay emission in the GeV-TeV gamma ray range.

2. Detection of supernova remnants in gamma-rays

The sample of supernova remnants detected in gamma-rays is now extremely large: it goes from evolved supernova remnants interacting with molecular clouds (MC) up to young shell-type supernova remnants and historical supernova remnants. The *Fermi*-LAT even detected one evolved supernova remnant without MC interaction, Cygnus loop. This section will review some of the main characteristics of detected SNRs. Complementary informations on GeV SNRs can be found in the description of Puppis A (J. W. Hewitt *et al.* (2012)) and in the *Fermi* Catalog of Supernova Remnants (T. J. Brandt (2013)).

2.1. *Historical supernova remnants*

Multi-PeV protons can be accelerated only during a relatively short period of the SNR evolution, namely, at the end of the free-expansion phase/beginning of the Sedov phase, when the shock velocity is high enough to allow sufficiently high acceleration rate. When the SNR enters the Sedov phase, the shock slows down leading to a decrease of the maximum energy of the particles that can be confined within the SNR. Therefore, historical supernova remnants are excellent targets for gamma-ray observations.

Two historical SNRs have been detected both at GeV and TeV energies: Cassiopeia A (Cas A) (A. A. Abdo *et al.* (2010), J. Albert *et al.* (2007), V. A. Acciari *et al.* (2010)) and Tycho (F. Giordano *et al.* (2012), V. A. Acciari *et al.* (2011)).

Cas A is the remnant of SN 1680. It is the brightest radio source in our Galaxy and its overall brightness across the electromagnetic spectrum makes it a unique laboratory for studying high-energy phenomena in SNRs. A multiwavelength modeling of Cas A does not allow a discrimination between the hadronic and leptonic scenarios. However, regardless of the origin of the observed gamma rays, this modeling implies that the total content of CRs accelerated in Cas A is $\sim(1-2)\times10^{49}$ erg, and the magnetic field amplified at the shock can be constrained as B ≈ 0.12 mG. Even though Cas A is considered to have entered the Sedov phase, the total amount of CRs accelerated in the remnant constitutes only a minor fraction ($\sim$ 2%) of the total kinetic energy of the supernova, which is well below the $\sim$ 10% commonly used to maintain the cosmic-ray energy density in the Galaxy.

Tycho's SNR (SN 1572) is classified as a Type Ia (thermonuclear explosion of a white dwarf) based on observations of the light-echo spectrum. Thanks to the large amount of data available at various wave bands, this remnant can be considered one of the most promising object where to test the shock acceleration theory and hence the CR – SNR connection. First, using the precise radio and X-ray observations of this SNR, G. Morlino & D. Caprioli (2012) have shown that the magnetic field at the shock has to be $> 200\,\mu$G to reproduce the data. Then, using multiwavenlength data, especially the GeV and TeV detections, they could infer that the gamma-ray emission detected from Tycho cannot be of leptonic origin, but has to be due to accelerated protons (this result is consistent with another modeling proposed in F. Giordano *et al.* (2012)). These protons are accelerated up to energies as large as $\sim$500 TeV, with a total energy converted into CRs estimated to be about 12% of the forward shock bulk kinetic energy. This is much more reasonable in the context of acceleration of Galactic cosmic-rays in SNRs.

2.2. *Young shell-type supernova remnants*

Four young shell-like SNRs with clear shell-type morphology resolved in VHE gamma-rays have been detected by H.E.S.S.: RX J1713.7-3946 (F. Aharonian *et al.* (2004), F. Aharonian *et al.* (2005)), RX J0852.04622 – also known as Vela Junior - (F. Aharonian *et al.* (2007)), SN 1006 (F. Acero *et al.* (2010)) and HESS J1731-347 (A. Abramowski *et al.* (2011)). A fifth case, RCW 86 (F. Aharonian *et al.* (2011)), might be added to this list although the TeV shell morphology has not yet been clearly proved. Two of them, RX J1713.7-946 (A. A. Abdo *et al.* (2011)) and Vela Junior (T. Tanaka *et al.* (2011)), have been detected by *Fermi*-LAT allowing direct investigation of young shell-type SNRs as sources of cosmic rays. Concerning RX J1713.7-3946, the *Fermi*-LAT spectrum is well described by a very hard power-law with a photon index of $\Gamma = 1.5\pm0.1$ that coincides in normalization with the steeper H.E.S.S.-detected gamma-ray spectrum at higher energies. The GeV measurements with *Fermi*-LAT do not agree with the expected fluxes around 1 GeV in most hadronic models published so far (e.g., E. G. Berezhko & H. J. Voelk (2010)) and requires an unrealistically large density of the medium. The agreement with the expected IC spectrum is better (as can be seen in Figure 1) but requires a very low magnetic field of $\sim 10\,\mu$G in comparison to the one measured in the thin filaments by X-ray observations. It is possible to reconcile a high magnetic field with the leptonic model if GeV gamma rays are radiated not only from the filamentary structures seen by Chandra, but also from other regions in the SNR where the magnetic field may be

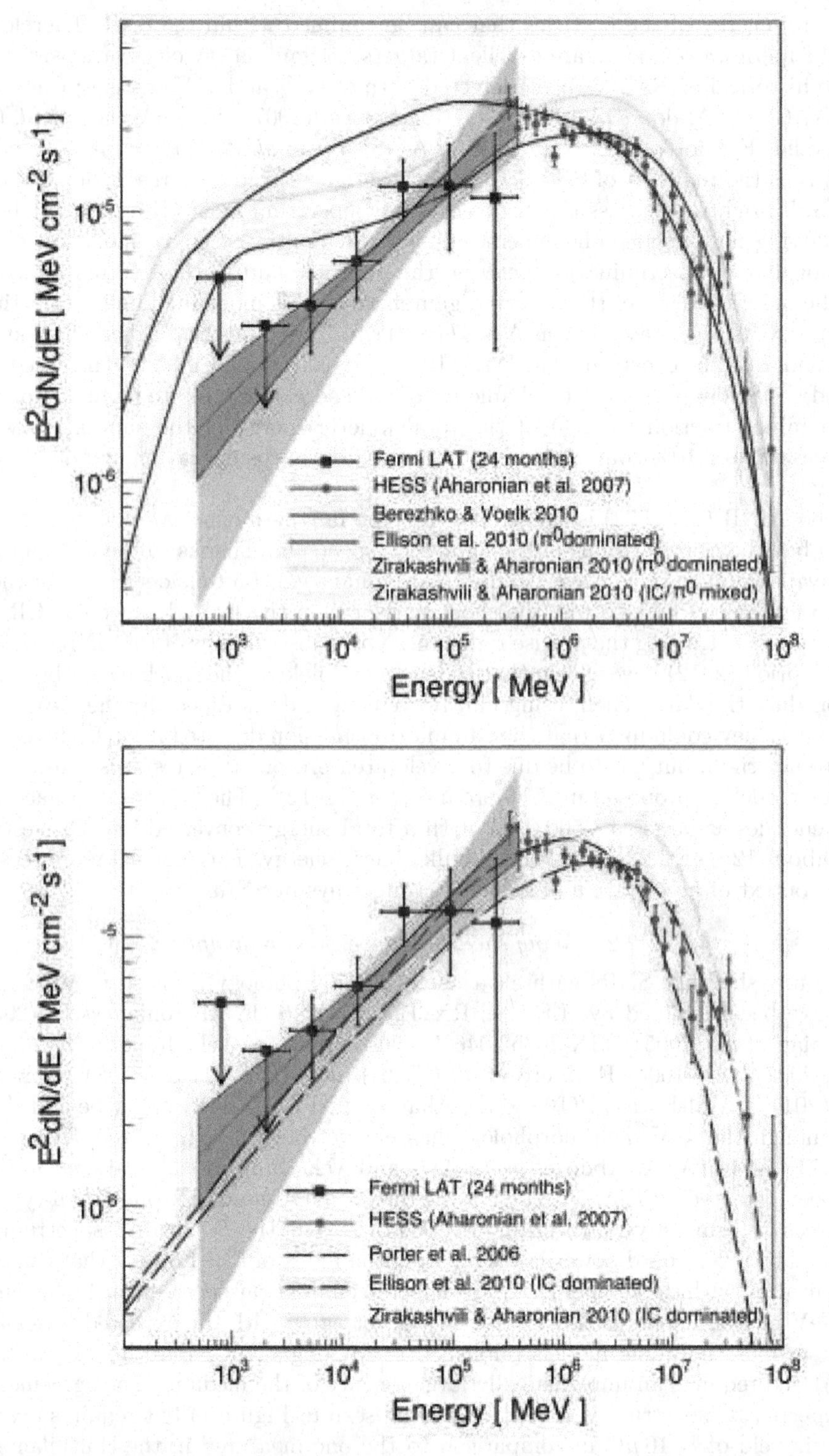

Figure 1. Energy spectrum of RX J1713.7-3946 in gamma rays. Shown is the *Fermi*-LAT (A. A. Abdo *et al.* (2011)) detected emission in combination with the energy spectrum detected by H.E.S.S. (F. Aharonian *et al.* (2005)). See A. A. Abdo *et al.* (2011) for more details.

weaker. Similar conclusions are reported for Vela Junior supernova remnant even though in this case the hadronic scenario can not be ruled out. However, being of hadronic or leptonic origin, the GeV-TeV gamma-ray detections imply a low maximal energy for the accelerated particles of $\sim$ 100 TeV, well below the knee of the cosmic-ray spectrum.

2.3. *Supernova remnants interacting with molecular clouds*

The *Fermi* LAT Collaboration has reported the discoveries of several middle aged ($\sim 10^4$ yrs) remnants interacting with molecular clouds: W51C (A. A. Abdo *et al.* (2010)), W44 (A. A. Abdo *et al.* (2010)), IC 443 (A. A. Abdo *et al.* (2010)), W49 (A. A. Abdo *et al.* (2010)) and W28 (A. A. Abdo *et al.* (2010)), being the most famous cases. Apart from W44, they have all been detected in the TeV regime as well. These SNRs are generally much brighter in GeV than in TeV in terms of energy flux (due to a spectral steepening arising at a few GeV), which emphasizes the importance of the GeV observations. The interaction with a molecular cloud provides the target material that allows to enhance the gamma-ray emission, either through bremsstrahlung by relativistic electrons or by pion-decay gamma-rays produced by high-energy protons. The observed large luminosity of the GeV gamma-ray emission precludes the inverse-Compton scattering off the CMB and interstellar radiation fields as the main emission mechanism since it would require an extremely low density (to suppress the bremsstrahlung and proton-proton interaction), a low magnetic field to enhance the gamma/X-ray flux ratio and an unrealistically large energy injected into protons. In addition, the break in the electron spectrum corresponding to the gamma-ray spectrum directly appears in the radio data leading to a bad modeling of the radio data and therefore disfavors the bremsstrahlung process. A model in which gamma-rays are produced via proton-proton interaction gives the most satisfactory explanation for the GeV gamma-rays observed in SNRs interacting with molecular gas as seen in Figure 2 for the case of W51C. However, the hadronic nature of the gamma-ray signal is mostly inferred from indirect arguments in such cases. Recently, an improved spectral analysis of W44 and IC 443 have been carried using four years of *Fermi*-LAT data above 60 MeV M. Ackermann *et al.* (2013). Both spectra are steeply rising below 200 MeV, showing a clear break at around 200 MeV (see the case of IC 443 in Figure 3). This spectral feature is often referred to as the "pion-decay bump" and uniquely identifies pion-decay gamma-rays. This detection thus provides a direct evidence that cosmic-ray protons are accelerated in SNRs W44 and IC 443.

There are two different types of hadronic scenarios to explain the GeV gamma-ray emission arising from such SNRs: the "Runaway CR" model (F. Aharonian & A. M. Atoyan (1996), Y. Ohira, K. Murase & R. Yamazaki (2011)) and the "Crushed Cloud" model (Y. Uchiyama *et al.* (2011)). The Runaway CR model considers gamma-ray emission from molecular clouds illuminated by runaway CRs that have escaped from their accelerators, whereas the Crushed Cloud model invokes a shocked molecular cloud into which cosmic-ray particles are adiabatically compressed and accelerated resulting in enhanced synchrotron and pion-decay gamma-ray emissions.

3. Where are the PeVatrons ?

The recent GeV and TeV detections of supernova remnants confirm the theoretical predictions that supernova remnants can operate as powerful cosmic ray accelerators. However, if these objects are responsible for the bulk of galactic cosmic rays, they should be able to accelerate protons and nuclei at least up to 10^{15} eV and therefore act as PeVatrons. S. Gabici & F. A. Aharonian (2007) have shown that the spectrum of nonthermal particles extends to PeV energies only during a relatively short period of the evolution of

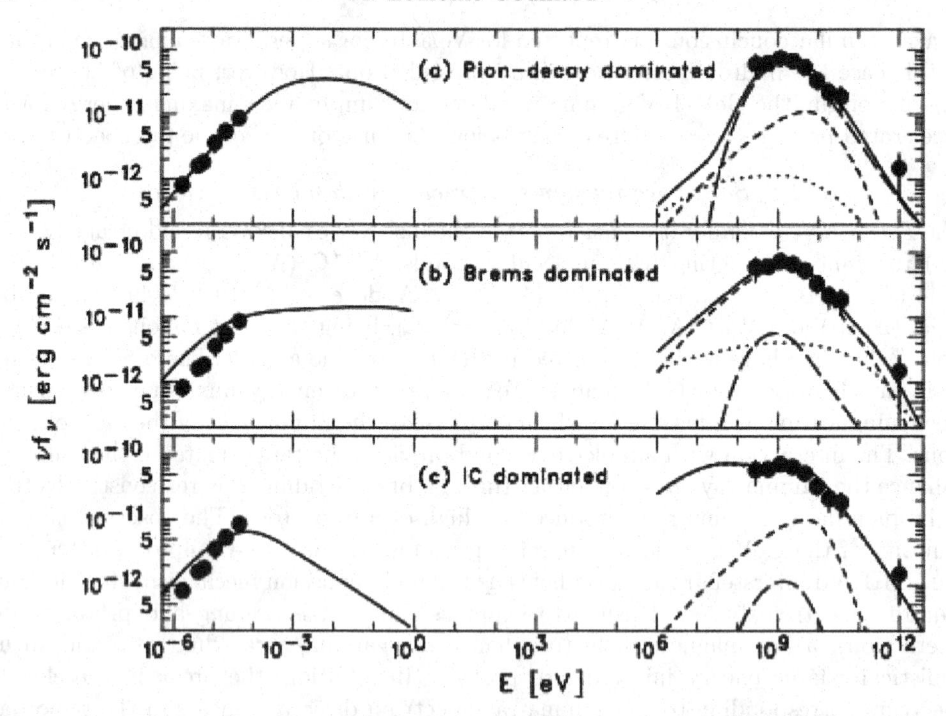

Figure 2. Different scenarios proposed for the multiwavelength modeling of W51C A. A. Abdo *et al.* (2010). The radio emission (from Moon & Koo 1994) is explained by synchrotron radiation, while the gamma-ray emission is modeled by different combinations of pion-decay (long-dashed curve), bremsstrahlung (dashed curve), and IC scattering (dotted curve). The sum of the three component is shown as a solid curve. See A. A. Abdo *et al.* (2010) for more details.

the remnant since high energy particles are the first to escape from the supernova remnant shock. For this reason one may expect spectra of secondary gamma-rays extending to energies beyond 10 TeV only from less than 1 kyr old supernova remnants. In this respect, Tycho could be considered as a half-PeVatron at least, since there is no evidence of a cut-off in the VERITAS data. One may wonder how many PeVatrons are expected to be detectable in our Galaxy. A simple estimate has been provided by Gabici and Aharonian (2007): assuming a rate of ~3 supernovae per century in our Galaxy, this directly implies that only a dozen of PeVatrons are present in the Galaxy on average and hence that they are likely to be distant and weak. This emphasizes the importance of TeV observations by the future generation of Cherenkov telescopes such as the Cherenkov Telescope Array (CTA) which will have a better effective area in the energy range already covered but that will also allow the observation up to 100 TeV of sources such as Tycho, therefore constraining the maximal energy at which protons are being accelerated in young SNRs. Monte-Carlo simulations of shell-type SNRs have already been carried out for different CTA array layouts, assuming a uniform exposure time of 20 hr everywhere along the Galactic Plane. In this purpose, the morphological and spectral characteristics of three SNRs (RX J1713.7-3946, Vela Junior and RCW 86), as measured with H.E.S.S., together with their respective distance estimates, have been used to simulate sources throughout the inner Galaxy (F. Acero *et al.* (2013), M. Renaud (2011)). This leads to ~20–70 detectable TeV SNRs, among which ~7–15 would be resolved with CTA (configurations I and D, optimized for providing the best sensitivity over the whole energy range or above 1 TeV, respectively). It is stricking to note that, thanks to its increased sensitivity, CTA will have the capability to detect SNRs as luminous as RX J1713.7−3946, Vela Junior,

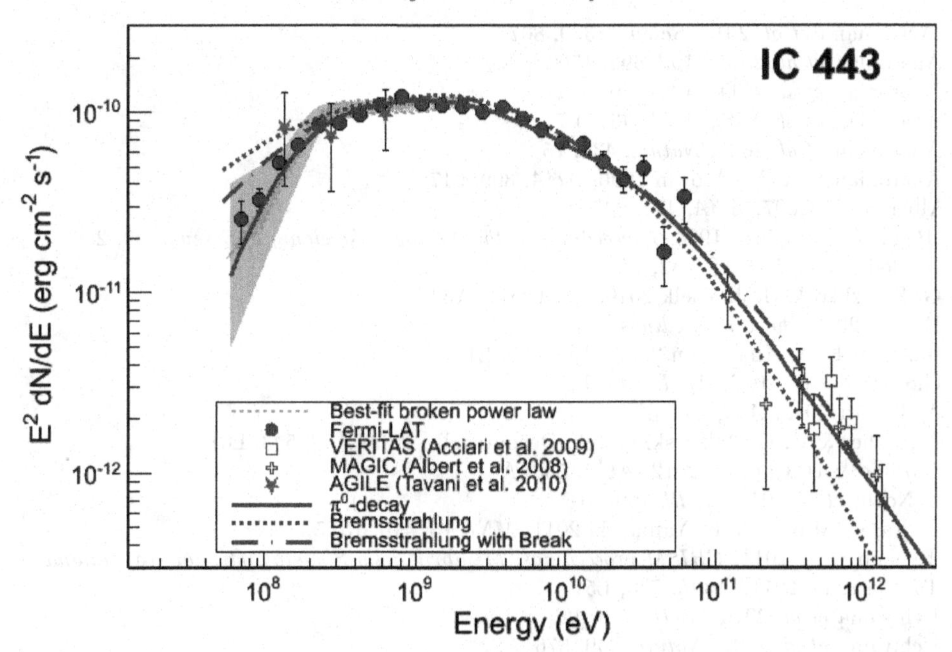

Figure 3. Gamma-ray spectra of IC 443 as measured with the *Fermi*-LAT. Color-shaded areas bound by dashed lines denote the best-fit broadband smooth broken power law (60 MeV to 2 GeV), gray-shaded bands show systematic errors below 2 GeV due mainly to imperfect modeling of the galactic diffuse emission. At the high-energy end, TeV spectral data points from MAGIC and VERITAS are shown. Solid lines denote the best-fit pion-decay gamma-ray spectra, dashed lines denote the best-fit bremsstrahlung spectra, and dash-dotted lines denote the best-fit bremsstrahlung spectra when including an ad hoc low-energy break at 300 MeV c^{-1} in the electron spectrum. See M. Ackermann *et al.* (2013) for more details.

or RCW 86 up to the other side of the Galaxy, providing a complete population study of our Galactic cosmic-ray accelerators.

Acknowledgements

I thank all the members of the *Fermi* GALACTIC and HESS SNR-PWN working groups for valuable discussion. I gratefully acknowledge funding from the European Community (contract ERC-StG-259391).

References

A. A. Abdo *et al.* 2011, *ApJ*, 734, 28
A. A. Abdo *et al.* 2010, *ApJL*, 710, L92
A. A. Abdo *et al.* 2010, *ApJ*, 722, 1303
A. A. Abdo *et al.* 2010, *ApJ*, 718, 348
A. A. Abdo *et al.* 2010, *ApJ*, 712, 459
A. A. Abdo *et al.* 2010, *Science*, 327, 1103
A. A. Abdo *et al.* 2009, *ApJL*, 706, 1
A. Abramowski *et al.* 2011, *A&A*, 531, A81
F. Acero *et al.* 2013, *Astroparticle Physics*, 43, 276
F. Acero *et al.* 2010, *A&A*, 512, A62
V. A. Acciari *et al.* 2011, *ApJL*, 730, L20
V. A. Acciari *et al.* 2010, *ApJ*, 714, 163

M. Ackermann *et al.* 2013, *Science*, 339, 807
F. Aharonian *et al.* 2011, *ApJ*, 692, 1500
F. Aharonian *et al.* 2011, *A&A*, 464, 235
F. Aharonian *et al.* 2005, *A&A*, 437, L7
F. Aharonian *et al.* 2004, *Nature*, 432, 75
F. Aharonian & A. M. Atoyan 1996, *A&A*, 309, 917
J. Albert *et al.* 2007, *A&A*, 474, 937
W. Baade & F. Zwicky 1934, *Proceedings of the National Academy of Science*, 20, 259
A. R. Bell 1978, *MNRAS*, 182, 147
E. G. Berezhko & H. J. Voelk 2010, *A&A*, 511, A34
T. Brandt 2013, *these proceedings*
S. Gabici & F. A. Aharonian 2007, *ApJ*, 665, L131
F. Giordano *et al.* 2012, *ApJL*, 744, L2
J. W. Hewitt 2012, *ApJ*, 759, 89
J. P. Hughes & C. E. Rakowski and A. Decourchelle 2009, *ApJ*, 543, L61
G. Morlino & D. Caprioli 2012, *A&A*, 538, A81
P. L. Nolan *et al.* 2012, *ApJS*, 199, 31
Y. Ohira, K. Murase & R. Yamazaki 2011, *MNRAS*, 410, 1577
M. Renaud *et al.* 2011, *CRISM proceedings, Memorie della Societa Astronomica Italiana*
T. Tanaka *et al.* 2011, *ApJL*, 740, L51
Y. Uchiyama *et al.* 2010, *ApJL*, 723, 122
Y. Uchiyama *et al.* 2007, *Nature*, 449, 576

Discussion

WANG: How many super bubbles have been detected as TeV sources?

LEMOINE-GOUMARD: There are two. But their IDs are still very uncertain.

Supernova Environmental Impacts
Proceedings IAU Symposium No. 296, 2013
A. Ray & R. A. McCray, eds.

doi:10.1017/S1743921313009617

Fermi-LAT and WMAP observations of the supernova remnant Puppis A

Marie-Hélène Grondin[1], John W. Hewitt[2], Marianne Lemoine-Goumard[3,4] & Thierry Reposeur[3], for the *Fermi*-LAT collaboration

[1]Institut de Recherche en Astrophysique et Planétologie, Université de Toulouse (UPS)/OMP, F-31028 Toulouse Cedex 4, France
E-mail: mgrondin@irap.omp.eu

[2]NASA Goddard Space Flight Center, Greenbelt, MD 20771, USA

[3]Centre dÉtudes Nucléaires de Bordeaux-Gradignan, Université Bordeaux 1, CNRS/IN2p3, F-33175 Gradignan, France

[4]Funded by contract ERC-StG-259391 from the European Community

Abstract. The supernova remnant (SNR) Puppis A (aka G260.4-3.4) is a middle-aged supernova remnant, which displays increasing X-ray surface brightness from West to East corresponding to an increasing density of the ambient interstellar medium at the Eastern and Northern shell. The dense IR photon field and the high ambient density around the remnant make it an ideal case to study in γ-rays. Gamma-ray studies based on three years of observations with the Large Area Telescope (LAT) aboard *Fermi* have revealed the high energy gamma-ray emission from SNR Puppis A. The γ-ray emission from the remnant is spatially extended, and nicely matches the radio and X-ray morphologies. Its γ-ray spectrum is well described by a simple power law with an index of ~ 2.1, and it is among the faintest supernova remnants yet detected at GeV energies. To constrain the relativistic electron population, seven years of *Wilkinson Microwave Anisotropy Probe (WMAP)* data were also analyzed, and enabled to extend the radio spectrum up to 93 GHz. The results obtained in the radio and γ-ray domains are described in detail, as well as the possible origins of the high energy γ-ray emission (Bremsstrahlung, Inverse Compton scattering by electrons or decay of neutral pions produced by proton interactions).

Keywords. ISM : supernova remnants, ISM: individual objects (Puppis A), cosmic rays

1. Introduction

Supernovae have long been thought responsible for accelerating protons to relativistic energies in our Galaxy. As reported in Brandt (2013) and Lemoine-Goumard (2013), the Large Area Telescope aboard the *Fermi Gamma-ray Space Telescope* has enabled the γ-ray detection of several middle-aged SNRs interacting with molecular clouds.

Among them, the supernova remnant (SNR) Puppis A (aka G260.4–3.4) is an important case to study, as it shows signs of recently encountering a higher ambient density in the vicinity of a nearby molecular cloud (Hwang *et al.* 2005). The remnant displays increasing X-ray surface brightness from West to East (Petre *et al.* 1982) corresponding to an increasing density of the ambient interstellar medium (ISM) at the Eastern and Northern shell (Dubner & Arnal, 1988). The proper motion of fast optical knots gives a dynamical age of 3700 ± 300 years, establishing that the SNR is in the Sedov-Taylor evolutionary phase. At an estimated distance of ~ 2 kpc (Reynoso *et al.* 1995), the diameter of Puppis A is ~ 30 pc, so the remnant is no longer interacting with the circumstellar medium of the progenitor, but with the surrounding ISM. There are two notable regions where the SNR shock has engulfed small denser clouds: the Bright Eastern Knot and the Northern Knot. Besides those, the shock has not yet become radiative for most of the SNR, consistent with the relatively young age and the low density of the surrounding medium.

In addition, SNR Puppis A hosts a central compact object recently identified as a pulsar, PSR J0821-4300, using *XMM-Newton* observations (Gotthelf & Halpern, 2009).

2. *Fermi*-LAT Observations and Data Analysis

The *Fermi*-LAT is an electron-positron pair conversion telescope, sensitive to γ-rays with energies from below 20 MeV to more than 300 GeV (Atwood *et al.* 2009). The following analysis was performed using 36 months of data collected from 2008 August 4 to 2011 August 20 within a $15^\circ \times 15^\circ$ region around the position of SNR Puppis A. Only events with zenith angles smaller than 100° were included to reduce contamination from the Earth limb. We used the P7V6 instrument response functions (IRFs), and selected the 'Source' events. Located at a distance of 3° from the SNR Puppis A, the Vela pulsar is the brightest steady γ-ray source in the sky, from which photons are observed up to 25 GeV (Abdo *et al.* 2010). To avoid any bias on the analysis of the SNR due to this bright nearby pulsar, the following analysis was performed in the off-pulse window of the Vela pulsar. For more details, please see Hewitt *et al.* (2012).

Morphology : Figure 1 shows the Test Statistic (TS) map for the region around Puppis A using photons with energies above 800 MeV. Significant emission coincident with Puppis A is clearly detected. The source extension was estimated using `pointlike` (Lande *et al.* 2012) with a uniform disk hypothesis for energies above 800 MeV. The fitted radius is $0.38^\circ \pm 0.04^\circ$, in good agreement with the size of the SNR as seen in the radio and X-rays. We have also examined the correspondence of the γ-ray emission from Puppis A with different source morphologies by using multi-frequency templates above 800 MeV. The radio template, the X-ray template and the uniform disk all produce improvements from the 2FGL 3-point source model, while having fewer degrees of freedom, the X-ray template providing the best log-likelihood of our fit.

Spectrum : Using the X-ray template (shown by green contours on Fig 1), we performed a spectral analysis between 200 MeV and 100 GeV using `gtlike`, a binned maximum likelihood method implemented in the *Science Tools* distributed by the *Fermi* Science Support Center (FSSC). The *Fermi*-LAT data are well described by a power-law with a flux above 200 MeV of $(1.6 \pm 0.2_{stat} \pm 0.2_{syst}) \times 10^{-8}$ ph cm^{-2} s^{-1} (renormalized to the whole phase interval) and a photon index of $2.09 \pm 0.07_{stat} \pm 0.09_{syst}$. Using radio template and the uniform disk (defined hereabove) yields consistent results within the statistical errors. Whatever the extended model considered for the spectral analysis, no evidence for cutoff or break is visible.

Temporal analysis : To check if the X-ray pulsar PSR J0821-4300 could contribute, at least partially, to the detected γ-ray emission, we folded the photon arrival times using an ephemeris from Gotthelf & Halpern (2009) and the Fermi plugin distributed with the `TEMPO2` software. No significant evidence of pulsation could be found. This confirms that with a small spin-down luminosity, PSR J0821-4300 is unlikely to be a γ-ray emitter.

3. WMAP Data Analysis

We used the 7-year all-sky data of the *Wilkinson Microwave Anisotropy Probe (WMAP)* to extend the radio spectrum of Puppis A to higher frequencies. Five bands were analyzed with effective central frequencies ($\nu_{\rm eff}$) of 23 to 93 GHz (Jarosik *et al.* 2011). To

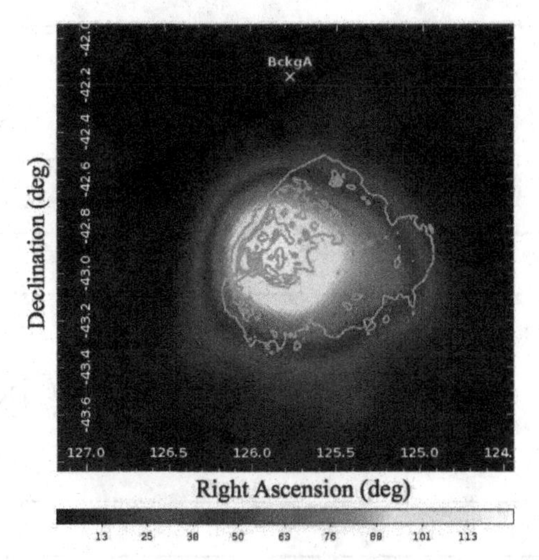

Figure 1. Test Statistic (TS) map in celestial coordinates of the SNR Puppis A. Only photons with energies above 800 MeV in the off-pulse window of the Vela pulsar were selected. The TS was evaluated by placing a point-source at the center of each pixel, Galactic diffuse emission and nearby sources being included in the background model. X-ray contours (green) are overlaid for comparison (Hewitt *et al.* 2012).

obtain *WMAP* flux densities we used template fitting of the flux-corrected 1.4 GHz radio image (Castelletti *et al.* 2006). This template image was smoothed with the *WMAP* beam profiles, and then fit in each band with a constant times the smoothed template plus a sloping planar baseline. To avoid contamination from the nearby Vela-X PWN, the template fit is restricted to a circular region within a 2° radius. To fit the radio spectrum, we exclude data below 300 MHz, which may suffer from low-frequency absorption by thermal electrons along the line of sight. We derived a best-fit 1 GHz flux density of 141 ± 4 Jy and a radio spectral index of -0.56 ± 0.01.

4. Discussion

We attempted to reproduce the *WMAP* and *Fermi*-LAT data by considering three scenarios dominated by each of the three plausible emission mechanisms : inverse Compton (IC) scattering, Bremsstrahlung and proton-proton interaction. We assumed similar injection spectra for protons and electrons, which implies that the spectral index of the particle spectrum below the cutoff energy is determined and fixed at ~ 2.1 by modeling the radio spectrum as synchrotron radiation by relativistic electrons. Modeled spectral energy distributions (SEDs) are presented in Figure 2 (see Hewitt *et al.* 2012 for more details). Simple constraints can be derived from these models. For instance, the electron-to-proton ratio needs to be larger than 0.1 for leptonic scenarios, and thus in excess of the ratio found for local cosmic-ray abundances, to inject a reasonable energy content in radiating electrons. In the same way, for the inverse-Compton dominated scenario, the average gas density needs to be lower than 0.3 cm^{-3} to reduce the Bremsstrahlung component. Although the hadronic scenario seems to be preferred, each emission mechanisms is able to fit the data with the current statistics. Future observations at high radio frequencies and at higher γ-ray energies will help to differentiate between leptonic and hadronic emission models.

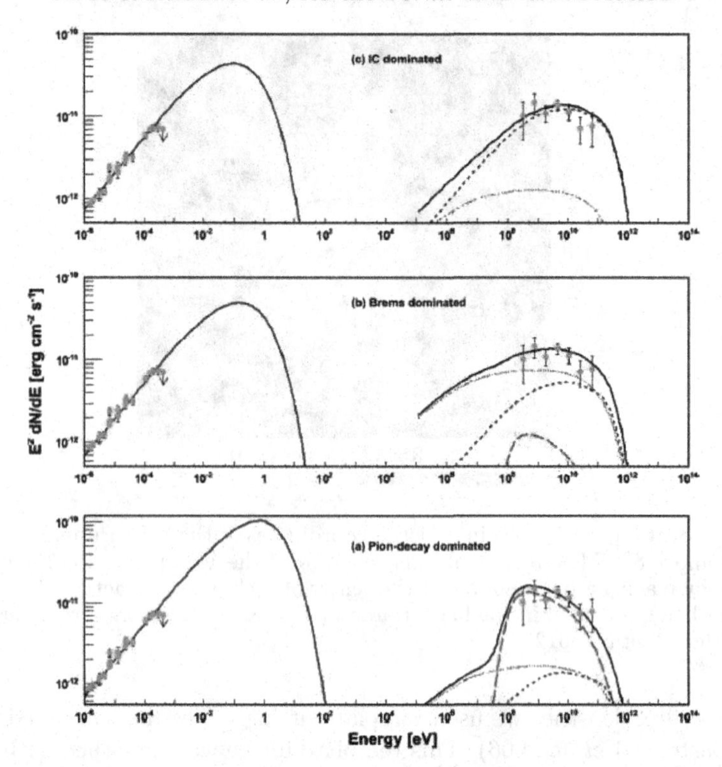

Figure 2. SED models for which IC (top), Bremsstrahlung (middle) and π^0-decay (bottom) are the dominant emission mechanism. In each model the radio data are fit with a synchrotron component. All models show the contributions of π^0-decay (long dashed, red), Bremsstrahlung (dotted, blue), and IC emission (dashed, blue) from CMB, IR dust photon field, and stellar optical photons. The sum of the three γ-ray components is shown as a solid black curve. The Fermi spectrum was renormalized to the whole phase interval.

The *Fermi* LAT Collaboration acknowledges support from a number of agencies and institutes for both development and the operation of the LAT as well as scientific data analysis. These include NASA and DOE in the United States, CEA/Irfu and IN2P3/CNRS in France, ASI and INFN in Italy, MEXT, KEK, and JAXA in Japan, and the K. A. Wallenberg Foundation, the Swedish Research Council and the National Space Board in Sweden. Additional support from INAF in Italy and CNES in France for science analysis during the operations phase is also gratefully acknowledged.

References

Abdo, A. A., *et al.* 2010, *ApJ*, 713, 154
Atwood, W. B., *et al.* 2009, *ApJ*, 697, 1071
Brandt *et al.* for the *Fermi-LAT* collaboration, 2013, *these proceedings*
Castelletti, G., Dubner, G. M., Golap, K., & Goss, W. M., 2006, *A&A*, 459, 535
Dubner, G. M. & Arnal, E. M., 1988, *A&AS*, 75, 363
Gotthelf, E. V. & Halpern, J. P., 2009, *ApJ*, 695, 35
Hewitt, J. W., *et al.* 2012, *ApJ*, 759, 89
Hwang, U., Flanagan, K. A., & Petre, R., 2005, *ApJ*, 635, 355
Jarosik, N., Bennett, C. L., Dunkley, J. *et al.* 2011, *ApJS*, 192, 14
Lande, J., *et al.* 2012, *ApJ*, 756, 5
Lemoine-Goumard, M., 2013, *these proceedings*
Petre, R., Kriss, G. A., Winkler, P. F., & Canizares, C. R., 1982, *ApJ*, 258, 22

Reynoso, E. M., Dubner, G. M., Goss, W. M., & Arnal, E. M., 1995, *AJ*, 110, 318

Discussion

SAHA L: What is the electron to proton ratio for Puppis A? Is it consistent with e/p ratio observed in the galaxy?

GRONDIN: The electron to proton ratio in the IC dominated and Bremsstrahlung dominated scenarios is estimated to be 1, which is significantly larger than what is observed in the galaxy. In the hadronic scenario, the e/p ratio is ~ 0.02 which seems a more reasonable value.

RAY: Will more data help distinguish different emission mechanisms in Pup A?

GRONDIN: Yes, first because it will decrease the statistical error bars. Then, TeV observations may also help to disentangle the different scenarios.

Supernova Environmental Impacts
Proceedings IAU Symposium No. 296, 2013
A. Ray & R. A. McCray, eds.

doi:10.1017/S1743921313009629

TeV γ-ray source MGRO J2019+37 : PWN or SNR?

Lab Saha and Pijushpani Bhattacharjee

Saha Institute of Nuclear Physics, Kolkata 700064, India
email: lab.saha@saha.ac.in

Abstract. Milagro has recently reported an extended TeV γ-ray source MGRO J2019+37 in the Cygnus region. It is the second brightest TeV source after Crab nebula in their source catalogue. No confirmed counterparts of this source are known although possible associations with several known sources have been suggested. We study leptonic as well as hadronic models of TeV emission within the context of Pulsar Wind Nebulae (PWN) and Supernova Remnant (SNR) type sources, using constraints from multi-wavelength data from observations made on sources around MGRO J2019+37. These include radio upper limit given by GMRT, GeV observations by Fermi-LAT, EGRET and AGILE and very high energy data taken from Milagro. We find that, within the PWN scenario, while both leptonic as well as hadronic models can explain the TeV flux from this source, the GMRT upper limit imposes a stringent upper limit on the size of the emission region in the case of leptonic model. In the SNR scenario, on the other hand, a purely leptonic origin of TeV flux is inconsistent with the GMRT upper limit. At the same time, a dominantly hadronic origin of the TeV flux is consistent with all observations, and the required hadronic energy budget is comparable to that of typical supernovae explosions.

Keywords. γ-rays, MGRO J2019+37, Observations.

---

1. Introduction

MGRO J2019+37 is one of the brightest sources in the Cygnus region and was first discovered by Milagro water Cherenkov telescope with very significant diffuse background. Milagro collaboration detected it with 10.9σ significance above isotropic background level (Abdo *et al.* 2007) and reported this to be a new source in this region . R.A. and decl. of the source are quoted to be $304.83 \pm 0.14_{stat} \pm 0.3_{sys}$ deg and $3.83 \pm 0.08_{stat} \pm 0.25_{sys}$ deg, respectively. Immediately after this discovery it drew attention of many people since it admits higher flux of γ-rays in TeV energies and then it has been associated with known sources in that region. For instance, MGRO J2019+37 has been associated with young Fermi pulsar PSR J2020.8+3649 and this pulsar is also considered to have possible association with one of the EGRET sources 3EG J2021+3716 (Roberts *et al.* 2002). GeV pulsation of this young pulsar was first detected by AGILE and was subsequently confirmed by Fermi observation. Radio observation by GMRT in this region shows no significant emission (Paredes *et al.* 2009) and they put some upper limit on the radio flux. Analysis of X-ray archival XRT data also shows no significant results. Lack of radio counterparts and presence of higher TeV γ-ray flux make it quite interesting source. In this article we built a model to explain the observed TeV flux considering multi-wavelength data which include the following viz., radio upper limit given by GMRT, X-ray upper limit from XRT data, GeV observation by Fermi-LAT (Abdo *et al.* 2009), EGRET (Hartman *et al.* 1999), AGILE (Halpern *et al.* 2008) and very high energy data taken from Milagro, in the framework of both pulsar wind nebula (PWN) and supernova remnant (SNR) scenario.

2. Models

We consider three photon emission processes :i) synchrotron radiation, ii) inverse Compton (IC) emission and iii) decay of neutral pions resulting from p-p collisions for both PWN and SNR scenarios. The spectra of electrons and protons in the emission volume are mentioned below.

2.1. *Pulsar Wind Nebula*

We assume that the nebula is filled with two types of relativistic electrons. First of this type is already cooled down by synchrotron radiation and is present in the nebula throughout the age of the pulsar. These low energetic electrons are called radio electron. The second population is freshly accelerated electrons in the wind and is called wind electrons. Total energy of all electrons are obtained from the rotational energy loss of the pulsar, resulting from its magnetic energy and the energy associated with relativistically charged particles. The spectral energy distribution of the two types of electrons are shown in Eqn. (2.1) and Eqn. (2.2), respectively. While the radio electrons follow single power law spectrum, the wind electrons have power law spectrum with an exponential cut-off at lower end as shown in Eqn. (2.2).

$$\begin{aligned}\frac{dN_r}{d\gamma} &= A_r \gamma^{-\alpha_r} \qquad (\gamma^r_{min} < \gamma < \gamma^r_{max}) \\ &= 0 \qquad \text{(otherwise)}\end{aligned} \tag{2.1}$$

$$\begin{aligned}\frac{dN_w}{d\gamma} &= A_w \gamma^{-\alpha_w}\, exp\left[-\frac{\gamma^w_{min}}{\gamma}\right] \qquad (\gamma < \gamma^w_{max}) \\ &= 0 \qquad \text{(otherwise)}\end{aligned} \tag{2.2}$$

Energy spectrum of proton is obtained by considering the injected rate of stripped Fe nuclei from the surface of the neutron star and by measuring their acceleration and propagation through the outer gap (Bednarek & Protheroe (1997), Bednarek & Bartosik (2003)). This spectrum may easily be fitted with the power-law spectrum with an exponential cut-off at very high energy. We simply consider that protons inside the nebula follow a power law spectrum with spectral index 2.3 with an exponential cut-off at 1000 TeV.

2.2. *Supernova Remnant*

We consider that the supernova remnant is spherically symmetric and homogeneous and is filled with hydrogen gas with number density n_H. It is widely believed that cosmic particles are accelerated by Fermi shock acceleration in supernovae remnants. Therefore we consider power law spectra for both electrons and protons with same spectral index -2 and with spectral cut-offs for protons and electrons at 500 TeV and 80 TeV, respectively.

3. Results

3.1. *PWN as a candidate*

As mentioned in Sec. 2.1, we consider two different populations of electrons in the PWN scenario and both radio and wind electrons having energy between $\gamma = 1$ and $\gamma = 10^{10}$ with exponential cut-off being 100 GeV for wind electrons, are contributing to synchrotron photon spectra. Using these synchrotron photons as target for IC process we fit the data at GeV energies. In addition to this, CMB photons with mean temperature 2.7K and thermal dust photons having gray body spectrum are also considered as target

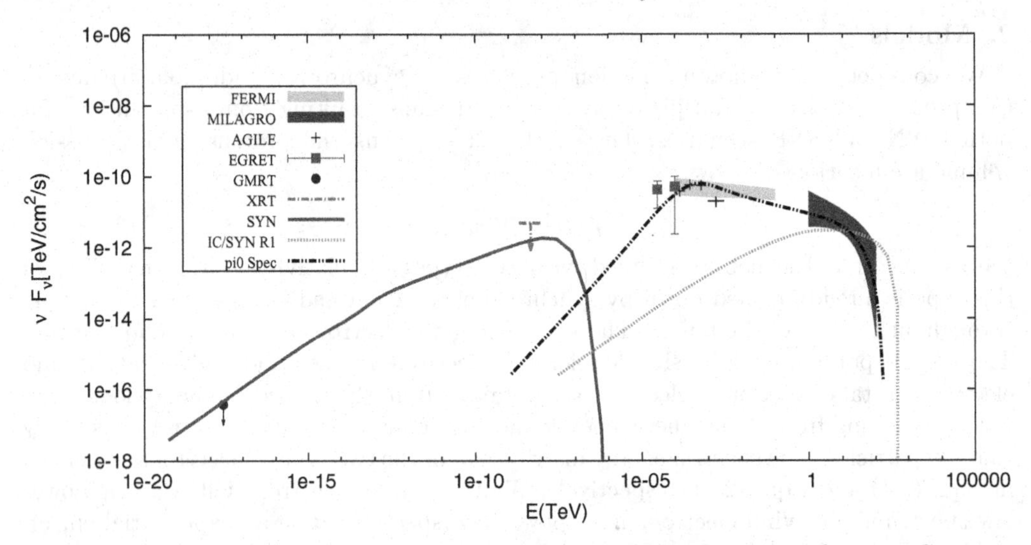

Figure 1. i) γ-ray energy spectrum from electrons for both Synchrotron and inverse Compton radiation processes (solid and dotted line, respectively). For inverse Compton radiation process, $r_{em} \sim 10^{-4}$ pc gives the best value for the fit. ii)(Double dot dashed line) γ-ray energy spectrum from decay of neutral pions ($\pi^0 \longrightarrow \gamma\gamma$) for the ambient proton density of 1 cm^{-3}.

photons. But, synchrotron photons contribute dominantly in this scenario. To calculate the flux from the IC emission we find out the density of target photons in the emission volume as follows: if L_ν is the differential luminosity and r_{em} is the radius of spherically symmetric emission volume, then the differential seed photon number density, n^ν_{seed}, is obtained as $L_\nu/4\pi ch\nu r^2_{em}$. In other words, L_ν is proportional to n^ν_{seed} r^2_{em}. But L_ν is bounded from above by F_{GMRT}/ν in the radio synchrotron range, where F_{GMRT} is the upper limit of radio emission from the position of MGRO J2019+37 in the Cygnus region. This implies an upper limit on n^ν_{seed} r^2_{em}. Since n^ν_{seed} is fixed by TeV flux, we get an upper limit on r_{em}. We estimate the radius of the emission volume to be about $10^{-4}pc$ which gives good fit to the observed TeV data as shown in Fig. 1.

For the case of pure hadronic contribution to γ-rays we consider ambient hydrogen density $< n_H >= 1/cm^3$, spectral index $\beta = 2.3$ and the distance to the source, $D = 3$ kpc. Normalising proportionality constant of proton energy spectrum to the Milagro data point at TeV energy, we compute the necessary energy that must be supplied to the protons as $E_p \sim 4.5 \times 10^{49}$ ergs. Fig. 1 shows the γ-ray spectrum from decay of π^0's fits well with the TeV data.

3.2. *SNR as a candidate*

For SNR scenario, we also consider that high energy photons are produced by synchrotron radiation, IC process and decay of neutral pions. CMB photons with mean temperature 2.7 k are considered here as dominant targets for IC emission process. For p-p collision, we consider ambient proton density to be $1/cm^3$ as we considered in PWN scenario. Fig. 2 shows the spectral energy distribution of γ-rays for the processes mention above. We see that, leptonic model in this scenario is unable to explain the observed TeV flux due to stringent upper limit on the observed radio flux. But hadronic model can explain the observed TeV flux as shown in Fig. 2. We estimate the total energy of protons to be 3.15×10^{50} which is comparable to that of typical supernovae explosions.

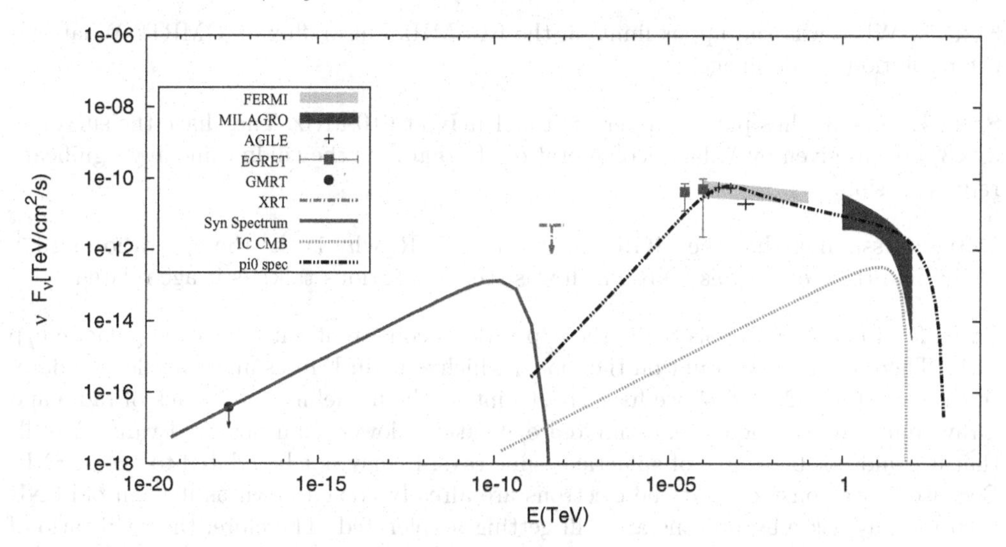

Figure 2. Spectral γ-ray energy distribution for three processes: i) Synchrotron emission spectrum (solid line), ii) IC with CMB photons (dotted line), iii) π^0 decay (double dot dashed line).

4. Conclusions

In this paper we carry out a study to find out the nature of the TeV γ-ray source MGRO J2019+37 as well as to explain the observed flux within the framework of PWN and SNR type sources. We find that for both PWN and SNR scenarios, hadronic contribution to γ-rays can explain the observed TeV flux. Purely leptonic origin of TeV flux can explain the observed flux for PWN and it also brings some constraint on the size of the emission volume. But, SNR scenario is inconsistent with the GMRT upper limit.

References

Abdo, A. A., *et al.* 2007, *ApJ*, 658, L33
Abdo, A. A., *et al.* 2009, *ApJS*, 183, 46
Roberts, M. S. E., *et al.* 2002, *ApJ*, 577, L19
Paredes, J. M., *et al.* 2009, *A&A*, 507, 241
Hartman, R. C., *et al.* 1999, *ApJS*, 123, 79
Halpern, J. P., *et al.* 2008, *ApJ*, 688, L33
Bednarek, W. & Protheroe, R. J. 1997, *PRL*, 79, 2616
Bednarek, W. & Bartosik, M. 2003, *A&A*, 405, 689

Discussion

CHAKRABORTI: Shouldn't the source be self absorbed if it is so compact?

SAHA L.: Yes, in general if the source is compact then there will be self absorption. We can consider the scenario in a little bit different way. We consider the density of synchrotron photon varies with distance from the center of the source. It follows gaussion function. Width of the gaussion function is being used to model the source. Using that we estimate the size of the remnant. So, there is still possibility that total size of the emitting volume is higher than estimated size. In that case there may not be that much absorption. We need obviously high resolution observation to unveil the nature of the source and also to get information about size of the emission volume.

SURNIS: What was the upper limit on the 610 MHz pulsar flux at GMRT? What was the resolution of the image?

SAHA L.: GMRT has put an upper limit of 1 mJy at 610 MHz. They have the survey of $3^o \times 3^o$ area as given by Milagro collaboration. In that area they didn't find any significant radio emission.

YADAV: Assuming that the MGRO source is a SNR, why is the the e/p ratio of 10^{-3} lower than that for Puppis SNR which was seen in previous talk? Is it age related?

SAHA L.: In case of Puppis SNR, the e/p ratio is consistent with observed galactic e/p ratio. There are no constraint on the model which explain Puppis multiwavelength data. In case of MGRO 2019+37, we have constraint on the modeling due to lack of radio and x-ray counterpart. Therefore, estimated e/p ratio is lower than observed value. It tells that it could not be source of galactic cosmic rays. It may not be related to age of SNR. Because, if we consider that the electrons are already cooled down as it is an old SNR then we may ask why protons are still getting accelerated. Therefore, that e/p ratio of 10^{-3} could not be related to age of the SNR.

Supernova Environmental Impacts
Proceedings IAU Symposium No. 296, 2013
A. Ray & R. A. McCray, eds.

doi:10.1017/S1743921313009630

Supernova remnants and the origin of cosmic rays

Jacco Vink
Astronomical Institute Anton Pannekoek & GRAPPA
Universiteit van Amsterdam
Postbus 94249
1090 GE Amsterdam, The Netherlands
email: j.vink@uva.nl

Abstract. Supernova remnants have long been considered to be the dominant sources of Galactic cosmic rays. For a long time the prime evidence consisted of radio synchrotron radiation from supernova remnants, indicating the presence of electrons with energies of several GeV. However, in order to explain the cosmic ray energy density and spectrum in the Galaxy supernova remnant should use 10% of the explosion energy to accelerate particles, and about 99% of the accelerated particles should be protons and other atomic nuclei.

Over the last decade a lot of progress has been made in providing evidence that supernova remnant can accelerate protons to very high energies. The evidence consists of, among others, X-ray synchrotron radiation from narrow regions close to supernova remnant shock fronts, indicating the presence of 10-100 TeV electrons, and providing evidence for amplified magnetic fields, gamma-ray emission from both young and mature supernova remnants. The high magnetic fields indicate that the condition for accelerating protons to $>10^{15}$ eV are there, whereas the gamma-ray emission from some mature remnants indicate that protons have been accelerated.

Keywords. acceleration of particles, magnetic fields, radiation mechanisms: nonthermal, shock waves, (stars:) supernovae: general, (ISM:) cosmic rays,(ISM:) supernova remnants

---

1. Introduction

The energy density of cosmic rays in the Galaxy is estimated to be 1-3 eV cm^{-3} (e.g. Webber 1998), which is similar to the internal gas density and magnetic field energy density in the interstellar medium (ISM). The idea that the bulk of these cosmic rays are associated with supernovae was first made by Baade & Zwicky (1934b), who also for the first time made a clear distinction between "*super* novae" and "common novae" (Baade & Zwicky 1934a).

The discovery of radio emission from supernova remnants (SNRs) and its identification as synchrotron radiation in the 1950ies (e.g. Ginzburg 1959) provided the first link between supernovae and particle acceleration. Of course, synchrotron radiation is primarily associated with relativistic electrons and positrons, whereas 99% of the cosmic rays observed on Earth consists of electrons. At the time it was not quite clear whether the acceleration occurred during the supernova explosion or in the SNR shells (Ginzburg & Syrovatskii 1964). The radio flux decline of the bright SNR Cassiopeia A, and the explanation by Shklovsky that this was caused by the adiabatic cooling of the relativistic electrons even suggested that electrons were no longer actively accelerated anymore (Shklovsky 1968).

The idea that the shocks at the outer boundaries of SNRs were the likely sites of cosmic ray acceleration became more prominent due to development of the theory of first order Fermi acceleration, also know as diffusive shock acceleration (DSA) (Axford *et al.* 1977; Krymskii 1977; Blandford & Ostriker 1978; Bell 1978). According to this theory particles gain momentum by repeatedly crossing the shock front as a result of particle diffusion. Each shock crossing leads to a fractional increase in momentum of order $\Delta p/p \approx \Delta v/c$, with Δv the difference in plasma velocity between both sides of the shock and c the particle velocity. The particle distribution becomes a power law, because the advection of plasma away from the shock in the downstream region (i.e. in the shock heated plasma), transports a fraction of $\sim v/(c\chi)$ (with χ the shock compression ratio) away from the shock region, preventing these particles from recrossing the shock again. The power law index of the energy spectrum, q, is for that reason a function of the compression ratio: $q = (\chi + 2)/(\chi - 1)$.

Although this theory made the case for shock acceleration in SNRs stronger, a number of problems remained in connecting the bulk of the cosmic rays with SNRs. For SNRs to be the origin of cosmic rays two criteria have to be met: 1) SNRs should be able to provide sufficient power to explain the cosmic ray energy density in the galaxy of 1 $\mathrm{eV/cm^{-3}}$, which requires that about 10% of the supernova explosion energy should be used to accelerate particles (Ginzburg & Syrovatskii 1964); 2) SNRs should be able to accelerate protons at least up to 3×10^{15} eV, which corresponds to a break in the observed cosmic ray spectrum, and probably marks the maximum energy that protons can be accelerated to in Galactic sources.

Applying shock acceleration theory to SNRs shows that it is difficult to fulfil the second criterion. The maximum energy to which particles can be accelerated by DSA depends on the time available for particle acceleration, the shock speed, and on the diffusion coefficient as a function of particle energy. The scattering causing the diffusion caused by turbulent magnetic fields. For the diffusion coefficient one often assumes that it is $D = \frac{1}{3}\eta c r_{\mathrm{g}}$, with $r_{\mathrm{g}} = E/eB$ the gyro radius radius of the particle, and η a parameter indicating how much the mean free path is larger than the gyro-radius. The smallest diffusion coefficient, which gives the fastest acceleration, is thought to be $\eta = 1$ (but see Reville & Bell 2013). The case $\eta = 1$ is referred to as "Bohm diffusion". In Lagage & Cesarsky (1983) it was shown that under some general assumptions it seems unlikely that SNRs can really accelerate particles to beyond 3×10^{15} eV. A more realistic estimate seemed to be 10^{14} eV, which even involved assuming the optimistic case that $\eta = 1$, and that the magnetic field is that of the ISM $B \approx 3\ \mu$G.

As for the first criterion, for a long time the estimate of the total energy in cosmic rays in SNRs had to rely on the observed synchrotron luminosity. Translating this into an particle energy density has to rely on assumptions on the magnetic field strength and the ratio between accelerated electron and proton density. For the magnetic field strength one usually relied on equipartition arguments, whereas for the electron/proton ratio it was usually assumed that electrons account for about 1% of the particles, based on the electron/proton ratio in the cosmic rays observed on Earth.

Over the last decade our understanding of cosmic rays inside SNRs has greatly improved, thanks to new X-ray observatories, like *Chandra*, *XMM-Newton*, and *Suzaku*, and thanks to new gamma-ray observatories like ground-based Cherenkov telescopes that cover the gamma-rays above ~ 0.1 TeV and the *Fermi* and *AGILE* satellites which observe gamma-rays in the $\sim 0.1 - 10$ GeV range.

I will cover here these recent advances, but would like to point out that more extended reviews on high energy emission from SNRs are Reynolds (2008), Hinton & Hofmann (2009), Vink (2012b), and Helder *et al.* (2012). In this volume there is overlap with the work presented by T. Brandt, S. Gabici, M. Lemoine-Goumard, on the gamma-ray emission, and with A. Decourchelle and M. Miceli on the X-ray emission.

2. X-ray synchrotron emission

Although synchrotron radiation only inform us about the accelerated electrons (leptonic cosmic rays), the discovery of X-ray synchrotron emission from young SNRs gives us important information about particle acceleration in general. Moreover, X-ray telescopes like *XMM-Newton*, and in particular, *Chandra* have a spatial resolution that is orders of magnitude better than that of gamma-ray observatories.

X-ray synchrotron radiation comes from the highest energy electrons. The characteristic frequency/photon energy of synchrotron radiation depends on electron energy as

$$h\nu_{\rm ch} = 13.9\Big(\frac{B_\perp}{100\,\mu{\rm G}}\Big)\Big(\frac{E}{100\,{\rm TeV}}\Big)^2\ {\rm keV}. \tag{2.1}$$

This tells us immediately that synchrotron radiation above 1 keV requires either high magnetic fields, or electrons with energies of $\sim 10^{13}$ eV have to be present.

For that reason the first discovery of X-ray synchrotron radiation from the limbs of the shell-type SNR SN1006 (Koyama *et al.* 1995) was very important, as it indicated that electrons could be accelerated to energies much higher than was inferred before. Moreover, electrons with TeV energies loose their energy relatively quickly, with a typical loss time scale of

$$\tau_{\rm syn} = \frac{E}{dE/dt} = 12.5\left(\frac{E}{100\ {\rm TeV}}\right)^{-1}\left(\frac{B_{\rm eff}}{100\mu{\rm G}}\right)^{-2}\ {\rm yr}. \tag{2.2}$$

So the presence of TeV electrons close to the shock front, also tell us that the acceleration is still taking place, or at least took place relatively recently. This clearly identifies the shock front of the SNR as a site of acceleration.

Finally, the time scale for acceleration to an energy E by DSA is estimated to be

$$\tau_{\rm acc} \approx 1.83\frac{D_2}{V_{\rm s}^2}\frac{3\chi^2}{\chi-1} = 124\eta B_{-4}^{-1}\Big(\frac{V_{\rm s}}{5000\ {\rm km\,s^{-1}}}\Big)^{-2}\Big(\frac{E}{100\,{\rm TeV}}\Big)\frac{\chi_4^2}{\chi_4-\frac{1}{4}}\ {\rm yr}, \tag{2.3}$$

with B_{-4} the downstream magnetic field in units of 100 μG and χ_4 the overall compression ratio in units of 4, and $V_{\rm s}$ the shock velocity. So in order to accelerate faster than the synchrotron loss time, for a 10 TeV electron, one needs a factor of η that cannot be too large compared to Bohm diffusion.

The maximum photon energy for synchrotron radiation can be calculated by making either the assumptions that the electron energy was limited by the time available to reach the maximum (Eq. 2.3), in which case the maximum electron energy must equal the maximum proton energy, or by assuming that the electron energy is limited by synchrotron losses. The latter can be calculated by setting $\tau_{\rm acc} = \tau_{\rm syn}$, which results in an equation that still depends on B. However, converting then electron energy to characteristic photon energy gives

$$h\nu_{\max} = 1.4\eta^{-1}\Big(\frac{\chi_4-\frac{1}{4}}{\chi_4^2}\Big)\Big(\frac{V_s}{5000\ {\rm km\,s^{-1}}}\Big)^2\ {\rm keV}, \tag{2.4}$$

which does not depend on B (Aharonian & Atoyan 1999; Zirakashvili & Aharonian 2007). If indeed the maximum electron energy is loss-limited the detection of X-ray synchrotron radiation indicates that $\eta \lesssim 10$.

The idea that for most young SNRs the synchrotron spectra are loss-limited is strengthened by the fact that the X-ray synchrotron emitting region is rather narrow (Helder *et al.* 2012, for an overview). In some cases, like Tycho's SNR (Fig. 1, Warren *et al.* 2005) and Cassiopeia A (Vink & Laming 2003), the synchrotron rims are only a few arcsec in width, and can only be resolved by *Chandra*. The easiest explanation is that the width is determined by the loss-time scale: while the electrons are advected away from the shock they loose energy and after some time $\tau_{\rm syn}$ they no longer emit X-rays. This would lead to the following expression for the rim width (Vink & Laming 2003)

$$l_{\rm adv} = \Delta v \tau_{\rm syn} = (V_{\rm s}/\chi)\tau_{\rm syn}. \tag{2.5}$$

Alternatively one assume that the rim width corresponds to the typical electron diffusion length scale (e.g. Berezhko *et al.* 2003; Bamba *et al.* 2004),

$$l_{\rm diff} \approx D/V_{\rm s}. \tag{2.6}$$

In fact the two approximations are both valid if the electrons are near their maximum energies, since then $\tau_{\rm syn} = \tau_{\rm acc}$, and $\tau_{\rm acc} \approx D/V_s^2$, which can be easily combined to give Eq. 2.5. Under the assumption that $\tau_{\rm syn} = \tau_{\rm acc}$ one can even find an expression for the magnetic field which is independent of the observed photon energy and electron energy:

$$B_2 \approx 26 \Big(\frac{l_{\rm adv}}{1.0 \times 10^{18} {\rm cm}}\Big)^{-2/3} \eta^{1/3} \Big(\chi_4 - \frac{1}{4}\Big)^{-1/3} \ \mu{\rm G}, \tag{2.7}$$

with B_2 the typical magnetic field strength in the shocked plasma.

The fact that using either Eq. 2.5 or Eq. 2.6 gives very similar magnetic field estimates (e.g. Ballet 2006), strengthens the case that the X-ray synchrotron emission is from loss-limited electron spectra. However, the low magnetic fields of some older X-ray synchrotron emitting SNRs may indicate the synchrotron spectrum is close to age-limited, or otherwise evolutionary effects should be taken into account. An example is RCW 86 for which $B_2 \approx 20 - 30\ \mu$G. Also for the youngest known Galactic SNR, G1.9+0.3, it is debatable whether the maximum electron energy is limited by the age of the SNR (~ 100 yr Carlton *et al.* 2011), or by radiative losses.

One of the surprising results that came out of the estimation of B_2 based on Eq. 2.7 or similar equations is that the magnetic field in young SNRs is much higher than anticipated, with typically $20\ \mu{\rm G} < B_2 < 600\ \mu$G (Vink & Laming 2003; Berezhko *et al.* 2003; Bamba *et al.* 2004; Ballet 2006). Moreover, there is a clear correlation with local circumstellar density and shock speed, which indicates that $B_2^2 \propto \rho_0 V_s^\alpha$, with α in the range 2-3 (Völk *et al.* 2005; Vink 2008; Helder *et al.* 2012). Note that all young SNRs have velocities in a narrow range (3000-6000 km s^{-1}), making α poorly constrained, whereas the dynamical contrast in ρ is much larger.

The magnetic fields are clearly larger than what can be expected from a compressed ISM magnetic field, which alleviates the problem discussed by Lagage & Cesarsky (1983), namely that the magnetic fields are not high enough to have protons accelerated to 3×10^{15} eV. The strong magnetic fields also support the idea that somehow magnetic field amplification near the shock is taking place, which in itself may be caused by the precursor resulting from efficient cosmic ray acceleration (e.g. Bell 2004; Drury & Downes 2012).

That magnetic field amplification may be very efficient is indicated by the narrow X-ray synchrotron filaments in the inside of Cas A, peaking in the west of the SNR (Helder & Vink 2008; Uchiyama & Aharonian 2008). Its location strongly suggest that these filaments are connected to the reverse shock of the SNR. Proper motion measurements in that region indicate that in our frame the reverse shock is almost at a stand still (Vink *et al.* 1998; Delaney & Rudnick 2003), indicating that the shock speed is almost equal to the free expansion velocity of the ejecta ~ 7000 km s^{-1}. What is surprising about acceleration and magnetic field amplification at the reverse shock is that the unshocked ejecta are likely to have a low magnetic field, because of the large expansion of the frozen

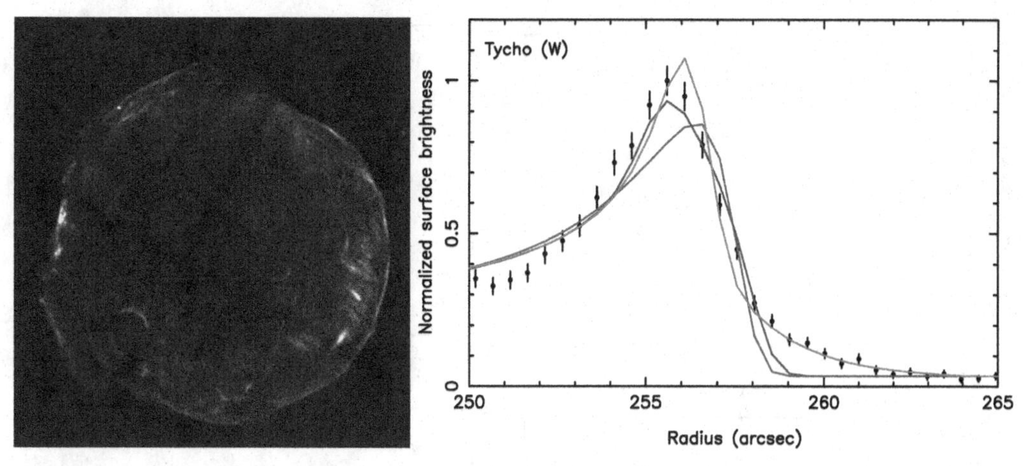

Figure 1. Left: *Chandra* X-ray image (2009) of Tycho SNR in the 4.6-6.1 keV band, which is dominated by X-ray synchrotron radiation. Note the narrow rim, in particular on the Western side, and the stripe-like pattern more inside the periphery in the West (Eriksen *et al.* 2011). Right: Emission profile from a 30° segment of the Western shock region in the 4.3-5.2 keV band. The blue line shows emission from a uniformly emitting shell with thickness l. The red and green lines show models in which the emissivity falls off behind the shock as $\exp(-(r_0 - r)/l)$, but the green line has an additional precursor component. The models take into account spherical projection and instrumental resolution (taken from Vink 2012a).

in magnetic field. Nevertheless, the width of the filaments indicate magnetic fields of the order of 100-500 μG.

The X-ray synchrotron emission from Cas A seems to decline rather rapidly (Patnaude *et al.* 2011), 1.5-2% yr^{-1}, which is much faster than the radio flux decline. This can be understood best by taking into account that the synchrotron cut-off energy is declining as a result of the slowing down of the shock. The decline is somewhat fast compared to the expectations based on a simple, Sedov-type shock speed evolution.

The other X-ray bright, young SNR, Tycho, also showed some surprising results. First of all, the ejecta seem to lie very close to the shock front (Warren *et al.* 2005; Cassam-Chenaï *et al.* 2007), which can be attributed to a very high compression ratio $\chi > 4$ as can be expected according to non-linear DSA theories (e.g. Decourchelle *et al.* 2001; Vink *et al.* 2010; Kosenko *et al.* 2011). However, there is currently a debate whether hydrodynamic instabilities may be another way to explain the appearance of ejecta close to the shock front (Orlando *et al.* 2012).

Another strange X-ray feature in Tycho's SNR is the appearance of a comb-like X-ray synchrotron structure, which has been interpreted as filamentation with a pattern-size matching the gyro-radius of the highest energy protons, which would then have to be close to 10^{15} eV (Eriksen *et al.* 2011; Bykov *et al.* 2011, Fig. 1). However, the coherence of the pattern, its orientation and its singular appearance need further clarification.

3. The maximum cosmic ray energies in SNRs

The X-ray synchrotron spectra indicate that magnetic fields in young SNRs can be as high as $B_2 \sim 500$ μG. The evidence that the magnetic field is amplified in the cosmic ray precursor, and that it scales with $B^2 \propto \rho_0 V_s^\alpha$, suggests that the highest energies may be reached in the early phase of the SNR evolution. Moreover, it helps if the SNR is evolving in a dense stellar wind. In that case $\rho_0 = \dot{M}/4\pi r^{-2} v_w$, which means that densities, and hence magnetic fields, are larger in the earliest phases of the SNR evolution (see Ptuskin

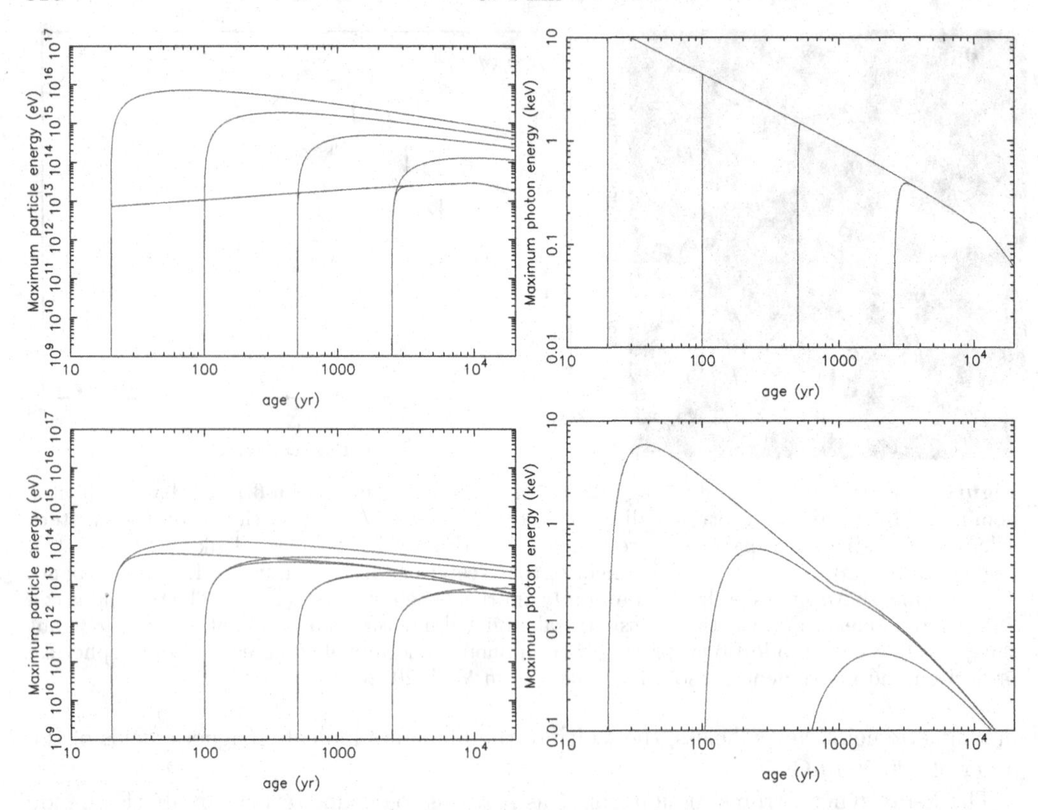

Figure 2. Top left: Maximum proton (top, red lines) and electron (blue) energies for particles that started to be accelerated at selected intervals. The parameters are chosen to match Cas A at the current epoch (at t = 330 yr: B_2 = 500 μG, η = 5,V_s = 5200 km s^{-1}, R_s = 2.6 pc, n_0 = 2 cm^{-2}, $R_s \propto t^{2/3}$). The magnetic field scales as $B^2 \propto \rho_0 V_s^3$, but is held constant once B_0 = 10 μG. Adiabatic losses are taken into account, and result in the downward slope of the proton spectra, once the maximum has been reached. Top right: The resulting characteristic maximum synchrotron photon energy. Although the maximum electron keeps going up with time till $t \approx 10^4$ yr, the photon energy declines rapidly because the magnetic field is declining. Bottom panels: similar to the top panels, but with a different magnetic field B_0 = 30 μG, a flat circumstellar medium, and $R_s \propto t^{0.5}$.

& Zirakashvili 2005; Schure *et al.* 2010). Moreover, the high early density results in more particles entering the shock front at early times.

The effects of magnetic field evolution on the maximum electron and proton energies is illustrated in Fig. 2. It shows that for a SNR with characteristics typical for Cas A the maximum energy that a proton that started to be accelerated at a SNR age of t = 20 yr could reach 10^{16} eV. The high magnetic field makes that the maximum electron energy for an electron injected at t = 20 yr is only 10^{13} eV. Note that the electrons are accelerated almost instantaneously, and are limited by synchrotron energy losses. Although the maximum electron energy goes up with age, the maximum characteristic photon energy declines with age, due to the decline of the magnetic field.

Note that adiabatic losses make that protons start losing energy after reaching the maximum energies. It is, due to the decline in magnetic field their diffusion length scales are rapidly increasing and become $l_{\rm diff} > 0.1 R_s$ shortly after reaching the peak energy. As a result they become detached from the SNRs and escape. The bottom panels of

Fig. 2 shows similar calculations, but now for a much lower magnetic field, resulting in an electron spectrum that is nearly age limited.

4. Gamma-ray emission from supernova remnants

EGRET (Esposito *et al.* 1996) identified several gamma-ray sources that appeared to be associated with SNRs. However, the big break-through in the field of gamma-ray studies of SNRs came with the coming of age of Cherenkov telescopes such as *HEGRA*, *H.E.S.S.*, *MAGIC*, and *Veritas*, which cover very high energy gamma-rays (VHE gamma-rays, $\sim 0.1 - 100$ TeV). The first detected shell-type SNR to be detected was Cas A by *HEGRA*(Aharonian *et al.* 2001), although it was not as bright as perhaps anticipated. A source that was very bright was the otherwise somewhat obscure SNR RX J1713.7-3946 (RXJ173, Aharonian *et al.* 2004, 2007), which has a VHE spectrum that extends up to 100 TeV, indicating primary particles with energies close to 10^{15} eV. Since these initial discoveries many more young SNRs have been detected in VHE gamma-rays, including Vela Jr (Aharonian *et al.* 2005), RCW 86 (Aharonian *et al.* 2009) and Tycho's SNR (Acciari *et al.* 2011).

However, what is not immediately clear from the VHE spectra alone is whether the emission is caused by accelerated electrons (leptonic cosmic rays) or from accelerated protons and other atomic nuclei (hadronic cosmic rays). Accelerated electrons cause gamma-ray emission either due to bremsstrahlung, which requires a sufficiently high local plasma density, or due to inverse Compton scattering of cosmic microwave background photons (or other strong photon fields). The mechanism by which accelerated protons cause gamma-ray emission is pion-decay: hadronic cosmic rays colliding with the background atoms result in the production of, among others, neutral pions, which decay emitting two photons.

In particular for RXJ1713 a strong debate ensued over the nature of the gamma-ray emission (Katz & Waxman 2008; Berezhko & Völk 2008; Acero *et al.* 2009; Berezhko & Völk 2010; Ellison *et al.* 2010). This debates has now settled somewhat in favor of inverse Compton scattering due to the detection of RXJ1713 by the *Fermi*-LAT instrument (Abdo & Fermi LAT Collaboration 2011), which indicates a rather hard spectrum that is more consistent with inverse Compton scattering than with pion decay. Also the detection of several other young SNRs by *Fermi* shows that in many cases the emission seems to be caused by inverse Compton scattering (e.g. RCW 86 Lemoine-Goumard *et al.* 2012). A likely exception among the young SNRs seems to be Tycho's SNR (Acciari *et al.* 2011; Giordano *et al.* 2012). For Tycho it also appears that the energy in cosmic rays inside the SNR is about 10% of the explosion energy, whereas for Cas A it seems to be only 4% (Abdo *et al.* 2010).

The case for pion-decay emission is more compelling for several mature SNRs. Many of the *Fermi* and *AGILE* detected mature SNRs fall in the class of the so-called mixed-morphology SNRs (see ?Helder *et al.* 2012, for a discusion), which have radio shells that are too cool to emit X-rays, but interiors that are hot enough to emit thermal X-ray emission. These SNRs are likely shaped by their high density environments, slowing down the shock front rapidly, resulting in cool SNR shells. The prevalence of the these SNRs in the *Fermi* source catalogue already tells us that their gamma-ray emission must be related to their high densities. The high densities also make it likely that the gamma-ray emission is due to either bremsstrahlung or pion-decay. A characteristic of pion-decay gamma-ray spectra is that their is cut-off in the spectrum at low energies, as the energy of the produced photon is $m_{\pi^0}/2 = 68$ MeV in the rest frame of the collision. This characteristic low energy cut-off has now been detected by *AGILE* (Giuliani *et al.* 2011) and *Fermi* (Ackermann *et al.* 2013) for the mixed-morphology SNRs W44 and IC 443.

The *Fermi* spectra of many mature SNRs show that the spectra are cut-off around 10 GeV, indicating that there are not many particles present with energies about a TeV. This suggests that these particles must have escaped the SNR in the past, since in young SNRs higher energy particles are present given the VHE gamma-ray emission. Interestingly, some mature SNRs do have associated TeV emission, but not immediately from the SNR shell, but coming from nearby molecular clouds. A case in point is the mixed-morphology SNR W28 (Aharonian *et al.* 2008). This indicates that escape is an important and in the past often neglected aspect of cosmic ray acceleration (see S. Gabici, 2013, this volume) . It has also to be taken into account when estimating the total energy in cosmic rays produced by a SNR, as some of the cosmic rays may have escaped, perhaps even at an early stage.

5. Conclusion

SNRs have for a long time been considered the dominant sources of Galactic cosmic rays. If true SNRs must be able to accelerate particles up to at least 3×10^{15} eV and have an efficiency of around 10% in transferring energy to accelerated particles.

The rapid development in X-ray and gamma-ray astronomy have helped to strengthen the case for cosmic ray acceleration by SNRs. The presence of X-ray synchrotron emission and the narrowness of the X-ray synchrotron emitting regions indicate that magnetic fields are amplified near young SNRs, which helps to create the right conditions for accelerating protons up to 3×10^{15} eV.

The gamma-ray emission from SNRs indicates that particles in young SNRs can be accelerated to very high energies, but for young SNRs it is not always clear whether the emission is caused by electrons or protons. However, for mature SNRs in dense regions there is now clear evidence for the presence of accelerated protons, although for these mature SNRs the highest energy protons seem to have escaped the SNR shells.

References

Abdo, A. A. & Fermi LAT Collaboration. 2011, ArXiv e-prints
Abdo, A. A., *et al.* 2010, *ApJ*, 710, L92
Acciari, V. A., *et al.* 2011, *ApJ*, 730, L20+
Acero, F., Ballet, J., Decourchelle, A., Lemoine-Goumard, M., Ortega, M., Giacani, E., Dubner, G., & Cassam-Chenaï, G. 2009, *A&A*, 505, 157
Ackermann, M., *et al.* 2013, *Science*, 339, 807
Aharonian, F. Akhperjanian, A., Barrio, J., *et al.* 2001, *A&A*, 370, 112
Aharonian, F., *et al.* 2005, *A&A*, 437, L7
—. 2007, *A&A*, 464, 235
—. 2008, *A&A*, 481, 401
—. 2009, *ApJ*, 692, 1500
Aharonian, F. A. & Atoyan, A. M. 1999, *A&A*, 351, 330
Aharonian, F. A., *et al.* 2004, *Nature*, 432, 75
Axford, W. I., Leer, E., & Skadron, G. 1977, in International Cosmic Ray Conference, Vol. 11, International Cosmic Ray Conference, 132–+
Baade, W. & Zwicky, F. 1934a, Contributions from the Mount Wilson Observatory, vol. 3, pp.73-78, 3, 73
—. 1934b, *Physical Review*, 46, 76
Ballet, J. 2006, *Advances in Space Research*, 37, 1902
Bamba, A., Ueno, M., Nakajima, H., & Koyama, K. 2004, *ApJ*, 602, 257

Bell, A. R. 1978, *MNRAS,*, 182, 147
—. 2004, *MNRAS,*, 353, 550
Berezhko, E. G., Ksenofontov, L. T., & Völk, H. J. 2003, *A&A*, 412, L11
Berezhko, E. G. & Völk, H. J. 2008, *A&A*, 492, 695
—. 2010, *A&A*, 511, A34+
Blandford, R. D. & Ostriker, J. P. 1978, *ApJ*, 221, L29
Bykov, A. M., Ellison, D. C., Osipov, S. M., Pavlov, G. G., & Uvarov, Y. A. 2011, *ApJ*, 735, L40
Carlton, A. K., Borkowski, K. J., Reynolds, S. P., Hwang, U., Petre, R., Green, D. A., Krishnamurthy, K., & Willett, R. 2011, ArXiv e-prints
Cassam-Chenaï, G., Hughes, J. P., Ballet, J., & Decourchelle, A. 2007, *ApJ*, 665, 315
Decourchelle, A., *et al.* 2001, *A&A*, 365, L218
Delaney, T. & Rudnick, L. 2003, *ApJ*, 589, 818
Drury, L. O. & Downes, T. P. 2012, *MNRAS*, 427, 2308
Ellison, D. C., Patnaude, D. J., Slane, P., & Raymond, J. 2010, *ApJ*, 712, 287
Eriksen, K. A., Hughes, J. P., Badenes, C., Fesen, R., Ghavamian, P., Moffett, D., Plucinksy, P. P., Rakowski, C. E., Reynoso, E. M., & Slane, P. 2011, *ApJ*, 728, L28
Esposito, J. A., Hunter, S. D., Kanbach, G., & Sreekumar, P. 1996, *ApJ*, 461, 820
Ginzburg, V. L. 1959, *in IAU Symposium, Vol. 9*, URSI Symp. 1: Paris Symposium on Radio Astronomy, ed. R. N. Bracewell, 589
Ginzburg, V. L. & Syrovatskii, S. I. 1964, *The Origin of Cosmic Rays*
Giordano, F., Naumann-Godo, M., Ballet, J., Bechtol, K., Funk, S., Lande, J., Mazziotta, M. N., Rainò, S., Tanaka, T., Tibolla, O., & Uchiyama, Y. 2012, *ApJ*, 744, L2
Giuliani, A., *et al.* 2011, *ApJ*, 742, L30
Helder, E., Vink, J., Bykov, A., Ohira, Y., Raymond, J., & Terrier, R. 2012, Space Science Rev., 1, 1
Helder, E. A. & Vink, J. 2008, *ApJ*, 686, 1094
Hinton, J. A. & Hofmann, W. 2009, *ARAA*, 47, 523
Katz, B. & Waxman, E. 2008, *Journal of Cosmology and Astro-Particle Physics*, 1, 18
Kosenko, D., Blinnikov, S. I., & Vink, J. 2011, ArXiv e-prints
Koyama, K., *et al.* 1995, *Nature*, 378, 255
Krymskii, G. F. 1977, *Soviet Physics Doklady*, 22, 327
Lagage, P. O. & Cesarsky, C. J. 1983, *A&A*, 125, 249
Lemoine-Goumard, M., Renaud, M., Vink, J., Allen, G. E., Bamba, A., Giordano, F., & Uchiyama, Y. 2012, *A&A*, 545, A28
Orlando, S., Bocchino, F., Miceli, M., Petruk, O., & Pumo, M. L. 2012, *ApJ*, 749, 156
Patnaude, D. J., Vink, J., Laming, J. M., & Fesen, R. A. 2011, *ApJ*, 729, L28+
Ptuskin, V. S. & Zirakashvili, V. N. 2005, *A&A*, 429, 755
Reville, B. & Bell, A. R. 2013, *MNRAS,*, 430, 2873
Reynolds, S. P. 2008, *ARAA*, 46, 89
Schure, K. M., Achterberg, A., Keppens, R., & Vink, J. 2010, *MNRAS,*, 406, 2633
Shklovsky, J. S. 1968, *Supernovae (Interscience Monographs and Texts in Physics and Astronomy*, London: Wiley, 1968)
Uchiyama, Y. & Aharonian, F. A. 2008, *ApJ*, 677, L105
Vink, J. 2008, *in American Institute of Physics Conference Series, Vol. 1085, American Institute of Physics Conference Series*, 169–180
Vink, J. 2012a, ArXiv e-prints
—. 2012b, *A&A*r, 20, 49
Vink, J., Bloemen, H., Kaastra, J. S., & Bleeker, J. A. M. 1998, *A&A*, 339, 201
Vink, J. & Laming, J. M. 2003, *ApJ*, 584, 758
Vink, J., Yamazaki, R., Helder, E. A., & Schure, K. M. 2010, *ApJ*, 722, 1727
Völk, H. J., Berezhko, E. G., & Ksenofontov, L. T. 2005, *A&A*, 433, 229
Warren, J. S., *et al.* 2005, *ApJ*, 634, 376
Webber, W. R. 1998, *ApJ*, 506, 329
Zirakashvili, V. N. & Aharonian, F. 2007, *A&A*, 465, 695

Discussion

KOO: Yesterday, we heard that as much as 90% of SNe may go off inside superbubbles. I guess the cosmic acceleration in such environment could be different from what we learned for single isolated SNRs. Could you comment on that?

VINK: Yes, if bubble is hot you may have for a longest time low Mach numbers, hampering efficient acceleration. However, Bykov e.g. have suggested that second order Fermi may take place in bubble, using Alfven waves excited by many SNRs together. 2nd order Fermi is slow, but the bubbles have longer lifetimes than SNRs.

Supernova Environmental Impacts
Proceedings IAU Symposium No. 296, 2013
A. Ray & R. A. McCray, eds.

doi:10.1017/S1743921313009642

Probing the effects of hadronic acceleration at the SN 1006 shock front

Marco Miceli[1], **F. Bocchino**[1], **A. Decourchelle**[2], **G. Maurin**[3], **J. Vink**[4], **S. Orlando**[1], **F. Reale**[5,1] **and S. Broersen**[4]

[1]INAF-Osservatorio Astronomico di Palermo, Piazza del Parlamento 1, 90134 Palermo, Italy
email: miceli@astropa.inaf.it

[2]Service d'Astrophysique/IRFU/DSM, CEA Saclay, Gif-sur-Yvette, France

[3]Université de Savoie, 27 rue Marcoz, BP 1107 73011-Chambery cedex, France

[4]Astronomical Institute "Anton Pannekoek", University of Amsterdam, P.O. Box 94249, 1090 GE Amsterdam, The Netherlands

[5]Dipartimento di Fisica e Chimica, Università di Palermo, Piazza del Parlamento 1, 90134 Palermo, Italy

Abstract. Supernova remnant shocks are strong candidates for being the source of energetic cosmic rays and hadron acceleration is expected to increase the shock compression ratio, providing higher post-shock densities. We exploited the deep observations of the XMM-Newton Large Program on SN 1006 to verify this prediction. Spatially resolved spectral analysis led us to detect X-ray emission from the shocked ambient medium in SN 1006 and to find that its density significantly increases in regions where particle acceleration is efficient. Our results provide evidence for the effects of acceleration of cosmic ray hadrons on the post-shock plasma in supernova remnants.

Keywords. X-rays: ISM – ISM: supernova remnants – ISM: individual object: SN 1006

1. Introduction

Supernova remnants (SNRs) shocks are efficient sites of particle acceleration and are candidates for being the main source of the observed spectrum of cosmic rays up to at least 3×10^{15} eV (Berezhko & Völk 2007). X-ray synchrotron emission from high energy electrons accelerated at the shock front up to TeV energies has been first observed in SN 1006 (Koyama *et al.* 1995) and then in other young SNRs as, for example, G1.9+0.3 (Reynolds *et al.* 2009), Vela Jr. (Slane *et al.* 2001, Bamba *et al.* 2005), and G353.6-0.7 (Tian *et al.* 2010). Recently, TeV emission has been detected in SN 1006 (Acero *et al.* 2010). The origin of the gamma-ray emission can be leptonic (i.e. inverse Compton from the accelerated electrons) or hadronic (i.e. proton-proton interactions with π^0 production and subsequent decay). The hadronic scenario would directly prove that SNRs can accelerate cosmic-rays up to PeV energies. Unfortunately, it is not easy to unambiguously ascertain the origin of the TeV emission. In fact, though a pure leptonic model is consistent with the SN 1006 observations, a mixed scenario that includes leptonic and hadronic components also provides a good fit to the data Acero *et al.* 2010.

An alternative way to reveal hadron acceleration in SNRs is to probe its effects on the shock dynamics. The loss of energy related to the acceleration of high-energy particles and their non-linear back-reaction on the background plasma are in fact predicted to modify the shock properties, by making the shock compression ratio higher and the post-shock temperature lower than that expected from the Rankine-Hugoniot equations (Berezhko & Ellison 1999, Decourchelle *et al.* 2000, Blasi 2002, Vink *et al.* 2010). This is known as "shock modification". The observational confirmation of these predictions requires accurate diagnostics of the thermal X-ray emission from the shocked interstellar medium (ISM). However, the ISM contribution in the X-ray spectra of remnants where particle acceleration is efficient is typically masked-out by the bright synchrotron emission and by the thermal emission from shocked ejecta, as in RXJ1713.7-3946 (Acero *et al.* 2009), Vela Jr. (Pannuti *et al.* 2010), G1.9+0.3 (Borkowski *et al.* 2010), and Tycho (Cassam-Chenaï *et al.* 2007).

SN 1006 exhibits a morphology characterized by two opposed radio, X-ray, and $\gamma-$ray bright limbs dominated by non-thermal emission, separated by a region at low surface brightness with soft, thermal X-ray emission. Thermal X-ray emission has been associated with shocked ejecta (Acero *et al.* 2007, Miceli *et al.* 2009). Indirect evidence for shock modification in SN 1006 has been obtained by measuring the distance, D_{BWCD}, between the shock front and the contact discontinuity that is expected to be smaller in non-thermal limbs, where particle acceleration is more efficient. D_{BWCD} is instead almost the same all over the shell (even in regions dominated by thermal emission), though it is much smaller than that expected from unmodified shocks (Cassam-Chenaï *et al.* 2008, Miceli *et al.* 2009). 3-D magneto-hydrodynamic simulations have recently shown that this small distance can be naturally explained by ejecta clumping, without invoking shock modification (Orlando *et al.* 2012). Therefore, the small value of D_{BWCD} is not a reliable indicator of hadronic acceleration.

We took advantage of new observations within the *XMM-Newton* SN 1006 Large Program (PI A. Decourchelle, 700 ks of total exposure time) to present the first robust detection of X-ray emission from shocked ISM in SN 1006 and show evidence for the effects of hadron acceleration on the post-shock plasma (see Miceli *et al.* 2012 for further details). The bilateral morphology of SN 1006 clearly reveals regions with high injection efficiency, η†, (i.e. the limbs), separated by regions with low η. We therefore expect to observe stronger shock modification near the non-thermal limbs than in the thermal regions.

2. Results

We first focused on the center of a thermal rim (region *e* in Fig. 1) where the contribution of the synchrotron emission is the smallest and we expected minimum shock modification. We have found that with the new data, a model with three components (ejecta, ISM, and synchrotron emission) fits the spectrum of region *e* significantly better than a model with only two components. For example, a thermal component for the ejecta (optically thin plasma in non-equilibrium of ionization with free O, Ne, Mg, and Si abundances, VPSHOCK model in XSPEC, Borkowski *et al.* 2001), plus the non-thermal SRCUT component (Reynolds & Keohane 1999), yields a $\chi^2 = 1199.8$ (with 658 d. o .f.), while by adding another VPSHOCK component with solar abundances, we get a much lower value ($\chi^2 = 1112.0$ with 655 d. o .f.). The normalization of the additional component is larger than zero at > 11 sigmas (more details of spectral analysis are provided in Miceli *et al.* 2012). The association of the additional thermal component with shocked ISM is supported by its parameters, derived from spectral analysis: i) the post-shock density $n_{ISM} = 0.14 \pm 0.01$ cm^{-3} indicates a pre-shock density <0.05 cm^{-3}, consistent with the high galactic latitude of the remnant and in agreement with the upper limit present in the literature (Acero *et al.* 2007); ii) the ionization parameter (i.e. the integral of density over time calculated from the impact with the shock front) $\tau_{ISM} \sim 7 \times 10^8$

† η is the fraction of particles injected in the acceleration process.

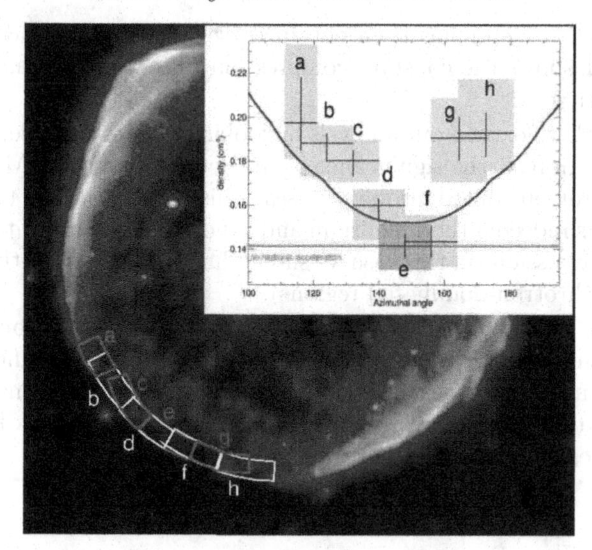

Figure 1. *XMM-Newton* image of SN 1006 in the 2 – 4.5 keV band (the bin size is 4″). The inset shows the azimuthal profile of the ISM density in the spectral regions. Error bars are at the 68% (crosses) and 90% (shaded areas) confidence levels. The blue curve shows the profile derived by assuming that η is proportional to the radio flux divided by $B^{3/2}$, and adopting the magnetic field model MF2 of Petruk *et al.* 2009 and by assuming the relationship between the injection efficiency and shock compression ratio adopted by Ferrand *et al.* 2010.

s cm^{-3} provides a very reasonable estimate for the time elapsed after the shock impact ($\sim$ 200 yr), considering the 1 kyr age of the remnant; and iii) the temperature ($kT_{ISM} \sim$ 1.4 keV) suits the expectations for the shocked ISM, as shown below. In collisionless shocks at high Mach number (shock velocity of the order of 10^3 km/s) the electron to proton temperature ratio is expected to be $T_e/T_p \ll 1$ (Ghavamian *et al.* 2007, Vink *et al.* 2003). In the north-western limb of SN 1006 the shock velocity is $v_{sh} \sim 3000$ km/s and it has been calculated that $T_e/T_p < 0.07$ (Ghavamian *et al.* 2002). Considering that in our south-eastern region the shock speed is $\sim$ 5000 km/s (Katsuda *et al.* 2009) and assuming the same ratio (and no or little shock modification) we get $kT_e < 3.5$ keV, in agreement with our findings.

If efficient hadron acceleration is at work, we expect to observe a higher shock compression ratio and a lower shock temperature in regions closer to the non-thermal limbs. The modeling of our spectra allows us to measure the electron temperature only and it is still unclear how to couple T_e with the shock temperature (Vink 2012). We therefore consider as the most reliable indicators of hadron acceleration the variations in the ISM post-shock density that can be directly derived through our spectral analysis. To verify the presence of shock modification, we then analyzed the spectra extracted from regions [a, b, c, d], and [f, g, h] (shown in Fig. 1) that are closer than e to the non-thermal limbs (more details in Miceli *et al.* 2012). In all these regions we confirmed the detection of the ISM component. We found that the post-shock density significantly increases in regions where particle acceleration is more efficient (see inset in Fig. 1). Considering that the pre-shock density is expected to be fairly uniform (Dubner *et al.* 2002), the density profile in Fig. 1 shows that *the shock compression ratio is higher near the non-thermal limbs*, thus providing a direct proof for the shock modification theory.

We also compared the observed ISM density profile with that obtained by assuming that η is proportional to the radio flux divided by $B^{3/2}$, and adopting the magnetic field model MF2 of Petruk *et al.* 2009 (the relationship between the injection efficiency and the shock compression factor has been obtained by following Ferrand *et al.* 2010). The green line in Fig. 1 shows the constant compression ratio scenario that would result for no shock modification.

It was not possible to extend our analysis by looking at regions closer to (or inside) the non-thermal limbs. In fact, the significance of the detection of the ISM component drops down as the synchrotron contribution increases. This result confirms that the additional ISM component has indeed a thermal origin and is not an artifact due to a misdescription of the synchrotron emission (in this case we should have detected it with higher statistical significance in synchrotron-dominated regions).

In conclusion, we have obtained a firm detection of the X-ray emission from the shocked ISM in SN 1006 and found evidence for the presence of shock modification induced by hadron acceleration at the shock front. Our results provide new constraints for future theories and prove that cosmic ray acceleration is present at the shock front and modifies the shock properties.

References

Acero, F., Aharonian, F., Akhperjanian, A. G., *et al.* 2010, *A&A*, 516, A62

Acero, F., Ballet, J., & Decourchelle, A. 2007, *A&A*, 475, 883

Acero, F., Ballet, J., Decourchelle, A., *et al.* 2009, *A&A*, 505, 157

Bamba, A., Yamazaki, R., & Hiraga, J. S. 2005, *ApJ*, 632, 294

Berezhko, E. G. & Ellison, D. C. 1999, *ApJ*, 526, 385

Berezhko, E. G. & Völk, H. J. 2007, *ApJl*, 661, L175

Blasi, P. 2002, *Astroparticle Physics*, 16, 429

Borkowski, K. J., Lyerly, W. J., & Reynolds, S. P. 2001, *ApJ*, 548, 820

Borkowski, K. J., Reynolds, S. P., Green, D. A., *et al.* 2010, *ApJ*, 724, L161

Cassam-Chenaï, G., Hughes, J. P., Ballet, J., & Decourchelle, A. 2007, *ApJ*, 665, 315

Cassam-Chenaï, G., Hughes, J. P., Reynoso, E. M., Badenes, C., & Moffett, D. 2008, *ApJ*, 680, 1180

Decourchelle, A., Ellison, D. C., & Ballet, J. 2000, *ApJl*, 543, L57

Dubner, G. M., Giacani, E. B., Goss, W. M., Green, A. J., & Nyman, L.-Å. 2002, *A&A*, 387, 1047

Ferrand, G., Decourchelle, A., Ballet, J., Teyssier, R., & Fraschetti, F. 2010, *A&A*, 509, L10+

Ghavamian, P., Laming, J. M., & Rakowski, C. E. 2007, *ApJ*, 654, L69

Ghavamian, P., Winkler, P. F., Raymond, J. C., & Long, K. S. 2002, *ApJ*, 572, 888

Katsuda, S., Petre, R., Long, K. S., *et al.* 2009, *ApJ*, 692, L105

Koyama, K., Petre, R., Gotthelf, E. V., *et al.* 1995, *Nature*, 378, 255

Miceli, M., Bocchino, F., Decourchelle, A., *et al.* 2012, *A&A*, 546, A66

Miceli, M., Bocchino, F., Iakubovskyi, D., *et al.* 2009, *A&A*, 501, 239

Orlando, S., Bocchino, F., Miceli, M., Petruk, O., & Pumo, M. L. 2012, *ApJ*, 749, 156

Pannuti, T. G., Allen, G. E., Filipović, M. D., *et al.* 2010, *ApJ*, 721, 1492

Petruk, O., Bocchino, F., Miceli, M., *et al.* 2009, extitMNRAS, 399, 157

Reynolds, S. P., Borkowski, K. J., Green, D. A., *et al.* 2009, *ApJl*, 695, L149

Reynolds, S. P. & Keohane, J. W. 1999, *ApJ*, 525, 368

Rothenflug, R., Ballet, J., Dubner, G., *et al.* 2004, *A&A*, 425, 121

Slane, P., Hughes, J. P., Edgar, R. J., *et al.* 2001, *ApJ*, 548, 814

Tian, W. W., Li, Z., Leahy, D. A., *et al.* 2010, *ApJ*, 712, 790

Vink, J. 2012, *A&Ar*, 20, 49

Vink, J., Laming, J. M., Gu, M. F., Rasmussen, A., & Kaastra, J. S. 2003, *ApJl*, 587, L31

Vink, J., Yamazaki, R., Helder, E. A., & Schure, K. M. 2010, *ApJ*, 722, 1727

Yamaguchi, H., Koyama, K., Katsuda, S., *et al.* 2008, *PASJ*, 60, 141

Discussion

EDITOR'S NOTE: The paper was presented by Fabrizio Bocchino. The discussion reproduced here took place after Bocchino's (separate) talk on "X-ray emission from ejecta fragments and protrusions in and around the SN1006 shell" which was on a related topic on the same SNR. No manuscript was received from Dr. Bocchino on his paper.

FOLATELLI: Do you have an explanation for the protrusions that are 'outliers' in terms of size as compared with your models?

BOCCHINO: We are looking into those. An expansion inside a more rarefied medium could be a possibility.

LU: Is there any difference along the lines from the center to the protrusions from the other regions of the remnant just inside the shock? (any sign of the fast moving fragments?)

BOCCHINO: We haven't investigated this issue yet. The limited-time span between XMM-Newton observations and the moderate spatial resolution make proper motion studies difficult.

BRANDT: What might we learn from regions of negative protrusions ?

BOCCHINO: "Negative Protrusions" are regions where the ejecta have not reached the main shock. They are useful in the sense that both "negative" and "regular" protrusion concur to establish the azimuthal periodicity seen in the data. They also concur to establish a mean R_{CD}/R_{BW} ratio which may be compared to models.

CHIOTELLIS: Do the ejecta protrusion affect the overall kinematics of the forward shock ? Do they also affect the shocked ejecta shell properties?

BOCCHINO: No, they don't, because perturbations of ejecta profiles are done maintaining the total kinetic energy constant, and moreover the density clumps are very few. So the overall dynamic is not significantly affected.

SLANE: The modeling here makes it seem clear that clumpy ejecta are required to get protrusions beyond the FS. However, we know there is CR acceleration too, so it isn't clear why this has been eliminated in the Final Modeling. Both effects must be playing a role in compressing the FS/CD ratio, right?

BOCCHINO: The poster of Orlando (already published on astro-ph) shows CR acceleration has only a second order effects on protrusions and R_{BW}/R_{CD} ratio. So it has been eliminated for simplicity in this work. Yes, CRs indeed do play a major role in compressibility and we show in the Miceli talk that this effect is measurable in SN1006.

SCHURE: On what is your initial clumpiness based and is there any information from explosion models that can shed light on this? It could well be that they are not homogeneously distributed?

BOCCHINO: The initial clumpiness is arbitrary, but realistic. We plan to improve it by coupling our simulation with explosion models having clumpiness. Yes, it could well be these are not homogeneously distributed.

Supernova Environmental Impacts
Proceedings IAU Symposium No. 296, 2013
A. Ray & R. A. McCray, eds.

doi:10.1017/S1743921313009654

Interaction of escaping cosmic rays with molecular clouds

Stefano Gabici

APC, AstroParticule et Cosmologie, Université Paris Diderot, CNRS, CEA, Observatoire de Paris, Sorbonne Paris Cité, France – email: `stefano.gabici@apc.univ-paris7.fr`

Abstract. The study of the gamma–ray radiation produced by cosmic rays that escape their accelerators is of paramount importance for (at least) two reasons: first, the detection of those gamma–ray photons can serve to identify the sources of cosmic rays and, second, the characteristics of that radiation give us constraints on the way in which cosmic rays propagate in the interstellar medium. This paper reviews the present status of the field.

Keywords. cosmic rays, supernova remnants, ISM: clouds

1. Introduction

The galactic disk is the sky's most prominent source of gamma rays in the GeV energy domain. Discovered in the late sixties (Clark *et al.* 1968), the galactic gamma–ray diffuse emission was soon interpreted as the decay of neutral pions (π^0) produced by cosmic rays (CRs) interacting with the interstellar gas (Stecker 1969). This confirmed a scenario first proposed by Hayakawa in 1952 (for early reviews see Fazio 1967 and Ginzburg & Syrovatskii 1964). During the past decades the diffuse gamma–ray emission from the galactic disk has been observed with constantly increasing accuracy by several space instruments (Fichtel *et al.* 1975, Mayer–Hasselwander *et al.* 1982, Hunter *et al.* 1997, Ackermann *et al.* 2012a), and the spatial distribution of CRs in the Galaxy could be extracted from such observations (Bloemen 1989 and references therein, Bertsch *et al.* 1993, Strong *et al.* 2004, Ackermann *et al.* 2012). It turned out that the distribution of CRs in the Galaxy is quite uniform and that, as an order of magnitude, the intensity of CRs measured in the solar system is representative of the intensity anywhere else in the galactic disk. This roughly uniform background is often referred to as the *cosmic ray sea.*

The π^0–decay gamma–ray emissivity $q_\gamma^0(> 100\ \mathrm{MeV})$ of the local atomic gas has been measured by the *Fermi* LAT and is equal to $q_\gamma^0(> 100\ \mathrm{MeV})/4\pi = 1.6\times10^{-26}$ ph/s/sr/H–atom (Abdo *et al.* 2009). Thus, an estimate of the intensity of the galactic diffuse emission from a specific direction in the sky can be obtained by integrating the gamma–ray emissivity of the gas $q_\gamma(> 100\ \mathrm{MeV}, l) \approx q_\gamma^0(> 100\ \mathrm{MeV})$ along the line of sight l. This straightforward procedure gives (e.g. Aharonian 2004):

$$J_\gamma(> 100\,\mathrm{MeV}) = \int \mathrm{d}l\ \frac{q_\gamma(> 100\ \mathrm{MeV}, l)}{4\pi}\ n_{gas} \approx 1.6\times10^{-4} \left(\frac{N_H}{10^{22}\ \mathrm{cm}^{-2}}\right) \mathrm{ph/s/cm^2/sr} \tag{1.1}$$

where N_H is a typical gas column density in the galactic disk.

In 1973 Black & Fazio suggested that the gamma–ray emission from individual massive molecular clouds (MCs) might be visible above the diffuse galactic emission. The gamma–ray flux from a MC of mass M_{cl}, distance d and embedded in the CR sea is:

$$F_\gamma(> 100\ \mathrm{MeV}) \approx \frac{q_\gamma^0(> 100\ \mathrm{MeV})\ (M_{cl}/m_p)}{4\pi\ d^2} \approx 2\times10^{-7}\left(\frac{M_5}{d_{kpc}^2}\right) \mathrm{ph/cm^2/s} \tag{1.2}$$

where M_5 is the mass of the cloud in units of $10^5 M_\odot$, d_{kpc} its distance in kiloparsecs, and m_p the proton mass. The detectability of massive MCs above the diffuse galactic emission follows from Equations (1.1) and (1.2) and from the fact that, for a typical cloud density of $n_{cl} \approx 1000\ n_3\ \text{cm}^{-3}$, the radius of the cloud is $R_{cl} \approx 10\ (M_5/n_3)^{1/3}$ pc and its angular extension is $\Omega_{cl} \approx 10^{-4}(M_5/n_3)^{2/3} d_{kpc}^{-1/2}$ sr.

Black & Fazio (1973) proposed that the masses of the MCs of known distance could be derived from the strength of their gamma–ray flux , under the assumption that the intensity of CRs is known throughout the Galaxy (i.e. the assumption of a uniform CR sea). Such an approach was used extensively to calibrate the methods for the determination of the mass of molecular and atomic gas in the Galaxy and also to check the (rough) spatial homogeneity of the intensity of CRs in the Galaxy over large spatial scales (e.g. Caraveo *et al.* 1980, Bloemen *et al.* 1984, Hunter *et al.* 1994, Digel *et al.* 1996, 1999, Abdo *et al.* 2010a, Ackermann *et al.* 2011, 2012b, 2012c, 2012d). However, the assumption of a uniform sea of CRs that permeates the whole Galaxy might be inappropriate in some circumstances, especially on small spatial scales (e.g. in the vicinity of CR sources). In these circumstances, the reasoning of Black & Fazio can be reversed and, if an estimate of the mass of the cloud is available (see e.g. Hartquist 1983 for caveats), one can use the gamma–ray observations of MCs to probe variations of the intensity of CRs in the Galaxy (Issa & Wolfendale 1981, Morfill *et al.* 1981, Aharonian 1991, 2001, Casanova *et al.* 2010). In this context, MCs serve as *cosmic ray barometers*, because from their gamma–ray flux it is possible to infer the intensity (and thus pressure) of the CRs.

The fact that the study of MCs could help in solving the problem of the origin of CRs became evident when it was realized that an association between MCs and CR sources is indeed to be expected (Montmerle 1979). This is because CRs are believed to be accelerated at supernova remnant (SNR) shocks (e.g. Hillas 2005) and thus are likely to be produced in star forming environments (where core–collapse supernovae explode), which are in turn expected to host massive MCs. In a seminal paper, Montmerle (1979) described a scenario in which CRs, after being accelerated at SNRs, could escape the acceleration site and diffuse to a nearby MC and produce there gamma rays via interactions with the gas. Due to the presence of these runaway CRs, the CR intensity is expected to be strongly enhanced in the vicinity of SNRs. In the same way, the gamma–ray emission from a MC illuminated by the runaway CRs will be much larger than the one derived in Equation (1.2), which refers to a MC embedded in the CR sea. It follows that the detection of gamma rays from MCs located in the vicinity of SNRs might constitute an hint for the fact that the nearby SNR is (or was, in the past) acting as a CR accelerator.

In this context, gamma–ray observations performed in the TeV domain are of great relevance, because: *i)* the expected gamma–ray emission for a MC illuminated by runaway CRs is expected to have a TeV flux which is within the reach of current Cherenkov instruments (Aharonian 1991, Aharonian & Atoyan 1996, Gabici *et al.* 2009), and *ii)* in the TeV energy domain the contribution from the CR sea (which has a steep spectrum) to the gamma–ray emission is virtually negligible, and thus any detection of MCs has necessarily to be interpreted as an excess of CRs above the sea at the location of the MC (Aharonian 1991). A spectacular example of this fact can be found in Aharonian *et al.* (2006), where the detection of a diffuse emission of TeV gamma rays was reported from a very massive MC complex located in the galactic centre region. These observations revealed an excess above the CR sea of a factor of $\approx$ 4...10 at TeV energies, and also an harder spectrum of CRs there (with slope ≈ 2.3). The excess indicates that a source (or more sources) of CRs might be present in the region (remarkably, runaway CRs from only one SNR would suffice to explain the observed gamma–ray emission).

A phenomenological description of the propagation of the CRs after their escape from the acceleration site has been developed in a pioneering paper by Aharonian & Atoyan (1996), who also discussed the expected radiative signatures (especially in gamma rays) due to the interactions of CRs in the ambient gas. Aharonian & Atoyan considered an isotropic and spatially homogeneous diffusion coefficient for CRs and computed the expected gamma–ray emission from the MC, and stressed the fact that the properties of such emission (intensity, spectral shape, duration in time, etc.) strongly depend on the value of the diffusion coefficient. This opens the possibility to constrain, from gamma–ray observations, the diffusion coefficient of CRs in the vicinity of their sources. This fact has a tremendous importance for CR studies, given that the diffusion coefficient is a very poorly determined quantity (both from an observational and theoretical point of view).

The multiwavelength emission resulting from the interactions of runaway CRs in a MC has been computed, for the specific case in which a SNR accelerates the CRs, by Gabici & Aharonian (2007) and Gabici *et al.* (2009). These studies have then been applied to specific situations in order to obtain constraints on the particle diffusion coefficient in the vicinity of SNRs (Gabici *et al.* 2010, Nava & Gabici 2013). Moreover, the TeV diffuse emission resulting from the interactions of runaway CRs in the diffuse interstellar medium (i.e. in the absence of a massive MC) has been predicted and found to be within the reach of future ground based instruments such as the Cherenkov Telescope Array, which is thus expected to play a crucial role in proving (or falsifying) the SNR paradigm for the origin of galactic CRs (Casanova *et al.* 2010, Acero *et al.* 2013). This paper is intended as a short review of these results. A more extended discussion can be found in Gabici (2013).

Though the effectiveness of the penetration of CRs into MCs still remains an open issue (Skilling & Strong 1976, Cesarsky & Völk 1978, Morfill 1982, Zweibel & Shull 1982, Everett & Zweibel 2011) I will assume in the following a full, unimpeded penetration of CRs in MCs. There is little doubt that TeV CRs can penetrate MCs (Gabici *et al.* 2007) and the penetration of GeV particles seems to be supported by gamma–ray observations, from the early ones by Lebrun & Paul (1978) to the ones by Ackermann *et al.* (2012b). For lower energies this remains an open issue, but this should not affect our considerations.

Here, I will not discuss the (yet not clear) way in which CRs escape their sources (see Gabici 2011 and references therein) nor the case in which the gamma–ray emission from the MC is the result of the interaction between the cloud itself and the SNR shock (numerous papers can be found in the literature, including: Blandford & Cowie 1982, Aharonian *et al.* 1996, Gaisser *et al.* 1998, Bykov *et al.* 2000, Fatuzzo & Melia 2005, Uchiyama *et al.* 2010, Malkov *et al.* 2011, Inoue *et al.* 2012, Fang & Zhang 2013)

2. Escape of cosmic rays from supernova remnants: isotropic diffusion

As an illustrative example I consider here the case of the SNR W28, which is an aged remnant ($t_{age} \approx 4 \times 10^4$ yr) located in the vicinity of three massive MCs (of total mass $\approx 10^5 M_{\odot}$). Gamma–ray emission has been detected from the MCs in both the GeV and TeV domain (Aharonian *et al.* 2008, Abdo *et al.* 2010b, Giuliani *et al.* 2010). Since most of the gamma–ray emission clearly comes from *outside of the SNR shell*, it seems natural to interpret it as the result of the interactions in the MCs of CRs that escaped the SNR (Fujita *et al.* 2009, Gabici *et al.* 2010, Li & Chen 2010, Ohira *et al.* 2011, Yan *et al.* 2012). This supports the idea that W28 was, in the past, an accelerator of CRs.

I provide now a simple argument to show how one can attempt to constrain the diffusion coefficient in the vicinity of the SNR W28 by using the above mentioned gamma–ray observations, especially the ones performed by H.E.S.S. in the TeV domain. The time elapsed since CRs with a given energy E escaped the SNR can be written as: $t_{diff} = t_{age} - t_{esc}$, where $t_{esc}(E)$ is the age of the SNR when CRs of energy E were released. This is a time dependent quantity, since the highest energy CRs are believed to be released first, and CRs with lower and lower energy are gradually released at later times (Gabici 2011 and references therein). However, for CRs with energies above 1 TeV (the ones responsible for the very high energy gamma–ray emission) one can assume $t_{esc} << t_{age}$ (i.e. high energy CRs are released when the SNR is much younger than it is now) and thus $t_{diff} \sim t_{age}$. At time t_{age} CRs have diffused over a distance $R_d \sim \sqrt{4\, D\, t_{age}}$. Within the diffusion radius R_d the spatial distribution of CRs, f_{CR}, is roughly constant, and proportional to $\eta E_{SN}/R_d^3$, where E_{SN} is the supernova explosion energy and η is the fraction of such energy converted into CRs. On the other hand, the observed gamma ray flux from each one of the three MCs detected in gamma rays is: $F_\gamma \propto f_{CR} M_{cl}/d^2$, where M_{cl} is the mass of the MC and d is the distance of the system. Note that in this expression F_γ is calculated at a photon energy E_γ, while f_{CR} is calculated at a CR energy $E_{CR} \sim 10 \times E_\gamma$, to account for the inelasticity of proton-proton interactions. By using the definitions of f_{CR} and R_d one can finally write the approximate equation, valid within a distance R_d from the SNR:

$$F_\gamma \propto \frac{\eta\, E_{SN}}{(\chi\, D_{gal}\, t_{age})^{3/2}} \left(\frac{M_{cl}}{d^2} \right).$$

Estimates can be obtained for all the physical quantities in the equation except for the CR acceleration efficiency η and the local diffusion coefficient D. By fitting the TeV data one can thus attempt to constrain, within the uncertainties given by the errors on the other measured quantities (namely, E_{SN}, t_{age}, M_{cl}, and d) and by the assumptions made (e.g. the CR injection spectrum is assumed to be E^{-2}, while the energy dependence of D is assumed to scale as a power law of index $\delta = 0.5$), a combination of these two parameters (namely $\eta/D^{3/2}$). The fact that the MCs have to be located within a distance R_d from the SNR can be verified a posteriori, and their exact location (unknown due to projection effects) can be tuned to match also the observed GeV emission. Given all the uncertainties above, our results have to be interpreted as a proof of concept of the fact that gamma ray observations of SNR/MC associations can serve as tools to estimate the CR diffusion coefficient. More detection of SNR/MC associations are needed in order to check whether the scenario described here applies to a whole class of objects and not only to a test-case as W28. Future observations from the Cherenkov Telescope Array will most likely solve this issue.

Fig. 1, from Gabici *et al.* 2010, shows a fit to the gamma–ray data for the three massive MCs in the W28 region. A simultaneous fit to all the three MCs is obtained by fixing a value for $\eta/D^{3/2}$, which implies that the diffusion coefficient of particle with energy 3 TeV (these are the particles that produce most of the emission observed by *HESS*) is:

$$D(3\ \text{TeV}) \approx 5 \times 10^{27} \left(\frac{\eta}{0.1} \right)^{2/3} \ \text{cm}^2/\text{s}\,. \tag{2.1}$$

This value is significantly smaller (more than an order of magnitude) than the one normally adopted to describe the diffusion of ~TeV CRs in the galactic disk, which is $\approx 10^{29}$ cm^2/s. This result remains valid (i.e. a suppression of the diffusion coefficient is indeed needed to fit data) even if a different value of the parameter δ is assumed, within the range 0.3...0.7 compatible with CR data.

The reason of this discrepancy between the average CR diffusion coefficient in the Galaxy and the one found in the vicinity of a SNR needs to be explained. A possible way to interpret these observations is given in the next Section.

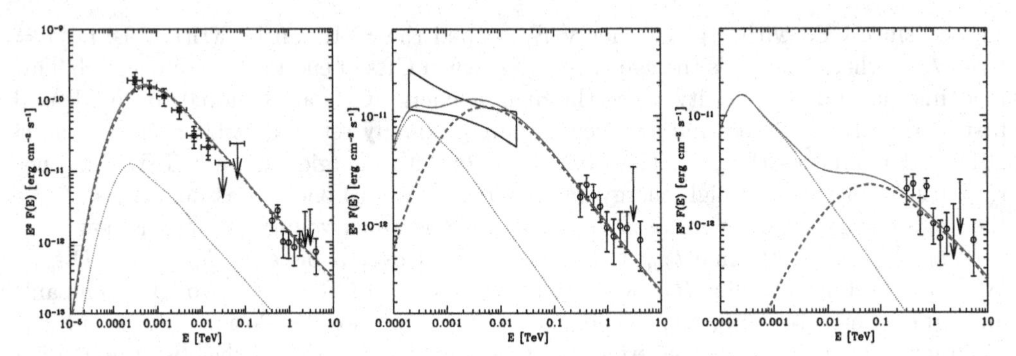

Figure 1. Broad band fit to the gamma ray emission detected by *FERMI* and *HESS* from the sources HESS J1801-233, HESS J1800-240 A and B (left to right), that coincide with three massive MCs. Dashed lines represent the contribution to the gamma–ray emission from CRs that escaped W28, dotted lines show the contribution from the CR sea, and solid lines the total emission. Distances to the SNR centre are 12, 65, and 32 pc (left to right). *FERMI* and *HESS* data points are plotted in black. No GeV emission has been detected from HESS J1800-240 A.

3. Anisotropic diffusion of runaway cosmic rays

Most of the studies aimed at predicting the gamma–ray emission from runaway CRs rely on the assumption of isotropic diffusion (see references in Sec. 2 and, e.g. Lee *et al.* 2008, Torres *et al.* 2008, Rodrguez Marrero *et al.* 2008, Torres *et al.* 2010, Ellison & Bykov 2011, Li & Chen 2012, Ellison *et al.* 22012, Telezhinsky *et al.* 2012). However, the validity of the assumption of isotropic diffusion of CRs, adopted in the previous section, needs to be discussed. In fact, if the intensity of the turbulent field δB on scales resonant with the Larmor radius of particles is significantly smaller than the mean large scale field B_0 (i.e. if $\delta B/B_0 \ll 1$), then *cosmic ray diffusion becomes anisotropic*, with particles diffusing preferentially along the magnetic field lines (e.g. Casse *et al.* 2002 and references therein). In the limiting (but still reasonable) case in which the perpendicular diffusion coefficient can be set equal to zero, the transport of CRs across the mean field is mainly due to the wandering of magnetic field lines (Jokipii & Parker 1969).

To give a qualitative idea of the role that anisotropic diffusion can play in the studies of the CRs that escaped SNRs, let us consider an idealized case in which the escaping particles diffuse along a magnetic flux tube characterized by a very long coherence length (i.e. the magnetic flux tube is preserved for a long distance). In this case, after a time t particle will diffuse up to a distance $R_d \approx \sqrt{2\,D_{\parallel} \times t}$ along the tube (here $D_{\parallel}$ is the *parallel* diffusion coefficient of cosmic rays, not to be confused with the isotropic diffusion coefficient D adopted in the previous Section), while their transverse distribution will be equal to the radius of the SNR shock at the time of their escape, R_{sh}, which is of the order of $\approx$ 1–10 pc. Thus, the CR density in the flux tube will be proportional to $n_{CR} \propto (R_d R_{sh}^2)^{-1}$ instead of $\propto R_d^{-3}$ as in the isotropic case (see previous Section). It is easy to see that the estimates of the diffusion coefficient based on the two opposite assumptions of isotropic and one–dimensional diffusion will differ by a factor of $\approx (R_d/R_{sh})^{4/3}$, which can be much larger than an order of magnitude! Thus, it is of paramount importance to investigate how the interpretation of gamma–ray observations depends on the assumptions made concerning CR diffusion.

As a first step, let us compare in Fig. 2 the results that are obtained if an isotropic diffusion coefficient is assumed, with the ones obtained for the anisotropic diffusion model considered in Nava & Gabici (2013). In both panels of Fig. 2, the SNR is located at the centre of the field and the color code refers to the excess of CRs with respect to the CR

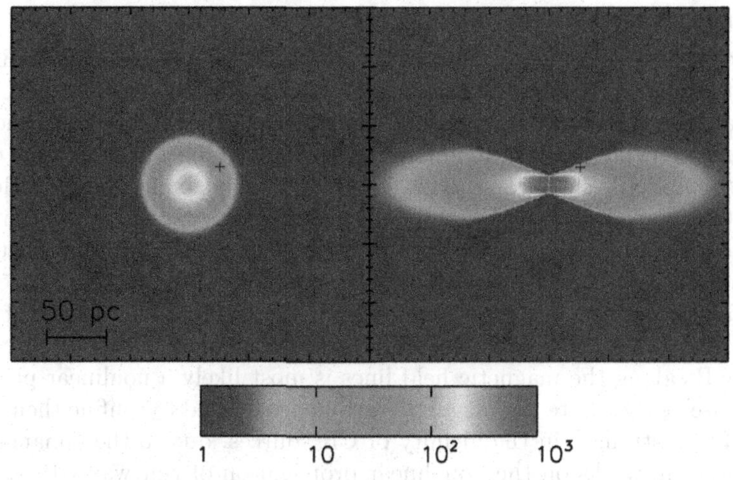

Figure 2. Cosmic ray over-density above the galactic background around a supernova remnant (located at the centre of the panels). A particle energy of $E = 1\,\text{TeV}$ and a time $t = 10\,\text{kyr}$ after the explosion are considered. The left panel refers to an isotropic diffusion coefficient of cosmic rays equal to $D = 5 \times 10^{26} (E/10\ \text{GeV})^{0.5}\ \text{cm}^2/\text{s}$, while the right panel refers to an anisotropic diffusion scenario with $D_{\parallel} = 10^{28} (E/10\ \text{GeV})^{0.5}\ \text{cm}^2/\text{s}$. The black cross marks the a position at which the CR over-density is equal in the two panels. Figure from Nava & Gabici (2013).

sea. Over-densities are plotted for a particle energy of $1\,\text{TeV}$ and for a time $t = 10\,\text{kyr}$ after the supernova explosion (see Nava & Gabici 2013 for more details on the model).

The spatial distribution of CRs is strikingly different in the two scenarios: spherically symmetric in the left panel, and strongly elongated in the direction of the magnetic field flux tube in the right panel. A filamentary diffusion of CRs was also found in the numerical simulations by Giacinti *et al.* (2012). The same parameters have been used to compute the over–densities in the two scenarios in Fig. 2, with the exception of the CR diffusion coefficient, which in the left panel has been assumed to be isotropic and equal to $D(1\ \text{TeV}) \approx 5 \times 10^{27}\ \text{cm}^2/\text{s}$, while in the right one is assumed to be strictly parallel (i.e. CRs diffuse only along field lines) and equal to $D_{\parallel} \approx 10^{29}\ \text{cm}^2/\text{s}$ (a value similar to the average CR diffusion coefficient in the Galaxy). The choice of two significantly different values for the diffusion coefficients, with $D \ll D_{\parallel}$ has been made in order to obtain the same level of CR over–density in the vicinity of the SNR. As an example, the black cross in Fig. 2 identifies a position, located 30 pc away from the centre of the explosion, where the CR over-density is identical in the two panels. To get comparable values for the CR over–density, a much smaller (isotropic) diffusion coefficient D is needed in order to compensate for the larger solid angle over which CRs can propagate. This fact must be taken into account when interpreting the gamma–ray observations of molecular clouds illuminated by CRs escaping from SNRs. For example, a fit to the gamma–ray data from the MCs in the W28 regions has been obtained by Nava & Gabici (2013) by assuming a *large* diffusion coefficient of $D_{\parallel}(1\ \text{TeV}) \approx 10^{29}\ \text{cm}^2/\text{s}$, much larger than the (isotropic) one adopted in the previous section (see Fig. 1). To conclude, the hint for a suppression of the diffusion coefficient in the vicinity of the SNR W28 obtained in the previous Section might depend on the assumption of isotropy of diffusion. If an anisotropic diffusion is adopted, a much larger diffusion coefficient can be assumed to fit gamma–ray data.

4. Conclusions and future perspectives

I have shown how gamma–ray observations of MCs located close to SNRs can serve to support the idea that SNRs are the sources of CRs. Information on the CR diffusion coefficient can also be extracted from such observations, though the conclusions of these studies strongly depend on the (still unknown) isotropic or anisotropic nature of diffusion. To date, only two SNRs show gamma–ray emission clearly coming from outside the SNR shell: W28 (see above) and W44 (Uchiyama *et al.* 2012). So, further observations are needed in order to obtain solid constraints on the CR diffusion coefficient. Future facilities as the Cherenkov Telescope Array will play a key role in this direction.

Theoretical studies are also needed in order to understand the details of CR propagation close to their sources and interpret correctly the gamma–ray observations. The diffusion of CRs along the magnetic field lines is most likely a nonlinear process, where the CRs themselves generate the magnetic turbulence needed to confine them. This effect is expected to be stronger in the vicinity of CR sources, due to the enhanced intensity of CRs. Pioneering works on the non–linear propagation of runaway CRs can be found in Skilling (1970) and Hartquist & Morfill (1994). These studies have been revived by the recent results obtained from the gamma–ray observations of SNR/MC associations (see e.g. Ptuskin *et al.* 2008, Malkov *et al.* 2013), and promise to become one of the most important developments in this field.

References

Abdo, A. A., *et al.* 2009, *ApJ*, 703, 1249
Abdo, A. A., *et al.* 2010a, *ApJ*, 710, 133
Abdo, A. A., *et al.* 2010b, *ApJ*, 718, 348
Acero, F., *et al.* 2013, *Astropart. Phys.*, in press – arXiv:1209.0582
Ackermann, M., *et al.* 2011, *ApJ*, 726, 81
Ackermann, M., *et al.* 2012a, *ApJ*, 750, 3
Ackermann, M., *et al.* 2012b, *ApJ*, 756, 4
Ackermann, M., *et al.* 2012c, *ApJ*, 755, 22
Ackermann, M., *et al.* 2012d, *A&A*, 538, A71
Aharonian, F. A. 1991, *Astrophys. Space Sci.*, 180, 305
Aharonian, F. A., Drury, L. O'C., & Völk, H. J. 1994, *A&A*, 285, 645
Aharonian, F. A. & Atoyan, A. M. 1996, *A&A*, 309, 917
Aharonian, F. A. 2001, *Space Sci. Rev.*, 99, 187
Aharonian, F. A. 2004, *Very high energy cosmic gamma radiation: a crucial window on the extreme Universe* (River Edge, NJ: World Scientific Publishing)
Aharonian, F. A., *et al.* 2006, *Nature*, 439, 695
Aharonian, F. A., *et al.* 2008, *A&A*, 481, 401
Bertsch, D. L., *et al.* 1993, *ApJ*, 416, 587
Black, J. H. & Fazio, G. G. 1973, *ApJ*, 185, L7
Blandford, R. D. & Cowie, L. L. 1982, *ApJ*, 260, 625
Bloemen, H., *et al.* 1984, *A&A*, 139, 37
Bloemen, H. 1989, *ARA&A*, 27, 469
Bykov, A. M., Chevalier, R. A., Ellison, D. C., & Uvarov, Yu. A. 2000, *ApJ*, 538, 203
Caraveo, P. A., *et al.* 1980, *A&A*, 91, L3
Casanova, S., *et al.* 2010, *PASJ*, 62, 769
Casanova, S., *et al.* 2010, *PASJ*, 62, 1127
Casse, F., Lemoine, M., & Pelletier, G. 2002, *Phys. Rev. D*, 65, 023002
Cesarsky, C. J. & Völk, H. J. 1978, *A&A*, 70, 367
Clark, G. W., Gamire, G. P., & Kraushaar, W. L. 1968, *ApJ*, 153, L203
Digel, S. W., Grenier, I. A., Heithausen, A., Hunter, S. D., & Thaddeus, P. 1996, *ApJ*, 463, 609

Digel, S. W., Aprile, E., Hunter, S. D., Mukherjee, R., & Xu, F. 1999, *ApJ*, 520, 196
Ellison, D. C. & Bykov, A. M. 2011, *ApJ*, 731, 87
Ellison, D. C., Slane, P., Patnaude, D. J., & Bykov, A. M. 2012, *ApJ*, 744, 39
Everett, J. E. & Zweibel, E. G. 2011, *ApJ*, 739, 60
Fang, J. & Zhang, L. 2013, *New Astron.*, 18, 35
Fatuzzo, M. & Melia, F. 2005, *ApJ*, 630, 321
Fazio, G. G. 1967, *ARA&A*, 5, 481
Fichtel, C. E., *et al.* 1975, *ApJ*, 198, 163
Fujita, Y., Ohira, Y., Tanaka, S. J., & Takahara, F. 2009, *ApJ*, 707, L179
Gabici, S., Aharonian, F. A., & Blasi, P. 2007, *Astrophys. Space Sci.*, 309, 365
Gabici, S. & Aharonian, F. A. 2007, *ApJ*, 665, L131
Gabici, S., Aharonian, F. A., & Casanova, S. 2009, *MNRAS*, 396, 1629
Gabici, S., *et al.* 2010, in: S. Boissier, M. Heydari–Malayeri, R. Samadi, & D. Valls–Gabaud (eds.) *SF2A-2010*, p. 313 - arXiv:1009.5291
Gabici, S. 2011, *MmSAI*, 82, 760
Gabici, S. 2013, in press, *San Cugat Forum on Astrophysics 2012* - arXiv:1208.4979
Gaisser, T. K., Protheroe, R. J., & Stanev, T. 1998, *ApJ*, 492, 219
Giacinti, G., Kachelriess, M., & Semikoz, D. V. 2012, *Phys. Rev. D*, 65, 023002
Ginzburg, V. L. & Syrovatskii, S. I. 1964, *The origin of cosmic rays* (New York: Macmillan)
Giuliani, A., *et al.* 2010, *A&A*, 516, L11
Hartquist, T. W. 1983, *Space Sci. Rev.*, 36, 41
Hartquist, T. W. & Morfill, G. E. 1994, *Astrophys. Space Sci.*, 216, 223
Hayakawa, S. 1952, *Prog. Theor. Phys*, 8, 571
Hillas, A. M. 2005, *J. Phys. G: Nucl. Part. Phys.*, 31, R95
Hunter, S. D., Digel, S. W., de Geus, E. J., & Kanbach, G. 1994, *ApJ*, 436, 216
Hunter, S. D., *et al.* 1997, *ApJ*, 481, 205
Inoue, T., Yamazaki, R., Inutsuka, S., & Fukui, Y. 2012, *ApJ*, 744, 71
Issa, M. R. & Wolfendale, A. W. 1981, *Nature*, 292, 430
Jokipii, J. R. & Parker, E. N. 1969, *ApJ*, 155, 777
Lebrun, F. & Paul, J. A. 1978, *A&A*, 65, 187
Lee, S. H., Kamae, T., & Ellison, D. C. 2008, *ApJ*, 686, 325
Li, H. & Chen, Y. 2010, *MNRAS*, 409, L35
Li, H. & Chen, Y. 2012, *MNRAS*, 421, 935
Malkov, M. A., Diamond, P. H., & Sagdeev, R. Z. 2011, *Nature Comm.*, 2, 194
Malkov, M. A. *et al.*(2013) - arXiv:1207.28
Mayer–Hasselwander, H. A., *et al.* 1982, *A&A*, 105, 164
Montmerle, T. 1979, *ApJ*, 231, 95
Morfill, G. E., *et al.* 1981, *ApJ*, 246, 810
Morfill, G. E. 1982, *ApJ*, 262, 749
Nava, L. & Gabici, S. 2013, *MNRAS*, 429, 1643
Ohira, Y., Murase, K., & Yamazaki, R. 2011, *MNRAS*, 410, 1577
Ptuskin, V. S., Zirakashvili, V. N., & Plesser, A. A. 2008, *Adv. Space Res.*, 42, 486
Rodrguez Marrero, A. Y., *et al.* 2008, *ApJ*, 689, 213
Skilling, J. 1970, *MNRAS*, 147, 1
Skilling, J. & Strong, A. W. 1976, *A&A*, 53, 253
Stecker, F. W. 1969, *Nature*, 222, 5196
Strong, A. W., Moskalenko, I. V., Reimer, O., Digel, S., & Dihel, R. 2004, *A&A*, 422, L47
Telezhinsky, I., Dwarkadas, V. V., & Pohl, M. 2012, *A&A*, 541, 153
Torres, D. F., Rodrguez Marrero, A. Y., & de Cea del Pozo, E. 2008, *MNRAS*, 387, L59
Torres, D. F., Rodrguez Marrero, A. Y., & de Cea del Pozo, E. 2010, *MNRAS*, 408, 1257
Uchiyama, Y., Blandford, R. D., Funk, S., Tajima, H., & Tanaka, T. 2010, *ApJ*, 723, L122
Uchiyama, Y., *et al.* 2012, *ApJ*, 749, L35
Yan, H., Lazarian, A., & Schlickeiser, R. 2012, *ApJ*, 745, 140
Zweibel, E. G. & Shull, J. M. 1982, *ApJ*, 567, 962

Supernova Environmental Impacts
Proceedings IAU Symposium No. 296, 2013
A. Ray & R. A. McCray, eds.

doi:10.1017/S1743921313009666

Simulating the Outer Nebula of SN 1987A

Ben Fitzpatrick, Thomas Morris and Philipp Podsiadlowski

Dept. of Astrophysics, University of Oxford,
Denys Wilkinson Building, Keble Road, Oxford, OX1 3RH, United Kingdom
email: podsi@astro.ox.ac.uk

Abstract. As has been shown previously, the triple-ring nebula around SN 1987A can be understood as a direct consequence of the merger of two stars, some 20,000 yr before the explosion. Here we present new SPH simulations that also include the pre-merger mass loss and show that this may be able to explain other structures observed around SN 1987A, such as Napoleon's hat and various light echoes.

Keywords. binaries (including multiple): close, stars: mass loss, supernovae: SN 1987A, supernova remnants

1. The Binary Merger Model for SN 1987A

SN 1987A was an unusual supernova. Its various anomalies, in particular, the blue-supergiant progenitor, the chemical anomalies in the ejecta and the complex triple-ring nebula surrounding it, are best explained by the merger of two massive stars, 20,000 yr before the explosion (Podsiadlowski *et al.* 2007). In this model (see diagram), the material that was ejected in the merger process was then swept up by the blue-supergiant wind to form the outer two rings, while the inner ring is the result of equatorial mass loss when the merged object shrank to become a blue supergiant. Morris & Podsiadlowski (2007) showed that this naturally reproduces all the features of the triple-ring nebula.

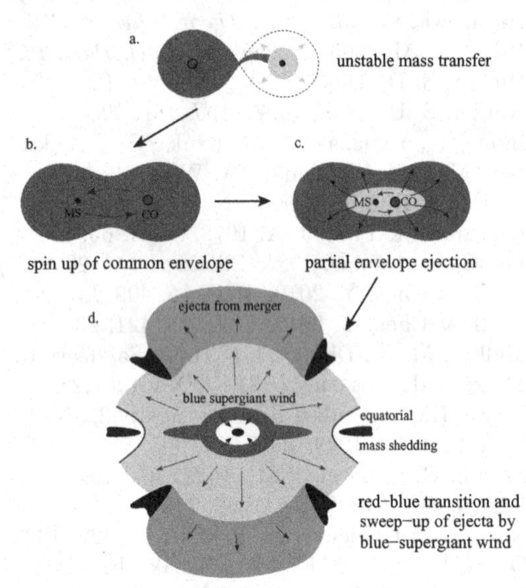

2. Modeling the Circumstellar Medium

Beyond the triple-ring nebula, there are other complex structures as seen with the NTT ('Napoleon's hat'; Wampler *et al.* 1990) and deduced from detailed light-echo studies (Sugerman *et al.* 2005). In order to understand their origin, we have performed detailed SPH simulations, using the GADGET-2 code (Springel 2005), that include the mass loss before the actual merger. Specifically, these include a slow red-supergiant wind (with $\dot{M}_{\rm RSG} = 2.0 \times 10^{-5} M_{\odot}{\rm yr}^{-1}$ and $v_{RSG} = 20\,{\rm km\,s}^{-1}$) and a bipolar outflow emitted from the accreting component in the early dynamically unstable mass-transfer phase (the first stage in the diagram above) with $\dot{M}_{\rm bi} = 6 \times 10^{-5}\, M_{\odot}{\rm yr}^{-1}$ and $v_{\rm bi} = 350\,{\rm km\,s}^{-1}$, lasting for 1000 yr.

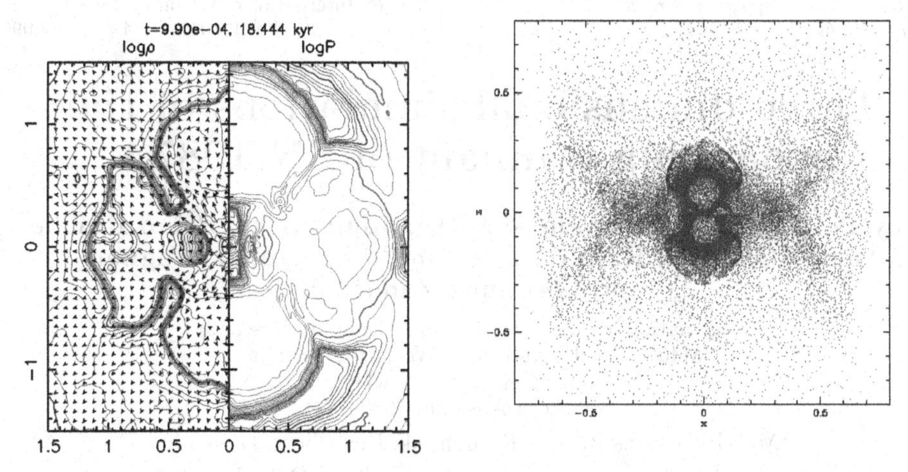

Figure 1. *Left:* Density/pressure distribution of the inner SN87A nebula from a supernova triple-ring simulation with a previous RSG wind (without bipolar jet). *Right:* Large-scale structure of the SN 87A nebula, including a RSG wind and a bipolar jet. The CE ejecta in the inner nebula are not properly resolved in this simulation. Structures similar to the 'spurs' in the Sugerman *et al.* (2005) reconstruction are found at the equatorial edges of the wind-compressed region. Axes are in units of 8 parsecs.

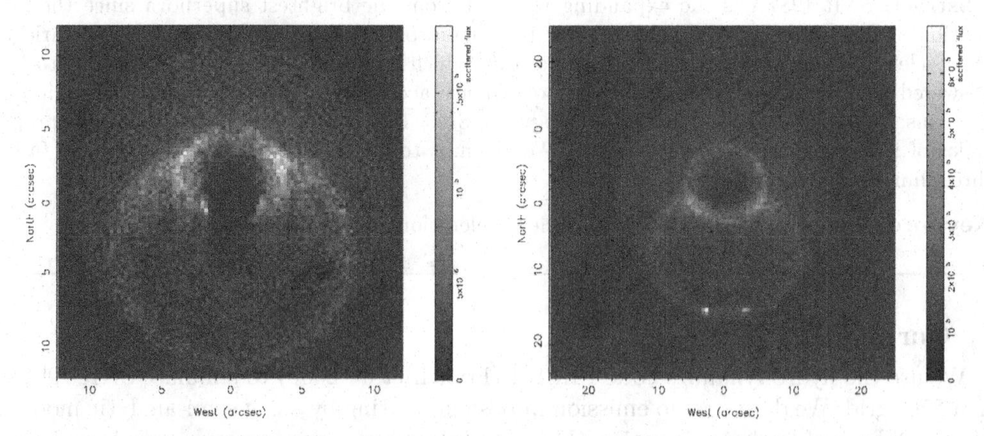

Figure 2. Simulated light echoes 1.8 years (left) and 4 years (right) after the supernova, assuming a constant-luminosity light curve of 0.3 years duration in a 24" by 24" field of view for the following parameters: jet velocity: 350 kms^{-1}, jet opening angle: 15°, and RSG wind velocity: 20 kms^{-1}.

Figure 1 present some the results of the SPH simulations, showing the large-scale structure and Figure 2 shows simulated light echos for our best-fit simulation.

References

Fitzpatrick, B., 2012, D. Phil. Thesis (Oxford University)

Morris, T. & Podsiadlowski, Ph., 2007, *Science*, 315, 1103

Podsiadlowski, Ph., Morris, T. S., & Ivanova, N., 2007, in *SN 1987A: 20 Years After*, AIP Conf.Proc., Vol. 937, p. 125

Springel, V. 2005, *MNRAS*, 37, 239

&Sugerman, B. E. K., *et al.* 2005, *ApJS*, 159, 60

Wampler, E. J., *et al.* 1990, *ApJL*, 362, 13

Supernova Environmental Impacts
Proceedings IAU Symposium No. 296, 2013
A. Ray & R. A. McCray, eds.

doi:10.1017/S1743921313009678

Three-dimensional simulations of the expanding remnant of SN 1987A

Toby Potter[1], Lister Staveley-Smith[1], John Kirk[2], Brian Reville[3], Geoff Bicknell[4], Ralph Sutherland[4], Alexander Wagner[4], and Giovanna Zanardo[1]

[1]International Centre for Radio Astronomy Research
M468, The University of Western Australia
35 Stirling Hwy, Crawley
Western Australia, 6009

[2]Max-Planck-Institut für Kernphysik, Heidelberg, Germany

[3]Department of Physics, University of Oxford

[4]Research School of Astronomy and Astrophysics,
Australian National University

email: tobympotter@gmail.com

Abstract. SNR 1987A is the expanding remnant from the brightest supernova since the invention of the telescope. The remnant has been monitored extensively in the radio at variety of wavelengths and provides a wealth of data on which to base a simulation. Questions to be answered include estimating the efficiency of particle acceleration at shock fronts, determining the cause of the one-sided radio morphology for SNR 1987A and investigating the gas properties of the pre-supernova environment. We attempt to address these questions using a fully three-dimensional model of SNR 1987A.

Keywords. Supernovae, SN 1987A, particle acceleration, three-dimensional, simulation

1. Our model

We use the hydrodynamics code FLASH (Fryxell *et al.* 2000) to simulate SNR 1987A in a 256^3 grid. We derive radio emission in post-processing by using semi-analytic models of sub-Diffusive Shock Acceleration (Kirk *et al.* 1996) to place inverse power-law phase-space distributions of electrons at the shock front. The magnetic field is assumed to be amplified at the shock using physics from Bell (2004). Both magnetic field and electron distributions are advected downstream using the velocity field from the hydrodynamics simulation. Features of the model include a custom Truelove and McKee (1999) progenitor with n = 9, $E = 1.5 \times 10^{44}$ J and $M = 10 M_\odot$; a free blue supergiant (BSG) wind; a shocked BSG wind, a central equatorial ring and the hourglass. Turbulence was excited in the expanding shock by perturbing the initial density by a random number chosen from a lognormal distribution with mean 1.0 and variance 5.0 (Sutherland and Bicknell 2007).

2. Results and conclusions of the study

Optimal fits to the asymmetric radio morphology were obtained if 1.55 times more kinetic energy was placed in the eastern hemisphere of the progenitor. The best-fit radius for the free BSG wind was 3.2×10^{15} m ($0.43''$) in the equatorial plane and 3.78×10^{15} m ($0.51''$) in the polar direction. For the HII region the optimal half-opening angle was $\theta_{\rm ho} = 15 \pm 5°$. Fits to the radius (see Figure 1) yield a peak HII region particle density of

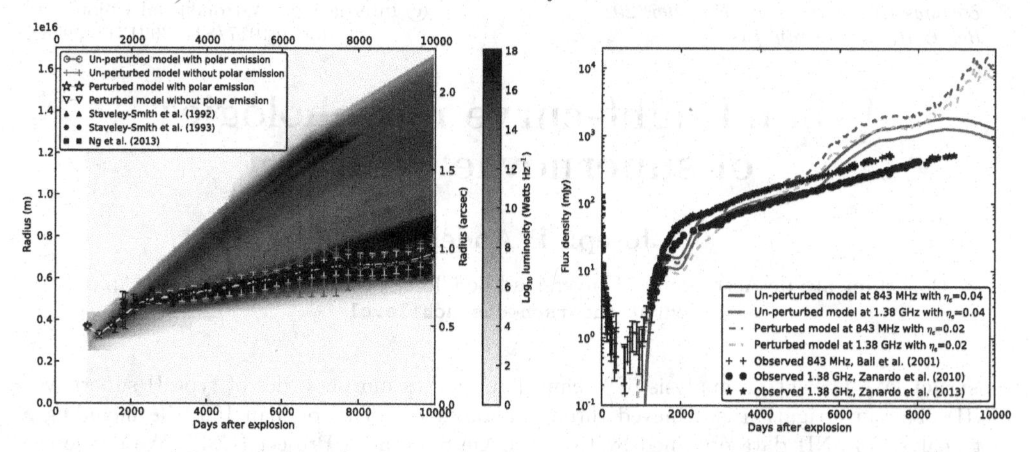

Figure 1. Radius and flux density from simulations of SN 1987A. The red background on the left is the time-varying radial distribution of simulated radio emission. Overlaid is the expectation of radius with error bars derived from expectation of variance. On the right the simulated flux density has been scaled by an injection efficiency n_e in the range (2-4%).

$(7.11 \pm 1.78) \times 10^7$ m^{-3}, given an abundance of 3.5 particles per hydrogen atom. These results are consistent to within errors of Zhekov (2010).

For the equatorial ring we used a thin crescent torus (Plait *et al.* 1995), with a peak particle density of 8.0×10^8 m^{-3} for the smooth component. The torus was azimuthally interspersed with 21 Gaussian clouds (Dewey *et al.* 2012) of radius 2.25×10^{14} m ($0.03''$), and a peak particle density of 3.14×10^{10} m^{-3}, consistent with Mattila *et al.* (2010). Radius measurements for both perturbed and unperturbed models agree well with the observational data from Ng *et al.* (2008, 2013) (see Figure 1). The simulations correctly model the apparent slowing in shockwave velocity seen around day 7500, and predict an accelerating expansion after day 9500. Flux density measurements (see Figure 1) from both un-perturbed and perturbed models show that the efficiency of particle acceleration for electrons is between 2 and 4%. Predicted radio images show that the asymmetric radio morphology will reverse by day 10000 as the shock leaves the eastern lobe first.

References

Bell, A. R. 2004, *MNRAS*, 353, 550-558

Dewey, D., Dwarkadas, V. V., Haberl, F., Sturm, R., & and Canizares, C. R. 2012, *ApJ*, 752, 103

Fryxell, B., Olson, K., Ricker, P., Timmes, F. X., Zingale, M., Lamb, D. Q., MacNeice, P., Rosner, R., Truran, J. W., & Tufo, H. 2000, *ApJS*, 131, 273

Kirk, J. G., Duffy, P., & Gallant, Y. A. 1996, *A&A*, 314, 1010–1016

Mattila, S., Lundqvist, P., Groningsson, P., Meikle, P., Stathakis, R., Fransson, C., & Cannon, R. 2010, *ApJ*, 717, 1140–1156

Ng, C. Y., Gaensler, B. M., Staveley-Smith, L., Manchester, R. N., Kesteven, M. J., Ball, L., & Tzioumis, A. K. 2008, *ApJ*, 684, 481–497

Ng, C. Y., *et al.* 2013 *in-preparation*

Plait, P. C., Lundqvist, P., Chevalier, R. A., & Kirshner, R. P. 1995, *ApJ*, 439, 730–751

Potter, T. M., Staveley-Smith, L., Ng, C. Y., Ball, L., Gaensler, B. M., Kesteven, M. J., Manchester, R. N. Tzioumis, A. K., & Zanardo, G. 2009, *ApJ*, 705, 261–271

Sutherland, R. S. & Bicknell, G. V. 2007, *ApJS*, 173, 37–69

Truelove, J. K. & McKee, C. F. 1999, *ApJS*, 120, 299–236

Zhekov, S. A. Park, S., McCray, R., Racusin, J. L., & Burrows, D. N. 2010, *MNRAS*, 407, 1157–1169

Supernova Environmental Impacts
Proceedings IAU Symposium No. 296, 2013
A. Ray & R. A. McCray, eds.

doi:10.1017/S174392131300968X

V-band light-curve morphologies of supernovae type II

Joseph P Anderson

Departamento de Astronomía, Universidad de Chile, Casilla 36-D, Santiago, Chile
email: anderson@das.uchile.cl

Abstract. We present an analysis of V-band light-curves morphologies of type II supernovae (SNII). This investigation is achieved through photometry of more than 100 SNe including a first analysis of SNII data obtained by the Carnegie Supernova Project (CSP). We define the important observables and present correlations between SNe absolute magnitudes and light-curve decline rates: we find that brighter SNII tend to have faster declining light-curves at all epochs.

Keywords. (stars:) supernovae: general

1. Introduction

The original spectroscopic classification of SNe into types I and II was based on the detection of hydrogen (Minkowski 1941). It is now believed that hydrogen rich SNII arise from the core-collapse of massive stars which explode with a significant fraction of their hydrogen envelopes retained. However, SNII show a wide range of light-curve properties from the steeply declining linear; SNIIL, to the more abundant plateau; SNIIP (Barbon, Ciatti & Rosino 1979). In these proceedings an initial analysis of the diversity of V-band light-curve morphologies is presented for a sample of 118 SNII obtained through the Carnegie Supernova Project (CSP), together with earlier follow-up campaigns.

2. SNII V-band light-curve correlations

In figure 1 (left panel) the defined light-curve measurements are displayed. Three magnitudes are defined: Mmax, the maximum magnitude; Mend, the magnitude at the end of the 'plateau'; and Mtail, the magnitude at the start of the radioactive tail. Three decline rates are also defined: s1, the initial decline from maximum; s2, the 'plateau' decline; and s3, the decline rate of the radioactive tail. Distributions of these values are built and correlations between the various parameters are investigated.

It is found that the magnitude at the end of the 'plateau', a value often used for SNII studies in the literature, does not corelate with decline rates. This is shown in the right panel of fig. 1. However, as shown in fig. 2, Mmax does correlate with SN decline rates: brighter SNe decline more quickly both during the 'plateau' and during the radioactive tail. This second correlation is intriguing as it implies that brighter SNe decline more quickly than that expected from the decay of ^{56}Co, which is believed to power the light-curve at late times. If this result is verified then it would constrain the mass and density of the ejecta of bright more linear SNe, to be significantly lower than that of dimmer SNIIP.

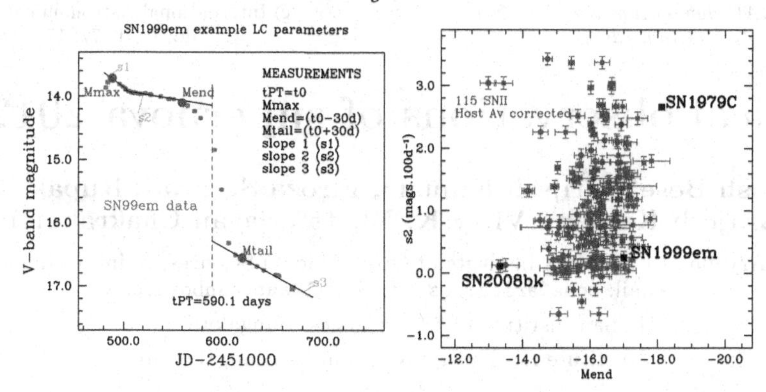

Figure 1. ***Left***: Schematic showing the defined SNII light-curve parameters for measurement: three absolute magnitudes; Mmax, Mend and Mtail, plus 3 decline rates; s1, s2 and s3. ***Right***: Correlation between Mend and s2.

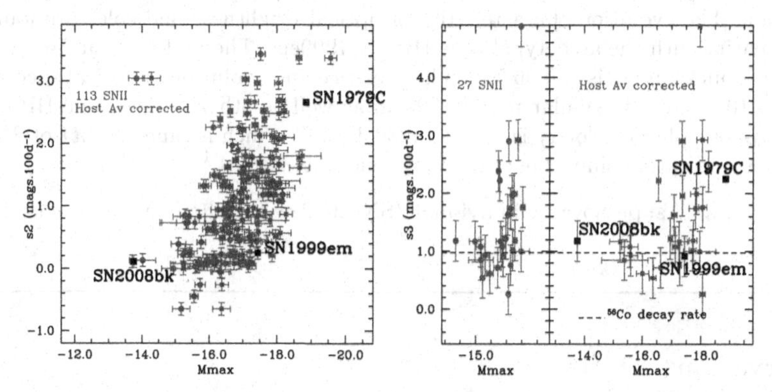

Figure 2. ***Left***: Correlation between Mmax and s2. ***Right***: Correlation between Mmax and s3.

3. Conclusions

In this initial study of SNII light-curve diversity it has been found that brighter SNe (at maximum) decline more quickly at all epochs. This can be rephrased to say SNIIL are in general more luminous than SNIIP, as has been suggested previously by Patat *et al.* (1994). However, here we vastly increase the statistics finding a continuum of events from low luminosity SNe which rise during s2, through to luminous SNIIL which decline quickly. The full analysis of this sample will provide further constraints on the diversity of SNII events, and the physical differences which drive differences in their transient evolution.

Acknowledgments: J.A. acknowledges support from FONDECYT grant 3110142, and grant ICM P10-064-F (Millennium Center for Supernova Science), with input from 'Fondo de Innovacin para la Competitividad, del Ministerio de Economa, Fomento y Turismo de Chile'.

References

Barbon, R, Ciatti, F & Rosino, L, 1979 *A&A*, 72, 287
Minkowski, R, 1941 *PASP*, 53, 224
Patat, F., *et al.* 1994 *A&A* 282, 731

Supernova Environmental Impacts
Proceedings IAU Symposium No. 296, 2013
A. Ray & R. A. McCray, eds.

doi:10.1017/S1743921313009691

Optical observations of supernova 2012aw

Subhash Bose[1], Brijesh Kumar[1], Firoza Sutaria[2], Rupak Roy[1], Brajesh Kumar[1], Vijay K. Bhatt[1], Sayan Chakraborti[3]

[1]Aryabhatta Research Institute of Observational Sciences, Nainital, India
email: `bose@aries.res.in`; `email@subhashbose.com`

[2]Indian Institute of Astrophysics, Bangalore, India

[3]Institute for Theory and Computation, Harvard

Abstract. We present optical *UBVRI* photometric and low-resolution spectroscopic follow-up observations of a type II SN 2012aw in a nearby (~10 Mpc) galaxy M95 during 4 to 270 days post-explosion. The evolution characteristics of optical brightness and color are found to have striking similarity with the archetypal type IIP SN 1999em. The mid-plateau M_V is −16.7 mag and the ejected nickel mass is ~ 0.06 M$_\odot$. The presence and evolution of optical spectral features during 7d to 104d are also similar to SN 1999em as well as other normal type IIP events. The mid-plateau photospheric velocity is around 4200 km s^{-1} which is same as that of SN 2004et at similar phases, indicating similar energy of explosion i.e. 2×10^{51} erg s^{-1}.

Keywords. (stars:) supernovae : individual (SN 2012aw); techniques: photometric, spectroscopic

1. Observations

The broadband photometric data in *UBVRI* Johnson-Cousins system are collected using the 104-cm Sampurnanand Telescope which is operated by the Aryabhatta Research Institute of Observational Sciences, Nainital, India (Sagar 2006). The data ranges from 4d to 270d. The long-slit low-resolution (~10Å) spectra in the visible range (4000 – 8000Å) were collected at 14 phases during 7d to 270d post explosion; nine from 2m IUCAA Girawali Observatory Telescope, Pune and five from 2m Himalayan Chandra Telescope, Hanle. Data reduction is done using IRAF in standard manner.

2. Preliminary results

The *UBVRI* light-curve of SN 2012aw is presented in Fig. 1. The plateau phase of about 100d duration is clearly visible and the light-curve shape matches well with SN 1999em. The optical bolometric light-curves of other well studied SNe are also overplotted. A comparison of nebular phase light-curve with SNe 1999em and 2004et suggests that the ^{56}Ni produced in 2012aw are similar to these events. Comparing with SN 1987, we derive the mass of ^{56}Ni for SN 2012aw to be ~ 0.058 M$_\odot$.

Fig. 2 [Left] shows low-resolution spectra at 14 phases between 7d and 104d. Using the Na I D absorption dips and employing empirical relation from Poznanski, Prochaska, & Bloom (2012), the total reddening $E(B-V)$ in direction to SN 2012aw has been estimated to be ~ 0.075 mag. The `SYNOW` modeling of spectra are done for all 14 spectra to identify lines and to determine photospheric velocities. Fig. 2 [Right] shows the `SYNOW` modeling for the phases 7d and 61d along with all identified spectral features. Striking similarity of spectral features with that of SN 1999em is noticed. The velocity profile of

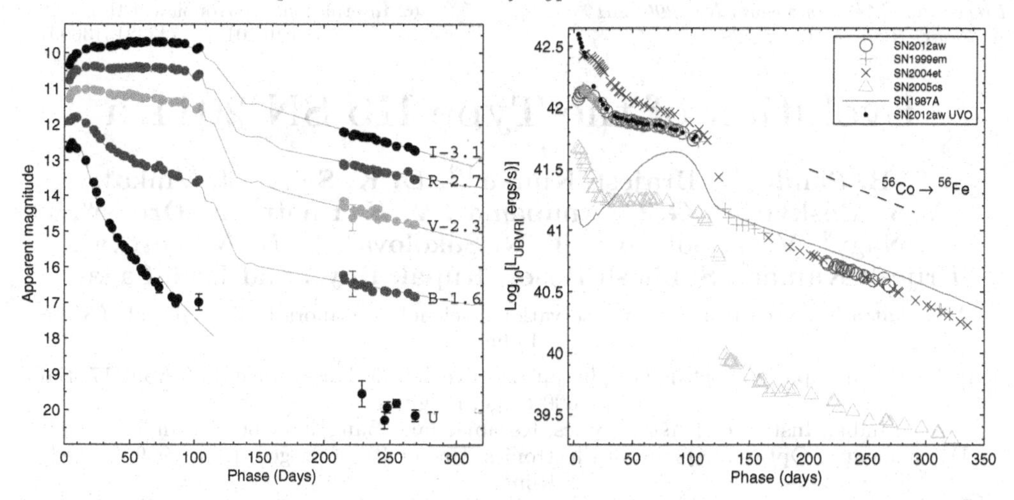

Figure 1. The apparent magnitude (left) and bolometric (right) light-curve of SN 2012aw. The apparent light curve of archetypal type IIP SN 1999em is shown in grey solid lines. The UV-optical bolometric curve for SN 2012aw is generated by incorporating UV data from Bayless *et al.* (2012)

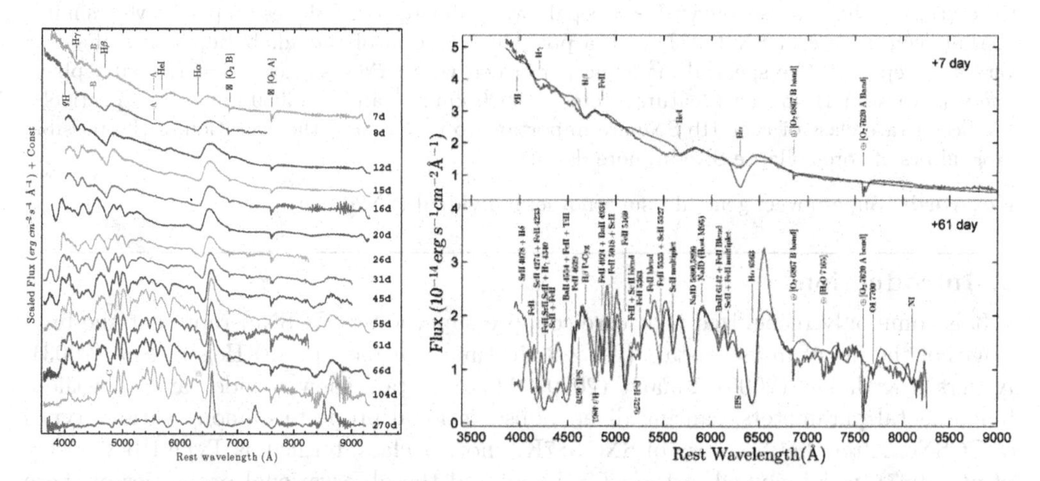

Figure 2. Left : The Doppler-corrected spectra of SN 2012aw are shown for 14 phases during 7d to 270d. Right : The SYNOW modeling is shown for 7d and 61d spectrum. Model spectra are shown with thick solid line, while the observed ones are in thin solid line.

SN 2012aw estimated using Fe II lines (4924, 5018, 5169Å) is found to be matching well with SN 2004et, though it is consistently higher than SN 1999em at all phases.

References

Bayless A. J., *et al.* 2012, *arXiv*, arXiv:1210.5496

Poznanski D., Prochaska J. X., & Bloom J. S. 2012, *MNRAS*, 426, 1465

Sagar R. 2006, *BASI*, 34, 37

Supernova Environmental Impacts
Proceedings IAU Symposium No. 296, 2013
A. Ray & R. A. McCray, eds.

doi:10.1017/S1743921313009708

Evolution of the Type IIb SN 2011fu

S. B. Pandey[1]†, Brajesh Kumar[1,2], D. K. Sahu[3], J. Vinko[4], A. S. Moskvitin[5], G. C. Anupama[3], V. K. Bhatt[1], A. Ordasi[6], A. Nagy[4], V. V. Sokolov[5], T. N. Sokolova[5], V. N. Komarova[5], Brijesh Kumar[1], Subhash Bose[1], Rupak Roy[1], and Ram Sagar[1]

[1]Aryabhatta Research Institute of observational sciencES, Manora Peak, Nainital 263129, India

[2]Institut d'Astrophysique et de Géophysique, Université de Liège, Allée du 6 Août 17, Bât B5c, 4000 Liège, Belgium

[3]Indian Institute of Astrophysics, Koramangala, Bangalore 560 034, India

[4]Department of Optics & Quantum Electronics, University of Szeged, Dóm tér 9, Szeged, Hungary

[5]Special Astrophysical Observatory, Nizhnij Arkhyz, Karachaevo-Cherkesia, 369167 Russia

[6]Department of Experimental Physics, University of Szeged, Dóm tér 9, Szeged, Hungary

Abstract. The *UBVRI* photometric follow-up of SN 2011fu has been initiated a few days after the explosion, shows a rise followed by steep decay in all bands and shares properties very similar to that seen in case of SN 1993J, with a possible detection of the adiabatic cooling phase at very early epochs. The spectral modeling performed with SYNOW suggests that the early-phase line velocities for H and Fe II features were ~ 16000 km s^{-1} and ~ 14000 km s^{-1}, respectively. Studies of rare class of type IIb SNe are important to understand the evolution of the possible progenitors of core-collapse SNe in more details.

Keywords. Supernovae: general - supernovae: individual (SN2011fu)

1. Introduction

It is commonly recognized that core-collapse supernovae (CCSNe) represent the final stages of the life of massive stars ($M > 8$–10 times to that of Sun) Heger *et al.* (2003), Anderson & James (2009), Smartt (2009). Massive stars show a wide variety in these fundamental parameters, causing diverse observational properties among various types of CCSNe. After the discovery of SN 1987K, another class, termed as Type IIb Woosley *et al.* (1987), was included in the CCSN zoo, and the observational properties of these SNe closely resemble those of Type II SNe during the early phases, while they are more similar to Type Ib/c events at later epochs. Type IIb and Type Ib/c SNe are collectively known as "stripped envelope" CCSNe as the outer envelopes of hydrogen and/or helium of their progenitors are partially or completely removed before the explosion. There are several studies about the discovery of the progenitors of Type IIb SNe but the debate about how they manage to keep only a thin layer of hydrogen, is still on and there are very few SNe of this class have been studied in great detail.

In this presentation, we discuss the results from photometric and spectroscopic monitoring of SN 2011fu starting shortly after the discovery and extending up to nebular phases Kumar *et al.* (2013).

2. Overview

The photometric and low-resolution spectroscopic monitoring of the Type IIb SN 2011fu, presented are the earliest ones reported for this event. The early photometric

† shashi@aries.res.in

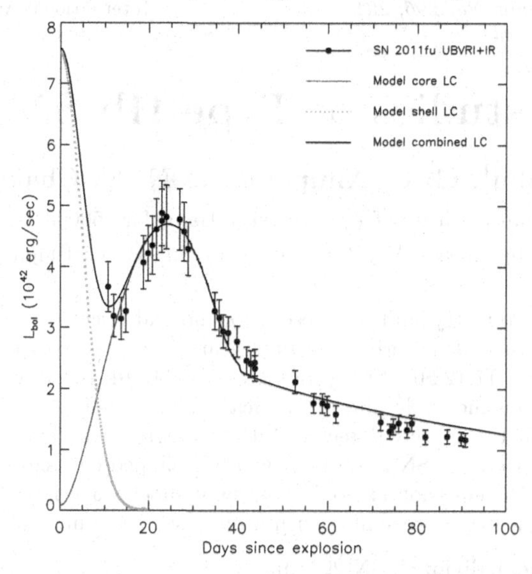

Figure 1. Comparison of the observed bolometric light curve (dots) with the best-fit two-component diffusion-recombination model. The red and green curves show the contribution from the He-rich core and the low-mass H-envelope, respectively, while the black line gives the combined light curve.

observations strongly suggest the presence of the early-time decline of the light curve (which is thought to be related to the shock break-out phase) as seen in case of SN 1993J. The color evolutions of SN 2011fu were studied using our *UBVRI* band observations. Our data showed that during the very early phases the $B - V$ color was very similar to that in SN 1993J. SN 2011fu seems to be the most luminous event in comparison to a sample of other well-observed type IIb SNe.

The quasi-bolometric light curve was computed by integrating the extinction-corrected flux values and the Infra-red contribution was approximated by assuming black-body flux distributions. The bolometric light curve was fitted by the semi-analytic light curve model of Arnett & Fu (1989), suggesting that the progenitor had an extended ($\sim 1 \times 10^{13}$ cm), low-mass ($\sim 0.1\ M_\odot$) H-rich envelope on top of a dense, compact ($\sim 2 \times 10^{11}$ cm), more massive ($\sim 1.1\ M_\odot$) He-rich core. The nickel mass synthesized during the explosion was found to be $\sim 0.21\ M_\odot$, slightly larger than seen in case of other Type IIb SNe.

The spectra of SN 2011fu taken at eight epochs were analyzed using the multi-parameter resonance scattering code `SYNOW`. The derived parameters describe the evolution of the velocities related to various atoms/ions and the variation of the black-body temperature of the pseudo-photosphere.

References

Anderson, J. P. & James, P. A. 2009, *MNRAS*, 399, 559

Arnett, W. D. & Fu, A. 1989, *ApJ*, 340, 396

Heger, A., Fryer, C. L., Woosley, S. E., Langer, N., & Hartmann, D. H. 2003, *ApJ*, 591, 288

Kumar, B., Pandey, S. B., Sahu, D. K., Vinko, J., Moskvitin, A. S., Anupama, G. C., Bhatt, V. K., Ordasi, A., Nagy, A., Sokolov, V. V., Sokolova, T. N., Komarova, V. N., Kumar, B., Bose, S., Roy, R., & Sagar, R. 2013, *Accepted to MNRAS*, arXiv:1301:6538

Smartt, S. J. 2009, *ARAA*, 47, 63

Woosley, S. E., Pinto, P. A., Martin, P. G., & Weaver, T. A. 1987, *ApJ*, 318, 664

Supernova Environmental Impacts
Proceedings IAU Symposium No. 296, 2013
A. Ray & R. A. McCray, eds.

doi:10.1017/S174392131300971X

Optical studies of Type IIb SN 2011dh

D. K. Sahu[1], G. C. Anupama[1] & N. K. Chakradhari[2]

[1]Indian Institute of Astrophysics, Bangalore 560034, India

[2]S.O.S. in Physics, Pt. Ravishankar Shukla Univ., Raipur, India

Abstract. UBVRI photometry and low resolution optical spectroscopy of the type IIb SN 2011dh in M51 are presented, covering the first year after the explosion. The peak absolute magnitude in V-band of -17.12±0.18 mag indicates SN 2011dh to be a normal bright type IIb event. The peak quasi-bolometric luminosity indicates that $\sim$ 0.06 M⊙ of ^{56}Ni was synthesized in the explosion. The He I lines were detected in the spectra much before the maximum light in B-band. The nebular spectra of SN 2011dh show a box shaped emission in the red wing of [OI] 6300, 6363 line due to Hα emission excited because of shock-wave interaction. The analysis of the nebular spectra indicates a progenitor with a main sequence mass of 10-15 M⊙.

Keywords. supernovae: individual: SN 2011dh

1. Introduction

SN 2011dh was discovered by A. Riou on 2011 June 01.89 in the nearby spiral galaxy M51 and was classified as a type IIb supernova (Silverman *et al.* 2011, Marion *et al.* 2011). SN 2011dh has been followed extensively from radio to X-rays (Arcavi *et al.* 2011; Soderberg *et al.* 2012). The search for progenitor star in the HST/ACS archive, led to the detection of a luminous star at the supernova location (Van Dyk *et al.* 2011).

2. Results

Imaging and spectroscopic observations of SN 2011dh were obtained, during one year after explosion, with the 2m Himalayan Chandra Telescope of Indian Astronomical Observatory, Hanle, India.

2.1. *Photometry*

The light curves of SN 2011dh in U, B, V, R and I bands are plotted in Figure 1(a). The maximum in B-band occurred on JD 2455732.6±0.35, ∼19.6 days after explosion, at an apparent magnitude of 13.39±0.02 mag. A significant steepening of the late time light curve in B-band is noticed. Light curve decline rate changes from 1.09±0.15 mag/100 days in the early phase to 1.71±0.13 mag/100 days during the late phase.

V-band peak absolute magnitude of SN 2011dh estimated adopting reddening $E(B-V) = 0.035$ and distance 8.4±0.7 Mpc is -17.12 ± 0.18 mag, which is ~ 1 mag fainter than the mean peak absolute magnitude of the entire sample of the stripped-envelope CCSNe, and is $\sim$ 0.3 mag fainter than the type IIb sample (Richardson *et al.* 2006). The peak bolometric luminosity is 1.267×10^{42} erg sec^{-1}, leading to an estimate of 0.06 M⊙ of ^{56}Ni synthesized in the explosion, using Arnett's rule (Arnett 1982).

2.2. *Spectroscopy*

The pre-maximum spectrum of SN 2011dh shows blue continuum with well developed P-Cygni absorption of hydrogen Balmer lines, CaII H&K, CaII NIR triplet (refer

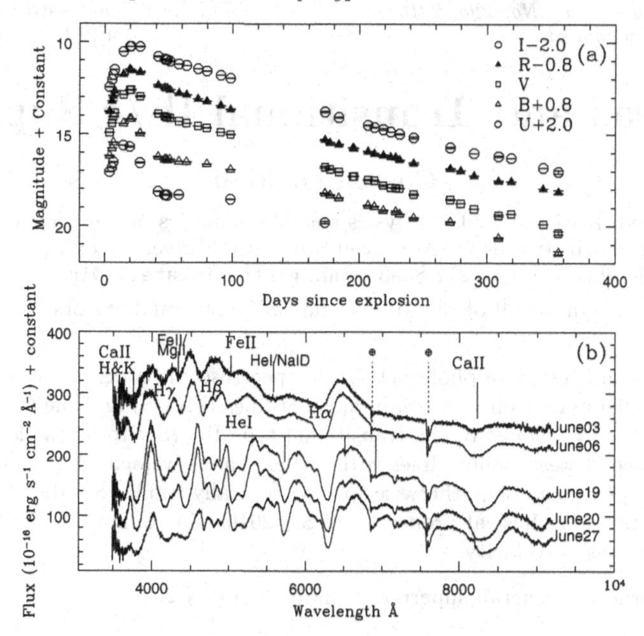

Figure 1. (a) UBVRI light curves of SN 2011dh, (b) Pre-maximum and early post-maximum spectra of SN 2011dh.

Figure 1(b)). The spectrum at B-band maximum shows that the supernova has already entered into a phase wherein the lines due to HeI become prominent. During the transitional phase the continuum becomes redder, the Balmer lines become sharper and lines due to HeI become stronger. The nebular phase spectra of SN 2011dh are dominated by emission lines of MgI], [OI], [CaII], OI, blend of [FeII] lines at $\sim$ 5000Å and CaII NIR triplet. A box-shaped emission in the red wing of [OI] 6300, 6364 line due to Hα is clearly identified. Presence of circumstellar material as revealed by X-ray (Campana & Immler 2012) and radio observations (Bietenholz *et al.* 2012), and high expansion velocity of the forward shock, indicate that shock wave interaction may be the most plausible mechanism for the observed Hα emission in the late phase.

The minimum mass of oxygen required to produce the observed [OI] emission in the nebular spectra is estimated as 0.22 M$\odot$. With metallicity a few tenths that of solar, the oxygen mass indicates the progenitor is a low-mass star of $\sim$13– 15 M$\odot$, in a binary system.

References

Arcavi, *et al.* 2011 *ApJL*, 742, 18
Arnett, W. D. 1982 *ApJ*, 253, 785
Bietenholz, *et al.* 2012 *ApJ*, 751, 125
Campana, S. & Immler, S. 2012 *MNRAS*, 427, 70
Marion, *et al.* 2011 *ATel*, 3435, 1
Richardson, D., Branch, D., Baron, E., 2006 *AJ*, 131, 2233
Silverman, *et al.* 2011 *ATel*, 3398, 1
Soderberg, *et al.* 2012 *ApJ*, 752, 78
Taubenberger, *et al.* 2011 *MNRAS*, 413, 2140
Van Dyk, *et al.* 2011 *ApJL*, 741, 28

Supernova Environmental Impacts
Proceedings IAU Symposium No. 296, 2013
A. Ray & R. A. McCray, eds.

doi:10.1017/S1743921313009721

SN 2010as and Transitional Ib/c Supernovae

Gastón Folatelli

Kavli Institute for the Physics and Mathematics of the Universe,
Todai Institutes for Advanced Study, the University of Tokyo,
Kashiwa, Japan 277-8583 email: gaston.folatelli@ipmu.jp

On behalf of the MCSS, the CSP, and collaborators

Abstract. We present intensive photometric and spectroscopic observations of SN 2010as carried out by the Millennium Center for Supernova Studies (MCSS) and the Carnegie Supernova Project (CSP). The SN belongs to the transitional type Ibc (SN Ibc) that is characterized by the slow appearance of weak helium lines with low expansion velocities. We find a wide variety of photometric properties among otherwise spectroscopically similar SN Ibc. A hydrodynamical model is used to provide physical properties of SN 2010as in comparison with the bolometric light curve and expansion velocity.

Keywords. supernovae: general, supernovae: individual (SN 2010as)

1. Introduction

SN 2010as was discovered in NGC 6000 on March 19.2 UT by the CHilean Automatic Supernova sEarch (CHASE; Maza *et al.* 2010) and classified as a type Ibc SN (Stritzinger *et al.* 2010). Here we compare its observed properties with those of other objects in this transitional class, namely SN 1999ex, SN 2005bf, and SN 2007Y.

2. Spectroscopic similarities among SN Ibc

A spectral time series between −10 and +309 days relative to maximum light was obtained for SN 2010as by the MCSS and CSP, with nearly daily coverage before maximum. The pre-maximum spectra are of type Ic, dominated by Ca II and Fe II lines, with possible Hα or Si II at ≈6200 Å. Then He I lines develop marking the transition to a SN Ib. Fig. 1 (left) compares the spectra with those of SNe 2005bf (Folatelli *et al.* 2006) and 2007Y (Stritzinger *et al.* 2009) at three epochs. The three SNe are very similar before maximum, but then SN 2007Y develops strong He I lines. The other two SNe remain similar, with SN 2010as showing stronger Ca II features. An additional similarity is the peculiarly low He I line velocities observed at all times (see Fig. 1, right).

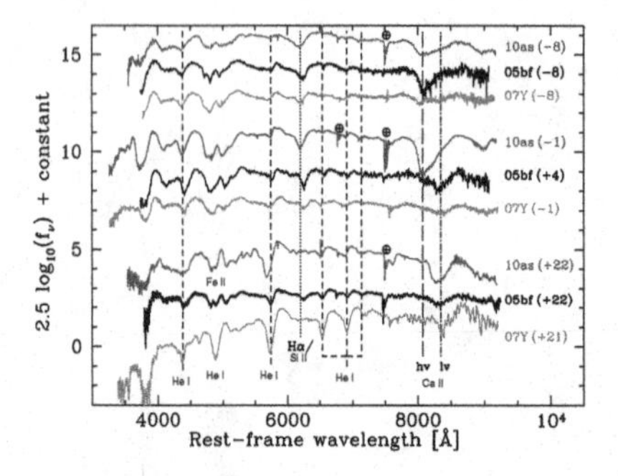

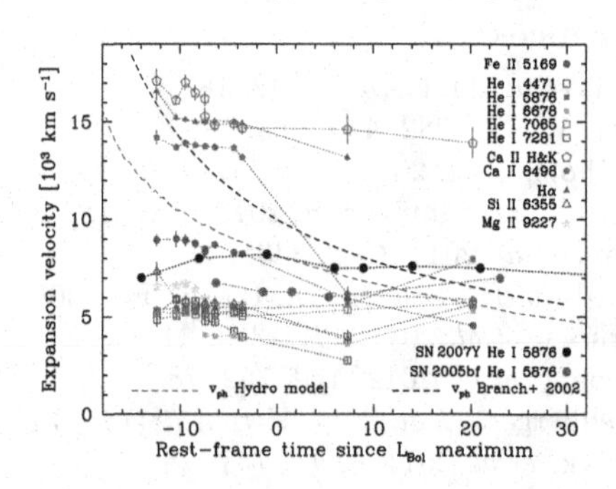

Figure 1. (Left) Spectral comparison with SNe Ibc 2005bf and 2007Y at three different epochs. (Right) Expansion velocities measured from spectral lines (dots) and models (lines).

Table 1. Absolute peak magnitudes of selected SN Ibc.

SN	M_B	M_V	M_J	M_K
2010as	-18.1 ± 0.5	-18.5 ± 0.4	-18.4 ± 0.3	-18.2 ± 0.3
1999ex	-17.4 ± 0.3	-17.9 ± 0.3		
2005bf	-18.5 ± 0.3	-18.6 ± 0.3		
2007Y	-16.2 ± 0.6	-16.4 ± 0.6	-16.7 ± 0.6	-17.0 ± 1.0

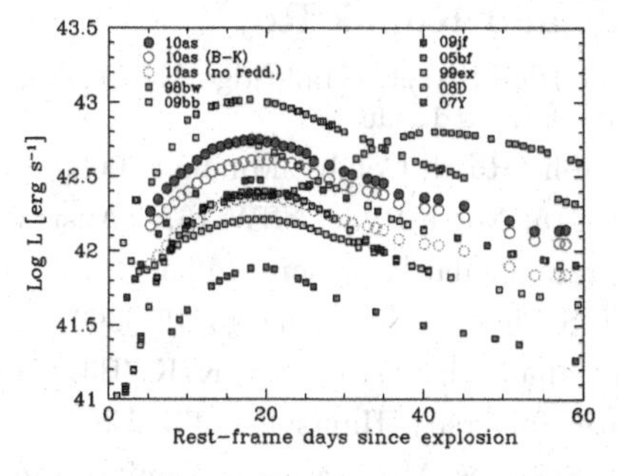

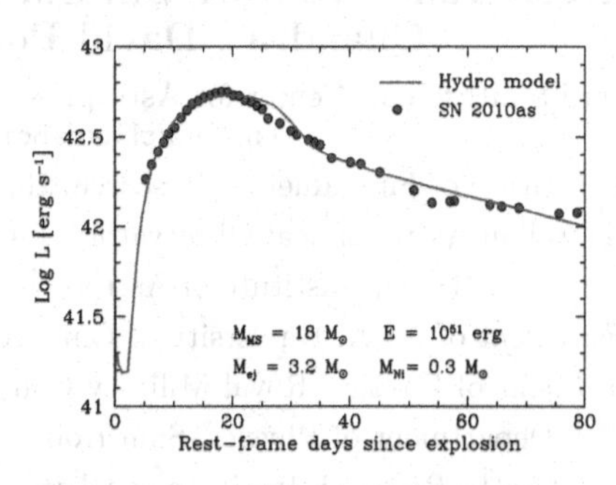

Figure 2. (Left) Bolometric light curves of core-collapse SNe including some SN Ibc. (Right) Bolometric luminosity for a model (solid line) with the indicated physical parameters, compared with the data (dots).

3. Photometric variety

Follow-up of SN 2010as was obtained in $BVRIg'r'i'z'$ bands at CTIO, and in JHK bands at ESO. The SN was caught 10 days before B-band maximum light and followed for over 100 days. Observed colors corrected for Galactic reddening were compared with those of a sample of SNe Ib and Ic from the CSP (Stritzinger *et al.*, in prep.) to derive a host-galaxy reddening of $E(B-V)_{\rm host} = 0.35$ mag†. Adopting a distance of 32.7±3.6 Mpc to the host (NED), reddening-free absolute peak magnitudes of SN 2010as are relatively luminous as compared with other SN Ibc (see Table 1).

A bolometric light curve was obtained by integrating the extinction-corrected optical flux and adding the extrapolation of a black-body fit toward the IR, and a straight line in the UV from the B-band point to zero flux at 2000 Å. The resulting peak luminosity of $L_{\rm Bol} = 6.3 \times 10^{42}$ erg s^{-1} is comparable with that of the peculiar SN 2005bf, although the total radiated energy is much larger for the latter SN. A hydrodynamical model (Bersten *et al.* 2011) was computed to reproduce the bolometric luminosity (see assumed physical parameters in the right panel of Fig. 2). A relatively large amount of ^{56}Ni, $M_{\rm Ni} = 0.3$ $M_\odot$, is required to explain the luminous peak. The explosion energy of $E = 10^{51}$ erg slightly overestimates the measured expansion velocities (see right panel of Fig. 1).

References

Bersten, M. C., *et al.* 2011, *ApJ*, 729, 61
Folatelli, G., *et al.* 2006, *ApJ*, 641, 1039
Maza, J., *et al.* 2010, *CBET*, 2215, 1
Stritzinger, M., *et al.* 2009, *ApJ*, 696, 713
Stritzinger, M., *et al.* 2010, *CBET*, 2221, 1
Stritzinger, M., *et al.* 2013 in prep.

† In agreement with the measured equivalent width of Na I D of 2.1 Å.

Supernova Environmental Impacts
Proceedings IAU Symposium No. 296, 2013
A. Ray & R. A. McCray, eds.

doi:10.1017/S1743921313009733

The strange case of SN 2011ja and its host

Sayan Chakraborti[1], Alak Ray[2], Randall Smith[1], Stuart Ryder[3], Naveen Yadav[2], Firoza Sutaria[4], Vikram V. Dwarkadas[5], Poonam Chandra[6], David Pooley[7], and Rupak Roy[8]

[1]Harvard Smithsonian Center for Astrophysics, 60 Garden Street, Cambridge MA 02138, USA
email: `schakraborti@fas.harvard.edu`

[2]Tata Institute of Fundamental Research, Homi Bhabha Road, Colaba, Mumbai 400005, India

[3]Australian Astronomical Observatory, P.O. Box 915, North Ryde, NSW 1670, Australia

[4]Indian Institute of Astrophysics, Koramangala, Bangalore, India

[5]Department of A&A, University of Chicago, 5640 S Ellis Avenue, Chicago, IL 60637, USA

[6]Department of Physics, Royal Military College of Canada, Kingston, ON, K7K 7B4, Canada

[7]Department of Physics, Sam Houston State University, Huntsville, TX, USA

[8]Aryabhatta Research Institute of Observational Sciences, Manora peak, Nainital, India

Abstract. SN 2001ja was observed twice in three months using the Chandra X-Ray Observatory. The X-ray flux could be due to interaction with the circumstellar medium, perhaps dominated by the reverse shock heated thermal plasma, or from inverse Compton scattering at the forward shock. In both cases, for a steady wind-like circumstellar density profile, the X-ray flux is expected to fall off as a power law or faster. But the flux from the position of SN 2011ja, increased by a factor of three between these observations. In this presentation, we investigated possible reasons, including contamination from other astrophysical sources such as a X-Ray Binary, within the Chandra's resolution, in the host galaxy using our observations, modelling and pre-explosion Chandra/XMM data.

Keywords. Stars: Mass Loss — Supernovae: Individual: SN 2011ja — shock waves — circumstellar matter — radio continuum: general — X-rays: general

---

1. Introduction

Type IIP supernovae have prominent P Cygni features of hydrogen at peak luminosity. Doggett and Branch (1985) show that their optical light curves have plateaus of ~ 100 days. The plateau arises as their progenitors retain extended hydrogen envelopes until the time of core collapse. Popov (1993) found that the duration of the plateau phase has a strong dependence on the mass of the hydrogen envelope. These arguments and direct pre-explosion imaging by Smartt *et al.* (2009) suggest that these stars exploded as red supergiants. Smith *et al.* (2011) found that half of the core collapse supernovae in their sample are type IIP. Red supergiants are found in the Local Group with masses up to 25 $M_{\odot}$, but Smartt *et al.* (2009) did not find any with masses greater than $17 M_{\odot}$ as progenitors of type IIP supernovae. Many solutions have been suggested for this *red supergiant problem.* O'Connor and Ott (2011) suggested that the ZAMS mass, metallicity, rotation and mass-loss prescription controls the compactness of the stellar core at bounce which determines whether a core-collapse supernova will fail. Walmswell and Eldridge (2012) have suggested circumstellar dust as a solution to the problem of the missing massive progenitors. In this situation, understanding the nature, amount and variability of mass loss from the progenitors of type IIP supernovae is crucial for resolving this puzzle.

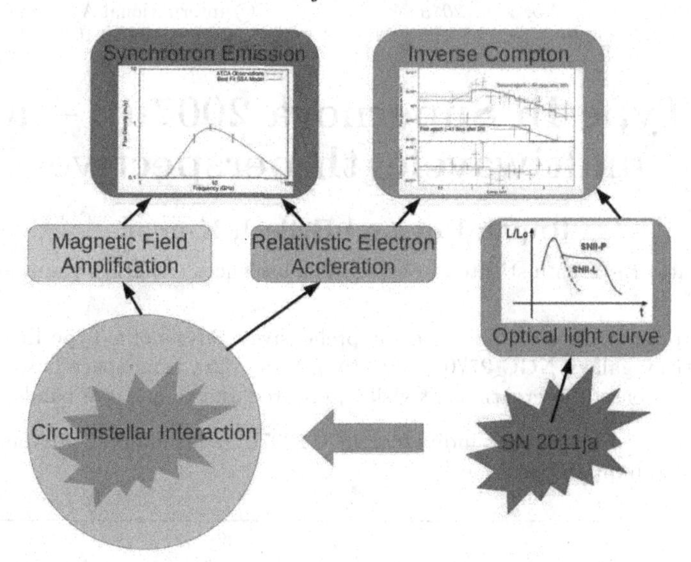

Figure 1. Radio synchrotron and X-ray inverse Compton emission from SN 2011ja.

2. Results

Supernova ejecta shocks circumstellar matter set up by stellar wind. Since the ejecta ($\sim 10^4$ km s^{-1}) moves about a thousand times faster than the stellar wind ($\sim$ 10 km s^{-1}), it probes a millennium of red supergiant mass loss history in a year. This interaction accelerates particles to relativistic energies, which then lose energy via synchrotron radiation in the shock-amplified magnetic fields and inverse Compton scattering against optical photons from the supernova (Fig 1). Chevalier *et al.* (2006) have shown that these processes produce separate signatures in the radio and X-rays. Chakraborti *et al.* (2012) have demonstrated that combining radio and X-ray spectra allows one to break the degeneracy between the efficiencies of shock acceleration and field amplification. Chakraborti *et al.* (2013) use X-rays observations of SN 2011ja from Chandra and radio observations from the Australia Telescope Compact Array (ATCA) to study the relative importance of particle acceleration and magnetic field amplification for producing the non thermal radiation from SN 2011ja. Chakraborti *et al.* (2013) use a multiple Chandra observation of SN 2011ja to establish variable mass loss from the progenitor.

References

J. B. Doggett & D. Branch, *AJ* **90**, 2303–2311 (1985).

D. V. Popov, *ApJ* **414**, 712–716 (1993).

S. J. Smartt, J. J. Eldridge, R. M. Crockett & J. R. Maund, *MNRAS* **395**, 1409–1437 (2009), 0809.0403.

N. Smith, W. Li, A. V. Filippenko & R. Chornock, *MNRAS* **412**, 1522–1538 (2011), 1006.3899.

E. O'Connor & C. D. Ott, *ApJ* **730**, 70 (2011), 1010.5550.

J. J. Walmswell & J. J. Eldridge, *MNRAS* **419**, 2054–2062 (2012), 1109.4637.

R. A. Chevalier, C. Fransson & T. K. Nymark, *ApJ* **641**, 1029–1038 (2006), arXiv:astro-ph/0509468.

S. Chakraborti, N. Yadav, A. Ray, R. Smith, P. Chandra, & D. Pooley, *ApJ* **761**, 100 (2012), 1206.4033.

S. Chakraborti, A. Ray, R. Smith, S. Ryder, N. Yadav, F. Sutaria, V. V. Dwarkadas, P. Chandra, D. Pooley, & R. Roy, *ArXiv e-prints* (2013), 1302.7067.

Supernova environmental impacts
Proceedings IAU Symposium No. 296, 2013
A. Ray & R. A. McCray, eds.

doi:10.1017/S1743921313009745

Type Ib Supernova 2007uy – a multiwavelegth perspective

Rupak Roy and Brijesh Kumar

Aryabhatta research institute of observational sciences (ARIES), Nainital, India

Abstract. We present the results from a comprehensive analysis of a Type Ib supernova (SN) 2007uy in a nearby galaxy NGC 2770 (∼30 Mpc), using data from space-based Swift/UVOT, along with ground-based observations at visible, infrared and radio wave bands.

Keywords. (Stars:) Supernovae: individual (SN 2007uy); techniques: photometric, spectroscopic; radio continuum: Supernovae

---

1. Introduction

The stripped-envelope supernovae of Type Ibc are rare and the detailed characteristics of explosion as well as the nature of progenitors have been studied only for a very few such events. Pre-SN imaging revealed that majority of Type II events are originated from supergiants. Recent investigation shows that progenitors of Type Ibc are probably Wolf Rayet stars, which are visually faint and hence challenging to get detected in comparison to low mass progenitors of Type II events (Yoon *et al.* 2012).

The supernova (SN) 2007uy is a type Ib event discovered on 31.7 UT December, 2007 by Yoji Hirose in the nearby spiral galaxy NGC 2770 (Nakano *et al.* 2008). Blondin & Calkins (2008), took the first spectrum of this event with FAST on the 1.5-m reflector at F. L. Whipple Observatory and its comparison with that of template type Ibc SNe spectra using the SNID tool (Blondin & Tonry 2007) revealed that SN 2007uy is young and the spectrum is similar to SN 2004gq at roughly one week before maximum.

2. Observations, data analysis and results

Optical photometric observation was performed using ground-based 1-m Sampurnanand Telescope (ST), 2.5-m NOT and Space-based *Swift*/UVOT. The NUV photometry was performed using *Swift*/UVOT, whereas NIR observations were conducted from 3.8-m UKIRT. Archival data from VLA have been used for radio follow-up. The spectroscopic data were acquired from 2.5-m NOT, 3.6-m NTT, 8.2-m VLT and 6.5-m MMT. For optical data, we performed template subtraction technique and applied standard stellar psf photometry (DAOPHOT), to get the calibrated magnitudes of the SN event.

Comparison of UVOIR light curves of SN 2007uy with other Type Ibc events shows that the supernova was discovered within a week after its explosion. We found that the reddening due to host is E(B−V)∼0.69 mag, which is substantially higher from the Galactic reddening (0.02 mag) along the line of sight, having contribution mainly from the highly inclined host galaxy. The UVOIR bolometric light curve has been computed and compared with different Type Ibc events (see left panel of fig. 1) and it is as luminous as GRB associated event SN 1998bw. After modeling the UVOIR bolometric light curve, we estimated the ejected mass $\sim 2\ M_\odot$, while amount of radioactive ^{56}Ni produced during this explosion is $\sim 0.4\ M_\odot$.

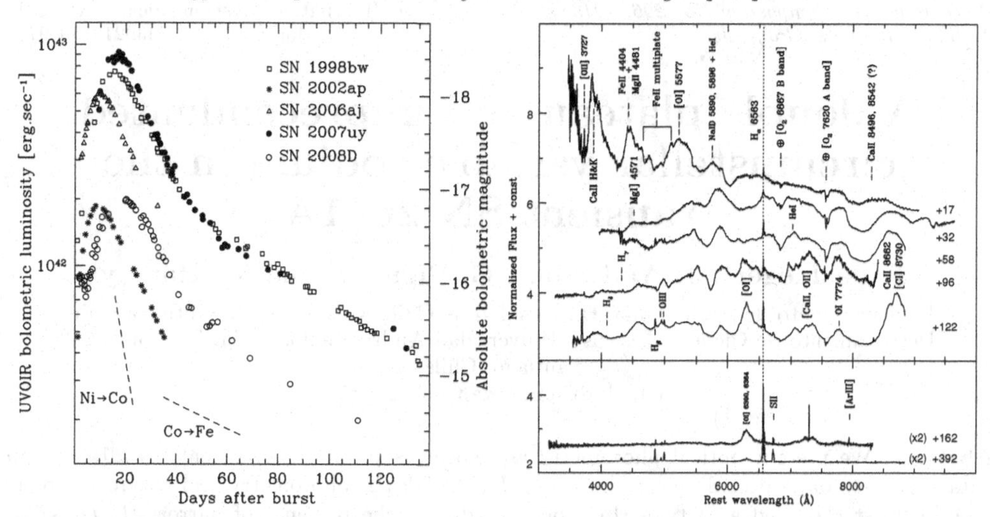

Figure 1. Left Panel: The UVOIR bolometric light curve of SN 2007uy. Comparison with other Type Ibc events. **Right Panel:** Spectroscopic evolution of SN 2007uy. All the spectra have been normalized with respect to the peak flux of the underling H_α feature and a constant offset has been applied to present them clearly. The +162d and +392d spectra have been multiplied by a factor of 2 to enlarge several tiny features. The dotted vertical line represents the position of H_α and confirms the wavelength calibration within the limits of the spectral resolution.

The event was monitored spectroscopically over an year using data from several 2 to 10-m class telescopes over the globe. The spectral evolution of the event is presented in right panel of fig. 1. The asymmetric evolution of different line profiles are clearly visible. The features are highly blended. The measured value of photospheric velocity around the peak is roughly around 15200 $\mathrm{km\,s^{-1}}$. We found that the temporal evolution of line profiles for different species are not similar to each other. This probably implies that there is no major effect of radioactive Ni on the evolution of different lines, at least for this particular case. The interaction of SN shock with the surrounding CSM generates radio waves. The archival radio data has been reduced, analyzed and modeled to find out several interaction parameters. According to radio data modeling the explosion happen nearly 4 days before the discovery of the event. This result is consistent with spectroscopic study. We measured the mass loss rate of this event during pre-SN phase and it is found to be more than 2.4×10^{-5} $\mathrm{M_\odot yr^{-1}}$.

Acknowledgements

This contribution is based on the work by Roy *et al.* (2013), which is under review.

References

Blondin S., Tonry J. L. 2007, *ApJ*, 6 66, 1024-1047
Blondin S. & Calkins M. 2008, *CBET*, 1191, 2
Nakano S., Kadota K., Itagaki K., & Corelli P. 2008, *CBET*, 1191, 1
Yoon S. C., Gräfener G., Vink J. S., Kozyreva A., & Izzard R. G. 2012, *A&A*, 544, L11

Supernova Environmental Impacts
Proceedings IAU Symposium No. 296, 2013
A. Ray & R. A. McCray, eds.

doi:10.1017/S1743921313009757

A double plateau and unprecendented circumstellar variable sodium in the transient SN 2011A

T. de Jaeger[1], J. Anderson[1], G. Pignata[2] and M. Hamuy[1]

[1]Departamento de Astronomía, Universidad de Chile, Casilla 36-D, Santiago, Chile
[2]Departamento de Ciencias Físicas - Universidad Andrés Bello, Avda. República 252, Santiago, Chile.
email: `dthomas@das.uchile.cl`

Abstract. We present optical photometry and spectrosopy of the transient SN 2011A. Our data spans 140 days after discovery including *BVRIu'g'r'i'z'* photometry and a sequence of 11 spectra. First classified as a type IIn supernova due to the presence of narrow H_α emission, this object shows exceptional characteristics. Firstly, the light curve shows a double plateau; a property only before observed in the impostor object SN 1997bs. Secondly SN 2011A has a very low luminosity for a type IIn supernova placing it between the type IIn supernovae and impostor classes in terms of luminosity. Thirdly, SN 2011A shows low velocity and high equivalent width sodium doublet absorption which increases with time and is most likely of circumstellar origin. This evolution is also accompanied by a change of line profile. When the absorption becomes stronger, a P-Cygni profile appears.

Keywords: circumstellar matter, supernovae: indivudual: 2011A, stars: mass loss,winds.

---

1. Introduction

The ejected material from a SN explosion is often seen to interact with surrounding circumstellar medium (CSM), left by progenitor mass-loss episodes prior to explosion (Chevalier 1981, Fransson 1982). When the CSM is dense, strong CSM-ejecta interaction can begin shortly after explosion; this is observed as type IIn supernova (Schlegel 1990, Chugai & Danziger 1994). These objects show narrow emission lines superimposed on broader emissions.

2. Photometry

In the light curve presented in Fig. 1 we see an initial plateau which lasts $\sim$ 15 days with a slope $\sim$ 0.42 mag 100 days^{-1} in V. After this first plateau, the light curve declines with a slope $\sim$ 3.4 mag 100 days^{-1}. Then there is a second plateau phase with a slope of 0.28 mag 100 days^{-1}, which is flatter and longer lasting at longer wavelengths. For the remaining observed epochs the light curve declines with a slope of 2.8 mag 100 days^{-1}. The presence of this rare double plateau leads us to speculate that the CSM is composed of two shells ejected by the pre supernova wind. During $\sim$ 15 days there is an initial plateau which corresponds to the interaction between the SN blast wave and the first shell. When the shock reaches the edge of the first shell the light curve drops. Then there is another weaker plateau; which lasts $\sim$ 15 days in the V-band.

Using a supernova distance = 37.70 Mpc we obtain an absolute magnitude without host exctinction of $M_V = -15.15$. Using the measurement of the equivalent width (EW) of NaI D absorption feature and we find an absolute magnitude of -16.4 in V-band, relatively low for SNeIIn but higher than impostors (Van Dyk *et al.* 2000).

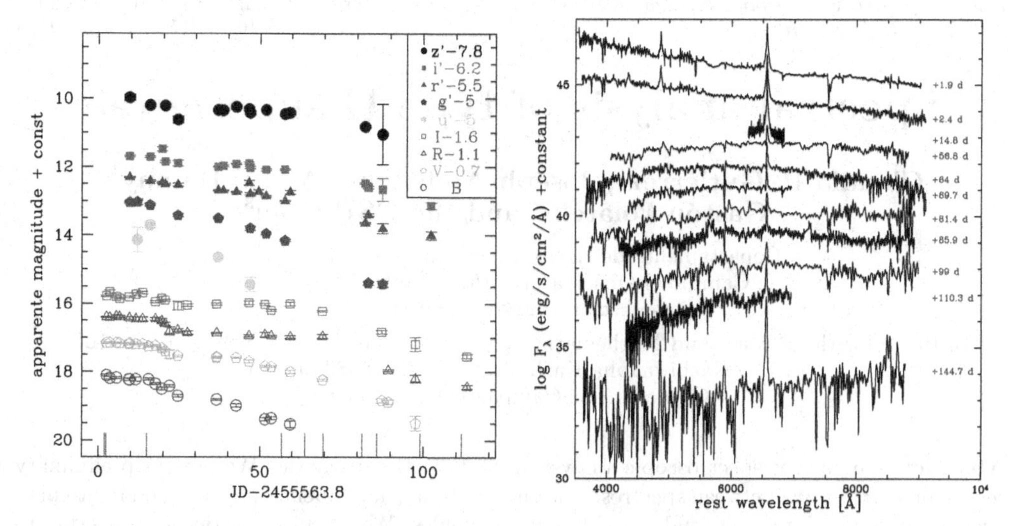

Figure 1. Left figure: *uBgVRrIiz light curve. Epochs are with respect to the discovery date (2011 Jan 2.30 UT= 0 days). Vertical red lines represent optical spectra epochs.* **Right figure**: *Optical spectra of SN 2011A. The epochs after discovery are indicated in red.*

3. Spectroscopy

Our spectral sequence (Figure 1., right figure) shows considerable H_α line evolution with time. From the discovery to 56.8 days after, our spectra are characterised by prominent broad H emission. Then a narrow P-Cygni absorption appears on the top of the broad component. We measured the blueshifted absorption velocity with respect to the host galaxy recession velocity and found low values, between 575 km s^{-1} to 1060 km s^{-1}. It is unlikely that these low velocities can be attributed to the ejecta, which is expected to have velocities of several thousand km s^{-1}. After days 85.9 we see again a H broad emission component. This is coherent with a CSM composing by two shells as discussed to explain the light curve evolution. Note that our spectrum taken 1.9 days after discovery fit perfectly a spectrum of SN 1994W taken 57 days after explosion. This allow us to constrain the SN 2011A explosion date to be $\sim$ 50 days after discovery.
We observe that the NaI D doublet absorption (λ5889.95, 5895.92) becomes stronger with time and has low velocity. Indeed the equivalent width is initially equal to $\sim$ 3 Å and increases to 10 Å after the first 70 days. At the same time, the NaI D profile evolves. First we see only an absorption line then a P-Cygni profile appears. The velocities measured from the doublet center, 5892.43 Å with respect to the host galaxy redshift never gets higher than 700 $km\ s^{-1}$ inconsistent with a ejecta origin, but consistent with a CSM origin interpretation.

References

Chevalier, R. A. 1981, *ApJ*, 251, 259–265

Chugai, N. N. & Danziger, I. J. 1994, *MNRAS*, 268, 173

Fransson, C. 1982, *A&A*, 111, 140–150

Schelegel, E. M 1990, *MNRAS*, 244, 296–271

Van Dyk, S. D., *et al.* 2000, *PASP*, 112, 1532–1541

Supernova Environmental Impacts
Proceedings IAU Symposium No. 296, 2013
A. Ray & R. A. McCray, eds.

doi:10.1017/S1743921313009769

Spectral analysis of type II supernovae

Claudia P. Gutiérrez[1], Joseph Anderson[1], Mario Hamuy[1], Gastón Folatelli[2] and the CSP team[3]

[1]Departamento de Astronomía Universidad de Chile,
Camino el Observatorio 1515, Santiago, Chile
email: cgutierr@das.uchile.cl

[2]Institute for the Physics and Mathematics of the Universe (IPMU) University of Tokyo,
515 Kashiwanoha Kashiwa, Chiba 2778583, Japan
[3]Carnegie/Las Campanas Observatories

Abstract. We present spectroscopic analysis of 63 type II supernovae. We present preliminary results on correlations between spectroscopic and photometric properties, focusing on light-curve decline rates, absolute magnitudes and H_α lines profiles. We found the ratio of absorption to emission of H_α P-Cygni profile as the dominant measured parameter as it has the highest median correlation with all other parameters.

Keywords. Supernova, Spectra, Photometry

1. Introduction

Type II Supernovae (SNe II) were initially classified in two subclasses depending to the shape of the light curve. SNe showing constant luminosity are called *plateau* (SN IIP), and SNe with linear decline are called *linear* (SN IIL; Barbon *et al.* 1979). While many individual analysis have been published (e.g. Hamuy 2003), few statistical analysis of SN II have been presented, to date . Patat *et al.* (1994) examined some properties of 51 SN II and concluded that SN IIL have large ratio of absorption to emission (a/e) of H_α P-Cygni profile values. Here, we show how the spectral and photometric properties are correlated using a large sample of high cadence and quality spectral sequences.

2. The sample and measurements

The sample of type II Supernovae (SNe II) employed in this study was obtained by CSP between 2004 and 2009 plus data from previous samples. From this database we selected a sub-sample of events with sufficient data to measure important spectral and photometric parameters. A large variety of SNe II are included in this sub-sample, which can be seen in the diversity of H_α. Figure 1 (left) shows this variety in SNe II focusing on the H_α P-Cygni profile.

For this initial study of SNe II spectral properties we choose to focus on H_α line profiles. We measure two spectral properties: the ejecta velocity via the FWHM of emission of H_α, and the ratio of EWs of absorption to emission of H_α, initially proposed by Patat *et al.* (1994). From photometry we measure properties of the V-band light-curves (see in this edition, Anderson 2013): s1: initial decline from maximum, s2: *'plateau'* decline rate (these are in V-band mags per 100 days), M_{max}: maximum absolute magnitude. All the spectral measurements are interpolated to the B_{inf}, defined as the time of transition between s1 and s2.

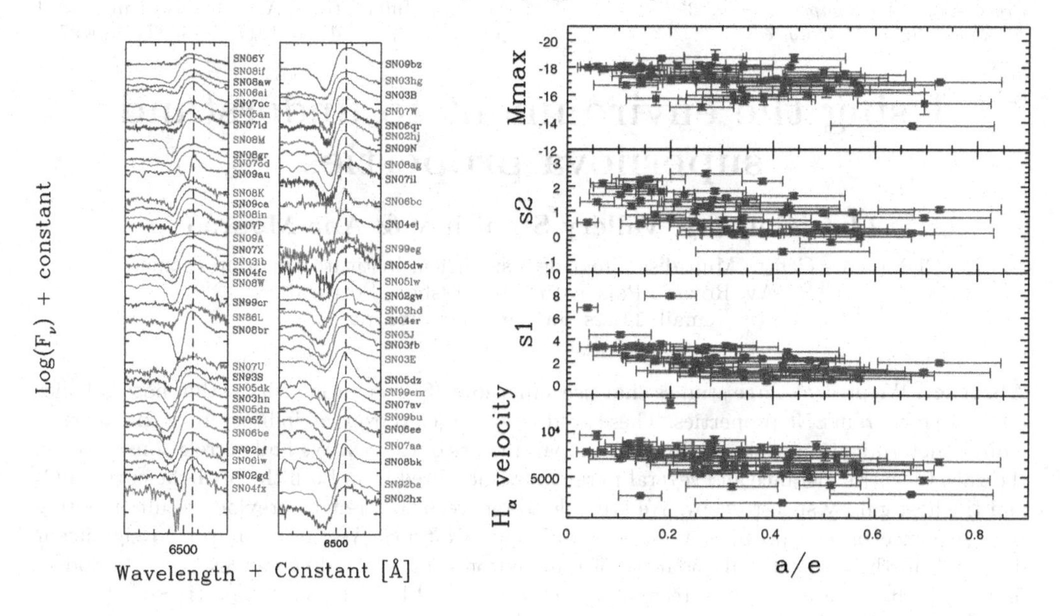

Figure 1. *Left:* Spectral sequences of 63 SNe focousing on the H_α PCygni profile near B_{inf}. The SNe are ordered in terms of a/e (ratio of absorption to emission in H_α) increasing. *Right:* : Correlations between (a/e) in H_α and ejecta velocity, s1, s2 and M_{max}.

3. Results

In Figure 1 (right) we correlate a/e with the ejecta velocity and three photometric parameters: s1, s2, M_{max} (corrected for A_v using NaD lines in spectra). a/e was defined as the dominant measured parameter as it has the highest median correlation with all other parameters. This plot shows that SNe with smaller (a/e) have higher H_α velocities, higher slope after the maximum and in the *'plateau'* phase and are brighter. While, SNe with higher (a/e) have smaller H_α velocities, smaller slope after the maximum and in the Plateau phase, and are dimmer.

The spectral diversity found in H_α P-Cygni profiles and their correlations with light curves properties could be interpreted as circumstellar material (CSM) interaction (Pastorello *et al.* (2006), Inserra *et al.* (2011), Roy *et al.* (2011), Inserra *et al.* (2012)). However, we can not rule differences in properties such as the changes in the mass, temperature and density of the ejecta.

Acknowledgments: We want to thank to he ALMA-CONICYT 31110018 fund: "Participation of Graduate Students in international Astronomy Meetings" and the IAU Grant Allocation.

References

Anderson, J. 2013. *Supernova environmental impacts*, Proc. IAU Symposium No. 296 (Kolkata)
Barbon, R., Ciatti, F., & Rosino, L. 1979, *A&A*, 72, 287
Hamuy, M. 2003, *ApJ*, 582, 905
Inserra, C., Turatto, M., Pastorello, A., *et al.* 2011, *MNRAS*, 417, 261
Inserra, C., Turatto, M., Pastorello, A., *et al.* 2012, *MNRAS*, 422, 1122
Pastorello, A., Zampieri, L., Turatto, M., *et al.* 2006, *MNRAS*, 347, 74
Patat, F., Barbon, R., Cappellaro, E., & Turatto, M. 1994, *A&A*, 282, 731
Roy, R., Kumar, B., Moskvitin, A. S., *et al.* 2011, *MNRAS*, 414, 167

Supernova Environmental Impacts
Proceedings IAU Symposium No. 296, 2013
A. Ray & R. A. McCray, eds.

doi:10.1017/S1743921313009770

Using the environment to understand supernova properties

Lluís Galbany, Vallery Stanishev & Ana Mourão

CENTRA - Centro Multidisciplinar de Astrofísica, Instituto Superior Técnico,
Av. Rovisco Pais 1, 1049-001 Lisbon, Portugal
email: `lluis.galbany@ist.utl.pt`

Abstract. We present three studies that use supernova (SN) environments within host galaxies (HGs) to constrain SNe properties. These studies are ordered from an indirect approximation to a direct determination of the environmental parameters of the SN. We find correlations between the galactocentric distance and several parameters measured from both the SN light-curve (LC) and the host galaxy spectroscopy. We are able to recover and strength previous results pointing to a sequence on the progenitor mass of different SN types. We also confirm no significant difference in the elemental abundances of the environment where different SN types exploded, measured with a more powerful technique such as Integral Field Spectroscopy (IFS).

Keywords. (stars:) supernovae: general, (ISM:) HII regions, galaxies: abundances, techniques: spectroscopic, surveys

---

1. Introduction

Current SNe surveys will increase drastically the number of events discovered and available to better understand both stellar evolution and cosmology. It is well known that the use of Type Ia SNe as a cosmological distance indicators requires a better control of the systematic errors since the statistical errors will be reduced by this new huge amount of events. One approximation is to account for environmental properties in the LC standardization. On the other hand, with this new data one will be able to make constrains on the properties of the progenitor stars that produce different types of SNe. Here we present a compilation of three studies that comprise a sequence from an indirect approximation to a direct determination of the environmental parameters of the SN.

2. Summary of the studies

GCD as a proxy for galaxy parameters. We use almost 200 SNe Ia at z $<$ 0.25 discovered by the SDSS-II SN Survey, to search for dependencies between SN Ia LC parameters and the projected distance to the HG center, using the distance as a proxy for local galaxy properties. We correlate the LC parameters with several definitions of the distance to the center of the HG, either normalized or not, and look for trends in the mean values of these parameters with increasing distance. We find several differences in the intrinsic color and LC stretch between SN exploding closer and further the galaxy core. More details can be found in Galbany *et al.* (2012).

Global galaxy parameters inferred from slit spectroscopy. The following step is to measure HG parameters, such as age, mass, metallicity, Hα emission, and specific star formation rate from the spectra of the HGs of those objects in SDSS-II SN sample that have the HG spectrum available in SDSS DR9. We look for correlations between these HG parameters and the redshift, SN LC parameters, and GCDs. We find that, for spiral

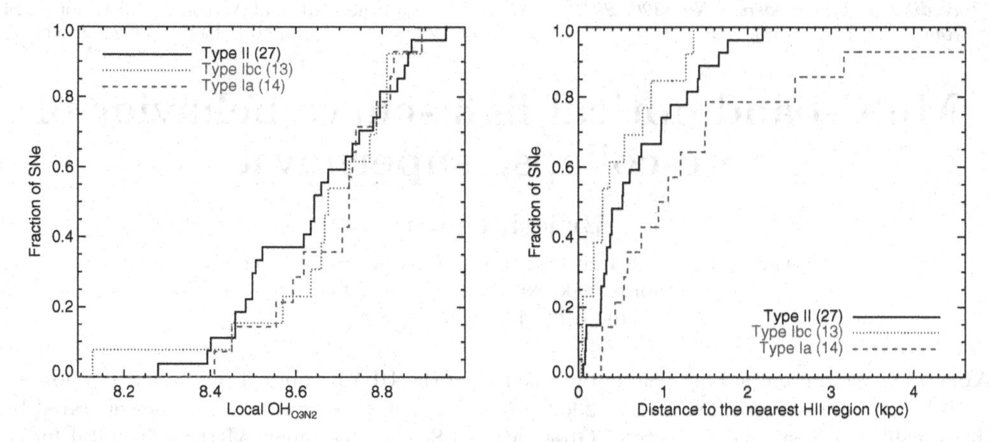

Figure 1. Fraction of SNe of different types that explode in an environment with an elemental abundance lower than each value (left), and at certain distance to the nearest HII region (right).

galaxies, SNe in more massive, old, and with high elemental abundances, tend to explode farther from the center.

Local galaxy parameters using IFS. Following Stanishev *et al.* (2012), we produce 2D maps of the gas emission lines and the stellar population parameters of nearby galaxies ($0.005 < z < 0.03$) from CALIFA and PINGS surveys that hosted an observed SN of any kind. We then look for differences among SN types that help to constrain properties of the progenitor star and environmental characteristics. With this approach we are going further than using simply an aperture spectrum centered at the galaxy core, or a spectrum from a slit positioned at the position of the SN explosion. Our sample consists of 54 SNe (27 II, 13 Ibc, 14 Ia). When correlating SN type with the local metallicity at SN position, we see that type Ibc and Ia SNe seem to explode in higher metallicity environments than type II SNe (Fig. 1, left plot). However there is a need for more statistics in order to make this difference significant. We also look for correlations between the SN type and the star formation rate traced by the Hα emission. The cumulative distributions of the distances from the SN location to the nearest HII region (right plot in Fig. 1) show a sequence from type Ibc, through type II, and type Ia SNe. The same ordering is obtained when correlating the SN type with the GCD. This result points to a difference in the lifetime and masses of their progenitor stars.

3. Conclusions

We have presented three studies that sequentially approach to a direct measurement of the local galaxy properties at the SN location. We see correlations between the LC parameters and the spectroscopic HG parameters to the GCD. We are also able to confirm and strength the differences in the progenitor mass of the main SN types, showing a decreasing sequence from type Ibc to type Ia SNe. We find no significant differences in the elemental abundance of the environments of different SN types.

References

Galbany, L., Miquel, R., Östman, *et al.* 2012, *ApJ*, 755, 125
Stanishev, V., Rodrigues, M., Mourão, A., & Flores, H., 2012, *A&A*, 545, A58

Supernova Environmental Impacts
Proceedings IAU Symposium No. 296, 2013
A. Ray & R. A. McCray, eds.

doi:10.1017/S1743921313009782

Multi-band optical light-curve behavior of core-collapse supernovae

Brijesh Kumar

Aryabhatta Research Institute of Observational Sciences,
Manora Peak, Nainital, 263 002, India
email: brij@aries.res.in

Abstract. We present survey results obtained from the UBVRI optical photometric follow-up of 19 bright core-collapse SNe during 2002-2012 using 1-m class optical telescopes operated by the Aryabhatta Research Institute of Observational Science (acronym ARIES), Nainital India. This homogeneous set of data have been used to study behavior of optical light/color curve, and to gain insight into objet-to-object peculiarity. We derive integrated luminosities for types IIP, Ibc and luminous SNe. Two peculiar type IIP events having photometric properties similar to normal IIP and spectroscopic properties similar to sub-lumnious IIP have been identified.

Keywords. (stars:) supernovae: general; techniques: photometric

1. Observations, data reduction and results

One-meter class optical telescopes equipped with a modern CCD detector are best suited for optical follow-up of supernovae until they fade below about 20 magnitude. During 2002-2012, we monitored nineteen bright (peak $V < 16$) SNe (see Table1) observable from Nainital -latitude $29^\circ 21' 39''$N. The observations are carried out using 104-cm Sampurnanand Telescope (ST) at Manora Peak in operation since 1972 and 130-cm Devasthal Fast Optical Telescope at Devasthal in operation since 2010. Both these telescopes are operated by ARIES, Nainital India. Both the telescopes are equipped with a 2kx2k CCD camera and at 104-cm ST, we have Johnson-Cousins UBVRI while at 130-cm, we also have SDSS ugriz filters. Both the telescopes are equipped with an auto-guider and the SNe observations are done in visitor mode. Instrumental magnitudes of SNe are estimated from profile fitting photometry using DAOPHOT. Landolt Standards are observed to calibrate the magnitudes of SNe. The galaxy-subtraction technique is used for SNe embedded in the galaxy background. The light-curves (apperant & bolometric) are shown in Fig. 1 and the derived physical properties are given in Table 1.

The mean of peak M_V for 6 type IIP events is -16.7 mag while for 6 type Ibc, it is -18.0 mag. The value of M_V for three luminous events lie between -19 to -22 mag. Mean values of peak luminosity ($\times 10^{42}$ erg s^{-1}) are 1, 3 and 90, respectively for IIP, Ibc and luminous events. Of six IIP events, the SNe 2008in and 2012A are found photometrically similar to normal type IIP events, while their spectral properties suggest them to be similar to low-luminosity SNe. The later are thought to originate from explosion of low-mass 8-10 $M_\odot$ progenitors. The total radiant energy for type II SNe is of the order of 5×10^{43} erg, while for type Ibc it is about 2 times higher. Color curve evolution of type Ibc and IIb SNe are quite similar.

We are thankful to the observatory staff of 104-cm and 130-cm optical telescope for their support during observations. We gratefully acknowledge Shashi B. Pandey, Kuntal Misra, Rupak Roy, Brajesh Kumar, Vijay K. Bhatt for providing data and Subhash Bose for data as well as the computer programs used in this work.

Table 1. Properties of the SNe sample. The time of explosion (t_0) in JD 2450000+, peak V-band magnitude V_p , reddening $E(B-V)$, distance modulus μ, peak absolute magnitude M_V, SNe types, peak $UBVRI/BVRI$ bolometric luminosity L_p, the time when luminosity peaks, t_p and total radiant energey E_t integrated over the observed period are given in successive columns. The reference for t_0, V_p, $E(B-V)$ and SNe type is given in the "Ref." column. Other parameters are derived in this work. Due to incomplete coverage of light-curve, the estimates for L_p and E_t are lower limits.

SN	t_0 day	V_p mag	$E(B-V)$ mag	μ mag	M_V mag	Type	Ref.	L_p 10^{42} erg s^{-1}	t_p day	E_t 10^{43} erg
SN 2002ap	2300.0	12.42	0.08	29.52	-17.35	Ic	1	1.2	14	4.4
SN 2004et	3270.5	12.50	0.41	28.70	-17.47	IIP	2	2.2	<5	21.0
SN 2006aj	3784.6	17.50	0.15	35.78	-18.75	Ibc	3	5.1	10	12.0
SN 2007uy	4462.0	15.74	0.71	32.37	-18.83	Ibc	4	6.2	22	16.0
SN 2008D	4475.1	17.39	0.80	32.38	-17.47	Ib	4	1.6	19	6.6
SN 2008gz	4694.0	15.50	0.07	32.03	-16.75	IIP	5	0.5	<87	2.5
SN 2008in	4825.6	15.07	0.10	30.60	-15.84	IIP	6	0.5	<5	2.8
SN 2009jf	5099.5	15.05	0.16	32.65	-18.10	Ib	7	3.0	23	9.7
CSS 100217	5170.0	16.41	0.02	38.94	-22.59	pisn	8	190.0	<125	930.0
SN 2010hq	5395.0	16.08	0.08	32.53	-16.70	IIP	9	0.6	<79	2.1
SN 2010jl	5479.1	13.80	0.06	33.50	-19.89	IIn	9	19.0	<28	200.0
SN 2010kd	5515.0	17.19	0.02	38.09	-20.96	pisn	4	55.0	40	470.0
SN 2011dh	5712.9	13.52	0.04	29.42	-16.02	IIb	10	0.9	>15	11.0
SN 2011fu	5825.5	16.92	0.22	34.46	-18.22	IIb	11	38.0	20	17.0
SN 2012A	5934.0	13.60	0.03	29.96	-16.45	IIP	12	0.6	<27	2.8
SN 2012P	5930.0	16.01	0.05	31.89	-16.04	IIb	13	0.5	<20	2.4
SN 2012aa	5957.0	18.00	0.20	37.61	-20.78	Ic	14	25.0	<12	78.0
SN 2012ap	5968.0	16.37	0.04	33.45	-17.20	Ic	13	1.1	11	2.9
SN 2012aw	6002.6	13.37	0.07	29.99	-16.84	IIP	14	1.4	8	12.0

1-Pandey *et al.* (2003, MNRAS 340, 375); 2-Misra *et al.* (2007, MNRAS 381, 280); 3-Misra (2007, PhD Thesis); 4-Roy & Kumar (2013, IAUS 296); 5-Roy *et al.* (2011, MNRAS 414, 167); 6-Roy *et al.* (2011, ApJ 776, 76); 7-Valenti *et al.* (2011, MNRAS 416, 318) 8-Drake *et al.* (2011, ApJ 735, 106); 9-Roy *et al.* (2011, ASInC 3, 124); 10-Van dyk *et al.* (2011, ApJ 741, 28); 11-Kumar *et al.* (2013, *MNRAS*, arXiv:1301.6538); 12- Roy *et al.* (2013, IAUS 296); 13-This work; 14-Bose *et al.* (2013, IAUS 296)

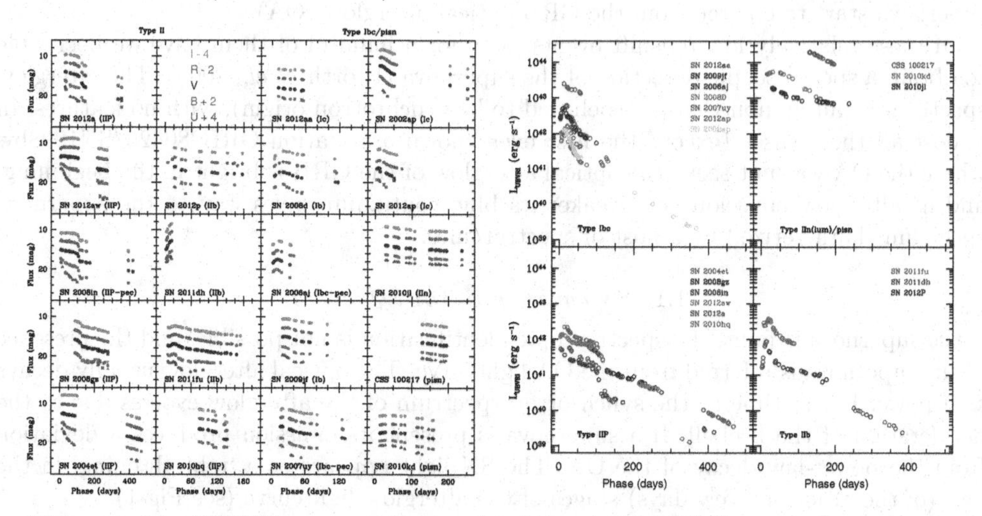

Figure 1. The $UBVRI$ apparent (left) and bolometric (right) ligh-curve of supernovae.

Supernova Environmental Impacts
Proceedings IAU Symposium No. 296, 2013
A. Ray & R. A. McCray, eds.

doi:10.1017/S1743921313009794

A comparative study of GRB-Supernovae

Lekshmi Resmi[1] and Kuntal Misra[2]

[1]Dept. of Earth & Space Sciences,
Indian Institute of Space Science & Technology,
Trivandrum-695547, India.
email: l.resmi@iist.ac.in

[2]Aryabhatta Research Institute of Observational Sciences (ARIES)
Manora Peak, Nainital-263129, India.
email: kuntal@aries.res.in

Abstract. Optical afterglow observations hold the indirect key to type-Ic Supernovae associated with Gamma Ray Bursts. In several cases where there is no spectroscopic confirmation available, presence of the supernova is inferred from the red bump seen in late afterglow light-curves. We do extensive afterglow modeling to extract the supernova contribution as residue. We compare the residual lightcurves of Supernovae associated with GRB041006, GRB030329, GRB050525A and GRB090618.

Keywords. gamma rays: bursts, supernovae

1. Introduction

Supernovae II/Ibc and Gamma Ray Bursts (GRBs) of the long duration kind are two manifestations of the end of massive stars. Not all massive stars end as long-GRBs. We believe, in certain conditions like if the star has high rotational angular momentum, the end stage will be a Gamma Ray Burst. However, it appears that in many if not all cases of the collapse that leads to a GRB, a type I-c supernova explosion also happens (see Woosley & Bloom, 2006 for a review). In several low redshift ($z < 1$) GRBs, the supernova start to emerge from the GRB optical afterglow (OA).

Of these relatively low redshift events, only for a handful of them have we been able to obtain a spectroscopic detection of the supernova (Hjorth *et al.*, 2003). The afterglow spectrum is purely non-thermal (believed to be synchrotron origin), with no features. In almost all these cases (except the first ever known association GRB980425/SN1998bw where the OA was not seen) the optical afterglow of the GRB is bright in the beginning, and as afterglow emission gets weaker its blue continuum gives way to the supernova spectrum characterized by a host of spectral lines.

1.1. *SN emission in OA lightcurve*

If the supernova is faint, the spectroscopic identification is not possible, and the presence of the supernova is inferred from the OA lightcurve. The optical afterglow usually decays as a power-law in time as the synchrotron spectrum of the afterglow evolves due to the deceleration of the fireball. If a supernova is present its emission produces a deviation from the power-law decay of the OA. The SN light appears as a 'rebrightening' in the late (of the order of a few days) stages of the afterglow lightcurve (see Fig-1).

Hence the late time optical emission will contain contributions from the optical afterglow, the supernova and the optical emission from the host galaxy of the progenitor star. We use afterglow physical models to subtract the OA and the host galaxy emission.

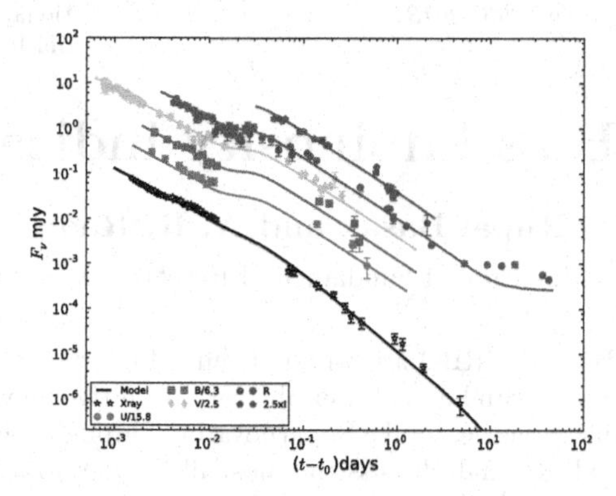

Figure 1. Optical afterglow with the multi-band model that takes care of the afterglow and host-galaxy contribution. In red band we can see the supernova contribution as a rebrightening (Resmi *et al.* 2012).

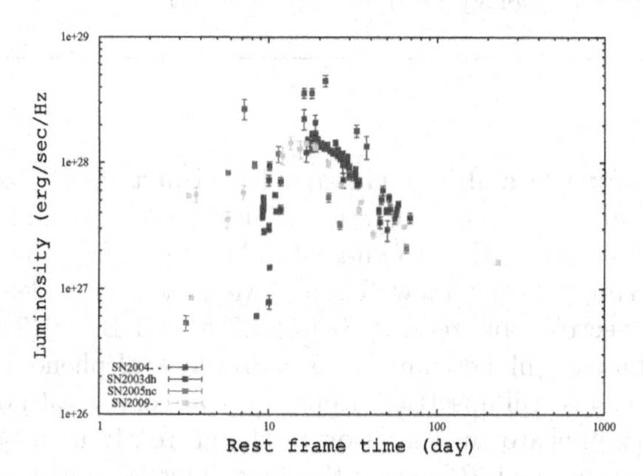

Figure 2. The Supernovae lightcurves in our sample. Luminosity vs. rest-frame time.

2. Sn lightcurves

We study four cases where sufficient data from x-ray to radio bands are available for doing a robust afterglow modeling. These are GRB030329/SN2003dh (z = 0.16), GRB041006 (z = 0.71), GRB050525A/SN2005nc (z = 0.606) and GRB090618 (z = 0.54). Of these, SN2003dh was identified spectroscopically too. We assume synchrotron emission from an adiabatically evolving fireball with a given isotropic equivalent energy and initial collimation angle. The flux also depends on the assumed ambient density profile and fractional energy content in non-thermal electrons and magnetic field (see Piran, 1999 for a review). We obtain the SN lightcurves as the residue and calculate the luminosity vs rest-frame time (shown in Fig-2).

Currently our sample contains four sources. A comparative study of the SN lightcurves is underway including more sources in the sample.

References

Hjorth, J., Sollerman, J., Moller, P., Johan, P. U., *et al.* 2003, *Nature*, 423, 847.

Piran, T. 1999, *Physics Reports*, 314, 575.

Resmi, L., Misra K., Johannesson, G., Castro-Tirado, A. J. *et al.*, 2012 *MNRAS*, 427, 288.

Woosley, S. E., & Bloom, J. S., 2006 *ARAA*, 44, 507.

Supernova environmental impacts
Proceedings IAU Symposium No. 296, 2013
A. Ray & R. A. McCray, eds.

doi:10.1017/S1743921313009800

GRB as luminosity indicator

Rupal Basak and A. R. Rao
Tata Institute of Fundamental Research, Mumbai

Abstract. Gamma Ray Bursts (GRBs) are found at much higher redshifts (z > 6) than Supernova Ia (z ∼ 1), and hence, they can be used to probe very primitive universe. However, radiation mechanism of GRB remains a puzzle, unlike Supernova Ia. Through comprehensive description, both empirical and physical, we shall discuss the most likely way to use the constituent pulses of a GRB to find the radiation mechanism as well as using the pulses as luminosity indicators.

Keywords. gamma-ray burst —general, method — data analysis, radiation mechanism — thermal and non-thermal, cosmology — luminosity indicator

---

1. Introduction

First discovered during late 1960's, Gamma Ray Burst (GRB) soon became one of the greatest puzzles in astrophysics in terms of its location, size and energetics. Many satellites have been flown since then to understand this puzzling phenomenon — *HETE-2*, *BATSE*, *Swift*, *Fermi*, to name a few. Though we have a clear idea about their cosmic origin (the highest spectroscopic redshift being 8.2 for GRB 090423) and a rough idea about their energy budget (highest among all astrophysical phenomena — $\sim 10^{52}$ erg), the emission mechanism is still unsettled. Hence, in spite of a great hope that GRB could be used as luminosity indicator in extension to the currently used ones, e.g., supernova Ia, one has to standardize the GRB energetics first. This, of course, serves two purposes — solving the GRB physics, which is not settled till now, and using GRB as luminosity indicator.

2. Methodology and Results

There exist certain empirical correlations of the peak energy (E_{peak}) of GRB spectrum with the energetics of GRB. These correlations are important as they can be used to independently measure a physical parameter, namely energy, using only prompt emission spectral data. Amati *et al.* (2002) showed that E_{peak} correlates with the isotropic equivalent energy ($E_{\gamma,iso}$). It is very important that this correlation should hold within a GRB, as that can prove the reality of such correlation and strongly refutes selection bias. But, Basak & Rao (2012b), using 9 GRBs with known redshift detected by *Fermi*/Gamma Ray Burst Monitor (GBM), have shown that this correlation breaks down if one uses the time-resolved data. The Pearson correlation, 0.80 drops to 0.37. They concluded that Amati correlation has no meaning in a time-resolved study. The situation is saved if one uses the constituent broad pulses (total 22 pulses), rather than intensity guided time cuts. Pulse-wise analysis not only restores the correlation, it improves that (0.89). They used the pulse description of Basak & Rao (2012a) and found that replacement of E_{peak} with a new quantity of their model, namely the peak energy at zero fluence ($E_{peak,0}$) improves the correlation even further (0.96).

In this study, we have enlarged our sample to 19 GRBs (43 pulses) having measured redshifts (z) from GBM catalog. We consider only pulse-wise Amati correlation here. The

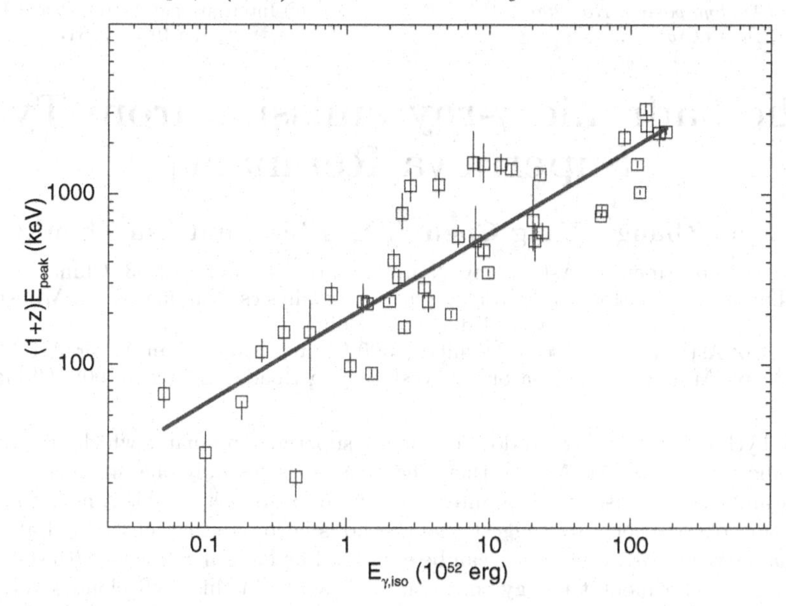

Figure 1. Pulse-wise Amati correlation. The thick line shows the correlation

Spearman rank correlation coefficient is 0.86 (see Figure 1). As we have a larger sample, we are able to divide the set of GRBs into various redshift (z) bins and study evolution of the correlation. A detailed analysis will be published later.

3. Discussions

The prompt emission spectrum of a GRB is generally fitted with Band model (Band *et al.* 1993). The empirical correlations, we have discussed, relies on the fact that the overall and instantaneous spectrum is Band like. Band model describes a non-thermal model. There are alternative models, e.g., black body with a powerlaw (Ryde 2004). Recently, we have analyzed the brightest GRBs, having separable pulses, namely GRB 081221 and GRB 090618, and found that the Band model is adequate in the falling part of a pulse. But, a different model is preferred in the rising part (Basak & Rao 2013). Hence, one should take into account these findings while describing the emission mechanism of GRB and thereby using the pulses for cosmological purpose.

References

Amati, L., Frontera, F., Tavani, M., in't Zand, J. J. M., Antonelli, A., Costa, E., Feroci, M., Guidorzi, C., Heise, J., Masetti, N., Montanari, E., Nicastro, L., Palazzi, E., Pian, E., Piro, L., & Soffitta, P. 2002, *A&A*, 390, 81

Band, D., Matteson, J., Ford, L., Schaefer, B., Palmer, D., Teegarden, B., Cline, T., Briggs, M., Paciesas, W., Pendleton, G., Fishman, G., Kouveliotou, C., Meegan, C., Wilson, R., & Lestrade, P. 1993, *ApJ*, 413, 281

Basak, R. & Rao, A. R. 2012a, *ApJ*, 745, 76

Basak, R. & Rao, A. R. 2012b, *ApJ*, 749, 132

Basak, R. & Rao, A. R. 2013, *arXiv*: 1302.6091

Ryde, F. 2004, *ApJ*, 614, 827

IAU Symposium 296: Supernova Environmental Impacts
Proceedings IAU Symposium No. 296, 2013
A. Ray & R. A. McCray, eds.

doi:10.1017/S1743921313009812

On the hadronic γ-ray emission from Tycho's Supernova Remnant†

Xiao Zhang[1], Yang Chen[1,2], Hui Li[3,1] and Xin Zhou[4,2]

[1]Department of Astronomy, Nanjing Univ., Nanjing 210093, China
[2]Key Laboratory of Modern Astronomy and Astrophysics, Nanjing Univ., Ministry of Education, China
[3]Department of Astronomy, Univ. of Michigan, 500 Church Street, Ann Arbor, MI 48109, USA
[4]Purple Mountain Observatory, 2 West Beijing Road, Nanjing 210008, China

Abstract. Tycho is one of nearly a dozen Galactic supernova remnants which are suggested to emit hadronic γ-ray emission. Among them, however, it is the only one in which the hadronic emission is proposed to arise from the interaction with low-density ambient medium. Based on the multi-band observations, we suggest that Tycho is encountering dense cloud at the north-eastern boundary. The γ-ray emissions can be explained by hadronic process with self-consistent parameters, such as a modest energy conversion efficiency. In this SNR-cloud association scenario, the distance can be estimated as $\sim$2.5 kpc.

Keywords. supernova remnants, ISM: individual (Tycho), radiation mechanisms: non-thermal

---

1. Introduction

Hadronic γ-ray emission from supernova remnants (SNRs) is an important tool to test shock acceleration of cosmic ray (CR) protons. For Tycho's SNR which is suggested evolving in a tenuous medium, an energy conversion efficiency 10–15% was invoked (e.g., Tang *et al.* 2011; Giordano *et al.* 2012; Morlino & Caprioli 2012). Although it is suggested that Tycho is a naked Ia SNR without any dense cloud (Tian & Leahy 2011), there seem to be signs and evidence in multiband showing the presence of dense clouds from the north to the east, which could be the target for bombardment of the accelerated protons and would thus reduce the needed converted energy. Constraints from γ-ray observation suggest that Cas A, another historical SNR, has only transferred a minor fraction ($\leqslant$ 2%) of the total kinetic energy to the accelerated particles (e.g., Araya & Cui 2010). Moreover, under the Bohm limit assumption, the conversion factor no more than 1.3% is predicted in kinetic models for an SNR at the Tycho's age, evolving in a pre-shock medium of density $\sim$0.3 cm^{-3} (Berezhko & Völk 1997). Therefore, the fraction of order $\sim$1% at Tycho's age deserves serious consideration.

2. Hadronic Gamma-ray emission

We assume that the energy spectrum of the accelerated particles (electrons and protons) is prescribed by a power law with high-energy cutoff. Some basic parameters are adopted as: distance 2.5 kpc, spectral index 2.3, explosion energy 6×10^{50} erg (Hughes 2000), the average downstream magnetic field 320 μG and the energy conversion efficiency $\eta \sim 10^{-2}$. By fitting the radio–γ-ray data (see the left panel in Fig. 1), we get

† Supported by the 973 Program grant 2009CB824800, NSFC grant 11233001, grant 20120091110048 from the Educational Ministry of China, and grant 2011M500963 from the China Postdoctoral Science Foundation .

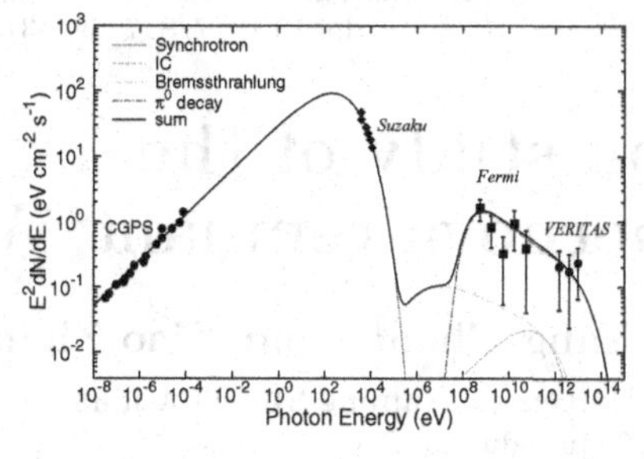

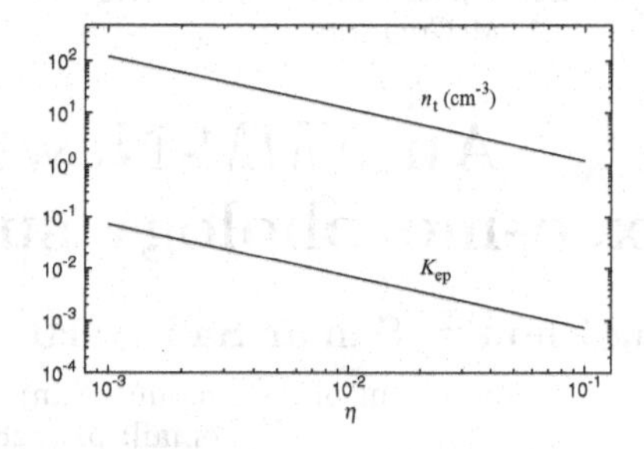

Figure 1. Left: Broadband SED of Tycho's SNR with the observed data in radio (Kothes *et al.* 2006), X-rays (Tamagawa *et al.* 2009) and γ-rays (*Fermi*: Giordano *et al.* 2012; *VERITAS*: Acciari *et al.* 2011). Right: Functional dependence of parameters $K_{\rm ep}$ and $n_{\rm t}$ on η.

the electron to proton number ratio $K_{\rm ep} \sim 0.7 \times 10^{-2}$ and the average density of the target protons $n_{\rm t} \sim 12$ cm^{-3} (see the right panel in Fig. 1 for functional dependence of $n_{\rm t}$ and $K_{\rm ep}$ on η). Considering the shock compression ratio, the average pre-shock proton density is in the range $\sim$ 4–12 cm^{-3}, which is much larger than the reported gas density $\sim 0.3\,{\rm cm}^{-3}$. This discrepancy can be reconciled if Tycho is encountering an intercloud medium. Indeed, there are some multi-band observational evidence of the presence of the surrounding dense medium from the north to the east: the expansion rate measured in Radio (Reynoso *et al.* 1997) and X-rays (Katsuda *et al.* 2010), CO-line observations (Lee *et al.* 2004; Cai *et al.* 2009; Xu *et al.* 2011), and infrared observations (Ishihara *et al.* 2010; Gomez *et al.* 2012).

The SNR-cloud association can be used to estimate the distance to Tycho's SNR. Due to the spiral shock, the local standard of rest velocity of the associated clouds allows two distances in two armed spiral shock model (Roberts 1972) for the longitude 120°. In combination with the line-of-sight HI absorption (Tian & Leahy 2011), the distance is determined to be ~2.5 kpc.

References

Acciari V. A., *et al.* 2011, *ApJ*(Letter), 730, L20
Araya M. & Cui W., 2010, *ApJ*, 720, 20
Berezhko E. G. & Völk H. J., 1997, *Astroparticle Physics*, 7, 183
Cai Z.-Y., Yang J., & Lu D.-R., 2009, *Chinese Astronomy and Astrophysics*, 33, 393
Giordano F., *et al.* 2012, *ApJ*(Letter), 744, L2
Gomez H. L., *et al.* 2012, *MNRAS*, 420, 3557
Hughes J. P., 2000, *ApJ*(Letter), 545, L53
Ishihara D., *et al.* 2010, *A&A*, 521, L61
Katsuda S., Petre R., Hughes J. P., Hwang U., Yamaguchi H., Hayato A., Mori K., & Tsunemi H., 2010, *ApJ*, 709, 1387
Kothes R., Fedotov K., Foster T. J., & Uyanıker B., 2006, *A&A*, 457, 1081
Lee J.-J., Koo B.-C., & Tatematsu K., 2004, *ApJ*, 605, L113
Morlino G. & Caprioli D., 2012, *A&A*, 538, A81
Reynoso E. M., Moffett D. A., Goss W. M., Dubner G. M., Dickel J. R., Reynolds S. P., & Giacani E. B., 1997, *ApJ*, 491, 816
Roberts W. W., 1972, *ApJ*, 173, 259
Tamagawa T., *et al.* 2009, *PASJ*, 61, 167
Tang Y.-Y., Fang J., & Zhang L., 2011, *Chinese Physics Letters*, 28, 109501
Tian W. W. & Leahy D. A., 2011, *ApJ*(Letter), 729, L15
Xu J.-L., Wang J.-J., & Miller M., 2011, *Research in Astronomy and Astrophysics*, 11, 537

Supernova Environmental Impacts
Proceedings IAU Symposium No. 296, 2013
A. Ray & R. A. McCray, eds.

doi:10.1017/S1743921313009824

An *XMM*-Newton study of the mixed-morphology supernova remnant W28

Ping Zhou[1,2], Samar Safi-Harb[2,3], Yang Chen[1,4] and Xiao Zhang[1]†

[1]Department of Astronomy, Nanjing University, Nanjing 210093, China
email: pingzhou@nju.edu.cn

[2]Department of Physics and Astronomy, University of Manitoba, Winnpeg R3T 2N2, Canada

[3]Canada Research Chair

[4]Key Laboratory of Modern Astronomy and Astrophysics, Nanjing University, Ministry of Education, China

Abstract. We perform an *XMM*-Newton study of the mixed-morphology supernova remnant (MMSNR) W28. The X-ray spectrum arising from the northeastern shell consists of a thermal component plus a non-thermal power-law component with a hard photon index (~ 1.5). Non-thermal bremsstrahlung is the most favourible origin of the hard X-ray emission. The gas in the SNR interior is centrally peaked and best described by a two-temperature thermal model. We found a non-uniform absorption column density and temperature profile for the central gas, indicating that the remnant is evolving in a non-uniform environment with denser material in the east. We argue that the cloudlet evaporation is an indispensable process to explain both the spectral properties and the clumpiness in the X-ray emission.

Keywords. Supernova remnants, ISM: individual (G6.4−0.1 = W28)

---

1. Introduction

Mixed-Morphology (or thermal composite) supernova remnants (MMSNRs) represent a class of SNRs that are shell-like in the radio but have a centrally-filled morphology in X-rays with a thermal spectrum (Jones *et al.* 1998; Rho & Petre 1998). W28 (G6.4−0.1) is an MMSNR with a double radio shell in its north. Two competitive scenarios were proposed to explain the X-ray emission in the SNR interior: thermal conduction (Cui & Cox 1992; Shelton *et al.* 1999) and cloudlet evaporation (White & Long 1991). However, the origin of the central X-ray emission is still not clear (Rho & Borkowski 2002).

W28 is interacting with molecular clouds (MCs) in the northeast, where GeV and TeV γ-rays have been detected (Aharonian *et al.* 2008; Abdo *et al.* 2010). Hadronic interaction of cosmic rays with the MCs are considered to produce the γ-rays (Li & Chen 2010) and several models are proposed to predict the broad-band spectrum generated in such sites (Bykov *et al.* 2000, Gabici *et al.* 2009). It is thus of great interest to explore in the X-ray band for the origin of the non-thermal emission from this particular shell.

2. Observation and results

Four archival *XMM*-Newton observations towards W28 were used for the analysis presented here. We here summarize briefly our results. A more detailed analysis and discussion will be presented in another paper (to be submitted to ApJ).

† Supported by NSFC grant 11233001, the 973 Program grant 2009CB824800, the grant from the Chinese Scholarship Council, the NSERC grant, and the grant 20120091110048 from the Educational Ministry of China.

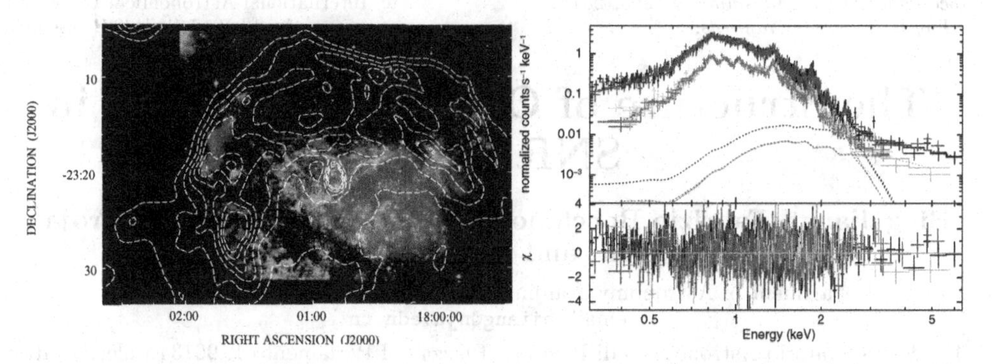

Figure 1. *Left panel*: Tri-color image of SNR W28. Red: The integrated JCMT intensity image of ^{12}CO $J = 3$-2 (-40–+40 km s^{-1}; Arikawa *et al.* 1999); Green: the Hα image from the archival SuperCOSMOS Hα Survey (Parker *et al.* 2005); Blue: *XMM*-Newton 0.3–7.0 keV X-ray map; and contours: 1.4 GHz radio continuum. *Right panel*: pn (upper) and MOS (lower) spectra and the fitted model (*vnei+power-law*, solid lines) of the northeastern shell. The short dashed lines show the components of the model.

As shown in the left panel of Fig. 1, the *XMM*-Newton image reveals blobby X-ray structures in the SNR interior and a deformed shell in the northeast. The remnant is evolving in a complicated environment with dense MCs in the east, explaining the difference in morphology between the northeast and south.

We have not found any evidence of ejecta inside the SNR. The X-ray spectra arising from the NE shell, where the shock-MC interaction is evident and γ-ray emission partly overlaps, consist of a thermal component with a temperature of $\sim$ 0.3 keV and a non-thermal component with a hard photon index of $\sim$ 1.5 (as shown in the right panel of Fig. 1). The non-thermal X-rays can not be explained by the secondary electrons from the hadronic interaction of cosmic-rays and the MCs Non-thermal bremsstrahlung from the cloud shock is the most favorable origin, at least in the view of the spectral slope.

The X-ray spectra in the central gas are well represented by a two-temperature thermal model *vnei+vmekal*. We performed a spatially resolved spectroscopy of the central gas and found variations of temperature, interstellar absorption and gas density across W28. The colder and denser gas are distributed to the north and east of the X-ray-brightness peak, where the X-rays suffer heavier absorption. We find that thermal conduction is not efficient in the SNR interior, while cloudlet evaporation is an indispensable process to explain both the clumpiness and some of the spectral properties of the X-ray emission.

References

Abdo, A. A., Ackermann, M., Ajello, M., *et al.* 2010b, *ApJ* 718, 348
Aharonian, F., Akhperjanian, A. G., Bazer-Bachi, A. R., *et al.* 2008, *A&A* 481, 401
Arikawa, Y., Tatematsu, K., Sekimoto, Y., & Takahashi, T. 1999, *PASJ* 51, L7
Bykov, A. M., Chevalier, R. A., Ellison, D. C., & Uvarov, Y. A. 2000, *ApJ* 538, 203
Cui, W. & Cox, D. P. 1992, *ApJ* 401, 206
Gabici, S., Aharonian, F. A. & Casanova, S. 2009, *MNRAS* 396, 1629
Jones, T. W., *et al.* 1998, *PASP* 110, 125
Li, H. & Chen, Y. 2010, *MNRAS* 409, L35
Parker, Q. A., Phillipps, S., Pierce, M. J., *et al.* 2005, *MNRAS* 362, 689
Rho, J. & Borkowski, K. J. 2002, *ApJ* 575, 201
Rho, J. & Petre, R. 1998, *ApJ* (Letters) 503, L167
Shelton, R. L., Cox, D. P., Maciejewski, W., *et al.* 1999, *ApJ* 524, 192
White, R. L. & Long, K. S. 1991, *ApJ* 373, 543

Supernova Environmental Impacts
Proceedings IAU Symposium No. 296, 2013
A. Ray & R. A. McCray, eds.

doi:10.1017/S1743921313009836

The Structure of Overionized Plasma in SNR IC 443

Bing Jiang[1], Fabrizio Bocchino[2], Marco Miceli[2], Eleonora Troja[3], Yang Chen[1] and Hiroya Yamaguchi[4] †

[1]Department of Astronomy, Nanjing University, Nanjing 210093, P.R.China
email: bjiang@nju.edu.cn

[2]INAF-Osservatorio Astronomico di Palermo, Piazza del Parlamento 1, 90134 Palermo, Italy
[3]NASA, Goddard Space Flight Center, Greenbelt, MD 20771, USA
[4]Harvard-Smithsonian Center for Astrophysics, 60 Garden Street, Cambridge, MA 02138, USA

Abstract. During the last few years, overionized (recombining) plasmas were unexpectedly discovered in a few supernova remnants, but the origin is still unclear. In this contribution, we present a preliminary spectroscopic analysis of the X-ray emission from the north central region of IC443, one of the "recombining" remnants. An overionized NEI plasma model can reproduce well the Ly-alpha lines and the recombination edges in the spectrum. The ionization temperatures for the metals Mg, Si and S are much higher than the electron temperatures. which is a strong indication of overionization of these elements. The different spectral features of the recombining plasma are characterized on scales of a few arcmin, such as the increasing trend of the pre-cooling temperature and the ionization time from south to north, which may imply a pre-heating direction.

Keywords. ISM: individual (IC 443 (G189.1+3.0)), radiation mechanisms: thermal, supernova remnants, X-rays:ISM

---

1. Introduction

In recent decade, the overionized (recombing) plasma was discovered in several supernova remnants (SNRs), such as IC 443 (Kawasaki *et al.* 2003, hereafter K03; Yamaguchi *et al.* 2009, hereafter Y09), W49B (Ozawa *et al.* 2009, Miceli *et al.* 2010), G359.1-0.5(Ohnishi *et al.* 2011), W44 (Uchida *et al.* 2012) and W28(Sawada & Koyama 2012), challenging the understanding of the evolution of SNRs and its origin is still controversial. IC 443 (G189.1+3.0), in which the overionization was first detected (K03), presents strong recombing emission with *Suzaku* (Y09) while the *XMM-Newton* observation suggested only marginal overionized plasma (Troja *et al.* 2008). With new *XMM-Newton* observation we investigated the difference and the spatial resolved structure of the plasma.

2. Data Analysis and Results

Our *XMM-Newton* observation of IC 443 was performed on 2010 March 07 to March 09 (Obs-ID = 0600110101, PI: E. Troja). The observation data files (ODF) were reprocessed using the *XMM-Newton* Science Analysis System (SAS) version 11.0.0 task following the standard procedures and were screened for the flares using the *XMM-Newton* Extended Source Analysis Software (*XMM*-ESAS) version 4.3.

† Supported by the NSFC grants 11203013 and 11233001, and the grants 20110091120001 and 20120091110048 from the Educational Ministry of China

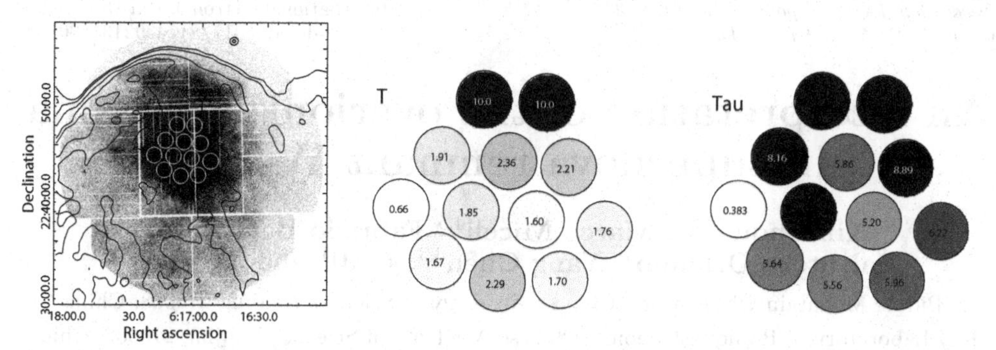

Figure 1. Left: Counts image of IC 443 (northern part) observed by *XMM-Newton* MOS2 in the 0.3-10 keV band, shown on an intensity color scale, superposed with VLA radio continuum at 1.4 GHz (contours) and the 12 regions (circles). Middel: the pre-cooling temperatures obtained from the overionized NEI plasma model for 12 regions in units of keV. Right: the same of the ionization timescales in units of $10^{11}\,\mathrm{s/cm^3}$.

We searched for overionized plasma and successfully found it in the north central area of the remnant, around where the overionization was previously detected by K03 and Y09. This area was then divided into 12 arcmin-scale regions for further analysis (Fig. 1).

For all 12 regions, an overionized NEI plasma model can reproduce well Ly-alpha lines and the recombination edges in the X-ray spectra. Using the collisional ionization equilibrium model with one temperature component or two temperature components, the spectra can also be fitted with good χ^2 value, however, the Ly-α lines, especially of Si and S, are failed to be explained. We obtained the ionization temperatures for different metals with that of Mg, Si and S much higher than the electron temperature of the gas, strongly indicating of overionization of these elements. These results are roughly consistent with that in Y09. The electron temperatures obtained from the overionized NEI model, are almost uniform, around 0.5 keV, over all regions, while the distributions of the pre-cooling temperatures (ionization temperatures) and the ionization timescales (see Fig. 1) both indicate an increasing trend from south to north, which is probably the pre-heating direction.

References

Kawasaki, M. T., Ozaki, M., Nagase, F., Masai, K., Ishida, M. & Petre, R. 2003, *ApJ*, 572, 897

Miceli, M., Bocchino, F., Decourchelle, A., Ballet, J., & Reale, F. 2010, *A&A* (Letters), 514, L2

Ohnishi, T., Koyama, K., Tsuru, T. G., Masai, K., Yamaguchi, H., & Ozawa, M. 2011, *PASJ*, 63, 527

Ozawa, M., Koyama, K., Yamaguchi, H., Masai, K., & Tamagawa, T. 2009, *ApJ* (Letters), 706, L71

Sawada, M. & Koyama, K. 2012, *PASJ*, 64, 81

Troja, E., Bocchino, F., Miceli, M., & Reale, F. 2008, *A&A*, 485, 777

Uchida, H., Koyama, K., Yamaguchi, H., Sawada, M., Ohnishi, T., Tsuru, T. G., Tanaka, T., Yoshiike, S., & Fukui, Y. 2012, *PASJ*, 64, 141

Yamaguchi, H., Ozawa, M., Koyama, K., Masai, K., Hiraga, J. S., Ozaki, M., & Yonetoku, D. 2009, *ApJ* (Letters), 705, L6

Supernova Environmental Impacts
Proceedings IAU Symposium No. 296, 2013
A. Ray & R. A. McCray, eds.

doi:10.1017/S1743921313009848

An interpretation of the overionized plasma in supernova remnant W49B

Xin Zhou[1,2,3]†, Marco Miceli[4,5] Fabrizio Bocchino[5], Salvatore Orlando[5], Yang Chen[6,3], Li Ji[1] and Ji Yang[1]

[1]Purple Mountain Observatory, Chinese Academy of Sciences Nanjing 210008, China

[2]Key Laboratory of Radio Astronomy, Chinese Academy of Sciences Nanjing 210008, China

[3]Key Laboratory of Modern Astronomy and Astrophysics (Nanjing University), Ministry of Education Nanjing 210093, China

[4]Dipartimento di Fisica, Universit di Palermo Piazza del Parlamento 1, I-90134 Palermo, Italy

[5]INAF-Osservatorio Astronomico di Palermo Piazza del Parlamento 1, I-90134 Palermo, Italy

[6]Department of Astronomy, Nanjing University Nanjing 210093, China

Abstract. W49B is a mixed-morphology supernova remnant (SNR) with the presence of enhanced abundances and overionization confirmed by X-ray observation. For the overionization, a strong radiative recombination continuum (RRC) has been detected and confirmed by *SUZAKU* and *XMM-Newton*. Here, we investigate these intriguing observational results through a multi-dimensional hydrodynamic model that takes into account, for the first time, the mixing of ejecta with the circumstellar and interstellar medium, thermal conduction, and non-equilibrium ionization. The model can reproduce the morphology and the overionization pattern of W49B. We found that the overionized plasma originates from the rapid cooling of the hot plasma originally heated by the shock reflected from the dense ring-like cloud. In addition, based on the most updated ATOMDB (v2.0.2), we calculated the spectrum of one cell in the overionized region from the simulation results at present. We got the overionized spectrum that is in agreement with the observational results. Thus, our primary result indicates that the model is consistent with the observations both spatially and spectrally.

Keywords. hydrodynamics; methods: numerical; ISM: individual: W49B(G43.3-0.2); supernova remnant; X-rays: ISM

Recently, strong radiative recombination continua (RRC) were detected by *SUZAKU* X-ray observations toward W49B (Ozawa *et al.* 2009) and several other supernova remnants (SNRs), which show the direct signatures of overionization in these SNRs. For W49B, Miceli *et al.* (2010) further confirmed the RRC feature by a spatially resolved spectral analysis of the *XMM-Newton* data and found the overionized plasma locate in the center of the remnant and in the western region, but not in the eastern region. The presence of such overionized recombining plasma is a contradictory to the traditional knowledge on SNRs where the plasma should be undergoing an ionizing process until reaching ionization equilibrium, and its origin is not well understood.

W49B is one of mixed-morphology (MM) SNRs (Rho & Petre 1998) with the presence of enhanced abundances. Its X-ray emission presents a central jet-like feature and a bright limb to the east (Miceli *et al.* 2006; Keohane *et al.* 2007). While surrounding the X-ray jet-like structure, a near-infrared barrel-shaped structure with coaxial rings was found (Keohane *et al.* 2007). It is also found that a strip of shocked molecular hydrogen

† Present address: West Beijing Road 2, Nanjing, China. email address: xinzhou@pmo.ac.cn
This work is supported by the 973 Program grant 2009CB824800, NSFC grant 11233001, 100 Talents Program of CAS, and ASI-INAF agreement No. I/009/10/0.

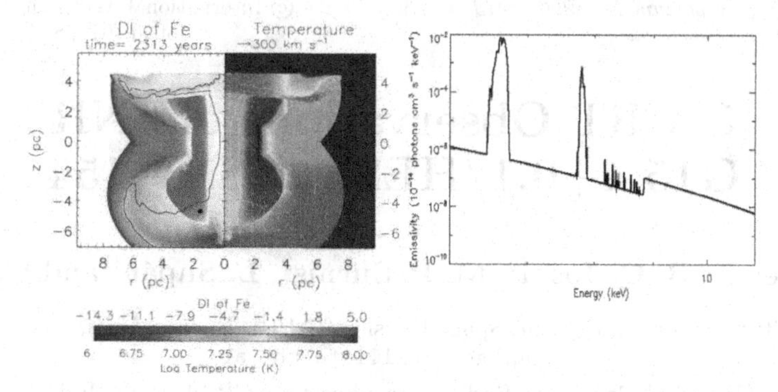

Figure 1. The left panel shows the differential ionization (DI; see Zhou *et al.* 2011 for the definition) and the temperature distributions in the numerical model, where the black point indicates the position for calculating the spectrum which is shown in the right panel.

is present in the eastern region just outside the shell, implying that the remnant is interacting with a molecular cloud there. These results indicate the remnant is encountering an inhomogeneous environment with particular structures, which are likely to be a remnant of the bipolar wind surrounding the massive progenitor star. In such interaction, the thermal conduction between the shocked molecular material and the inner part of the remnant could be efficient.

In Zhou *et al.* (2011) we proposed a possible scenario for the origin of overionized plasma in W49B, which is basically consistent with multi-band observational results (Keohane *et al.* 2007, Ozawa *et al.* 2009, Miceli *et al.* 2010). The overionized plasma can be the results of the rapid cooling of the hot plasma originally heated by the shock reflected from a dense ring-like cloud. In particular, we found two different ways for the rapid cooling of plasma to appear: (i) the mixing of relatively cold and dense material evaporated from the ring with the hot shocked plasma and (ii) the rapid adiabatic expansion of the ejecta. Here we show new results on the calculation of X-ray spectra from the numerical model, and we are trying to evaluate this scenario by comparing the numerical model to the observational results spatially and spectrally.

We calculate the X-ray spectrum from a single cell of the numerical model based on the most updated ATOMDB (v2.0.2), which locates in the overionized region with enhanced abundances (see Figure 1). Overionized plasma detected by *SUZAKU* and *XMM-Newton* probably locates in this region. The calculated spectrum is consistent with the spectra observed by *SUZAKU* (see Figure 3 in Ozawa *et al.* 2009) in many respects, they have corresponding (i) series recombination lines and (ii) prominent RRC feature. We got, for the first time, the model which is spectrally consistent with the observation. These results indicate that our scenario is consistent with the observations both spatially and spectrally. We will further examine the spectral features over the entire remnant in our model for a thorough comparison.

References

Keohane, J. W., Reach, W. T., Rho, J., & Jarrett, T. H. 2007, *ApJ*, 654, 938

Miceli, M., Bocchino, F., Decourchelle, A., Ballet, J., & Reale, F. 2010, *A&A*, 514, L2

Miceli, M., Decourchelle, A., Ballet, J., Bocchino, F., Hughes, J. P., Hwang, U., & Petre, R. 2006, *A&A*, 453, 567

Rho, J. & Petre, R. 1998, *ApJ*, 503, L167

Ozawa, M., Koyama, K., Yamaguchi, H., Masai, K., & Tamagawa, T. 2009, *ApJ*, 706, L71

Zhou, X., Miceli, M., Bocchino, F., Orlando, S., & Chen, Y. 2011, *MNRAS*, 415, 244

Supernova Environmental Impacts
Proceedings IAU Symposium No. 296, 2013
A. Ray & R. A. McCray, eds.

doi:10.1017/S174392131300985X

GMRT Observations of SNR G15.4+0.1/HESS J1818−154

G. Castelletti[1], B. C. Joshi[2], M. P. Surnis[2], L. Supán[1] and G. Dubner[1]

[1]Institute of Astronomy and Space Physics (IAFE), Buenos Aires, Argentina.
email: gcastell@iafe.uba.ar

[2]National Centre for Radio Astrophysics (NCRA), Pune, India.

Abstract. We report here on the first dedicated simultaneous imaging and pulsar observations towards the supernova remnant (SNR) G15.4+0.1, the possible counterpart of the very high energy (VHE) source HESS J1818−154. The observations were carried out using the Giant Metrewave Radio Telescope (GMRT) at 610 and 1400 MHz. Preliminary analysis of data suggests absence of pulsations towards the centroid of HESS J1818−154, with upper limits of 0.6 and 0.3 mJy at 610 and 1400 MHz, respectively. Analysis of data with a larger beam is in progress, which may confirm the presence of a putative pulsar and its wind nebula if it is offset from the centroid of HESS J1818−154.

Keywords. (ISM:) supernova remnants, (stars:) pulsars: general, (stars:) pulsars: individual (J1645−0317, J1939+2134, J1901−0906)

1. Introduction

VHE γ-ray emission was recently detected from SNR G15.4+0.1 (Hofverberg *et al.* 2011). The new source was identified as HESS J1818−154. The VHE emission is center dominated with an extent of about 8'.5, smaller than the radio shell of the SNR (~10'). This fact led to the speculation that the VHE radiation is originating from a yet unknown pulsar wind nebula (PWN), making G15.4+0.1 a probable candidate for a composite SNR.

We have carried out simultaneous imaging and pulsar observations over 32 MHz bandwidth towards SNR G15.4+0.1 at 610 and 1400 MHz with the GMRT, which has the unique capability for such observations. The centroid of the source HESS J1818−154 (α_{J2000} = 18h 18m 3.4s, δ_{J2000} = -15° 27' 54"), marked with a cross symbol in Figure 1, was used as the phase centre for our radio observations. Total integration time on the source was 240 minutes and 260 minutes at 610 and 1400 MHz, respectively. Imaging data were sampled every 16 s, while the pulsar data were sampled every 61 μs. Pulsar data were obtained with a narrow beam phased array (PA) as well as wide beam incoherent array (IA). Three known pulsars, PSRs J1645−0317, J1939+2134 and J1901−0906 were also observed for estimating the sensitivity of pulsar observations.

The imaging data were analysed using the NRAO Astronomical Image Processing Software (AIPS) (www.aips.nrao.edu). The pulsar search was carried out using SIGPROC (www.sigproc.sourceforge.net) on a high performance computing cluster having 64 dual core processors at NCRA.

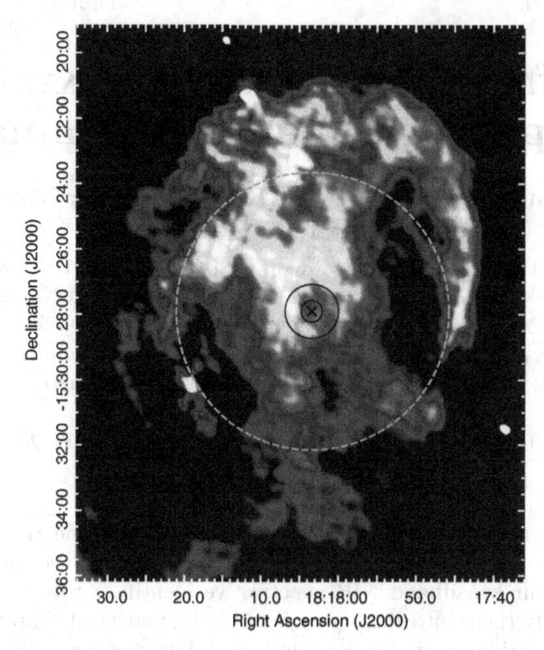

Figure 1. GMRT image of SNR G15.4+0.1 obtained at 610 MHz. The synthesized beam size is 5". The cross and the outermost dashed circle indicate the centroid and the extent of HESS J1818−154, respectively. Inner smaller and larger black circles denote the PA beam sizes at 1400 (40") and 610 MHz (100"), respectively.

2. Results and Discussion

A deep high resolution image of SNR G15.4+0.1 obtained from new GMRT data at 610 MHz is presented for the first time in Figure 1. This image reveals a morphology similar to the previously published lower resolution images at 330 and 1400 MHz (Brogan *et al.* 2006, Helfand *et al.* 2006). Higher resolution and more sensitivity have brought out greater details of radio emission from this SNR. Our preliminary pulsar search did not detect pulsations near the centroid of the VHE source. We estimate the flux density of a putative pulsar to be lower than 0.6 and 0.3 mJy at 610 and 1400 MHz, respectively, assuming a 10% duty cycle, in 1' radius from the centroid of the VHE source. IA observations cover the entire SNR and pulsar search in these data is currently in progress.

Multi-band observations of PWNs are useful to better understand the emission processes responsible for VHE emission and extend our comprehension of the complex acceleration processes involved. In particular, radio measurements are very important to constrain the proposed dynamic and evolutionary models of a PWN.

References

Brogan, C., Gelfand, J. D., Gaensler, B. M., Kassim, N. E. & Lazio, J. W. 2006, *ApJ*, 639, L25

Helfand, D. J., Becker, R. H., White, R. L., Fallon, A. & Tuttle, S. 2006, *AJ*, 131, 2525H

Hofverberg, P., Chaves, R. C. G., Méhault, J. & de Naurois, M. for the H. E. S. S. Collaboration 2011, *32nd International Cosmic Ray Conference*, 7, 248

Supernova Environmental Impacts
Proceedings IAU Symposium No. 296, 2013
A. Ray & R. A. McCray, eds.

doi:10.1017/S1743921313009861

Near-Infrared Study of Iron Knots in Cassiopeia A Supernova Remnant

Yong-Hyun Lee[1], Bon-Chul Koo[1], Dae-Sik Moon[2], and Michael G. Burton[3]

[1]Department of Physics and Astronomy, Seoul National University, 1 Gwanak-ro, Gwanak-gu, Seoul 151-742, Republic of Korea
email: yhlee@astro.snu.ac.kr, koo@astro.snu.ac.kr

[2]Department of Astronomy and Astrophysics, University of Toronto, 50 St. George Street, Toronto, ON M5S 3H4, Canada
email: moon@astro.utoronto.ca

[3]School of Physics, University of New South Wales, Sydney, NSW 2052, Australia
email: m.burton@unsw.edu.au

Abstract. We present the results of near-infrared (NIR) imaging and spectroscopic observations of the Galactic supernova remnant Cassiopeia A (Cas A). Applying the method of Principal Component Analysis to our broadband NIR spectra, we identify a total of 61 NIR emission knots of Cas A and classify them into three groups of distinct spectral characteristics: Helium-rich, Sulfur-rich, and Iron-rich groups. The first and second groups are of the circumstellar and supernova ejecta origin, respectively. The third group, which has enhanced iron emission, is of particular interests since it shows intermediate characteristics between the former two groups. We suggest that the Iron-rich group is knots of swept-up circumstellar medium around the contact discontinuity in Cas A and/or supernova ejecta from deep layers of its progenitor star which have recently encountered a reverse shock in the remnant.

Keywords. ISM: individual (Cassiopeia A), ISM: supernova remnants, infrared: ISM

---

1. Introduction and Observations

Cas A, at the age of ~340 yrs and distance of ~3.4 kpc, is one of the brightest and youngest known supernova remnants (SNRs) in our Galaxy which has provided invaluable information about the dynamical and chemical evolutions of core-collapse SNRs in an early stage. An important observational result of Cas A is discoveries of numerous emission knots scattered around the remnant in the visible wavebands: (1) fast-moving knots (FMKs) of the supernova ejecta origin abundant with nucleosynthetic metallic elements; (2) quasi-stationary flocculi (QSFs) from circumstellar material of its progenitor star; and (3) fast-moving flocculi (FMFs) (or nitrogen-knots) thought to originate from expelled photospheric layers.

In contrast to the studies in the visible wavebands, NIR observations of Cas A have been limited to spectroscopic confirmations of a few previously-identified FMKs and QSFs (Gerardy & Fesen 2001), together with morphological studies of its [Fe II] emission, which is exceptionally bright in its southwestern rim (Rho *et al.* 2003). Given the importance and usefulness of NIR observations of young SNRs, especially broadband spectroscopy enclosing [Fe II] lines for investigating ejecta from deep layers of a progenitor star and shocked circumstellar material (e.g., Koo *et al.* 2007; Moon *et al.* 2009), it is important to conduct such observations for Cas A as we report here.

We obtained intermediate-resolution ($R \simeq 2{,}700$), broadband (0.95–2.46 μm) NIR long-slit spectra of eight positions of bright [Fe II] (1.64 μm) emission distributed around the

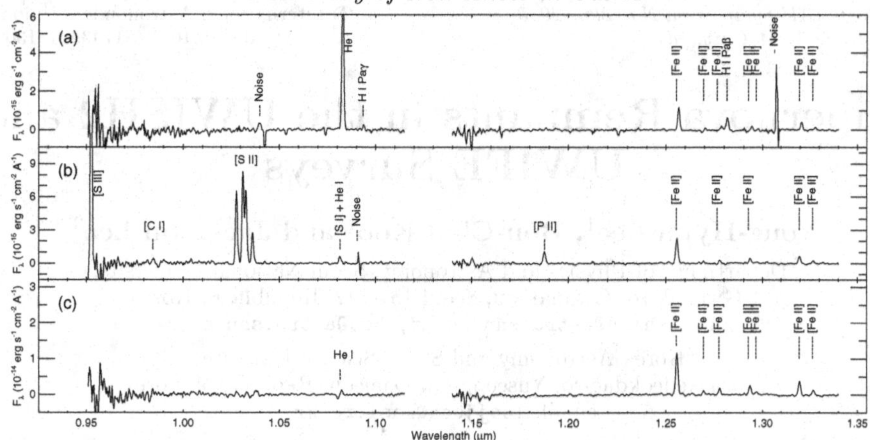

Figure 1. Sample *J*-band spectra of (a) Helium-rich, (b) Sulfur-rich, and (c) Iron-rich knots.

rim of Cas A with the Triplespec spectrograph of the Palomar 5-m telescope. Using the clump-finding algorithm of Williams *et al.* (1994), we have identified a total of 61 knots from the spectra showing 44 emission lines of H, He, and other heavy elements in various ionized stages (see Fig. 1).

2. Results of Principle Component Analysis

The identified knots, 61 in total, show multivariate similarities and differences among them in their spectral characteristics such as identified lines and their intensities and velocities. In order to effectively classify the knots based on their spectral patterns, we adopt the method of Principal Component Analysis (PCA) which leads us to categorize the knots into three distinctive groups: Helium-rich, Sulfur-rich, and Iron-rich groups.

The characteristics of the knots in the Helium-rich group, which have radial velocities within ±100 km s^{-1}, are almost the same as those of the known QSFs, indicating that they are of the circumstellar nature. The knots in this group have strong emission of He I (1.08 μm) and numerous [Fe II] line transitions, but without any evidence of Si, P, and S line emission. The knots in the Sulfur-rich group, on the other hand, have strong [Si VI] (1.96 μm), [P II] (1.19 μm), [S II] (1.03 μm), [S III] (0.95 μm) lines, and show increased radial velocities around +800 km s^{-1} which sometimes extends to +2,000 km s^{-1}. These are consistent with the observed characteristics of FMKs (Gerardy & Fesen 2001). Finally, the knots in the Iron-rich group show both the QSF-like spectral features (e.g., bright [Fe II] emission accompanied only by He I emission) and FMK-like ones (e.g., high radial velocity of several thousand km s^{-1}). We, therefore, suggest that they are either swept-up circumstellar material around the contact discontinuity in Cas A and/or ejecta material from deep layers, near the iron core, of the progenitor which have recently encountered a reverse shock in the SNR.

References

Gerardy, C. L. & Fesen, R. A. 2001, *AJ*, 121, 2781
Koo, B.-C., Moon, D.-S., Lee, H.-G., Lee, J.-J., & Matthews, K. 2007, *ApJ*, 657, 308
Moon, D.-S., Koo, B.-C., Lee, H.-G., *et al.* 2009, *ApJL*, 703, L81
Rho, J., Reynolds, S. P., Reach, W. T., *et al.* 2003, *ApJ*, 592, 299
Williams, J. P., de Geus, E. J., & Blitz, L.1994, *ApJ*, 428, 693

Supernova Environmental Impacts
Proceedings IAU Symposium No. 296, 2013
A. Ray & R. A. McCray, eds.

doi:10.1017/S1743921313009873

Supernova Remnants in the UWISH2 and UWIFE Surveys

Yong-Hyun Lee[1], Bon-Chul Koo[1] and Jae-Joon Lee[2]

[1]Department of Physics and Astronomy, Seoul National University,
1 Gwanak-ro, Gwanak-gu, Seoul 151-742, Republic of Korea
email: yhlee@astro.snu.ac.kr, koo@astro.snu.ac.kr

[2]Korea Astronomy and Space Science Institute,
776, Daedeokdae-ro, Yuseong-gu, Daejeon, Republic of Korea
email: leejjoon@kasi.re.kr

Abstract. We report the preliminary results for the detection of H_2 and [Fe II] line features around the Galactic supernova remnants (SNRs) from the UWISH2 and UWIFE surveys that cover the first galactic quadrant of $7° < l < 65°$ and $|b| < 1.3°$. By this time, we have found a total of 17 H_2-emitting and 14 [Fe II]-emitting SNRs in the coverage, and more than a half of them are detected in both H_2 and [Fe II] emissions, which implies that the environment of these SNRs might be complex and composed of multi-phase medium. In this paper, we present our identification strategy and some preliminary results including H_2 and [Fe II] luminosity distributions.

Keywords. surveys, ISM: supernova remnants, infrared: ISM

1. Introduction

The UWISH2/UWIFE (UKIRT Wide-field Infrared Survey for H_2/[Fe II]) is near-infrared (NIR) H_2-2.12 μm/[Fe II]-1.64 μm narrow-band imaging survey for the first Galactic quadrant ($7° < l < 65°$; $|b| < 1.3°$) that have been performed with the Wide-Field Camera at the United Kingdom Infrared Telescope (Froebrich *et al.* 2011; Lee *et al.*, in prep.). Since the H_2 and [Fe II] lines trace the shocked and/or fluorescently excited regions of molecular and atomic gases associated with jets/outflows around star formation regions, photo-dominated regions, planetary nebulae, evolved massive stars, and supernova remnants, the combination studies of the H_2 and [Fe II] are one of the valuable tools for studying the formation and death of stars.

We have searched for the H_2 and [Fe II] line features around the Galactic SNRs from the survey data. A total of 77 SNRs are falling in the survey area among the currently known 274 Galactic SNRs (Green 2009). These two complementary NIR H_2 and [Fe II] imaging surveys can help us to understand not only the environment and evolution of the individual remnants but also statistical properties of the Galactic SNRs.

2. Observation & Continuum Subtraction

The H_2 survey had been completed in August 2011, while the [Fe II] survey is expected to finish in September 2013. The pixel scale of both surveys is 0.2 arcsec by using 2×2 microstepping sequence, and the median seeings are 0.73 arcsec and 0.80 arcsec for H_2 and [Fe II] surveys, respectively. The total per-pixel integration time is 720 s in both surveys, and the 5σ detection limits of point sources in H- and K-bands are ~ 19 and ~ 18 magnitude, respectively. That limits may show diffuse structures as deep as the surface brightness of $\sim 10^{-19}$ W m^{-2} arcsec^{-2}.

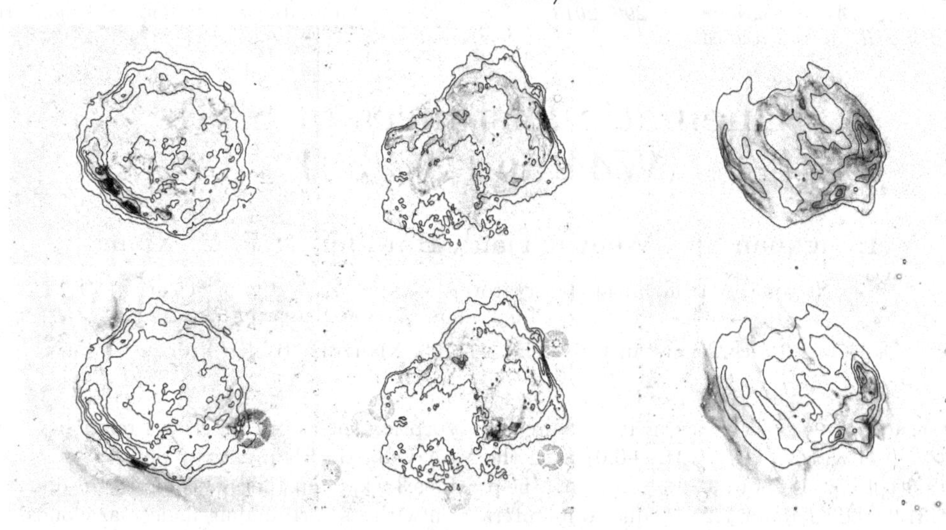

Figure 1. Continuum-subtracted images of G11.2-0.3 (left), 3C 391 (middle), and W49B (right). The upper-pannel shows [Fe II] in UWIFE, while the lower-pannel presents H_2 in UWISH2. Radio continuum contours are superposed on each pannel.

In order to identify the narrow-band emission features, we subtract the *H* and *K*-band continuum images obtained as part of the UKIDSS GPS (UKIRT Infrared Deep Sky Survey of the Galactic Plane) from the H_2 and [Fe II] narrow-band images, respectively. For this, we developed an IDL-based automatic program for dealing with complex and space-variant Point Spread Functions (PSFs) using the subroutines of *Starfinder* code (Diolaiti *et al.* 2000).

3. Preliminary Results

By this time, we have found 17 H_2-emitting and 14 [Fe II]-emitting SNRs corresponding to 22% (17 out of 77) and 26% (14 out of 56) in detection rates, respectively. More than a half of them are newly-confirmed H_2/[Fe II]-emitting SNRs that have never been reported in previous studies. This is likely to increase in future as we inspect the images in more detail. About 60% of the H_2-emitting SNRs have [Fe II] features as well, thus the environment of these SNRs might be complex and composed of multi-phase medium.

Fig. 1 shows three SNRs (G11.2-0.3, 3C 391, and W49B) that radiate both strong H_2 and [Fe II] emissions. Note that the [Fe II] emission features are well correlated with the radio morphologies, while the H_2 emissions are often found outside of the radio boundary.

According to our preliminary results, the extinction-corrected luminosities ranges from $2.4\ L_\odot$ to $2.1 \times 10^3\ L_\odot$ in H_2, and from $2.3\ L_\odot$ to $8.6 \times 10^3\ L_\odot$ in [Fe II], and the brightest SNR is W49B in both H_2 and [Fe II]. This luminosity range is comparable to that of the LMC SNRs, but slightly lower than that of the starburst galaxies, such as M82 and NGC 253 (Alonso-Herrero *et al.* 2003).

References

Alonso-Herrero, A., Rieke, G. H., Rieke, M. J., & Kelly, D. M. 2003, *AJ*, 125, 1210
Diolaiti, E., Bendinelli, O., Bonaccini, D., *et al.* 2000, *A&AS*, 147, 335
Froebrich, D., Davis, C. J., Ioannidis, G., *et al.* 2011, *MNRAS*, 413, 480
Green, D. A. 2009, *Bulletin of the Astronomical Society of India*, 37, 45

Supernova Environmental Impacts
Proceedings IAU Symposium No. 296, 2013
A. Ray & R. A. McCray, eds.

doi:10.1017/S1743921313009885

Kinematic Distances of SNRs W44 and 3C 391

Hongquan Su[1], Wenwu Tian[1], Hui Zhu[1] & F. Y. Xiang[2]

[1]National Astronomical Observatories, Chinese Academy of Sciences, 100012, Beijing, China. email: hq_su@bao.ac.cn

[2]Dept. of Physics, Xiangtan University, 411105, Xiangtan, Hunan Province, China

Abstract. We extracted the neutral hydrogen absorption spectra of supernova remnants W44 (G34.7-0.4) and 3C 391 (G31.9+0.0) from the VLA Galactic Plane Survey data. The revised distance of W41 is about 3.3 kpc replacing the previous 3.0 kpc. Further, we confined the distance of G31.9+0.0 to about 7.2 kpc due to its interaction with its surrounding molecular clouds.

Keywords. methods: data analysis, techniques: radial velocities, ISM: supernova remnants

---

1. Introduction

Distance of supernova remnants (SNRs) is not only a basic parameter of finding their properties such as radius and explosion energy, but also a key factor to research the related high energy phenomena. Recently, the kinematic distances of more than 10 SNRs have been revised by analyzing their H I absorption spectra and ^{13}CO spectra (Tian & Leahy 2011). Here we report our measurement on SNRs W44 and 3C391 which are related with high energy phenomena and interacting with their surrounding environment. Our measurements are based on the assumption of the basic Galactic parameters $V_{\odot}$ = 220 km s^{-1}, $R_{\odot}$ = 8.5 kpc.

2. W44

W44 (also named G34.7-0.4, 3C392) has a shell-like radio-continuum morphology (Figure 1.) with a centrally peaked distribution of thermal X-ray surface brightness (Cox *et al.* 1999). It lies in a complex region which is close to several molecular clouds and a giant molecular cloud (GMC) complex (Dame *et al.* 1986).

We have obtained its H I absorption spectrum (Figure 1.) from the VLA Galactic Plane Survey (VGPS) data. This spectrum shows the highest absorption velocity at 50 km s^{-1} much less than the velocity of tangent point in the direction of W44. This gives a lower limit distance of 3.3 kpc for W44. The H I spectra of W44 show H I emission but no associated absorption beyond the velocity of ~50 km s^{-1}. This constraints W44 at a distance of 3.3 kpc. Caswell *et al.* (1975) found the highest velocity absorption feature of 42 km s^{-1} then concluded that the distance of W44 was 3 kpc ($R_{\odot}$ ~ 10 kpc was adopted) with low velocity resolution data. The VGPS observations provide the strongest evidence to date that the W44 is no further than 3.3 kpc distant; in turn, this supports earlier conclusions that the OH (1720 MHz) maser emission, which peaks at a velocity of 46.9 km s^{-1} (Claussen *et al.* 1999), is physically associated with the W44.

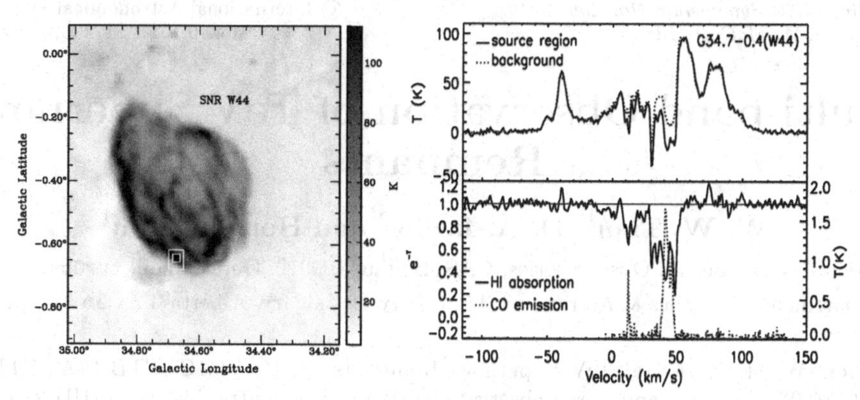

Figure 1. The 1420 MHz continuum image of W44 (left). The H I and CO emission apectra and H I absorption spectrum of W44 (right).

3. 3C 391

3C 391 (also named G31.9+0.0) is a thermal composite remnant, with center-brightened thermal X-ray emission (Rho & Petre 1996). The radio morphology of 3C 391 is a half shell of radius 5′, with another half extend out the open end of the shell which suggests it breakouts into a region of significantly lower density (Renolds & Moffett 1993).

New H I absorption spectrum of 3C 391 reveals the highest absorption velocity at 108 km s^{-1}, i.e. at tangent point, this implies a lower limit distance of ~7.2 kpc for 3C 391. There is H I emission peak at 70 km s^{-1} but lack of absorption, which hints the H I cloud lies behind the continuum source 3C 391. So we give an upper limit distance of 10.4 kpc.

This conclusion is supported by several H I emission peaks without associated H I absorptions in the velocity from 30 to 50 km s^{-1}. Previous studies showed that this SNR has an interaction with clouds at the tangent point (Wilner *et al.* 1998). We conclude that 3C 391 is most likely at a distance of 7.2 ± 0.3 kpc. The uncertainty of 0.3 kpc is from the Galactic parameter's uncertainty and the measurement errors.

Caswell *et al.* (1971) suggested a distance of 8.5−13.4 kpc for 3C 391 because of the H I absorption velocity at 105 km s^{-1} and the lack of absorption at 35−60 km s^{-1} (for a Galactocentric distance of 10 kpc) by Parkes 21cm line observations.

4. Acknowledgement

We acknowledge supports from the NSFC program (011241001, 11261140641, 11273022). It is also supported by a grant of from the John Templeton Foundation and NAOC.

References

Caswell, J. L., Dulk, G. A., Goss, W. M., Radhakrishnan, V., & Green, A. J. 1971, *A&A*, 12, 271

Caswell, J. L., Murray, J. D., Roger, R. S., Cole, D. J., & Cooke, D. J. 1975, *A&A*, 45, 239

Claussen, M. J., Goss, W. M., Frail, D. A., & Desai, K. 1999, *ApJ*, 522, 349

Cox, D. P., Shelton, R. L., Maciejewski, W., Smith, R. K., Plewa, T., Pawl, Andrew., & Ryczka, M. 2011, *ApJ*, 524, 179

Dame, T. M., Elmegreen, B. G., Cohen, R. S., & Thaddeus, P. 1986, *ApJ*, 305, 892

Reynolds, S. P. & Moffett, D. A. 1993, *AJ*, 105, 2226

Rho, J.-H. & Peter, R. 1996, *ApJ*, 467, 698

Tian, W. W. & Leahy, D. A. 2011, *ApJ*, 729, L15

Wilner, D. J., Reynolds, S. P., & Moffett, D. A. 1998, *AJ*, 115, 247

Supernova Environmental Impacts
Proceedings IAU Symposium No. 296, 2013
A. Ray & R. A. McCray, eds.

doi:10.1017/S1743921313009897

Multi-band Observation of TeV Supernova Remnants

W. W. Tian[1], D. A. Leahy[2] and Hongquan Su[1]

[1]National Astronomical Observatories, CAS, Beijing 100012, China email: tww@bao.ac.cn
[2]Department of Physics & Astronomy, University of Calgary, Alberta T2N 1N4, Canada

Abstract. We study several TeV Supernova Remnants (SNRs W51C, CTB 37A, CTB 37B and G353.6-0.7) by radio and X-ray observations. We utilize neutral hydrogen (HI) 21 cm line data to measure their kinematic distances, and use the CO line survey sensitive to molecular hydrogen clouds to validate these distance measurements and understand their relation to the TeV SNRs. Our study show that the TeV γ-ray emission from W51C should not be associated with the high-velocity HI clouds; CTB 37A and CTB 37B are at different distances and are only by chance nearby each other on the sky; the extended TeV emission from G353.6-0.7 possibly originates from the interaction between the SNR shock and the adjacent CO clouds.

Keywords. ISM:supernova remnants - ISM:HII regions - ISM:lines and bands - cosmic rays

---

1. Background

Galactic cosmic Rays(CRs) have been believed to originate from supernova remnants (SNRs). Recent very-high-energy observations show that more than 10 Galactic Supernova Remnants (SNRs) emit TeV γ-rays (http://tevcat.uchicago.edu/). However, it is still unclear whether the charged particles emitting the TeV γ-rays are accelerated protons or electrons. Since 2007, we have studied TeV SNRs (e.g. W51C, CTB 37A/B and G353.6-0.7) by multi-band observation which has proven to be robust in shedding light on γ-ray sources' nature. We have utilized recently-released neutral hydrogen (HI) 21 cm line data from international Galactic plane survey (Stil *et al.*2006) and the ^{13}CO line data from the Galactic ring survey (Jackson *et al.* 2006) to understand their relation to the TeV SNRs.

2. Results

2.1. *W51C/HESS J1923+141*

W51C is located in the strong radio source complex W51. W51 contains components of thermal emission from HII regions (W51A/B) and non-thermal emission from SNR W51C (Koo *et al.* 2005). The recently-detected TeV γ-ray source HESS J1923+141 coincides with Supernova Remnant (SNR) W51C and the star forming region W51B (Aleksic *et al.* 2012). W51C was suggested to be posssibly interacting with nearby hydrogen molecular/atomic clouds at radial velocity of $\sim$ 70 km s^{-1} surrounding the SNR's north-western shell (Koo & Moon 1997). Two compact 1720 MHz OH masers have been detected at W51B and the north-western edge of the W51C shell (Hewitt *et al.* 2008). The masers are caused by shocks either from the SNR W51C or from the star forming region W51B.

Our observation detects high-velocity (HV) HI clouds (>83 km s^{-1}) which coincide with W51B. We find that the clouds are behind W51B and W51C is possibly in front of W51B. This argues against previous claims that the SNR has shocked the HV HI clouds.

In addition, we think that the observed nearby OH masers from dense molecular gas are likely associated with the star forming region in W51B but not W51C. These new results show clear absence of evidence of molecular or atomic gas interacting with W51C, thus casting doubt on a hadronic origin for the GeV/TeV emission from W51C. We think W51C is similar to Tycho SN 1572 (Tian & Leahy 2011) which is not interacting with HI clouds. We also suggest a distance of about 4.3 kpc for W51C, smaller than the tangent point distance of 5.5 kpc in that direction, but still in the Sagittarius spiral arm.

2.2. *CTB 37A/HESS J1714-385 and CTB 37B/HESS J1713-381*

SNRs CTB 37A and CTB 37B are associated with the TeV source HESS J1714-385 and HESS J1713-381 respectively. We use the 1420-MHz radio continuum and 21-cm HI data from the Southern Galactic Plane Survey (SGPS, Haverkorn *et al.* 2006) and build 21-cm HI absorption spectrum to constrain kinematic distances to the SNRs. Our measurements show that CTB 37A has a distance in the range 6.3-9.5 kpc (previously $\sim$11.3 kpc) and an HI column density of $\sim 7.1 \times 10^{21}$ cm^2. CTB 37B is at a distance of $\sim$13.2 kpc (previously 5-9 kpc) and has an HI column density of $\sim 8.3 \times 10^{21}$ cm^2. CTB 37A and CTB 37B are at different distances and are only by chance nearby each other on the sky (Tian & Leahy 2012).

2.3. *G353.6-0.7/HESS J1731-347*

There is a rare class of TeV-emitting SNRs which have non-thermal X-ray and TeV gamma-ray shells (i.e. SNRs G266.2-1.2, G347.3-0.5, G315.2-2.3 and SN1006). SNR G353.6-0.7/HESS 1731-347 is the newest one of the class (Tian *et al.* 2008, Abramowski *et al.* 2011). We study the system by using radio data from the SGPS and Delinha CO observation, X-ray data from XMM-Newton, Suzaku and Chandra observations. We detect the extended hard non-thermal X-ray emission which is coincident with HESS J1731-347 and the shell of the radio remnant, and no thermal X-ray emission from the SNR at a significant level (Bamba *et al.* 2012). Based on the probable association between the hard X-ray and γ-ray emissions and likely association between the CO cloud and the SNR, we conclude that the extended TeV emission likely originates from the interaction between the SNR shock and adjacent CO cloud.

Acknowledgements

We thank supports from the NSFC(011241001, 211381001), BaiRen program of the CAS(034031001), and China Ministry of Science and Technology's Program (2013CB837901, 2012CB821800). This publication was partly supported by a grant from the John Templeton Foundation and National Astronomical Observatories of the CAS.

References

Abramowski, A., Acero, F., Aharonian, F., *et al.* 2011, *ApJ*, 735, 12
Aleksic, J., Alvarez, E. A., Antonelli, L. A., *et al.* 2012, *A&A*, 541, 13
Bamba, A., Pühlhofer, G., Acero, F., *et al.* 2012, *ApJ*, 756, 149
Hewitt, J. W., Yusef-Zadeh, F., & Wardle, M. 2008, *ApJ*, 683, 189
Haverkorn, M., Gaensler, B. M., McClure-Griffiths, N. M., *et al.* 2006, *ApJS*, 167, 230
Jackson, J. M., Rathborne, J. M., Shah, R. H., *et al.* 2012, *ApJ*, 163, 145
Koo, B. C., Moon,D. S. 1997, *ApJ*, 475, 194
Koo, B. C., Kim, K. T., Seward, F. D., & Lee, J. J. 2005, *ApJ*, 633, 946
Stil, J. M., Taylor, A. R., Dickey, J. M. *et al.* 2006, *AJ*, 132, 1158
Tian, W. W., Leahy, D. A., Haverkorn, M., & Jiang, B. 2008, *ApJ*, 679, L85
Tian, W. W. & Leahy, D. A. 2011, *ApJL*, 729, 215
Tian, W. W. & Leahy, D. A. 2012, *MNRAS*, 421, 2593

Supernova Environmental Impacts
Proceedings IAU Symposium No. 296, 2013
A. Ray & R. A. McCray, eds.

doi:10.1017/S1743921313009903

A combined GMRT/CLFST image of IC443 at 150 MHz

D. Mitra[1], D. A. Green[2] and A. Pramesh Rao[1]

[1]National Centre for Radio Astrophysics, Tata Institute of Fundamental Research, Pune University Campus, Post Bag 3, Ganeshkhind Pune 411007, India
email: dmitra@ncra.tifr.res.in, pramesh@ncra.tifr.res.in

[2]Cavendish Laboratory, 19 J. J. Thomson Ave., Cambridge, CB3 0HE, U.K.
email: dag@mrao.cam.ac.uk

Abstract. IC443 is a relatively large Galactic (≈45 arcmin) SNR with a high radio surface brightness. It has fine scale structure down to arcsec scales, and so is difficult to image on all angular scales with a single instrument. Here observations of IC443 at 151 MHz made with both the GMRT and the CLFST are combined to give a composite image of IC443 on all scales from > 45 arcmin down to ≈ 20 arcsec.

1. Background

IC443 (=G189.1+3.0) is a relatively bright SNR in the Galactic anti-centre, where the Galactic background emission is relatively faint. It is ≈ 45 arcmin in diameter, with brighter emission to the northeast, and fainter emission (with a somewhat larger out radius) in the southwest. Structure is seen at radio wavelengths down to scales of arcsec (e.g. Dickel *et al.* 1989; Wood *et al.* 1991). Various observations show that IC443 is interacting with a surrounding molecular cloud (e.g. Rosado *et al.* 2007), and radio spectral index studies reveal a region of flatter spectrum emission in the east (see Green 1986).

2. Observations and Results

We have combined observations made of IC443 at 151 MHz from two telescopes (see Table 1), in order to cover a wide range of angular scales. The Giant Metrewave Radio Telescope (GMRT) – see Pramesh Rao (2002) – is a synthesis telescope that provides baselines up to ~25 km, but lacks good *uv*-plane coverage on baselines less than a few hundred metres. The GMRT observations of IC443 miss about 20% of the total expect flux density of the SNR (≈280 Jy at 151 MHz, Green 1986), due to the missing

Table 1. Parameters of the GMRT and the CLFST.

	GMRT	CLFST
number of antennas	30	60
antenna type	45-m dish	4 × 10-element yagi
number of baselines	435	776
longest baseline	~ 25 km	~ 4.6 km
shortest baseline	~ 100 m	~ 12 m
array layout	14 in central 'square' 16 in 3 arms	~ east–west
frequency	153 MHz	151.5 MHz
bandwidth	6 MHz	0.8 MHz
primary beam	~ 3°	~ 17°

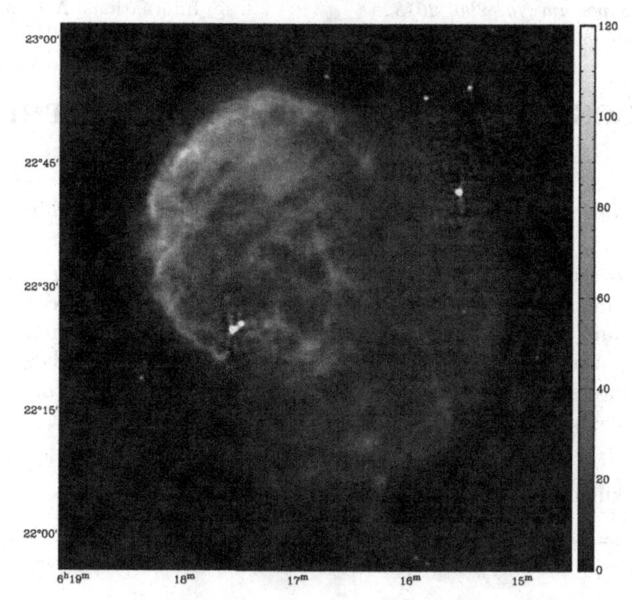

Figure 1. Combined GMRT plus CLFST image of IC443 with a resolution of $24'' \times 19''$ arcsec^2 (at PA of 61°). The scale is 0 to 120 mJy beam^{-1} (using the 'CUBEHELIX' colour scheme of Green 2011). The peak emission is $\approx$2.1 Jy beam^{-1}, from the background source near the NE edge of the remnant, and the noise is $\approx$2.0 mJy beam^{-1}.

short baselines. The Cambridge Low-Frequency Synthesis Telescope (CLFST) – see Rees (1990) – was an (approximately) E–W synthesis telescope that provides good coverage of the *uv*-plane for the small baselines missed by the GMRT. The smoothed 151-MHz image of IC443, from observations made in 1983/84, as used in Green (1986), with a resolution of 5.′4 × 2.′1 arcmin^2 (NS×EW) covers baselines up to $\sim$ 1 km at all position angles. Figure 1 shows the combined GMRT plus CLFST image of IC443 at 151 MHz. This includes emission on a wide range of scales from $\approx$20 arcsec to 45 arcmin. This was made using the `IMERG` task in AIPS which takes the larger/smaller scale structure from the CLFST/GMRT images respectively, gradually merging the contributions on the intermediate scales in both images.

Acknowledgements

We thank staff of the Mullard Radio Astronomy Observatory and the GMRT. GMRT is run by National Centre for Radio Astrophysics of TIFR.

References

Dickel, J. R., Williamson, C. E., Mufson, S. L., & Wood, C. A., 1989, *AJ*, 98, 1363
Green, D. A., 1986, *MNRAS*, 221, 473
Green, D. A., 2011, *BASI*, 39, 289
McGilchrist, M. M., Baldwin, J. E., Riley, J. M., Titterington, D. J., Waldram, E. M., & Warner, P. J., 1990, *MNRAS*, 246, 110
Pramesh Rao, A., 2002, in: A. Pramesh Rao, G. Swarup & Gopal-Krishna (eds), *The Universe at Low Radio Frequencies*, Proc. IAU Symposium No. 199, (San Francisco: ASP), p. 439
Rees, N., 1990, *MNRAS*, 244, 233
Rosado, M., Arias, L., & Ambrocio-Cruz, P., 2007, *AJ*, 133, 89
Wood, C. A., Mufson, S. L., & Dickel, J. R., 1991, *AJ*, 102, 224

Supernova Environmental Impacts
Proceedings IAU Symposium No. 296, 2013
A. Ray & R. A. McCray, eds.

doi:10.1017/S1743921313009915

Distances of Galactic supernova remnants

Hui Zhu and Wenwu Tian
National Astronomical Observatories, CAS, Beijing, China

Abstract. Supernova remnants (SNRs) play a key role in understanding supernovae explosion mechanisms, exploring the likely sources of Galactic cosmic rays and the chemical enrichment of interstellar medium (ISM). Reliable distance determinations to Galactic SNRs are key to obtain their basic parameters, such as size, age, explosion energy, which helps us to study their environment and interstellar medium. We review the methods to determine the distances to SNRs and highlight the kinematic distance measurement by HI absorption and CO emission observations.

Keywords. ISM: kinematics and dynamics, ISM: supernova remnants

1. The empirical surface brightness(Σ) - diameter(D) relation

The theoretical and observational studies indicate that there should be a relationship between surface brightness and diameter of an SNR, $\Sigma = aD^{\beta}$. However, the validity of this relation has not yet been universally accepted. The error in using this method can reach about 40% of the measured distance (Case & Bhattacharya 1998). The reasons to explain dispersions in $\Sigma - D$ relation (Guseinov *et al.* 2003) include: 1) The wide range of explosion energies of supernovae. 2) The influence of ISM around SNRs. Normally, denser ISM will lead to higher surface brightness. 3) The influence of neutron star within an SNR. 4) The distance and diameters uncertainty in calibrators.

A frequently used relation was derived by analyzing 36 Galactic shell SNRs (Case & Bhattacharya 1998). The derived slope was =−2.38 . Guseinov *et al.* (2003) showed a two slopes $\Sigma - D$ relation with a turning point at D = 36.5 pc. For D > 36.5 pc, $\beta = -5.99$. For D < 36.5 pc, $\beta = -2.47$. Pavlović *et al.* (2013) used orthogonal regression to replace the previous used vertical regression and the newly calculated slope, $\beta = -4.8$, is significantly steeper than those in previous works.

2. The kinematic distance

Kinematic distance measurement is based on an assumption that rotation curve of the Milk Way is flat outside the bulge region. For objects that follow the rotation curve well, we can derive their distances from the observed line-of-sight velocity (e.g. HI clouds). 21 cm line observation toward SNRs is used to build HI absorption spectra. Cold gases between observer and SNRs will lead to absorption, gases behind an SNR will not be seen in this SNR's absorption spectrum (see Figure 1). Therefore, the HI absorption features are used to give a lower distance restriction and the emissions without absorptions are used to derive an upper distance restriction. For SNRs showing continuum emission and absorption features, their positions are given by the velocity at which the continuous absorption stops. Tian *et al.* (2007) presented an improved method (combining HI and CO lines's analysis) to build 21 cm HI absorption spectrum against a background extended radio source. In comparison with the traditional method-selecting background beside the source, they let the background surround the source directly which can minimize

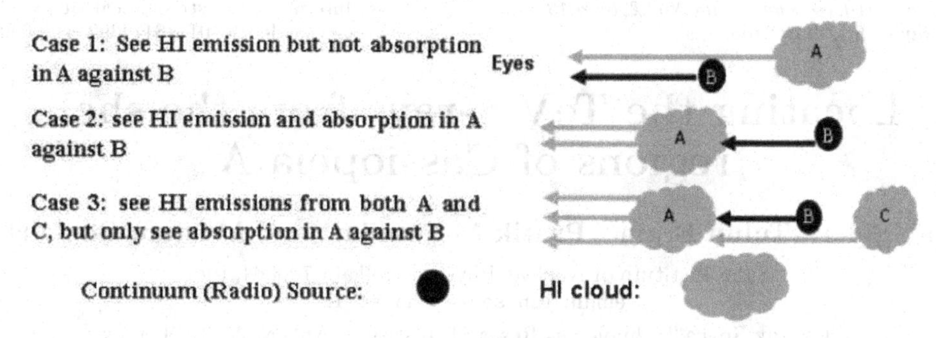

Figure 1. Three cases of relative positions between continuum source and HI clouds.

the possibility of false absorption spectrum. In addition, they use CO spectrum toward target source and the HI absorption spectra of other nearby bright continuum sources to understand absorption spectrum of an SNR better. Using this method, they showed new or revised distances for SNRs, e.g., G18.8+0.3, Kes 69, 73 & 75, W41, CTB 37 & 109, Tycho's SNR (Tian & Leahy 2008a, 2008b, 2011, 2013; Leahy & Tian 2008).

3. Other methods

Kassim *et al.* (1994) found that X-ray observations could be used to calculated the Sedov distance of shell type SNRs which are in the adiabatic expansion phase. Based on the X-ray data from ROSAT observations, they derived Sedov distances to 10 SNRs successfully. Some SNRs are associated with distance-known objects, such as pulsar, HII region, CO cloud (e.g. Vela, Green 1984). Errors in this method are caused by not only the distance's errors of the associated objects, but also the uncertainty of such association. For nearby SNRs (e.g. Crab nebula), proper motions and radial velocity can be used to derive an expansion distance (Green 1984). Extinction measurement based on red clump stars has been used to measure distance (Güver *et al.* 2010). For infrared or optical bright SNRs, extinction A_V can be measured by intensity ratio of H_α and H_β, the SII multiplet ratios at ~1032.0 nm and ~406.8 nm, IR transitions of Fe[II] at 1.6435 and 1.2567 μm etc. (Güver & Özel 2009). For X-ray bright SNRs, we can derive the column density of hydrogen by X-ray absorption. Then hydrogen column density (N_H) can be converted to extinction by the N_H–A_V relation.

References

Case, G. L. & Bhattacharya, D. 1998, *ApJ*, 30, 490
Green, D. A. 1984, *MNRAS*, 209, 449
Guseinov, O. H., Ankay, A., Sezer, A., & Tagieva, S. O. 2003, *A&AT*, 22, 273
Güver, T. & Özel, F. 2009, *MNRAS*, 400, 2050
Güver, T., Özel, F., Cabrera-Lavers, A., & Wroblewski, P., 2009, *ApJ*, 712, 964
Kassim, N. E., Hertz, P., van Dyk, S. D., & Weiler, K. W. 1994, *ApJ*, 427, 95
Leahy, D. A. & Tian, W. W. 2008, *AJ*, 135, 167
Pavlović, M. Z., Urošević, D., Vukotić, B., Arbutina, B., & Göker, Ü. D. 2013, *ApJS*, 204, 4
Tian, W. W., Leahy, D. A., & Wang, Q. D. 2007, *A&A*, 474, 541
Tian, W. W. & Leahy, D. A. 2008a, *ApJ*, 679, 85
Tian, W. W. & Leahy, D. A. 2008b, *MNRAS*, 391, 54
Tian, W. W. & Leahy, D. A. 2011, *ApJ*, 729, 15
Tian, W. W. & Leahy, D. A. 2013, *MNRAS*, 421, 2593

Supernova Environmental Impacts
Proceedings IAU Symposium No. 296, 2013
A. Ray & R. A. McCray, eds.

doi:10.1017/S1743921313009927

Locating the TeV γ-rays from the shell regions of Cassiopeia A

Lab Saha[1], **Tulun Ergin**[2], **Pratik Majumdar**[1] **and Mustafa Bozkurt**[3]

[1]Saha Institute of Nuclear Physics, Kolkata 700064, India,
email: `lab.saha@saha.ac.in`

[2]Tubitak Space Technologies Research Institute, Ankara 06531, Turkey

[3]Bogazici University, Physics Department, Istanbul 34134, Turkey

Abstract. We have analyzed Chandra X-ray data from different parts of the shell of young supernova remnant (SNR) in the energy range of 0.7 - 8 keV. We observed that X-ray flux level varies over different shell regions of the source. Implications of X-ray observation will be discussed here. We also analyzed Fermi-LAT data in the energy range 0.5 - 50 GeV for the source. The differential spectrum obtained in this way fits with simple power-law. We also present here multi-wavelength modeling of the source considering archival radio and TeV data along with Chandra and Fermi-LAT data.

Keywords. Cassiopeia A, X-rays, Observations

---

1. Introduction

Cassiopeia is a young shell type supernova remnant and is observed in radio, infra red, X-rays and even further to GeV-TeV high energy γ-rays (Atoyan *et al.* (2000); Aharonian *et al.* 2001; Albert *et al.* 2007; Abdo *et al.* (2010); Araya *et al.* (2010)). Here we have focussed on different shell regions of Cas A instead of the radio knots which have been observed many times. We have analysed X-ray data from different shell regions and have seen that there are significant differences in number of high energy photons from different regions of the shell. Based on the different level of fluxes we see the region of the shell which will be responsible for the production of very high energy γ-rays. We have also estimated contribution to GeV-TeV energy spectra as well as associated magnetic field from different shell regions of the CasA based on the X-ray data.

2. Leptonic model

For multi-wavelength modeling of CasA we use TeV data from MAGIC (Albert *et al.* 2007), GeV data from Fermi-LAT, X-ray data from Chandra and radio data given by Baars *et al.*, 1977.

We have estimated X-ray fluxes from 4 different regions of the Cas A. To be consistent with the magnetic field B = 80 -160 μG estimated by Vink & Laming (2003), we considered magnetic field 90 μG for one region (R1). The magnetic field for other regions (R2, R3 and R4) are estimated w.r.t the magnetic field considered in R1. We then fit the X-ray data for this region and estimate other model parameters as shown in Table 1. The fitted spectrum is then scaled such that it can explain the total X-ray flux from the remnant. The parameters obtained in that way is used to get the inverse Compton spectrum for the whole remnant. Figure 1 shows the fit to the X-ray data from different parts of the remnant by synchrotron spectra and corresponding inverse Compton spectra. We also estimate the contribution of bremsstrahlung process to the TeV energies and see

Table 1. Parameters for Synchrotron spectra for different regions of the filaments.

Region	Magnetic Field (μG)	spectral index	γ_{min}	γ_{max}	Distance (kpc)	Normalization Constant
Region 1 (R1)	90	2.54	1.0	5.5×10^7	3.4	4.1×10^{53}
Region 2 (R2)	170	2.54	1.0	5.5×10^7	3.4	4.1×10^{53}
Region 3 (R3)	120	2.54	1.0	5.5×10^7	3.4	4.1×10^{53}
Region 4 (R4)	90	2.54	1.0	5.5×10^7	3.4	4.1×10^{53}
Region 5 (R5)	100	2.54	1.0	5.5×10^7	3.4	4.1×10^{53}

that bremsstrahlung process cannot explain the TeV data for ambient proton density of $10/cm^3$.

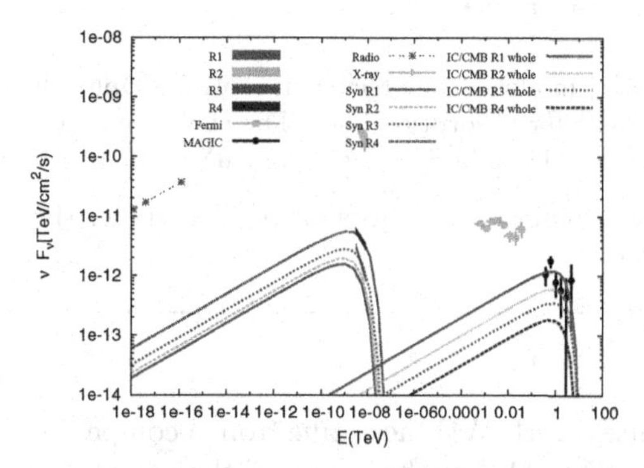

Figure 1. Synchrotron spectra and inverse Compton spectra for different regions of the Cassiopeia A along with the observed multi--wavelength data.

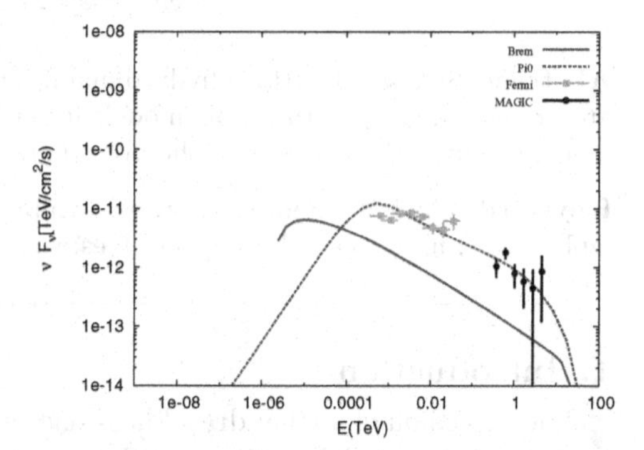

Figure 2. Bremsstrahlung spectrum (solid line) and spectrum due to decay of neutral pions (dotted line) of Cassiopeia A for ambient proton density 10 cm^{-3}.

3. Hadronic model

For hadronic contribution to the γ-ray flux through decay of neutral pions (π^0's), we have considered ambient proton density to be $10/cm^3$ as we have considered for bremsstrahlung process. Accelerated proton spectrum was considered as $dN/dE_p \propto E^{-2.35}$ with an exponential cutoff at 80 TeV. Figure 2 shows the contribution to γ-rays from decay of π^0's. The γ-ray spectrum fits well with the observed data in GeV -TeV range. As we know that the about 10% of the explosion energy of SNR is converted to the energy of relativistic particles, therefore the total explosion energy of the supernova is estimated to be 10^{51} ergs.

4. Conclusion

We show that different shell regions of Cassiopeia differ in X-ray fluxes which can help us to identify the region which is more bright in high energy γ-rays. Apart from that we show that leptonic model cannot explain GeV data. We need to invoke haronic model to explain GeV-TeV data.

References

Atoyan *et al.*, 2000, *A&A*, 354, 915
Aharonian, F., *et al.* 2001, *A&A*, 370, 112
Albert, J., *et al.* 2007, *A&A*, 474,937
Abdo, A., *et al.* 2010, *ApJL*, 710, L92
Araya, M., *et al.* 2010,*ApJ*, 720, 20
Baars, J. W. M., *et al.* 1977, *A&A*, 61, 99
Vink, J., *et al.* 2003, *ApJ*, 584, 758

Supernova Environmental Impacts
Proceedings IAU Symposium No. 296, 2013
A. Ray & R. A. McCray, eds.

doi:10.1017/S1743921313009939

Simulations of RS Oph and the CSM in Type Ia Supernovae

Richard A. Booth[1], Shazrene Mohamed[2] and Philipp Podsiadlowski[1]

[1]Dept. of Astrophysics, University of Oxford,
Denys Wilkinson Building, Keble Road, Oxford, OX1 3RH, United Kingdom
email: richard.booth@astro.ox.ac.uk, podsi@astro.ox.ac.uk

[2]South African Astronomical Observatory
Observatory Road, Observatory, 7925, South Africa
email: shazrene@saao.ac.za

Abstract. Smoothed particle hydrodynamics (SPH) models of the recurrent nova RS Ophiuchi are presented, along with simple models for circumstellar absorption lines. The evolution of the model sodium line is similar to the behaviour in some Type Ia SNe, e.g. SN2006X.

Keywords. hydrodynamics, binaries: symbiotic, circumstellar matter, stars: individual (RS Oph), stars: novae, cataclysmic variables, supernovae

---

1. Introduction

The explosion of a Chandrasekhar-mass white dwarf (WD) accreting from a companion is one popular model for Type Ia Supernovae (SNe Ia), but the nature of the companion is still being debated. The circumstellar material (CSM) constrains the mass accretion process and therefore provides potential insight on the nature of the companion. In the case of a red gaint (RG) companion, there is an equatorial wind and the possibility of recurrent novae, as is observed in the Galactic symbiotic binary RS Oph. SPH simulations are used to model the CSM in RS Oph and investigate the CSM seen in some SN Ia.

2. Circumstellar model

The interaction of the RG wind with the binary potential is modelled with the 3D SPH code GADGET-2 (Springel 2005), which has been modified as described in Mohamed (2010). The binary parameters are the same as in Mohamed *et al.* (2013). The RG and WD masses are 0.8 and $1.38 M_\odot$, and the binary separation is 1.48 AU. The wind mass-loss rate and velocity are $10^{-7} M_\odot \, \mathrm{yr}^{-1}$ and $20 \, \mathrm{km \, s}^{-1}$, respectively. The binary potential focuses the wind into the equatorial plane and forms an Archimedian spiral shock.

Every 20 years a nova is ejected into the system. The nova shell is spherically symmetric with a mass of $2 \times 10^{-7} M_\odot$, and velocity of $4000 \, \mathrm{km \, s}^{-1}$. The equatorial wind constrains the nova, which results in a bipolar structure. The evolution of the novae is followed for 80 years, including the ejection of three novae and their interaction with the previous novae. Fig. 1 (left) shows the system roughly 2 years after the 3rd nova. Components from all three novae can be seen. There is also a smooth equatorial component, which is material from the 1st nova that has been shock heated by the 2nd nova.

3. Circumstellar Sodium Absorption in SN Ia

The CSM 20 years after the 3rd nova is used to model sodium absorption as seen in some SN Ia. The material is assumed to be fully ionized for the first 17 days after

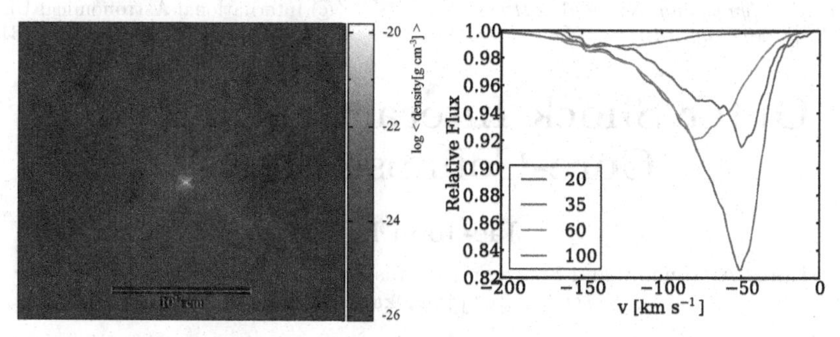

Figure 1. *Left*: Density Profile for a model of RS Oph 2 years after the 3rd nova. *Right*: Sodium absorption line evolution against a SN, for an observer aligned with the plane of the orbit. Recombination gives rise to the initial deepening of the line after the UV flux has decreased. The SN sweeping up the CSM and the decreasing covering factor of the densest regions causes the lines to weaken.

the explosion, after which it recombines on timescales of 10–15 days. This leads to an initial deepening of the line, until the supernova begins to sweep up the material, and the photosphere becomes larger than the equatorial waist. The line begins to weaken thereafter, with the lowest velocity components disappearing first. Fig. 1, (right) shows the evolution, which has features similar to those seen in SN2006X (Patat *et al.* 2007) and SN2007af (Simon *et al.* 2009).

Densities of $10^5\,\mathrm{cm}^{-3}$ are required in the nova shell to reproduce the 10–15 day recombination time scale, however the simulations produce densities of a few $100\,\mathrm{cm}^{-3}$. The formation of optically thin micro-clumps may be able to explain the discrepancy, as may be necessary to explain the apparently super-Eddington mass loss rates from massive stars (Gräfener 2008). Small-scale clumping increases the recombination rate (which depends on the square of the density), without increasing the mean density. An alternative explanation is shell thickness. The simulations currently produce nova shell thickness $\Delta r/r \approx 1$, which maybe much thicker than the shells produced in nature. For example, Patat *et al.* (2007) suggest a thickness $\Delta r/r \approx 0.01$ for the shells in SN2006X. An investigation into nova shells is now being conducted.

4. Conclusions

The detection of CSM in some SN Ia provides a way to test the progenitor channels of at least some of the supernovae. Preliminary models of sodium absorption lines from 3D SPH models of the CSM show some of the features seen in SN2006X. A systematic investigation into the line profiles is being conducted to compare them to a broader range of SNe Ia.

References

Gräfener G., 2008, in: W.-R. Hamann, A. Feldmeier, L. M Oskinova (eds.), *Clumping in Hot-Star Winds*, p. 103

Mohamed, S., 2010, *Ph.D. thesis, Univ.* Oxford

Mohamed, S., Booth, R., & Podsiadlowski, P., 2013, in: R. Di Stefano, M. Orio, & M. Moe (eds.), *Binary Paths to Type Ia Supernovae Explosions*, Proc. IAU Symposium No. 281, (Cambridge), p. 195

Patat, F., Chandra, P., Chevalier, R., Justham, S., *et al.* 2007, *Science*, 317, 924

Simon, J. D., Gal-Yam, A., Gnat, O., *et al.* 2009, *ApJ*, 702, 1157

Springel, V. 2005, *MNRAS*, 37, 239

Supernova Environmental Impacts
Proceedings IAU Symposium No. 296, 2013
A. Ray & R. A. McCray, eds.

doi:10.1017/S1743921313009940

Outer Shock Interaction in Young Core-Collapse SNRs

Jae-Joon Lee

Korea Astronomy and Space Science Institute, Daejeon 305-348, Korea
email: leejjoon@kasi.re.kr

Abstract. Studying the environments in which core-collapse supernovae (SNe) explode and then subsequently evolve is essential to establish the nature of the mass loss and the explosion of the progenitor star. The spatial structure of the outer shock in young core-collapse SNRs provides an opportunity to study the nature of the medium into which the remnant has been expanding. We present our X-ray study of the outer shocks in young core-collapse SNRs in our Galaxy. For Cas A and G292.0+1.8, we find that both remnants have been likely interacting with dense red supergiant winds. For other remnants with bright thermal X-ray emission from the shell, we suggest that they are interacting with pre-existing circumstellar structure. We discuss the nature of the winds and the progenitor stars.

Keywords. ISM: individual (Cas A) — supernova remnants — X-rays: ISM —stars: winds

---

1. Young Core-Collapse SNRs expanding inside their RSG Winds.

Stellar mass is the most important parameter that determines the structure, the evolution and the fate of stars. Observations have established that stars with initial mass of $8 - 15\ M_{\odot}$ undergo core collapse during their red supergiant (RSG) phases and produce Type II-P SNe (Smartt *et al.* 2009). On the other hand, the fate of more massive stars is still not well established (see the discussion in Smith *et al.* 2011). For example, stars in $15 - 30\ M_{\odot}$ may explode during the RSG phase, producing Type II-L, IIb, or IIn SNe while stars $> 30\ M_{\odot}$ may evolve into Wolf-Rayet stars and produce Type Ib or Ic SNe. But observational constraints are rather indirect and weak. Also, their fate likely become complicated by their binary interaction.

For a star that undergoes SN explosion during the RSG phase (or soon after it evolved past the RSG phase), its remnant will primarily interact with the surrounding RSG wind. Existence of such an RSG wind has been suggested for SN 1987A (Chevalier & Emmering 1989; Sugerman *et al.* 2005). The extent of the RSG wind would be determined by the point where the ram pressure of the wind equals the pressure of the surrounding medium. A radius of 5 pc is estimated for a canonical case (Chevalier 2005). Thus, the interaction of the SNR with the RSG wind can last for thousands of years.

Galactic core-collapse SNRs Cas A and G292.0+1.8 are likely examples that are currently interacting with the RSG wind of their progenitor (Lee *et al.* 2010, Lee *et al.* in prep.). By analyzing the X-ray emission from the outer shock regions of these SNRs with the *Chandra* X-ray observatory, we found that the remnants have been expanding inside a wind. For G292.0+1.8, the estimated wind density ($n_{\rm H} = 0.1 \sim 0.3\ {\rm cm}^{-3}$) at the current outer radius (~ 7.7 pc) of the remnant is consistent with a slow wind from a RSG star. For Cas A, we estimate a wind density $n_{\rm H} \sim 0.9\,(\pm 0.3)\,{\rm cm}^{-3}$ at the current outer radius of the remnant (~ 3 pc), which we interpret as a dense slow wind from a red supergiant (RSG) star. Our results suggest that the progenitor star of Cas A had an initial mass around $\sim 16\ M_{\odot}$, and its mass before the explosion was about $5\ M_{\odot}$.

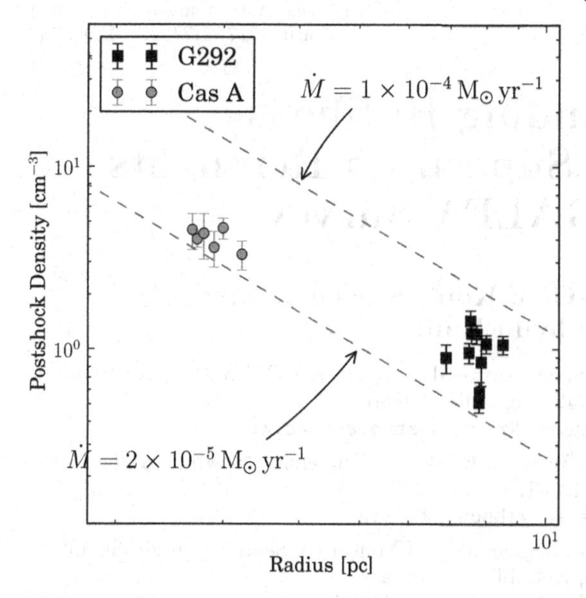

Figure 1. Densities estimated from the postshock X-ray spectra of outermost regions in Cas A and G292.0+1.8. The lines are expected postshock density for SNRs expanding inside the wind of given mass loss rate (wind velocity of $\sim$ 10 km s^{-1} is assumed).

Furthermore, the results suggest that, among the mass lost from the progenitor star ($\sim$11 M$_\odot$), a significant amount (more than 6 M$_\odot$) could have been via its RSG wind.

2. Other Young Core-Collpases SNRs

For Young SNRs interacting with their dense RSG winds, we may expect strong X-ray emission from the reverse-shocked metal rich ejecta. This seems true for Cas A and G292.0+1.8, but in other young SNRs, even though they are bright in X-ray, the ejecta signature seems weak or insignificant. A few examples are G11.2-0.3, Kes 73 and RCW 103. All 3 SNRs show NIR [Fe II] 1.64 μm lines likely from their interaction with dense CSM (Koo B.-C., this Volume). An intriguing possibility is that the SNRs are interacting with dense structure around the progenitor star that were created by the star itself before the SN explosion. The nature of these SNRs needs further investigation.

References

Chevalier, R. A. 2005, *ApJ*, 619, 839
Chevalier, R. A. & Emmering, R. T. 1989, *ApJ*, 342, L75
Lee, J., Park, S., Hughes, J. P., Slane, P. O., Gaensler, B. M., Ghavamian, P., & Burrows, D. N. 2010, *ApJ*, 711, 861
Smartt, S. J., Eldridge, J. J., Crockett, R. M., & Maund, J. R. 2009, *Mon. Not. R. Astron. Soc.*, 395, 1409
Smith, N., Li, W., Filippenko, A. V., & Chornock, R. 2011, *Mon. Not. R. Astron. Soc.*, 412, 1522
Sugerman, B. E. K., Crotts, A. P. S., Kunkel, W. E., Heathcote, S. R., & Lawrence, S. S. 2005, *ApJ*, 627, 888

Supernova Environmental Impacts
Proceedings IAU Symposium No. 296, 2013
A. Ray & R. A. McCray, eds.

doi:10.1017/S1743921313009952

Fast-Expanding HI Shells Associated with Supernova Remnants in the I-GALFA Survey

Geumsook Park[1], Bon-Chul Koo[1], Steven J. Gibson[2], and Ji-hyun Kang[3,4]

[1]Department of Physics and Astronomy, Seoul National University, 1 Gwanak-ro, Gwanak-gu, Seoul 151-742, Republic of Korea
email: pgs@astro.snu.ac.kr, koo@astro.snu.ac.kr

[2]Department of Physics and Astronomy, Western Kentucky University, Bowling Green, KY 42101, USA
email: steven.gibson@wku.edu

[3] Yonsei University Observatory, Yonsei University, 50 Yonsei-ro, Seodaemun-gu, Seoul 120-749, Republic of Korea

[4] Korea Astronomy and Space Science Institute, 776 Daedeokdae-ro, Yuseong-gu, Daejeon 305-348, Republic of Korea
email: jkang@kasi.re.kr

Abstract. We examine excess emission at high positive and negative velocities toward known Galactic supernova remnants (SNRs) in the "Inner-Galaxy Arecibo L-band Feed Array (I-GALFA)" HI 21-cm survey data. The I-GALFA survey covers $\ell = 32°$ to $77°$, and has a velocity range of ± 700 km s^{-1} with high angular and velocity resolutions ($4'$ and 0.18 km s^{-1}, respectively) and good sensitivity (0.2 K). The excess emission which is thought to be part of a fast-expanding HI shell of a SNR is detected from four among 39 SNRs in the I-GALFA area: W44, G54.4 − 0.3, W51C, and CTB 80. Although the HI shells of the four SNRs were already reported in low-resolution studies, the first detection of both sides of an expanding HI shell associated with W44 is very inspiring. We discuss physical properties of these four SNRs and their statistical nature.

Keywords. ISM: supernova remnants, radio lines: ISM

---

1. Introduction

Supernova (SN) explosion is one of important sources to supply energy and matter to the interstellar medium (ISM). In the process, the ISM is swept up by supersonic shock waves, and produce expanding shells. The expanding shells are initially ionized, and over time cool down and neutralize. The neutral shells can be observed in HI emission line at 21 cm, but the identification is difficult because of Galactic foreground and background HI emission. Nevertheless, shells with a large expanding velocity, larger than the maximum velocity allowed by the Galactic rotation in that direction, can be detected.

The I-GALFA survey observed the sky of $\ell = 32°$ to $77°$ and $b \lesssim 10°$ with high resolution ($4'$) and high sensitivity (0.2 K) (See Peek *et al.* 2010, 2011 for more details about the survey.). There are known 275 SNRs in our Galaxy (Green 2009a,b), and 39 among them are located in the I-GALFA area. We have done a systematic study of the HI shells toward the Galactic SNRs using the I-GALFA data.

2. Results and Discussion

We detected fast-moving H I emission toward 4 of 39 SNRs in the I-GALFA area at high positive and/or negative velocities: G34.7 − 0.4 (W44), G49.2 − 0.7 (W51C), G54.4 − 0.3 (HC40), and G69.0 + 2.7 (CTB 80). The excess emission is well constrained within each SNR area in radio continuum, so it is thought that the emission is part of an atomic expanding shell of each SNR. The presence of high-velocity H I shells in these SNRs was reported before by previous studies (e.g., Koo & Heiles 1991, Koo *et al.* 1990, Koo & Heiles 1995, Koo & Moon 1997a), but it is the first time to observe both sides of an expanding shell associated with an SNR (W44) at high negative and positive velocities in H I line. Also, in G54.4 − 0.3, the high-resolution I-GALFA map reveals a well-defined circular symmetric shell at high positive velocities.

We have derived physical parameters, such as an expansion velocity, ambient density, SN energy, age, and so on, by fitting the average H I profiles from two SNRs, W44 and G54.4 − 0.3. The other two SNRs had been studied well in previous researches (Koo & Moon 1997a, Koo *et al.* 1990, Koo & Moon 1997b), and we adopt their parameters which are consistent with the I-GALFA results. The expansion velocities of the H I shells range from 59 to 135 km s^{-1}, and the ages of the SNRs are $1.8 - 9.5 \times 10^4$ yrs. Interestingly, all four SNRs with fast-expanding H I shells (H I SNRs) are all middle-aged SNRs, and their derived ambient densities (0.8 – 400 cm^{-3}) are higher than that of warm (~ 0.1 cm^{-3}) or hot ($\sim 10^{-3}$ cm^{-3}) ISM. We suggest that the SNRs are probably core-collapse SN origin interacting with relatively dense medium.

To explore the fact that only a few of known SNRs in the I-GALFA area are detected as H I SNRs, we examined statistics of SNRs with respect to a model showing the visibility of fast expanding H I shells in the Galactic plane. There are many SNRs where H I shells have not been detected in spite of their favorable locations, i.e., galactic longitudes that require relatively small velocities for expanding shells to be detected. These SNRs might be either too young or too old to have fast-expanding radiative H I shells.

References

Green, D. A. 2009a, *Bulletin of the Astronomical Society of India*, 37, 45

Green, D. A. 2009b *MNRAS*, 399, 177

Koo, B.-C. & Heiles, C. 1991, *ApJ*, 382, 204

Koo, B.-C. & Heiles, C. 1995, *ApJ*, 442, 679

Koo, B.-C. & Moon, D.-S. 1997a, *ApJ*, 475, 194

Koo, B.-C. & Moon, D.-S. 1997b, *ApJ*, 485, 263

Koo, B.-C., Reach, W. T., Heiles, C., Fesen, R. A., & Shull, J. M. 1990, *ApJ*, 364, 178

Peek, J. E. G., Begum, A., Douglas, K. A., *et al.* 2010, *Astronomical Society of the Pacific Conference Series*, 438, 393

Peek, J. E. G., Heiles, C., Douglas, K. A., *et al.* 2011, *ApJS*, 194, 20

Supernova Environmental Impacts
Proceedings IAU Symposium No. 296, 2013
A. Ray & R. A. McCray, eds.

doi:10.1017/S1743921313009964

Observations of O VI Absorption from the Superbubbles of the Large Magellanic Cloud

Ananta C. Pradhan[1], Amit Pathak[2], Jayant Murthy[3] and D. K. Ojha[1]

[1]Tata Institute of Fundamental Research, Homi Bhabha Road, Mumbai 400005, India,
email: acp@tifr.res.in

[2]Department of Physics, Tezpur University, Tezpur 784028, India

[3]Indian Institute of Astrophysics, Koramangala II block, Bangalore 560034, India

Abstract. We have presented the observations of O VI absorption at 1032 Å towards 22 sightlines in 10 superbubbles (SBs) of the Large Magellanic Cloud (LMC) using the data obtained from the *Far Ultraviolet Spectroscopic Explorer (FUSE)*. The estimated abundance of O VI in the SBs varies from a minimum of $(1.09 \pm 0.22)\times 10^{14}$ atoms/cm^2 in SB N206 to a maximum of $(3.71 \pm 0.23)\times 10^{14}$ atoms/cm^2 in SB N70. We find about a 46% excess in the abundance of O VI in the SBs compared to the non-SB lines of sight. Even inside a SB, O VI column density (N(O VI)) varies by about a factor of 2 to 2.5. These data are useful in understanding the nature of the hot gas in SBs.

Keywords. Galaxies: Magellanic Clouds, Ultraviolet: general, ISM: abundances

1. Introduction

SBs are cavities across the interstellar medium (ISM) created by cataclysmic processes such as strong stellar winds and supernova explosions associated with massive stars in an OB association. The swept up material by stellar winds and supernova remnants produce an outer dense shell with an interior filled by low-density hot gas. The gas at the center of the SB is bright in X-rays, the outer shell formed by dust and cool gas is observed in the infrared while the interface between the cool shell and the hot interior is seen in the ultraviolet (UV). O VI absorption lines at 1032 Å and 1037.6 Å are diagnostic of the energetic processes of the interface environments of the SBs which are produced by shock heating and are collisionally ionized. The LMC hosts more than 20 SBs which have been observed in X-rays and UV due to their proximity (∼50 kpc). *FUSE* has observed O VI doublet along many sightlines using early type stars as background objects. We have presented a study of O VI absorption at 1032 Å in the SBs of the LMC in order to assess the properties of intermediate environments in them.

2. Measurement of O VI Column Density

We selected high resolution *FUSE* spectra from the archival data which were observed in the SB lines of sight in the LMC and had well pronounced O VI profiles. We found 22 good O VI observations covering 10 SBs in the LMC. The stellar continuum of O VI profiles are of low order Legendre polynomials ($\leqslant 5$). The measurement of the optical depth of O VI and the equivalent width for the LMC component are done by the apparent optical depth method following the procedures of Savage & Sembach (1991), Sembach & Savage (1992) and Howk *et al.* (2002). The results are listed in Table 1. The 1σ error in the measurements were estimated from the fitting procedures using the uncertainties in the *FUSE* data. The Milky Way ($v \lesssim 150$ km/s) and LMC ($v \gtrsim 150$ km/s) absorption components are well resolved by *FUSE* for most of the sightlines. The average value of the equivalent width of O VI profiles of the SBs in the LMC comes out to be 214±19 mÅ.

Table 1. Details of O VI observations in the SBs of the LMC.

SBs	Size of SBs	*FUSE* Targets	RA (J2000) hr min sec	Dec (J2000) deg min sec	FWHM (mÅ)	Integration limit(km/s)	N(O VI)/ (10^{14} atoms/cm^2)
N11	10′×7′	PGMW-3070	04 56 43.25	-66 25 02.0	132±23	180, 345	$1.27^{+0.43}_{-0.43}$
		LH103102	04 56 45.40	-66 24 45.9	132±23	180, 330	$1.46^{+0.10}_{-0.13}$
		LH91486	04 56 55.58	-66 28 58.0	266±29	175, 385	$2.96^{+0.68}_{-0.68}$
		PGMW-3223	04 57 00.80	-66 24 25.3	129±15	175, 315	$1.27^{+0.18}_{-0.18}$
N51	12′×10′	Sk-67D106	05 26 15.20	-67 29 58.3	180±7	175, 345	$2.01^{+0.50}_{-0.53}$
		Sk-67D107	05 26 20.67	-67 29 55.4	254±10	160, 360	$2.85^{+0.27}_{-0.27}$
		Sk-67D111	05 26 47.95	-67 29 29.9	214±19	175, 365	$2.20^{+0.29}_{-0.24}$
N57	12′×7′	Sk-67D166	05 31 44.31	-67 38 00.6	206±9	165, 390	$2.09^{+0.25}_{-0.20}$
N70	8′×7′	SK-67D250	05 43 15.48	-67 51 09.6	316±33	165, 375	$3.71^{+0.23}_{-0.23}$
		D301-NW8	05 43 15.96	-67 49 51.0	228±30	175, 365	$2.60^{+0.09}_{-0.13}$
		D301-1005	05 43 08.30	-67 50 52.4	284±57	165, 385	$3.37^{+0.17}_{-0.22}$
N144	13′×12′	HD36521	05 26 30.32	-68 50 25.4	126±9	175, 340	$1.18^{+0.28}_{-0.28}$
		Sk-68D80	05 26 30.43	-68 50 26.6	303±16	145, 335	$3.61^{+0.30}_{-0.26}$
N204	14′×13′	Sk-70D91	05 27 33.74	-70 36 48.3	256±7	160, 365	$2.69^{+0.24}_{-0.17}$
N206	9′×15′	BI184	05 30 30.60	-71 02 31.3	118±10	165, 330	$1.09^{+0.22}_{-0.22}$
		Sk-71D45	05 31 15.55	-71 04 08.9	194±9	160, 345	$1.80^{+0.24}_{-0.21}$
N154	12′×8′	SK-69D191	05 34 19.39	-69 45 10.0	185±25	165, 340	$1.65^{+0.26}_{-0.23}$
N158	8′×7′	HDE269927	05 38 58.25	-69 29 19.1	245±8	160, 320	$2.62^{+0.16}_{-0.21}$
30DOR C	7′×6′	MK42	05 38 42.10	-69 05 54.7	228±24	160, 330	$2.60^{+0.46}_{-0.45}$
		SK-69D243	05 38 42.57	-69 06 03.2	307±15	150, 345	$3.63^{+0.40}_{-0.45}$
		30DOR-S-R136	05 38 51.70	-69 06 00.0	185±25	165, 320	$1.77^{+0.20}_{-0.24}$
		SK-69D246	05 38 53.50	-69 02 00.7	211±7	155, 325	$2.36^{+0.18}_{-0.14}$

3. Results and Discussions

N(O VI) of SBs varies from $(1.09\pm0.22)\times10^{14}$ to $(3.71\pm0.23)\times10^{14}$ atoms/cm^2. The mean N(O VI) for the SBs is found to be $(3.07\pm0.62)\times10^{14}$ atoms/cm^2 while the mean N(O VI) for the non-SBs lines of sight is $(2.10\pm0.50)\times10^{14}$ atoms/cm^2 (Pathak *et al.* 2011). Thus, the SBs in the LMC show an excess O VI abundance of about 46% in comparison to non-SB regions. Studies for SB N70 (Danforth *et al.* 2006) found similar results with 60% more O VI than the non-SB targets. Considering N(O VI) inside the individual SBs (e.g., N11 and 30 Dor C), a variation of about a factor of 2 to 2.5 in N(O VI) was obtained. The thermal conduction between the interior hot, X-ray producing gas and the cool, photoionized shell of SBs is found to be the most favourable mechanism for the production of O VI in the SBs (Danforth *et al.* 2006) as the thermal conduction models (Weaver *et al.* 1977 and Borkowski *et al.* 1990) have found N(O VI) $\approx$ few $\times$ 10^{13} atoms/cm^2, which is a close approximation of the observed N(O VI). O VI in the SBs does not show any correlation with the ISM morphologies such as Hα and X-ray surface brightnesses. The temperature estimated from the line width of O VI comes to be $\sim10^6$ K which represents slightly higher FWHM than expected for O VI absorption.

References

Borkowski, K. J., Balbus, S. A., & Fristrom, C. C. 1990, *ApJ*, 355, 501

Gaustad, J. E., McCullough, P. R., Rosing, W., & Van Buren, Dave 2001, *PASP*, 113, 1326

Danforth, C. W., Shull, J. M., Rosenberg, J. L., & Stocke, J. T. 2006,*ApJ*, 640, 716

Howk, J. C., Sembach, K. R., Savage, B. D., Massa, D., Friedman, S. D., & Fullerton, A. W. 2002, *ApJ*, 569, 214

Pathak, Amit, Pradhan, A. C. Murthy, Jayant, & Sujatha, N. V. 2011, *MNRAS*, 412, 1105

Savage, B. D. & Sembach, K. R. 1991, *ApJ*, 379, 245

Sembach, K. R. & Savage, B. D. 1992, *ApJS*, 83, 147

Weaver, R., McCray, R., Castor, J., Shapiro, P., & Moore, R. 1977, *ApJ*, 218,377

Supernova Environmental Impacts
Proceedings IAU Symposium No. 296, 2013
A. Ray & R. A. McCray, eds.

doi:10.1017/S1743921313009976

Heavy Elements Produced in Supernova Explosion and thier Propulsion in the ISM

Rulee Baruah
Department of Physics, HRH The Prince of Wales Institute of Engineering and Technology, Jorhat-785001, India
email: ruleeb@yahoo.com

Abstract. We study the r-process path at temperatures from $1.0 - 3.0 \times 10^9$K and neutron number density from 10^{20}-10^{30} cm^{-3}. At low density of 10^{20} cm^{-3} and $T_9 = 2.0$, the path contains all the elements as given by experimental data of Wapstra *et al.* (2003). The element $_{98}Cf^{254}$ shown by supernova light curves is found in our results. We take iron (Z = 26) as seed for calculation of abundances for supernova.

Keywords. r-process, light curves, abundances

Supernova Explosion

Due to successive fusion, once the inner core is converted to iron/nickel, no more energy is available from the fusion process and the inner core collapses catastrophically to form a neutron star or black hole. In the resulting explosion, the outer layers of the star are blown out into space with a velocity of upto 15,000 km s^{-1}. Here the r-process nucleosynthesis takes place producing the heavy elements as shown. The enormous amount of energy released in a supernova explosion has major effects on the interstellar medium.

Under the assumption that in a steady state the abundances of elements in the neutron capture path are proportional to β^- decay lifetimes, the computed abundances of all the r-process elements are compared with observed universal abundances taken from Lodders (2003) for different values of Q_n i.e. different T_9 and n_n.

Results

- Peaks at A $\simeq$ 80 and A $\simeq$ 130 are well produced; which corresponds to isotones N = 50 and N = 82, the neutron magic numbers (Z = 28,29,30 and Z = 47,48,49,50). A peak at A $\simeq$ 195 is seen, quite in agreement with the observed data.

Table 1. Chemical elements at the r-process site

Element	$T_9\,(10^9$ **K**$)$	$n_n\,(cm^{-3})$
$_{56}Ba^{137}$	2.5	10^{20}
$_{82}Pb^{207}$	2.5	10^{22}
$_{92}U^{236}$	3.0	10^{22}
$_{98}Cf^{254}$	1.9	10^{20}
For double magic nuclei		
$_{28}Ni^{78}{}_{50}$	1.0	10^{20}
	1.1	10^{22}
	1.2	10^{24}
	1.4	10^{26}
	2.0	10^{28}
$_{50}Sn^{132}{}_{82}$	1.7	10^{20}
	1.9	10^{22}

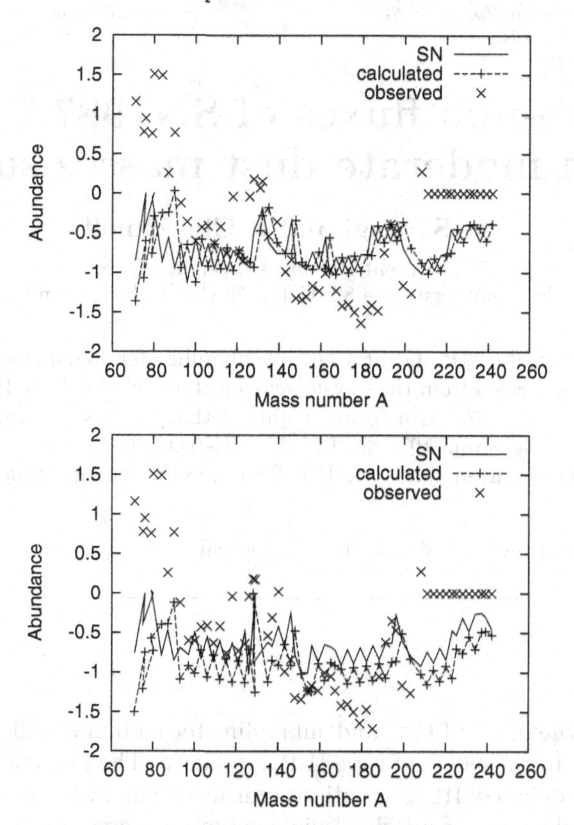

Figure 1. Global r-abundance curve, with waiting point and steady flow approximation at (a) $T_9 = 1.98$ and $n_n = 1.28 \times 10^{26} cm^3$; (b) $T_9 = 1.6$ and $n_n = 1.25 \times 10^{24} cm^3$

- A small peak at A $\simeq$ 210 is noticed; which may be proposed as due to neutron magic isotones N = 126 (Z = 79,80,81,82,83,84,85,86). A peak at A $\simeq$ 235 is observed; which is beyond the scope of discussion as not much nuclear data is available beyond A > 209. We want to propose that this peak is justified considering the fact that U^{235} is a stable element found in nature.

Discussion

In the calculation, we note an element of mass 273 as a termination element of an r-process path, corresponding to atomic number 115. Experimentally some new elements were synthesized at the Lawrence Berkley laboratory e.g. element with Z = 116, 118 etc. As the high density conditions do not show much of the experimentally observed elements, we propose that the heavy elements produced during extreme conditions of SN explosion instantly undergo photodisintegration. In the later expansion stages, they were distributed all over the universe.

References

Audi, G., Wapstra, A. H. & Thibault, C. 2003, *Nucl. Phys. A*, 729, 337

Burbidge, E. M., Burbidge, G. R., Fowler, W .A. & Hoyle, F., 1957, *Rev. Mod. Phys.*, 29, 547

Lodders, K., 2003, *ApJ*, 591, 1220

Swiategki, W. J., Wilczynska, K. S. & Wilczynski, J., 2005, *Phys. Rev. C.*, 71, 014602

Supernova Environmental Impacts
Proceedings IAU Symposium No. 296, 2013
A. Ray & R. A. McCray, eds.

doi:10.1017/S1743921313009988

IR and sub-mm fluxes of SN1987A revisited: when moderate dust masses suffice

A. Sarangi and I. Cherchneff

Department Physik, Universität Basel
Klingelbergstrasse 82, CH-4056 Basel, Switzerland

Abstract. We model the fluxes in the infrared and submillimeter domain using the dust chemical composition and mass derived from the physico-chemical model of a Type II-P supernova ejecta with stellar progenitor of 19 $M_\odot$. Our results highlight that the dust mass predicted to rise over time in our chemical models from 10^{-2} to 10^{-1} $M_\odot$satisfactorily reproduce the infrared and sub millimeter fluxes. They confirm that type II-P SNe are efficient but moderate dust makers in galaxies.

Keywords. supernovae: general, dust, infrared: general

1. Introduction

Observations in the infrared (IR) and submillimeter (submm) indicate the presence of molecules and dust in the ejecta of type II-P supernova. The mass of dust formed in the ejecta is still highly debated: IR observations indicate small dust mass (10^{-5} to 10^{-3} $M_\odot$) formed before 500 days post-explosion, while submm observations with Herschel reveal large reservoirs of cool dust (10^{-2} to 0.7 $M_\odot$) in supernova remnants. The chemistry and time evolution of the ejecta of a typical type II-P supernova with solar metallicity has been studied (Sarangi & Cherchneff 2013). The synthesis of dust grain was modelled with a bottom-up approach through formation of molecules and small clusters in the gas phase from day 100 to 2000 post-explosion. The upper limits on dust masses in the ejecta was estimated for different progenitors (Cherchneff & Sarangi, this volume). We then model the fluxes emitted by dust at different epochs and in different wavelength ranges using the 3D radiative code MOCASSIN (Ercolano *et al.* 2007) and reproduce the Spitzer and Herschel data.

2. Dust mass

The ejecta parameters correspond to a typical 19 $M_\odot$ model (Sarangi & Cherchneff 2013) and the elemental yields are those of Woosley (1988). We model the formation of small dust clusters that include dimers of forsterite ($Mg_4Si_2O_8$), silica, carbon ring (C_{10}), alumina, silicon carbide, iron and magnesium sulphide, and metallic clusters of Fe, Mg and Si. The cluster masses represent an upper limit to the total dust mass and increases from 10^{-4} $M_\odot$ at day 400 to ~ 0.1 $M_\odot$ at day 2000 post-explosion. The prevalent dust clusters are carbon (0.08 $M_\odot$), forsterite (0.014 $M_\odot$) and alumina (6×10^{-3} $M_\odot$).

3. Modelling of IR & submm fluxes from SN1987A

Based on the dust chemical composition and mass from the model, the dust contribution to the flux at different epochs was assessed. The observed IR flux from SN1987A at

Table 1. Input parameters used in MOCASSIN to model the IR flux from SN1987A

Time (days)	Rin (10^{15} cm)	Rout (10^{15} cm)	Luminosity ($L_\odot$)	Blackbody (K)	Mass (10^{-2} $M_\odot$)	Composition Silicate/Alumina (%)
615	9.5	95	6.7e5	6200	1.0	99/1

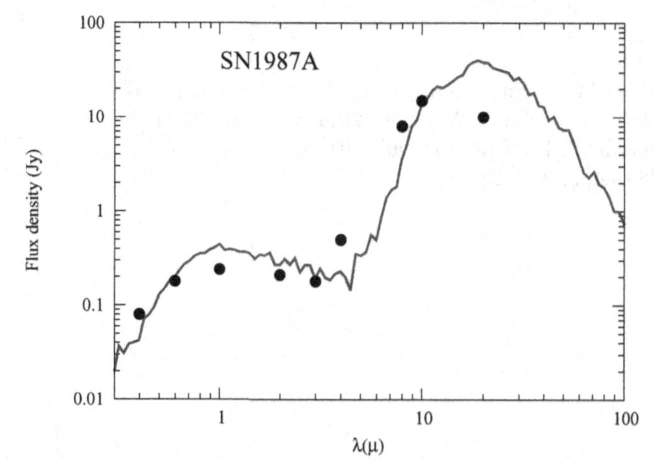

Figure 1. IR flux emitted by our modelled dust composition and mass for a 19 $M_\odot$ progenitor at day 615 post-explosion. The IR flux data of SN1987A at 615 days are shown.

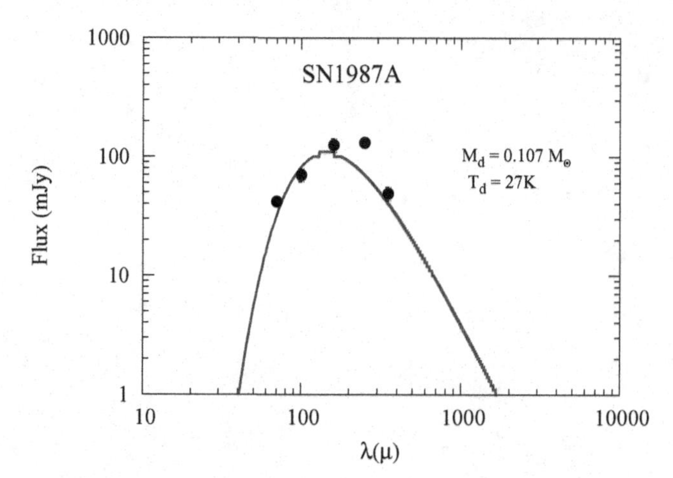

Figure 2. Submm flux emitted by our modelled dust composition and mass for a 19 $M_\odot$ progenitor. The dust temperature is 27 K. The submm dust flux data from Herschel are also shown.

615 days was modelled with the 3D radiative transfer code MOCASSIN (Ercolano *et al.* 2007). A smooth model (i.e., without clumps) with a grain size distribution ranging from 0.005 to 0.05 μm was considered. The input parameters are listed in Table 1. The fitting of the IR data with our modelled dust composition and mass at 615 days is presented in Fig. 1. The submm data of SN1987A from the Herschel space telescope (Matsuura *et al.* 2011) is fitted by the dust composition and mass derived at day 2000 and an absorption coefficient dependant black body fit. The best fit is achieved for a dust temperature of 27 K (Fig. 2).

4. Implication

Our results highlight that a moderate mass of dust (10^{-2} to 0.1 $M_{\odot}$) is sufficient to reproduce the IR flux of the supernova SN1987A and the submm flux measured in the young remnant. They indicate that type II-P SNe are efficient but moderate dust makers in galaxies.

References

Ercolano, B., Barlow, M. J., Sugerman, B. E. K., 2007 *MNRAS*, 375, 753
Matsuura, M., Dwek, E., Meixner, M. *et al.* 2011 *Science*, 333, 1258
Sarangi, A. & Cherchneff, I., 2013 *ApJ*, submitted
Woosley, S. E, 1988, *ApJ*, 330, 218

Supernova Environmental Impacts
Proceedings IAU Symposium No. 296, 2013
A. Ray & R. A. McCray, eds.

doi:10.1017/S174392131300999X

Radio studies of relativistic SN 2009bb

Alak Ray[1], Naveen Yadav[1], Sayan Chakraborti[2], Alicia Soderberg[2] and Poonam Chandra[3]

[1]Tata Institute of Fundamental Research, Mumbai 400005, India
email: akr@tifr.res.in, nyadav@tifr.res.in

[2]Harvard-Smithsonian Center for Astrophysics, Cambridge, MA 02138, USA

[3]National Centre for Radio Astrophysics, TIFR, Pune 411007, India

Abstract. A local sub-population of type Ib/c supernovae (stripped envelope SNe) with mildly relativistic outflows have been detected as sub-energetic Gamma Ray Bursts (GRBs) or X-ray Flashes (XRFs) and as radio afterglows without detected GRB counterpart. SN 2009bb belongs to the last class of objects. The long term radio observations with (J)VLA and GMRT of this SN map the dynamics of the relativistic ejecta characteristic of Central Engines associated with GRBs. We present here GMRT observations of this SN from October 2009 onwards.

Keywords. shock waves, supernovae, radio continuum: general, circumstellar matter

---

Relativistic bulk motion of matter is implied (Piran 1999, Piran 2004) in long duration gamma ray bursts (GRBs) that are linked to core-collapse explosion of a stripped envelope massive star (Paczynski 1986, Woosley 1993). In the collapsar model (MacFadyen & Woosley 1999), matter flows towards a newly formed black hole or rapidly spinning, highly magnetized neutron star, constituting the central engine from which powerful jets are launched along the spin axis which plow through the collapsing star eventually attaining relativistic speeds and producing the GRB. The association of energetic core collapse SN 1998bw with the underluminous low redshift ($z = 0.0085$) GRB 980425, as well as spectroscopic identification of SN features well after the GRB event, established the SN-GRB connection (Hjorth & Bloom 2012 and references therein). These SNe belong to a rare subclass of type Ibc SNe called the broad-line Ic's. The central engine driven objects have been discovered mainly through their concomitant gamma-ray emission, but the discovery of luminous radio emission from the type Ibc SN 2009bb that was undetected in gamma-rays despite extensive search led to the measurement of a substantial relativistic outflow that must be powered by the central engine (Soderberg *et al.* 2010). The outflow speed was measured by combining observations of SN 2009bb from the VLA and the GMRT at multiple epochs. These are well fitted by Synchrotron Self Absorption (SSA) models which can lead to radiosphere locations interpreted from a combination of spectral peak frequencies and fluxes.

We report GMRT observations of SN 2009bb from October 2009. The GMRT and earlier VLA data are plotted in Fig 1.

The radius evolution with time measured from the SSA spectrum showed the SN had mildly relativistic ejecta (Soderberg *et al.* 2010, Chakraborti & Ray 2011). The magnetic field amplified by the shock and how it evolves with time and the role of such central engine driven explosions as accelerators of Ultra High Energy Cosmic Ray accelerators are described in Chakraborti *et al.* 2011.

Table 1. GMRT observations of SN 2009bb

Observation Dates	Days after explosion	Frequency (MHz)	Flux density (mJy)
22Oct–01Nov 2009	218–223	325	12.0±0.26
		610	6.3±0.70
		1280	4.4±0.30
09Feb–18Feb 2010	328–337	325	7.2±0.29
		610	3.6±0.05
		1280	1.2±0.07
17Apr–25Apr 2010	395–403	325	7.4±0.34
		610	3.6±0.09
		1280	1.9±0.05
26Dec–31Dec 2010	648–653	325	6.1±0.12
		610	1.9±0.05
		1280	0.9±0.05
26Apr–05May 2011	769–778	325	4.6±0.19
		610	1.4±0.08
		1280	0.7±0.05
04Jan–13Jan 2012	1022–1031	325	5.9±0.18
		610	1.4±0.06
		1280	0.7±0.03
17Jul–20Jul 2012	1217–1220	325	4.4±0.15
		610	2.0±0.04

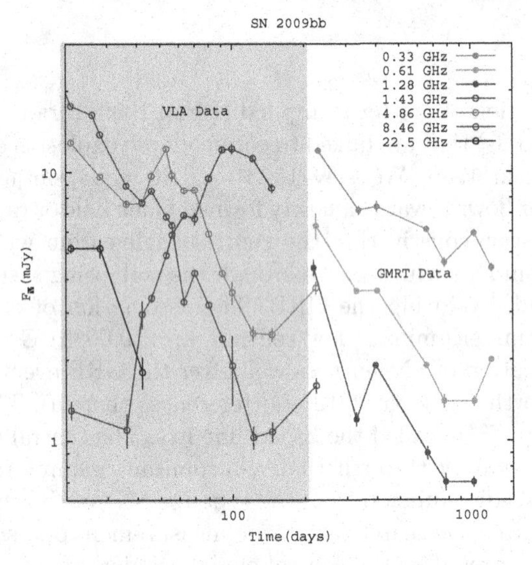

Figure 1. Radio light curves of SN 2009bb at indicated frequencies

References

Chakraborti, S. & Ray, A. 2011, *ApJ*, 729, 57

Chakraborti, S., Ray, A., Soderberg, A. M., & Loeb, A. 2011, *Nature Communications*, 2, 175

Hjorth, J. & Bloom, J. S. 2012, *in "Gamma-Ray Bursts", Cambridge Astrophysics Series 51, eds. C. Kouveliotou, et al., Cambridge University Press (Cambridge)*, p. 169

MacFadyen, A. I. & Woosley, S. E. 1999, *Ap J*, 424, 262

Paczynski, B. 1986, *Ap J*, 308, L43

Piran, T. 1999, *Phys. Reports*, 314, 575

Piran, T. 2004, *Rev. Mod. Phys*, 76, 1143

Soderberg, A. M., Chakraborti, S., *et al.* 2010, *Nature*, 463, 513

Woosley, S. E. 1993, *ApJ*, 405, 273

Supernova Environmental Impacts
Proceedings IAU Symposium No. 296, 2013
A. Ray & R. A. McCray, eds.

doi:10.1017/S1743921313010004

Clumping of ejecta and accelerated cosmic rays in the evolution of type Ia SNRs

S. Orlando[1], F. Bocchino[1], M. Miceli[1], O. Petruk[2] and M. L. Pumo[3]

[1]INAF - Osservatorio Astronomico di Palermo, Piazza del Parlamento 1, 90134 Palermo, Italy
[2]Inst. for Appl. Problems in Mechanics and Mathematics, Naukova St. 3-b Lviv 79060, Ukraine
[3]INAF - Osservatorio Astronomico di Padova, Vicolo dell'Osservatorio 5, 35122 Padova, Italy

Abstract. We investigate the role played by initial clumping of ejecta and by efficient acceleration of cosmic rays (CRs) in determining the density structure of the post-shock region of a Type Ia supernova remnant (SNR) through detailed 3D MHD modeling. Our model describes the expansion of a SNR through a magnetized interstellar medium (ISM), including the initial clumping of ejecta and the effects on shock dynamics due to back-reaction of accelerated CRs. The model predictions are compared to the observations of SN 1006. We found that the back-reaction of accelerated CRs alone cannot reproduce the observed separation between the forward shock (FS) and the contact discontinuity (CD) unless the energy losses through CR acceleration and escape are very large and independent of the obliquity angle. On the contrary, the clumping of ejecta can naturally reproduce the observed small separation and the occurrence of protrusions observed in SN 1006, even without the need of accelerated CRs. We conclude that FS-CD separation is a probe of the ejecta structure at the time of explosion rather than a probe of the efficiency of CR acceleration in young SNRs.

Keywords. (magnetohydrodynamics:) MHD, (ISM:) cosmic rays, (ISM:) supernova remnants

---

1. Introduction

Current multi-dimensional models of supernova remnants (SNRs) predict an average distance D between the contact discontinuity (CD) and the forward shock (FS) that is much larger than that observed in many young SNRs (e.g. SN1006) and cannot explain the high number of knots observed to protrude ahead of the shock. A possible cause invoked to explain these features has been the back-reaction of accelerated cosmic rays (CRs) at the FS. However, this mechanism cannot explain the evidence that the distance D is lower and the occurrence of protrusions is higher than predicted even in regions where the local CRs acceleration efficiency is very low (e.g. Miceli *et al.* (2009)).

Nowadays, there is a growing consensus that density clumping of ejecta may be intrinsic at early phases of the remnant evolution (e.g. Maeda *et al.* (2010)). An important question is: can the ejecta clumping enhance the growth of hydrodynamic instabilities up to a level that allows them to reach and possibly overtake the FS? We investigate this issue by developing a 3D MHD model of SNR expanding through a magnetized medium, including consistently both the initial ejecta clumping and the effects on shock dynamics due to back-reaction of accelerated CRs (see Orlando *et al.* (2012) for a detailed description of the model and an extended discussion of the results).

2. The MHD model and the results

Our model describes the expansion of a SNR through a non-uniform magnetized medium and is described by the time-dependent MHD equations (Orlando *et al.* (2007), Orlando *et al.* (2012)). The cosmic ray back-reaction is taken into account by means of

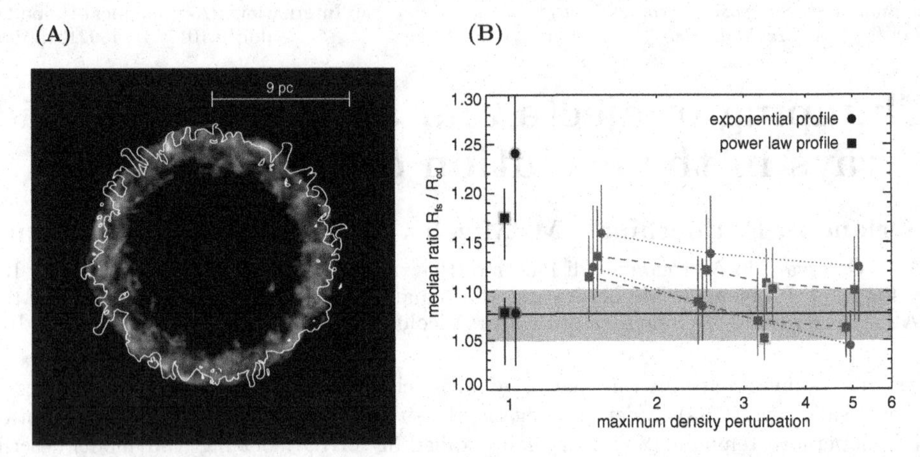

Figure 1. **(A)** 3D rendering of mass density for a model with an initial clumpy structure of the ejecta and no shock modification by CRs. The white contour encloses the original ejecta material. **(B)** Median values of the ratio of the forward shock radius to the contact discontinuity radius R_{fs}/R_{cd} versus the maximum density perturbation for models either with (red and blue symbols for initial size of the clumps $\approx 1\%$ and $\approx 2\%$ of the remnant diameter, respectively) or without (black symbols) ejecta clumping, and for models including the back-reaction of accelerated CRs (symbols with a cyan or magenta halo). The grey region marks the range of values observed in SN1006 (Miceli *et al.* (2009)).

a space- and time-dependent effective adiabatic index γ_{eff} by following the approach of Ferrand *et al.* (2010). The calculations are performed with the FLASH code (Fryxell *et al.* (2000)). As initial conditions we adopted parameters appropriate to describe the SNR SN1006. We also assume that the initial ejecta has a clumpy structure. The clumps have been modelled as per-cell random density perturbations (Orlando *et al.* (2012)).

We performed simulations either with or without the backreaction of accelerated CRs and either with or without the clumping of ejecta. For each simulation, we derived the median values of the ratio between the FS and the CD radii R_{fs}/R_{cd} versus the maximum density perturbation as reported in Fig. 1. We found that the back-reaction of accelerated CRs alone cannot reproduce the observations in SN1006 unless the CRs energy losses are extreme (i.e. $\gamma_{eff} \approx 1.1$) and independent on the obliquity angle. An initial clumping of ejecta turns out to be a fundamental ingredient to reproduce the observed values of R_{fs}/R_{cd} in SN1006 and its obliquity dependence. We conclude therefore that, in general, the separation between the FS and the CD is not a reliable diagnostic tool for studying the CR shock modification (see Orlando *et al.* (2012) for a full discussion of the results).

Acknowledgements

The software used in this work was in part developed by the DOE-supported ASC / Alliance Center for Astrophysical Thermonuclear Flashes at the University of Chicago.

References

Ferrand, G., Decourchelle, A., Ballet, J., Teyssier, R., & Fraschetti, F. 2010, *A&A*, 509, L10
Fryxell, B., *et al.* 2000, *ApJS*, 131, 273
Maeda, K., *et al.* 2010, *Nature*, 466, 82
Miceli, M., *et al.* 2009, *A&A*, 501, 239
Orlando, S., Bocchino, F., Reale, F., Peres, G., & Petruk, O. 2007, *A&A*, 470, 927
Orlando, S., Bocchino, F., Miceli, M., Petruk, O., & Pumo, M. L. 2012, *ApJ*, 749, 156

Author Index

Printed in the United States
by Baker & Taylor Publisher Services

Printed in the United States
by Baker & Taylor Publisher Services